Stereotactic Radiosurgery and Stereotactic Body Radiation Therapy

IMAGING IN MEDICAL DIAGNOSIS AND THERAPY

William R. Hendee, Series Editor

Published titles

Quality and Safety in Radiotherapy
Todd Pawlicki, Peter B. Dunscombe, Arno J. Mundt, and Pierre Scalliet, Editors
ISBN: 978-1-4398-0436-0

Adaptive Radiation Therapy
X. Allen Li, Editor
ISBN: 978-1-4398-1634-9

Quantitative MRI in Cancer
Thomas E. Yankeelov, David R. Pickens, and Ronald R. Price, Editors
ISBN: 978-1-4398-2057-5

Informatics in Medical Imaging
George C. Kagadis and Steve G. Langer, Editors
ISBN: 978-1-4398-3124-3

Adaptive Motion Compensation in Radiotherapy
Martin J. Murphy, Editor
ISBN: 978-1-4398-2193-0

Image-Guided Radiation Therapy
Daniel J. Bourland, Editor
ISBN: 978-1-4398-0273-1

Targeted Molecular Imaging
Michael J. Welch and William C. Eckelman, Editors
ISBN: 978-1-4398-4195-0

Proton and Carbon Ion Therapy
C.-M. Charlie Ma and Tony Lomax, Editors
ISBN: 978-1-4398-1607-3

Comprehensive Brachytherapy: Physical and Clinical Aspects
Jack Venselaar, Dimos Baltas, Peter J. Hoskin, and Ali Soleimani-Meigooni, Editors
ISBN: 978-1-4398-4498-4

Physics of Mammographic Imaging
Mia K. Markey, Editor
ISBN: 978-1-4398-7544-5

Physics of Thermal Therapy: Fundamentals and Clinical Applications
Eduardo Moros, Editor
ISBN: 978-1-4398-4890-6

Emerging Imaging Technologies in Medicine
Mark A. Anastasio and Patrick La Riviere, Editors
ISBN: 978-1-4398-8041-8

Cancer Nanotechnology: Principles and Applications in Radiation Oncology
Sang Hyun Cho and Sunil Krishnan, Editors
ISBN: 978-1-4398-7875-0

Monte Carlo Techniques in Radiation Therapy
Joao Seco and Frank Verhaegen, Editors
ISBN: 978-1-4665-0792-0

Image Processing in Radiation Therapy
Kristy Kay Brock, Editor
ISBN: 978-1-4398-3017-8

Informatics in Radiation Oncology
George Starkschall and R. Alfredo C. Siochi, Editors
ISBN: 978-1-4398-2582-2

Cone Beam Computed Tomography
Chris C. Shaw, Editor
ISBN: 978-1-4398-4626-1

Tomosynthesis Imaging
Ingrid Reiser and Stephen Glick, Editors
ISBN: 978-1-4398-7870-5

Stereotactic Radiosurgery and Stereotactic Body Radiation Therapy
Stanley H. Benedict, David J. Schlesinger, Steven J. Goetsch, and Brian D. Kavanagh, Editors
ISBN: 978-1-4398-4197-6

IMAGING IN MEDICAL DIAGNOSIS AND THERAPY

William R. Hendee, Series Editor

Forthcoming titles

Computer-Aided Detection and Diagnosis in Medical Imaging
Qiang Li and Robert M. Nishikawa, Editors

Handbook of Small Animal Imaging: Preclinical Imaging, Therapy, and Applications
George Kagadis, Nancy L. Ford, George K. Loudos, and Dimitrios Karnabatidis, Editors

Physics of Cardiovascular and Neurovascular Imaging
Carlo Cavedon and Stephen Rudin, Editors

Ultrasound Imaging and Therapy
Aaron Fenster and James C. Lacefield, Editors

Physics of PET Imaging
Magnus Dahlbom, Editor

Hybrid Imaging in Cardiovascular Medicine
Yi-Hwa Liu and Albert Sinusas, Editors

Scintillation Dosimetry
Sam Beddar and Luc Beaulieu, Editors

Stereotactic Radiosurgery and Stereotactic Body Radiation Therapy

Edited by
Stanley H. Benedict, PhD
David J. Schlesinger, PhD
Steven J. Goetsch, PhD
Brian D. Kavanagh, MD, MPH

LONDON AND NEW YORK

First published 2015 by CRC Press

2 Park Square, Milton Park, Abingdon, Oxon OX14 4RN
711 Third Avenue, New York, NY 10017, USA

Routledge is an imprint of the Taylor & Francis Group, an informa business

First issued in paperback 2016

Version Date: 20140304

ISBN 13: 978-1-4398-4197-6 (hbk)
ISBN 13: 978-1-138-19854-8 (pbk)

Library of Congress Cataloging-in-Publication Data

Stereotactic radiosurgery and stereotactic body radiation therapy / editors, Stanley H. Benedict, David J. Schlesinger, Steven J. Goetsch, Brian D. Kavanagh.
p. ; cm. -- (Imaging in medical diagnosis and therapy)
Includes bibliographical references and index.
ISBN 978-1-4398-4197-6 (hardcover : alk. paper)
I. Benedict, Stanley H., editor. II. Schlesinger, David J., editor. III. Goetsch, Steven J., editor. IV. Kavanagh, Brian D., editor. V. Series: Imaging in medical diagnosis and therapy.
[DNLM: 1. Radiosurgery. WN 250.5.R15]

RC78
616.07'57--dc23 2014007009

I would like to dedicate this book to my many clinician and scientist friends in radiation oncology. I have had the benefit of many mentors over the years—too many to list them all here—but would like to thank James Smathers (UCLA), who taught me the wonders of medical physics; Rupert Schmidt-Ullrich (MCVH-VCU), the value of clinical research and clinical trials; and James Larner (UVA), the importance of sound business and management practices. Finally, I would like to dedicate this to my loving wife, Lori, and our daughters, Erin & Noelle; may you stay forever young.

Stanley Benedict

In memory of Robert and Marguerite Palmer. We will never forget the strength and dignity with which you fought.

David Schlesinger

I would like to dedicate this book to my mentors in radiological physics, medical physics, and radiation biology: Herman Cember, Herb Attix, Paul De Luca Jr., and Jack Fowler. There are too many colleagues to begin thanking them all. I would also like to thank my wife Mona and son David for putting up with me during trying times.

Steven J. Goetsch

I am deeply indebted to the abiding love and support I receive from my family (Julia, Clare, and Thomas), without which I could not possibly achieve any meaningful success.

Brian Kavanagh

Contents

Series preface

Advances in the science and technology of medical imaging and radiation therapy are more profound and rapid than ever before since their inception more than a century ago. Further, the disciplines are increasingly cross-linked as imaging methods become more widely used to plan, guide, monitor, and assess treatments in radiation therapy. Today, the technologies of medical imaging and radiation therapy are so complex and so computer-driven that it is difficult for the persons (physicians and technologists) responsible for their clinical use to know exactly what is happening at the point of care when a patient is being examined or treated. The persons best equipped to understand the technologies and their applications are medical physicists, and these individuals are assuming greater responsibilities in the clinical arena to ensure that what is intended for the patient is actually delivered in a safe and effective manner.

The growing responsibilities of medical physicists in the clinical arenas of medical imaging and radiation therapy are not without their challenges, however. Most medical physicists are knowledgeable in either radiation therapy or medical imaging and expert in one or a small number of areas within their discipline. They sustain their expertise in these areas by reading scientific articles and attending scientific talks at meetings. In contrast, their responsibilities increasingly extend beyond their specific areas of expertise. To meet these responsibilities, medical physicists periodically must refresh their knowledge of advances in medical imaging or radiation therapy, and they must be prepared to function at the intersection of these two fields. How to accomplish these objectives is a challenge.

At the 2007 annual meeting of the American Association of Physicists in Medicine in Minneapolis, Minnesota, this challenge was the topic of conversation during a lunch hosted by Taylor & Francis Publishers and involving a group of senior medical physicists (Arthur L. Boyer, Joseph O. Deasy, C.-M. Charlie Ma, Todd A. Pawlicki, Ervin B. Podgorsak, Elke Reitzel, Anthony B. Wolbarst, and Ellen D. Yorke). The conclusion of this discussion was that a book series should be launched under the Taylor & Francis banner with each volume in the series addressing a rapidly advancing area of medical imaging or radiation therapy of importance to medical physicists. The aim would be for each volume to provide medical physicists with the information needed to understand technologies driving a rapid advance and their applications to safe and effective delivery of patient care.

Each volume in the series is edited by one or more individuals with recognized expertise in the technological area encompassed by the book. The editors are responsible for selecting the authors of individual chapters and ensuring that the chapters are comprehensive and intelligible to someone without such expertise. The enthusiasm of volume editors and chapter authors has been gratifying and reinforces the conclusion of the Minneapolis luncheon that this series of books addresses a major need of medical physicists.

Imaging in Medical Diagnosis and Therapy would not have been possible without the encouragement and support of the series manager, Ms. Luna Han of Taylor & Francis Publishers. The editors and authors are and, most of all, I am indebted to her steady guidance of the entire project.

William R. Hendee, Series Editor
Rochester, Minnesota

Foreword

Stereotactic radiosurgery (SRS) has substantially progressed since it was conceived of by Professor Lars Leksell in 1951. As a concept markedly distinct from then-current radiation therapy and open neurosurgery approaches, radiosurgery was met with skepticism and resistance by some. Ultimately, the disruptive technology has become the preferred treatment strategy for many benign and malignant conditions. Advances in manufacturing and control system engineering have led to breakthroughs in linear accelerator (linac) technology that allow treatment of multiple small targets at extremely high doses. Improvements in imaging science have translated into tremendous advances in the accuracy and precision of targeting both static and moving targets. The accumulation of clinical evidence for the efficacy of radiosurgery has broadened its reach beyond its initial home in the brain to locations throughout the body. Radiosurgery technology continues to evolve and improve even today.

The University of Virginia (UVA) was fortunate to have one of the first Gamma Knife installations in North America in 1989. The intervening years have brought changes that mirror the evolution of the field as a whole. Gamma Knife technology at UVA saw a significant improvement with the installation of the Perfexion model Gamma Knife in 2007, which enhanced workflow and improved treatment efficiency. Linac-based radiosurgery for lung and spine tumors began to gain traction at about the same time, and they are now both busy clinical programs at UVA using our Varian TrueBeam and TomoTherapy HD treatment systems. In addition, we continue to explore new minimally invasive approaches for treating small, well-localized targets, demonstrated by our program for MR-guide–focused ultrasound surgery.

In this text, the editors and authors have taken on a new strategy in their approach to covering the topic of SRS and SBRT. While other works have focused, in large part, on identifying specific anatomical sites and the application of SRT techniques to disease-specific clinical applications, this book explores the historical developments of the Gamma Knife and linac-based SRS, the physics and radiobiology of SRS and SRT using photons and charged particles, dedicated SRS and SBRT devices, and patient positioning and immobilization strategies as well as a comprehensive clinical chapter on the various disease sites with relevant ongoing and completed protocols. This refreshing approach for assessing SRS and SBRT thereby allows the reader to explore well-referenced chapters on many individual aspects of the developing field.

We trust the physicians, physicists, and other professionals involved in radiation oncology will find this book very valuable for their practices of SRS and SBRT.

James Larner, MD
Chairman, University of Virginia Department of Radiation Oncology

Jason Sheehan, MD, PhD
Professor and Vice-Chair, University of Virginia Department of Neurological Surgery

Preface

Since its inception as a field more than 50 years ago, stereotactic radiation surgery (SRS) has developed into arguably the most sophisticated external beam radiation therapy delivery available today. SRS incorporates the most rigorously developed strategies for patient simulation and immobilization, treatment planning, and dose calculation algorithms. Additionally, it requires multimodality image coregistration, thoughtful radiobiological considerations, and careful attention to clinical and anatomic details. In this book, we address the field of SRS and stereotactic body radiation therapy (SBRT) in a comprehensive manner, including a brief history, major technological developments, and practical clinical and radiobiological considerations.

The book begins with an overview of the history of SRS/SRT/SBRT, thereby setting the stage for a review of the current status of the technology and clinical treatments and finally closing with a brief insight into the anticipated future of this treatment delivery.

We hope that readers will benefit by obtaining a detailed understanding of the development of the field coupled with an overview on the current status of the technology and clinical status and a comprehensive bibliography. Among the target audience are physicians, physicists, and a wide range of associated professionals with interests in radiation oncology, including medical oncologists, neurosurgeons, and other specialists whose patients may benefit from this treatment–delivery approach. The editors and authors include physicians and physicists by design in order to ensure that the contents meet the needs of clinicians and all others in related fields. The book is also intended as a comprehensive reference for those involved in research, teaching, management, or administration in the field of radiation oncology.

Acknowledgments

The editors wish to express their sincere appreciation to Manuscript Editor Patricia Dunn, without whose tireless efforts this book would not have made it into print. Her patience, expertise, wisdom, and humor led four benighted scientists through the inscrutable mysteries of the publishing process.

Thanks, Pat!

Editors

Dr. Stanley Benedict received his PhD in biomedical physics from UCLA and holds a certification in radiological therapeutic physics from the American Board of Radiology. His first faculty appointment was at the Medical College of Virginia Hospitals of the Virginia Commonwealth University in Richmond, Virginia, where he rose to be Chief of Clinical Physics. Dr. Benedict later joined the faculty at the University of Virginia, Charlottesville, Virginia, where he served as Professor and Director of Radiological Physics. Dr. Benedict is currently Professor and Vice Chair of Clinical Physics in the Department of Radiation Oncology at the University of California at Davis, in Sacramento, California, where he is responsible for supervising the physics, dosimetry, and clinical engineering activities of the faculty and staff and overseeing the technological resources invested in the simulation, planning, and delivery of radiation therapy for the department and its satellite partners. His professional achievements include many scientific peer-reviewed publications, several book chapters, a wide array of peer-reviewed scientific abstracts at international symposia, and serving as Chairman of the AAPM Task Group on Stereotactic Body Radiation Therapy. Dr. Benedict has been PI, co-PI, co-investigator, and collaborator on several public and privately funded clinical medical physics research projects involving image-guided radiation therapy, stereotactic radiosurgery, and radiation biology. He lives in Davis, California, with his wife Lori, and their daughters Erin and Noelle.

Dr. David Schlesinger is currently an Associate Professor of Radiation Oncology and Neurological Surgery at the University of Virginia and chief medical physicist at UVA's Lars Leksell Center for Gamma Surgery. Dr. Schlesinger earned his MS and PhD in biomedical engineering at the University of Virginia. After several years working in the industry, he rejoined the UVA faculty in the Department of Neurological Surgery in 2004 and the medical physics faculty within the Department of Radiation Oncology in 2009. He earned his board certification in therapeutic medical physics in 2010 from the American Board of Radiology. Dr. Schlesinger's primary clinical expertise is in stereotactic radiosurgery, and he is involved in all aspects of treatment planning, treatment delivery, quality assurance, and radiation safety at the UVA Gamma Knife Center. Dr. Schlesinger also helps to direct the operation of the UVA MR-guided Focused Ultrasound Surgery Center, helps to supervise several joint radiosurgery partnerships between UVA and regional medical centers, and is active in a variety of ongoing research projects within the Department of Radiation Oncology. Dr. Schlesinger has authored or coauthored more than 40 publications in the field of radiosurgery and is a frequent speaker at national and international conferences and training courses. He actively participates in national societies, such as the American Association of Physicists in Medicine, for which he has served as cochair of a task group on MR-guided focused ultrasound surgery and served as the secretary and treasurer of the society's Mid-Atlantic Chapter. He is also a frequent referee for peer-reviewed publications in radiation oncology, medical physics, and neurosurgery. Dr. Schlesinger lives in Charlottesville, Virginia, with his wife Michelle, son Aidan, and daughter Maya; they share space with their two Saint Bernards, Kahuna and Omaha.

Dr. Steven J. Goetsch has served as Director of Medical Physics at the San Diego Gamma Knife Center since 1994. He is presently a Professor of Medical Physics at Radiological Technologies University in South Bend, Indiana, and has previously been a faculty member at the University of Wisconsin, University of California Los Angeles, and San Diego State University. He earned his MS in health physics from Northwestern University and his PhD in medical physics from the University of Wisconsin. He directed the University of Wisconsin Accredited Dosimetry Calibration Laboratory for seven years before becoming a hospital physicist at UCLA Medical Center. Dr. Goetsch has authored or coauthored 40 publications in the fields of Gamma Knife dosimetry, small-field dosimetry, health physics, quality assurance, patient safety in radiation therapy, and calibration of radiation sources. He has served on the International Commission of Radiation Units and Measurements and serves as chair of a task group on Gamma Stereotactic Radiosurgery Dosimetry and Quality Assurance of the American Association of Physicists in Medicine. Dr. Goetsch is a reviewer and guest editor for several refereed publications in medical physics and radiation oncology. He has chaired dozens of radiation therapy conferences for medical physicists, medical dosimetrists, and radiation therapists. He lives in Solana Beach, California, with his wife Mona and has one son, David.

Dr. Brian Kavanagh graduated summa cum laude with honors in biomedical engineering from the Tulane University School of Engineering and then remained at Tulane to receive his MD and MPH degrees from the School of Medicine and School of Public Health. He completed an internship at the Charity Hospital of Louisiana before entering a radiation oncology residency at Duke University, where he was an American Cancer Society Clinical Oncology Fellow. Following his first faculty appointment at the Medical College of Virginia, he moved to the University of Colorado. His current research interests include laboratory studies of how to enhance the effectiveness of radiotherapy by blocking the action of macrophages that stimulate tumor regrowth after treatment, and he has a special clinical interest in stereotactic radiosurgery for brain tumors and stereotactic body radiation therapy (SBRT) for tumors elsewhere in the body. His recent publication list includes numerous peer-reviewed papers, chapters, and monographs related to SBRT.

Contributors

Igor Barani
University of California
San Francisco, California

Marc R. Bussière
Massachusetts General Hospital
Boston, Massachusetts

Edwin Crandley
University of Virginia
Charlottesville, Virginia

Megan E. Daly
University of California
Sacramento, California

Indra J. Das
Indiana University School of Medicine
Indianapolis, Indiana

Sonja Dieterich
University of California
Sacramento, California

Toufik Djemil
Cleveland Clinic
Cleveland, Ohio

Joshua Evans
Virginia Commonwealth University
Richmond, Virginia

Ruben Fragoso
University of California
Sacramento, California

Rohini George
University of Maryland
Baltimore, Maryland

Steven J. Goetsch
San Diego Gamma Knife Center
San Diego, California

and

Radiological Technologies University, VT
South Bend, Indiana

William H. Hinson
Wake Forest School of Medicine
Winston-Salem, North Carolina

Bassel Kassas
York Cancer Center
York, Pennsylvania

Richard P. Levy
Scripps Proton Therapy Center
San Diego, California

Patricia Lindsay
Princess Margaret Hospital
Ontario, Canada

D. Michael Lovelock
Memorial Sloan-Kettering Cancer Center
New York, New York

Ryan McMahon
St. Vincent Hospital
Green Bay, Wisconsin

Anthony L. Michaud
University of California
Sacramento, California

Moyed Miften
University of Colorado School of Medicine
Aurora, Colorado

Michael F. Moyers
Shanghai Proton and Heavy Ion
Shanghai, China

Malika Ouzidane
Radiation Oncology Cleveland Clinic
Cleveland, Ohio

Mihaela Rosu
Virginia Commonwealth University
Richmond, Virginia

Kyle E. Rusthoven
Coastal Carolina Radiation Oncology
Wilmington, North Carolina

Habeeb Saleh
University of Kansas Cancer Center
Kansas City, Kansas

Bill J. Salter
University of Utah
Salt Lake City, Utah

Vikren Sarkar
Radiation Oncology University of Utah
Salt Lake City, Utah

Zachary Seymour
University of California
San Francisco, California

Ke Sheng
University of California
Los Angeles, California

Timothy D. Solberg
Hospital of the University of Pennsylvania
Perelman Center for Advanced Medicine
Philadelphia, Pennsylvania

Andrew Vaughan
University of California
Sacramento, California

Yevgeniy Vinogradskiy
University of Colorado School of Medicine
Aurora, Colorado

Brian Wang
University of Louisville
Louisville, Kentucky

Krishni Wijesooriya
University of Virginia
Charlottesville, Virginia

David Wilson
University of Virginia
Charlottesville, Virginia

John Wong
Johns Hopkins University
Baltimore, Maryland

Michael Wright
Varian Medical Systems
Palo Alto, California

Kamil M. Yenice
University of Chicago
Chicago, Illinois

Fang-Fang Yin
Duke University Medical Center
Durham, North Carolina

Cedric Yu
University of Maryland School of Medicine
Baltimore, Maryland

Part I

Early history of stereotactic radiation therapy

This book details the history and practice of two relatively recently developed complimentary techniques: stereotactic radiosurgery and stereotactic radiotherapy. Each has made extraordinary contributions to the practice of medicine, especially in the field of neuro-oncology.

Archeologists have found evidence from Neanderthal skulls that medical practices in humans date back at least 36,000 years. Remarkably, the origins of neurosurgery are known to date back at least 5500 years, making it the oldest known form of human surgery. Trephinations (the practice of cutting through the skull) have been found in skeletal remains across the planet in many cultures, practiced for unknown reasons. From these primitive roots sprang the pseudoscience of phrenology, which tried to associate bony skull protuberances with intellectual properties and characteristics. This popular cultural phenomenon lead to real brain mapping, in which different regions of the human brain were found to be associated with different skills and capabilities, which varied little morphologically from one person to another. Stereotactic radiosurgery and its evolutionary offspring, stereotactic radiation therapy, have had an enormous impact on the fields of neurological surgery, radiation oncology, and, more recently, oncological surgery. From a modest beginning in a single site in Stockholm, the concept of stereotactic radiosurgery evolved from the closely related field of stereotactic neurosurgery. The Gamma Unit incorporated stereotactic technology, incorporating the extremely narrow beams of proton radiosurgery as well as the compactness and affordability of the recently developed cobalt-60 teletherapy devices invented in Canada by Harold

Johns and others. After 14 years of use by one single center, the Gamma Unit began to proliferate around the world, with hundreds of Elekta Gamma Knife units and also many rotating gamma stereotactic radiosurgery units now in use worldwide.

Lars Leksell invented radiosurgery in 1951 and rapidly moved from x-ray technology to proton therapy. The concept of intracranial radiosurgery advanced rapidly as the technology evolved. John Lawrence (brother of E. O. Lawrence, the inventor of the cyclotron) and Cornelius Tobias at the Lawrence Berkeley National Laboratory in California preceded Leksell into particle beam therapy in 1954. These clinicians expanded the concept of radiosurgery to include light ion charged particle beams, some of which have never been duplicated. The restless intellects of Dr. Leksell and his colleague physicist Borje Larsson lead them to invent a brilliant hospital-based device, the Leksell Gamma Unit (now the Gamma Knife) in 1967. The proliferation of the Gamma Knife and the success of the light ion clinical programs located at a handful of high-energy physics research centers gave rise to ingenious adaptations of the widely employed linear accelerator in the mid 1980s. After about 10 years of this adaptive development, new dedicated linear accelerator units, such as the Accuray CyberKnife and the hybrid Varian/Brainlab Novalis, were developed. Many previously intractable diseases are now being far better controlled with radiosurgery and stereotactic radiation therapy.

1 Historic development of stereotactic radiosurgery and stereotactic body radiation therapy

Steven J. Goetsch

Contents

1.1 INTRODUCTION

The history of brain surgery (Table 1.1) goes back about 7000 years or more to the practice of trephination, or opening of the skull, often with primitive (e.g., flint) tools (Piek et al. 1999). This is the oldest surgical procedure known to have been performed on human beings (Figure 1.1). The practice has been noted in thousands of human skulls with more than 450 dating from the early and late Neolithic Age in Europe (5000 BCE). The practice was discovered all across the globe, particularly in Mesoamerica. Skulls were determined to be from both men and women of various ages and never with any sign of associated trauma

Table 1.1 Timeline of early neuroscience and neurosurgery

ERA	DATE	EVENT
Prehistory	5500 BCE	Trephination with stone tools
17th century	1618	Paracelsus introduces laudanum
18th century	1796	Gall proposes phrenology
Early 19th century	1804	Morphine purified
Mid 19th century	1846	Ether used as anesthetic
Mid 19th century	1859	Cocaine used as local anesthetic
Late 19th century	1870s	Lister proposes antisepsis
Late 19th century	1880s	Elective craniotomies begin
Late 19th century	1896	Radiography invented
Early 20th century	1908	Beginnings of stereotaxis

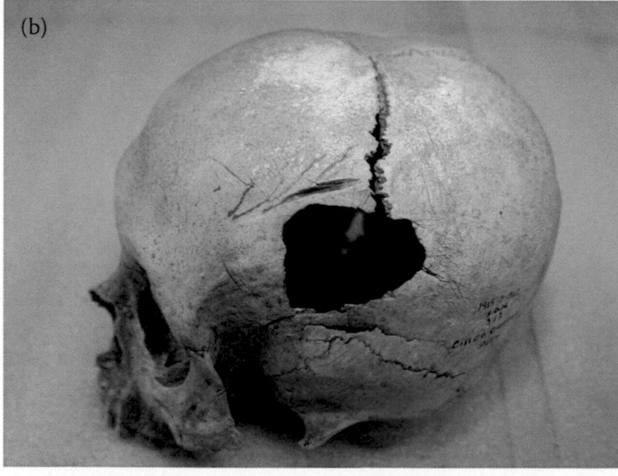

Figure 1.1 (a) Prehistoric female skull, San Damian, Peru, showing trephination marks. Aperture approximately 3.0 × 1.7 cm. (b) Prehistoric female skull, Cinco Cerros, Peru, left frontal and parietal incision with no signs of healing. (Courtesy of San Diego Museum of Man, San Diego, CA.)

or disease. One skull was found with two openings, each in the 9-cm range, and another skull was found with three openings. Images exist on ancient pottery of Egyptian physicians in 1400 BCE using a drill (perforating rod) powered by a back-and-forth bow to more easily open the skull. Hippocrates, father of modern medicine, wrote a book in the fifth century BC on the merits and methods of performing surgery for brain injury that was referred to for nearly 2000 years. No known reason for this practice has ever been established, and it continues in isolated primitive societies to this day.

The brain was recognized by serious philosophers at the time of the American Revolution to be the home of intellectual and character functions. This led to the rise of the pseudoscience of *phrenology* (from the Greek language for "mind" and "knowledge"), based on a presumed relationship between surface contours and bumps on the skull with character attributes. The "science" of the relationship of skull size and shape to human intellect and abilities was created by German physician Franz Joseph Gall in 1796 and was popular worldwide until about 1840 (Greenblatt 1995). The term *phrenology* is traced to Benjamin Rush, patriot, physician, signer of the Declaration of Independence, and founder of American psychiatry. Rush believed in a relationship between behavior and the physical state of the brain and was the first to hospitalize a mentally ill patient. An extremely important medical event occurred at the Massachusetts General Hospital in 1846 when ether was given to a patient by inhalation so that a neck surgery could be performed (Barker 1993). This innovation, combined with the adoption of antiseptic techniques advocated by the English surgeon Joseph Lister (spraying the operating room with carbolic acid vapor), helped advance neurosurgery from the rare attempt to relieve a traumatic skull fracture or serious brain infection to the planned trephination of the skull to relieve symptoms of injury or to remove a tumor.

As the 19th century progressed, neurosurgeons began to attempt bolder brain operations. Victor Horsley (later knighted) began performing craniotomies at the National Hospital for the Paralyzed and Epileptic in London in 1896 (Tan and Black 2002). The introduction of surgical anesthetics, such as cocaine and morphine, allowed longer and more difficult operations to be performed although surgical morbidity and mortality was a huge problem. One of the early advocates and practitioners of elective brain surgery was Harvey Cushing, who trained under pioneering American surgeon William Halsted at the new Johns Hopkins Medical School and Hospital (Voorhees et al. 2009). In partnership with another neurosurgery pioneer, Walter Dandy, Cushing performed 29 craniotomies in 1910 alone. Cushing moved on to Harvard's new Peter Bent Brigham Hospital in 1911, where he stayed for 22 years except for the time spent in military service during World War I. He is renowned for the description of Cushing's disease, development of the electrocautery tool with physicist William Bovie, and leadership in the development of brain surgery.

Walter Dandy stayed at Johns Hopkins from 1918 to 1946, the only neurosurgeon on the staff, until his untimely death at age 60. He pioneered the techniques of *ventriculography* and *pneumoencephalography* (air contrast injected into the spinal canal to make the ventricles of the brain visible on a radiograph) for brain imaging, which was the key to the development of image-guided brain surgery. He also wrote 10 classic papers on hydrocephalus. Dandy pioneered an innovative technique for neurosurgery for tic douloureux as well as vastly improving the results for the resection of acoustic neuromas. These are both now popular indications for modern stereotactic radiosurgery (SRS) (Campbell 1951).

1.2 ORIGINS OF STEREOTAXIS

1.2.1 HISTORIC FRAME OF HORSLEY AND CLARKE

Surgical pioneer Sir Victor Horsley was quite an inventor as well as a physician as many of the innovators of later radiosurgery would also be (Sachs 1958; Rahman, Murad, and Mocco 2009). In collaboration with mathematician Robert Clarke, he coined the term *stereotaxis* (from the Greek words for "solid" and "orderly arrangement") (Horsley and Clarke 1908). Together, the two investigators fabricated an apparatus that could be rigidly clamped to the skull of a living experimental animal with a tiny probe entering into the cerebellum (Figure 1.2). They reported lesioning targets in the monkey brain. Shortly afterward, they had a disagreement, which ended their collaboration.

Figure 1.2 Horsley and Clarke original apparatus. (Courtesy of London Science Museum, online image.)

Canadian neurophysiologist Aubrey Mussen purchased a used Horsley-Clarke apparatus in 1908 and later designed and built a modified version, which he took back with him to McGill University in Montreal in 1918. He used his own stereotactic frame to guide lesioning electrodes into the brains of experimental animals. This work led to publication of an atlas of the brain stem (Mussen 1922–1923). Another Horsley-Clarke frame was made for the American surgeon Ernest Sachs, who trained with Horsley and returned with this frame to Washington University in St. Louis (Fodstad, Hariz, and Ljunggren 1991). A copy of the Horsley-Clarke apparatus was later rebuilt from diagrams at the Institute of Neurology of Northwestern University Medical School in Chicago by famed neuroanatomist Stephen Walter Ranson, who had seen the frame when he was at Washington University, and colleague W. R. Ingram (Ingram et al. 1932). They used it to perform a classic series of animal investigations on the reticular formation, midbrain, and hypothalamus from 1929 to 1932.

1.2.2 CLINICAL APPLICATIONS OF STEREOTACTIC FRAMES: SPIEGEL AND WYCIS

A critical part of stereotaxis was reliance on brain imaging and the use of maps of human or animal brains. In the early 19th century, *craniometry* (cranial measurement) began to develop. Generations of anatomists mapped out the human brain in great detail although they were not generally aware of the function of each brain region. Neurologist Ernest Spiegel began to do similar research in Vienna, Austria, and began using pneumoencephalograms for image guidance. He coined the term *stereoencephalotomy* to describe three-dimensional (3-D) image-guided surgery on the human brain. Spiegel moved to Temple University in Philadelphia where he collaborated with neurosurgeon Henry Wycis in modifying a Horsley-Clarke apparatus for use in human brain surgery (Spiegel et al. 1947). They mounted the apparatus on a ring, which was suspended from a plaster of Paris cap made individually for each patient (Figure 1.3). The Spiegel-Wycis frame went through five innovations with the Model V frame used in their practice for many years. The Model V Spiegel-Wycis stereotactic apparatus required surgical placement of four countersunk screws into the patient's skull, which permitted removal of the apparatus and repositioning over several days. Pneumoencephalography was an agonizingly painful procedure requiring a spinal tap so that pressurized air could be introduced into the cerebrospinal fluid space. This created two large air bubbles filling the patient's ventricles so that they could be imaged radiographically. The patient was suspended from the ceiling by a full body harness, which was rotated in three dimensions to assist the introduction of air into the brain. Mercifully, this procedure became obsolete when computed tomography (CT) was introduced in the mid-1970s.

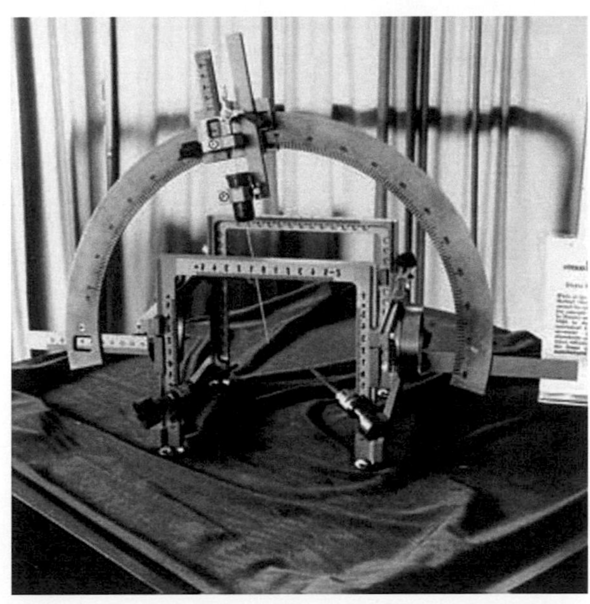

Figure 1.3 Spiegel and Wycis apparatus. (Courtesy of Cyber Museum of Neurosurgery, online image.)

1.2.3 PROLIFERATION OF STEREOTACTIC FRAMES

Neuronavigation was defined as the practice of identifying objects in the brain radiographically prior to surgery. Although early primitive radiological procedures could identify the air-filled ventricles, tumors and other abnormalities were not visible, and no identifying landmarks inside the skull were visible. Jean Talairach in Paris was the first neurosurgeon to impose a coordinate system inside the brain that could be used with landmarks defined from radiographic procedures (Kelly 2004). His rectilinear stereotactic frame was designed with small holes at 1-mm intervals along the vertical plates so that an x-ray beam could demonstrate both the external frame markers (fiducials) and the interior anatomy. Talairach defined a line from the anterior commissure (AC, a bundle of fibers at the midpoint between the two hemispheres, just anterior to the fornix) and the posterior commissure (PC, a band of fibers on the dorsal aspect of the cerebral aqueduct). Talairach used this AC-PC line to navigate to other nearby predefined anatomical structures (Figure 1.4). A brain atlas of well-localized structures based on the detailed dissection of one human brain was published by Talairach in 1949 (Talairach, David, and Tournoux 1957). Talairach's brain atlas set off a profusion of interest in brain mapping with the Schaltenbrand and Wahren work generally regarded as definitive (Schaltenbrand and Bailey 1959). The Society of Brain Mapping and Therapeutics was founded in 2003.

Stereotaxis enjoyed an explosive development in the years following the Second World War. A Swedish neurosurgeon and prolific surgical inventor, Lars Leksell, developed a stereotactic frame exclusively for human patients (Leksell 1950). Leksell's innovation was the introduction of the arc-centered concept, which replaced the dependence of the Spiegel and Wycis device on internal and external bony landmarks, such as the auditory canal. The location of the desired target was first determined from radiographic procedures (with the Leksell frame attached) and then translated into the X (lateral), Y (anterior-posterior), and Z (inferior-superior) "Leksell coordinates." A pointer following an arc, which could be used to attach a surgical probe or to guide a surgical microscope or radiosurgical device (such as an x-ray tube), had wide freedom of movement but always pointed at the exact coordinates of the selected intracerebral target. The neurosurgeon could then use his or her knowledge of brain anatomy to select the least harmful approach to the target for biopsy or lesion creation. This frame is still in use today as the Leksell Model G frame and is supplied with the Gamma Knife Perfexion. Traugott Riechert and Fritz Mundinger created a frame in Freiburg, Germany, that was popular in Europe (Riechert and Mundinger 1955). Bertrand Gilles at McGill

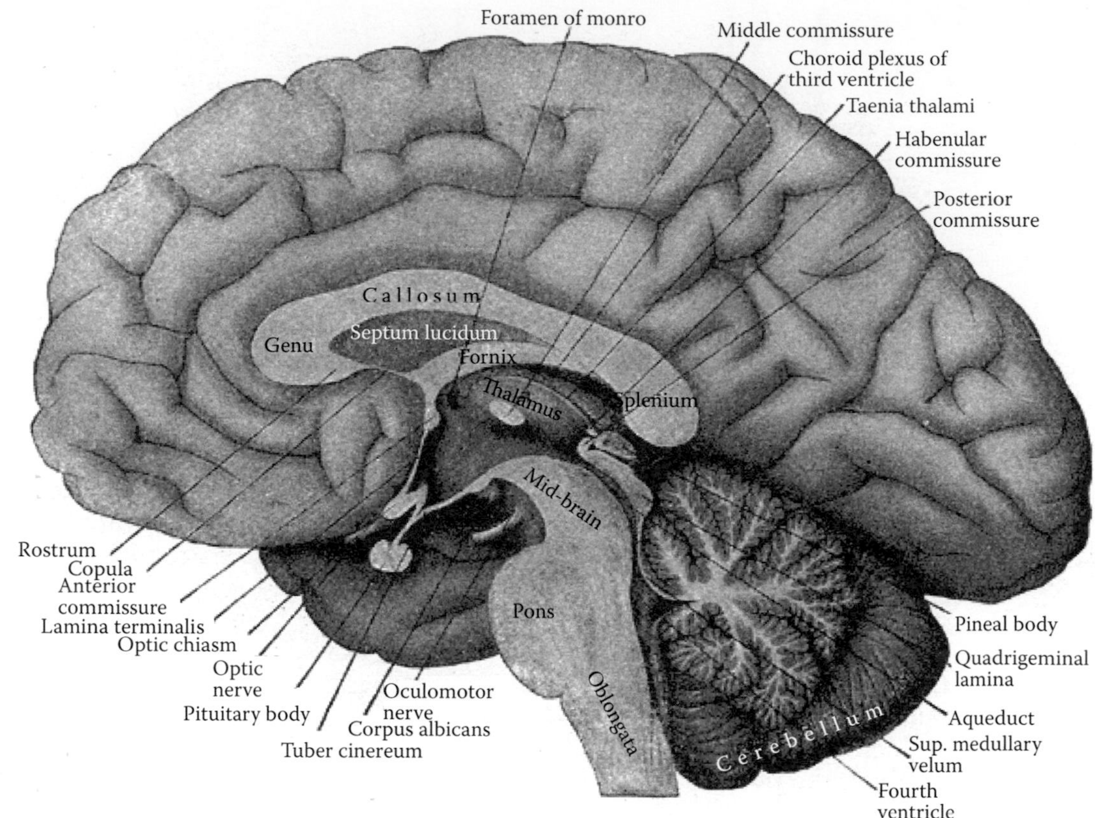

Figure 1.4 Brain image: AC PC line (online image).

University in Montreal modified a Leksell frame to permit room for lateral radiographs in 1963 (Bertrand 2004). Gerard Guiot in France created his own stereotactic frame (1962) to permit microelectrode recordings of the stimulation of the internal capsule (Rahman et al. 2009).

1.3 EARLY HISTORY OF X-RAYS AND RADIOACTIVITY

1.3.1 PHYSICAL DISCOVERIES

The famous mathematical physicist William Thomson (Lord Kelvin) was quoted late in the 19th century as saying, "There is nothing new to be discovered in physics now. All that remains is more and more precise measurement." Events that occurred less than 6 months apart and only 558 km away from each other in 1895 and 1896 would prove him totally wrong (Table 1.2). Serendipity may be defined as the fortunate discovery of something when one is looking for something else. Louis Pasteur remarked, "Science favors the prepared mind." Professor Wilhelm C. Röntgen of the University of Wurzburg, Germany, was working with a Hittorf-Crookes tube, a device that created a stream of electrons in a clear glass evacuated tube, when he noticed something odd (del Regato 1975). He saw a glow emanating from a piece of paper painted with barium platinocyanide, which fortunately happened to be on his desktop near the Crookes tube. Astonishingly, nothing else in the room glowed, and the glow remained when Röntgen interposed opaque objects in the path between the tube and the piece of metal. Röntgen spent weeks investigating this new phenomenon. He assessed the penetration of these mysterious rays (which he named "x-rays") through objects, such as a thick book and various thicknesses of aluminum, copper, silver, platinum, and lead. He deduced that the fluorescence was inversely proportional to the square of the distance from the tube. Röntgen also observed the shadow of the bones of his finger, thereby creating the world's first radiograph!

Table 1.2 Timeline of significant milestones in radiation therapy and stereotactic radiosurgery

DATE	CLINICIAN OR PHYSICIST	EVENT
1895	Röntgen	Discovery of x-rays
1896	Becquerel	Discovery of radioactivity
1896	Many physicians	Diagnostic radiology begins
1896	Grubbé	First radiation therapy
1898	Marie and Pierre Curie	Discovery of radium and polonium
1913	Coolidge	Introduction of hot filament x-ray tube
1945	Kerst	Invention of betatron
1951	Leksell	Radiosurgery named and performed
1952	Johns	Cobalt-60 teletherapy
1954	Lawrence	Treatment of patients with charged particles
1967	Leksell	First Gamma Knife used to treat patients
1970	Steiner	First AVM patients treated with Gamma Knife
1980	Fabrikant	Helium ions used to treat patients
1982	Barcia-Solario	Cobalt-60 teletherapy radiosurgery
1982	Betti, Derechinsky	Linac modified for radiosurgery
1985	Winston, Lutz	SRS program at Joint Center, Boston
1987	Lunsford, Flickinger, Wu	First commercial Gamma Knife Model U at University of Pittsburgh
1997	De Salles, Solberg, Selch	First Novalis installed at UCLA Med Center

He also observed that these shadows could be recorded on photographic film, and on December 22, 1895, he published the now-famous portrait of his wife's hand.

Röntgen delivered a manuscript to the Secretary of the Wurzburg Physical and Medical Society on December 28, 1895, titled "On a New Kind of Rays." The *Vienna Free Press* published a front-page story on January 5, 1896, which rapidly spread around the world by telegraph. Undoubtedly, many other scientists had observed some of these phenomena prior to this (Nikola Tesla, Sir William Crookes, and Wilhelm Hittorf among them), but they never deduced the significance of what they noticed. Röntgen received the first Nobel Prize in Physics in 1901 for his discovery. Over 1000 scientific papers were published about x-rays worldwide in 1896.

Röntgen's astonishing discovery caused an equally momentous discovery only a few months later at the University of Paris (Tubiana, Dutreiz, and Pierquin 1996). Antoine Henri Becquerel was the third member of his family to occupy the chair of physics professor at the Muséum National d'Histoire Naturelle. Becquerel had long studied phosphorescent materials, such as uranium salts, and thought they might give off x-rays when stimulated by bright sunlight. He coated photographic plates with uranium and exposed the covered plates to sunlight to stimulate the phosphorescence, which worked reasonably well. However, one day he developed plates that had not been exposed to sunlight and found that the film had been equally darkly exposed. He announced the discovery of natural radioactivity on March 2, 1896.

At this point a newly married graduate student in Becquerel's laboratory, Marie Sklodowksa Curie (who had married physicist Pierre Curie in July 1985), decided to conduct her doctoral thesis work on this new phenomena (del Regato 1976b). Marie's work depended critically on the invention of a piezoelectric electrometer (also known as a gold-leaf electroscope) fabricated by Pierre and Paul-Jacques Curie. She found that radiation from natural radioactivity could cause the air to be somewhat conductive through the process of ionization. Marie Curie obtained tons of pitchblende ore from her native Poland and painstakingly began separating it by chemical means to isolate the portion that was emitting radiation. Her work was interrupted in 1897 by the birth of her daughter Irene (who would later join her in important collaborations). Pierre became so intrigued by Marie's work that he joined her, beginning in 1898. In this torturous, hazardous way, they isolated two compounds far more active than uranium. Marie's results were presented by her former professor Gabriel Lippmann on April 12, 1898. Later that year, they announced two new elements: radium and polonium. Pierre Curie, Marie Curie, and Henri Becquerel shared the 1903 Nobel Prize in Physics. Tragedy struck in 1906 when Pierre Curie was killed by a horse-drawn carriage while crossing a busy street. The next year, Marie Curie founded the Radium Institute at the University of Paris (with the help of the Pasteur Foundation and American philanthropist Andrew Carnegie) to both honor Pierre and continue their work.

1.3.2 EARLY ADVERSE EFFECTS AND ORIGIN OF RADIATION PROTECTION

Becquerel and the Curies noticed that the new radiation coming from these materials could be quite harmful to their skin. They all noticed lesions from handling the material, and Becquerel received a burn on the skin of his chest from carrying a vial of radium chloride in his coat pocket on a lecture tour of England. He published a report on his injury in 1901. Early x-ray workers also noticed that the skin of their hands suffered from dermatitis after holding photographic film in front of the x-ray tube. This led to the origin of radiation protection, a science in its own right. Röntgen himself warned a London instrument manufacturer of the harmful effects of x-rays, showing him photographs of skin burns (Mould 1993). The manufacturer reported back to a meeting of the British Roentgen Society on these hazards in April 1898. X-rays were rapidly used almost immediately, and the new science of *roentgenography* had proliferated worldwide by the end of 1896. A total of 23 cases of radiation dermatitis were reported by the end of 1896, and Thomas Edison's assistant Clarence Dally died of cancer in 1904 following severe radiation dermatitis (Bushong 1995). The eminent American scientist Nikola Tesla (one of the inventors of radio technology) warned of x-ray hazards by June 1896. The English engineer Elihu Thomson (later one of the founders of the General Electric Company) in late 1896 deliberately exposed his left little finger to an x-ray tube for 30 min per day, inducing a blistering lesion 1 week later (Kathren 1962). A court settlement in the United States in 1897 of $10,000 for a radiation necrosis injury stimulated the development of standards for both patients and early radiographers. Dentist and x-ray inventor William Rollins advocated in 1902 heavy-metal shields for x-ray housings to reduce stray radiation, the earliest known example of radiation shielding. Rollins demonstrated the lethality of x-rays on guinea pigs in 1898. The use of fluoroscopic plates made direct observation of radiographs possible to supplement the use of photographic film. Unfortunately, most radiologists placed themselves behind the plates in direct line with the x-rays traversing the patient! The first attempt at calibration of x-ray tube output was for the radiologist to test their device personally by exposing his or her own hand to radiation and watching for skin-reddening within a few days. The skin erythema dose (SED) was in widespread use as a qualitative standard for assessing the output of an x-ray tube. Harvey Cushing also fell victim to the practice of testing his own medicine and became addicted to cocaine, which he used on his patients during craniotomies.

Creation of scientific standards in radiation protection and measurement had to await the development of more precise radiation instrumentation. A 1915 conference of the British Roentgen Society recommended a series of strict x-ray precautions (Brodsky, Kathren, and Willis 1995). This report, delayed by the outbreak of World War I, was published in 1921. American physicist Arthur Mutscheller published a recommendation in 1925 that the level of exposure in x-ray centers be limited to 1/100th of the SED in a 30-day period (Mutscheller 1925). Mutscheller estimated the SED by multiplying the tube current times the exposure length divided by the square of the distance from the tube multiplied by an arbitrary constant. Later physicists estimated this "tolerance dose" as about 60–70 rem per year. Swedish physicist Rolf Sievert also recommended a tolerance dose of 1/10th of the SED per year, close to Mutscheller's recommendation. So many radiologists had been maimed or killed by overexposure to x-rays that in the 1930s the German Roentgen Society constructed a monument at a hospital in Hamburg to 136 x-ray "martyrs" and later added several hundred more names.

A huge conceptual breakthrough occurred with the formal definition of a measurement unit (named in honor of the discoverer of x-rays), the *Roentgen* (or R) by the Second International Congress on Radiology in Stockholm in 1928 (Brodsky, Kathren, and Willis 1995). It was defined as the amount of "corpuscular radiation" (measured in electrostatic units) liberated per volume of dry air at a standard temperature and pressure. The definition was modified several times, but international bodies later accepted that one Roentgen liberated one electrostatic unit of charge per 1 cm^3 of dry air. Accompanying this watershed event was the introduction of the "condenser R meter" measurement device by John Victoreen in 1929. This reliable, durable, easy-to-use device could measure x-rays reasonably accurately and was in use for another 50 years. The significance of this international standard is difficult to overstate: Both radiation protection and the use of ionizing radiation in medicine could be put on a firm quantitative physical basis.

The National Bureau of Standards (NBS) in the United States established what became the National Council on Radiation Protection and Measurement in 1929. This committee published a report in 1931 recommending an exposure level for radiation workers not to exceed 0.2 R per day (NBS 1931). This recommendation (roughly equivalent to 125 rem per year) has been lowered many times and now stands at a maximum of 5 rem per year. The preferred unit for specification of radiation absorbed dose in terms of energy absorbed per unit mass of absorber was the *rad*, which was adopted by the International Commission on Radiation Units and Measurements (ICRU) in 1953 and is equivalent to 100 ergs per gram (ICRU 1959). Ultimately, the preferred measure of radiation protection became the *rem* (Roentgen equivalent man). This standard was adopted by the International Commission on Radiological Protection in 1962.

1.3.3 EARLY HISTORY OF DIAGNOSTIC RADIOLOGY

As described above, the new technology for producing x-rays was widely available and was placed into use in medicine within weeks of being reported electronically around the world. Early x-ray tubes were very inefficient and of low energy, thought to be about 50 kVp with a mean energy of as little as 9 keV. Much of the reason for erythema was the unfiltered spectrum clogged with wasteful low-energy photons that contributed nothing to image formation and merely increased skin dose. The idea of adding filtration was attributed to Rollins as early as 1903.

The first major technological breakthrough in x-ray technology was the invention by American physicist William D. Coolidge in 1913 of the hot filament cathode copper anode and tungsten target x-ray tube with higher voltages and more stable operation than previous units (Seibert 1995). Earlier gas tubes relied on ionization of residual gas in the x-ray tube to produce a stream of electrons, but the Coolidge tube had a much harder vacuum. Coolidge worked at the General Electric research laboratory in Schenectady, New York, where he continued to work for nearly 40 years and then consulted with GE nearly until his death at age 102. Because far more than 99% of x-ray tube energy is converted into heat in the target, the high temperatures generated were a constant problem. Gustav Bucky in Germany created the first x-ray grid in 1913, and it reduced scattered radiation from reaching the imaging plane. Chicago radiologist H. E. Potter (1920) successfully created a way of moving the grid simultaneously with motion of the x-ray target, which drastically increased image quality (Webster 1995). The Potter-Bucky grid remained in use for decades. Further imaging gains were achieved with field-limiting apertures (about 1921) that simultaneously shielded the patient from extraneous radiation while limiting scattered radiation, which contributed to image unsharpness.

The next major technological breakthrough in x-ray tube design was achieved by Albert Bouwers of the Philips X-ray Research Laboratory in Eindhoven, Netherlands, in 1930 when he created a rotating anode tube (Webster 1995). This major advance increased tolerable power levels by about 900%, allowing for much shorter, sharper exposures and reducing motion blurring. No major watershed developments took place in diagnostic radiology until the Digital Revolution of the 1970s and beyond.

1.3.4 EARLY HISTORY OF RADIATION THERAPY

The twin bombshell discoveries of x-rays in 1895 and natural radioactivity in 1896 led to the rise of both diagnostic radiology and radiation therapy. It is astonishing how quickly these developments were put into medical practice. A radiologist in Birmingham, England, located a needle stuck in a patient's hand with a radiograph taken on January 11, 1896. The first use of radiation in medical therapy is generally attributed to Emil Grubbé, a manufacturer living in Chicago who manufactured incandescent lamps and Crookes tubes (Lederman 1981). He had enrolled as a medical student at Hahnemann Medical College two years earlier and would go on to a long and distinguished career as a radiologist (Grubbé 1933). Grubbé tested his own x-ray tubes by placing his hand in the beam path (years later, he had many disfiguring surgeries from radiation injuries) and was seen by a physician who appreciated the power of this new radiation. His own physician, a Dr. Gilman, sent him a patient with breast cancer whom Grubbé treated in his factory on January 29, 1896, only 24 days after the *Vienna Free Press* story on Röntgen's invention was published worldwide. Grubbé, by his own account, treated a second patient for lupus a short while later. Thus was *teletherapy* (from the Greek for "distance" and "treatment") invented. Other x-ray treatments took place in Germany in February 1896 and France in July 1896. A patient with a large squamous cell tumor under the left eye was apparently cured with x-ray therapy performed in Sweden in 1899. A prominent young Parisian physician, Henri Beclere, shifted his practice from infectious diseases to diagnostic and therapeutic radiology and became the father of French radiology (del Regato 1978). A gifted French physician and painter, Georges Chicotot, painted a portrait of himself treating a young woman with breast cancer in about 1906 (Figure 1.5).

A parallel track of radiation therapy for cancer developed simultaneously. The adverse effects of radium were almost immediately noticed by the discoverers, and the possibility for treatment of human disease with this potent modality was immediately obvious. The practice of treatment of cancer (and some benign lesions) with radium was first called *Curietherapy* in honor of the Curies, but later became known as *brachytherapy* (from the Greek for "close" and "treatment"). Shortly after Becquerel's famous chest burn, the Curies and Becquerel presented some of their radium to physicians at St. Louis Hospital, where a dermatologist treated a patient with lupus, and radium therapy began (Grubbé 1933). Radium was very precious and very expensive, so mesothorium (decay products of thorium, typically a radium or actinium isotope) or radon gas, which emanates constantly from radium-226, was often used in hospitals. Sometimes, a great deal of radium was gathered together in the form of a "radium bomb," and external treatments took place, blurring the distinction between brachytherapy and teletherapy. After the Roentgen was defined in 1928, American physicist Edith Quimby of Memorial Hospital in New York

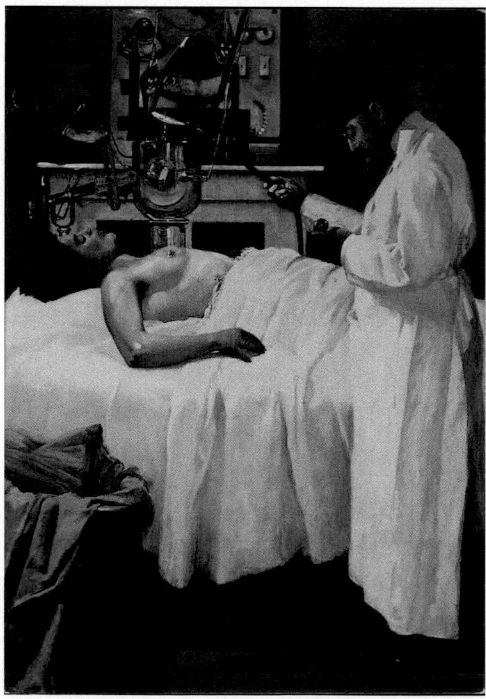

Figure 1.5 "The First Attempt to Treat Breast Cancer with X-Rays." (Self-portrait by Dr. Georges Chicotot, 1906.)

City systematized the relationship between milligrams of radium and dose to patients. Over the next few decades, brachytherapy systems would be created at Memorial Hospital in New York, in Stockholm, in Paris, and in Manchester (Mould 1993). These and other systems persisted in some form into the computerized treatment-planning era.

External-beam x-ray therapy made a significant advance when higher voltage apparatus was made available in the 1920s with so called *x-ray cannons* (Orton 1995). These well-shielded devices (which resembled an artillery piece) attained high voltages of 150 kVp to as much as 300 kVp. This therapy went by several names, including *orthovoltage* and *deep x-ray therapy*. Although D_{max} for these units was still nearly at skin level, early treatment patterns of cross-firing beams now became somewhat more successful although deep tumors could really only be palpated, and their relationship to adjacent critical organs remained speculative. Radiation therapy in this x-ray energy range continues to this very day, primarily for basal cell and squamous cell skin cancers. Several innovative devices, such as Van de Graaf generators, one million volt x-ray machines, and the betatron (Kerst 1975), were introduced in the 1930s and 1940s but did not gain widespread use (Orton 1995).

The technological inventions stimulated by World War II, namely the nuclear reactor and radar, led to the introduction of two high-energy treatment devices. The first operational linear accelerator, based on the magnetron microwave power unit, was built by Metropolitan Vickers and installed in Hammersmith Hospital in London (Orton 1995). It was a 3-m-long 8-MeV standing wave stationary device and treated its first patient in August 1953. A 4-MeV linac was installed in Newcastle in 1954 by Mullard Equipment Division of Philips Medical, which was later purchased by Elekta, Inc. Varian Associates in Palo Alto, California, installed a traveling wave linear accelerator based on their patented klystron microwave power unit at the request of Dr. Henry Kaplan at Stanford University in 1954 (Figure 1.6). Another Varian linear accelerator was their first 6-MV device and was also the first isocentric linear accelerator, capable of 360-degree rotation. It was installed at UCLA Medical Center in 1962, one of only 15 medical linacs in the world at that time, plus an estimated 50 betatrons (Thwaites and Tuohy 2006).

Access to byproduct material of nuclear reactors made the production of long-lived (5.25-year half-life), high-energy (1.17 and 1.33 MeV gamma rays in cascade), cobalt-60 plentiful. This led rival medical groups in Canada to create megavoltage cobalt-60 teletherapy units in 1951 (Figure 1.7). The first group was at Royal University Hospital Saskatoon, Saskatchewan, and was led by Harold Johns (Johns, Bates, and Watson 1952). This 1000-Curie unit was installed in November 1951. However, a rival claim by a cancer center in London, Ontario, was put forward with the first patient treatment reported in October 1951. This machine was manufactured by Atomic Energy of Canada Limited and launched a proliferation of such devices with 1120 cobalt-60 teletherapy units worldwide by 1961 (Mould 1993). Henry Kaplan published a manuscript in 1977 summarizing the previous 25 years of radiation therapy and concluded that survival

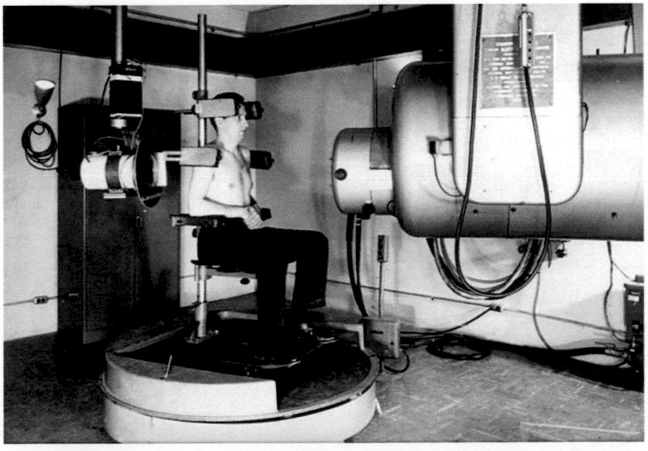

Figure 1.6 First Varian linear accelerator at Stanford University Medical Center.

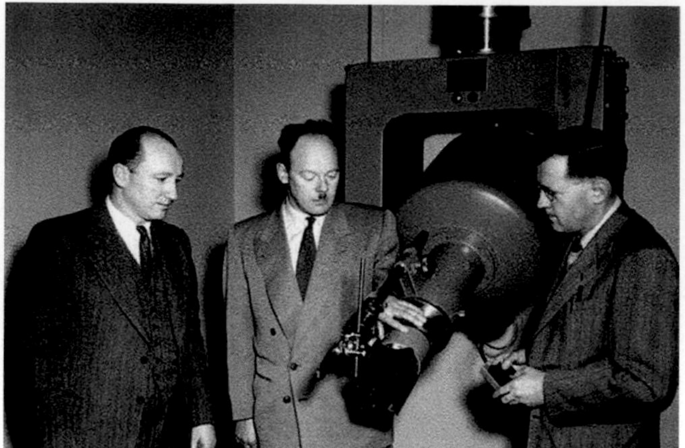

Figure 1.7 First cobalt-60 teletherapy source at Saskatoon Cancer Centre, with Sandy Watson, John MacKay, and Harold Johns. (From the University of Saskatchewan web page.)

of patients treated with megavoltage beams was a factor of two to four times greater for most cancers than those treated with orthovoltage x-rays (Das and Kase 1992). The stage was now set for the SRS revolution.

1.4 STEREOTACTIC RADIOSURGERY: A SYNTHESIS OF MANY DISCIPLINES

1.4.1 PIONEERING WORK OF LARS LEKSELL

Swedish neurosurgeon and inventor Lars Leksell was a Renaissance man. Tall and distinguished, he left behind such a prolific trail of inventions that his sons Lawrence and Daniel gathered them together to form Elekta, now one of the largest medical corporations in the world. Leksell's earliest invention was the arc-centered stereotactic frame, still in use more than 60 years later. He also invented surgical instruments and was one of the pioneers of radiofrequency rhizotomy, which he used to create controlled lesions in precisely selected brain centers. His most revolutionary achievements came from a synthesis of many ideas. Leksell published a paper in 1951 titled "The Stereotactic Method and Radiosurgery of the Brain," which is the beginning of the entire field of SRS (Leksell 1951). Leksell was one of the inventors of minimally invasive surgery and decided that some of the lesions he was creating in the brain could be created with x-rays, thereby eliminating the risk of opening the brain with a twist drill and inserting a catheter (Figure 1.8). Leksell used an orthovoltage x-ray in combination with his stereotactic frame to treat two trigeminal neuralgia patients, both of whom reported relief of pain for many years.

Leksell abandoned the medium-energy x-ray machine for the 185-MeV cyclotron at nearby Uppsala University, about one hour's drive from Stockholm (Larsson et al. 1958). Together with his longtime collaborator, radiation biologist and physicist Borge Larsson, they treated 76 patients with this device between 1957 and 1976, despite the long distance from the Karolinska Hospital and the disadvantages of treatment in a high-energy physics laboratory (Figure 1.9).

1.4.2 EARLY CHARGED-PARTICLE THERAPY ERA

The cyclotron was invented by University of California physicist E. O. Lawrence in 1931 (for which he received the Nobel Prize in Physics in 1939). Together with his brother John (a physician) as well as physicians Robert Stone and John Larkin, he treated some 226 patients with advanced cancers at his cyclotron laboratory with neutron therapy between 1938 and 1939. Results were very poor, and the long-term side effects on the few survivors were catastrophic (Stone 1948). It would be years before radiation biologists learned that the relative biological effectiveness of fast neutrons is close to 2.5, accounting for the terrible radiation overdoses these patients suffered.

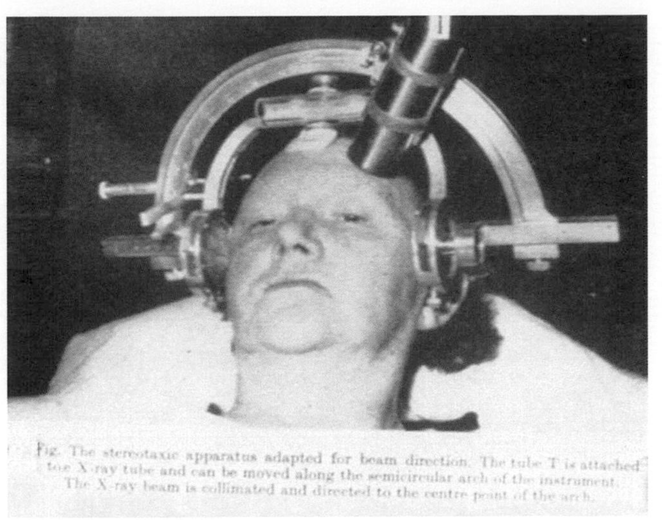

Figure 1.8 First Leksell radiosurgery patient. (Courtesy of Elekta, officially released picture.)

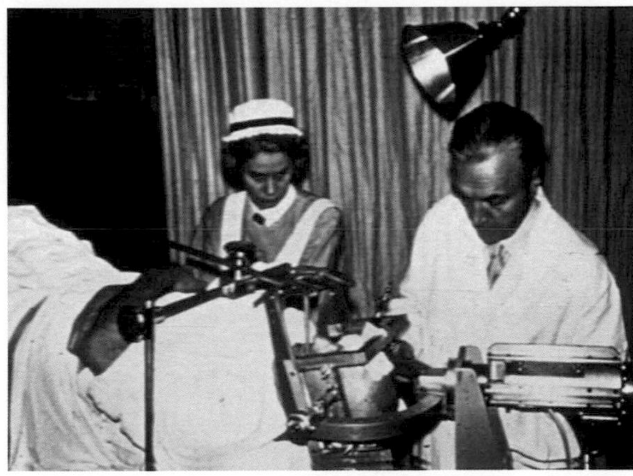

Figure 1.9 Leksell at Uppsala cyclotron facility.

Attention then turned to use of the cyclotron to produce protons, which had been suggested by physicist Robert Wilson in 1947 (Wilson 1947). The Berkeley laboratory began treating patients for pituitary tumors in 1954 with high-energy protons, treating 30 patients in three years. An energy upgrade on the cyclotron caused them to switch to heavy charged particles (910-MeV helium ions), and more than 2000 patients were treated by Jacob Fabrikant and others between 1957 and 1992 (Tobias, Anger, and Lawrence 1952; Fabrikant, Lyman, and Hosobuchi 1984). A 160-MV cyclotron located in the physics department at Harvard University was used beginning in 1961 until 2002 by neurosurgeon Ray Kjellberg to treat arteriovenous malformations (AVMs) and other intracranial conditions (Kjellberg et al. 1968) as well as by Herman Suit and others for extracranial disease (Suit et al. 1977). Kjellberg's pioneering work was critical to the development of radiosurgery. He was the first to work out risk curves relating single fraction doses to the total volume of brain tissue irradiated (Kjellberg et al. 1983). Kjellberg's work, in turn, borrowed from the radiation biology research of Swedish radiation oncologist Magnus Strandqvist at Radiumhemmet (Swedish for "radium home") Hospital in Stockholm (del Regato 1989). Based on a suggestion by his doctoral thesis advisor, Strandqvist plotted the logarithm of the dose administered to patients treated at his center versus the logarithm of the total number of days in which the doses were administered. He found that the successful cases could be linked by a straight line with adverse effects above that line and recurrences falling below the line. Kjellberg extended the concept, plotting the single dose (in rads) on

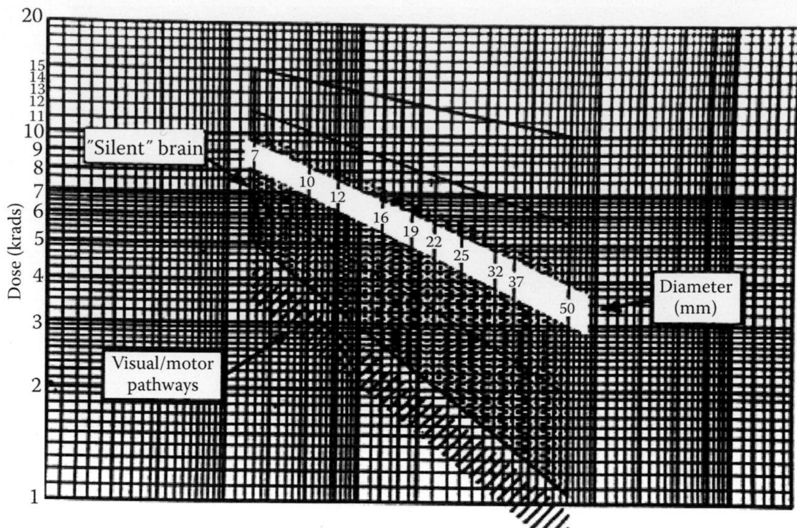

Figure 1.10 Raymond Kjellberg dose versus total volume risk curve for proton therapy of AVM. (From *New England Journal of Medicine* abstract online.)

the ordinate versus proton beam diameter (Figure 1.10). His publication in the *New England Journal of Medicine* in 1983 came just before the proliferation of photon SRS and gave early investigators a method to assess risk of previously untried treatments on a wide variety of intracranial disease (Kjellberg et al. 1983).

Three more centers began treating patients with protons in Russia, and two more centers (one in France, one in the United States) began the treatment of human patients with helium ions by 1977 (Boone et al. 1977). Three centers began treatment with pi-meson therapy, including TRIUMF in Vancouver in 1979, Los Alamos National Laboratory in 1974, and Paul Scherrer Institute in Switzerland in 1980. All pi-meson therapy worldwide was discontinued by 1994. New emphasis in recent years has been placed on the use of heavier ions, such as carbon.

1.4.3 ORIGIN OF THE LEKSELL GAMMA KNIFE

Neurosurgeon Lars Leksell formed a successful partnership with radiation biologist Borge Larsson on a number of projects, including treatment of neurosurgical patients with proton therapy at Uppsala University from 1957 to 1976. Both scientists, however, sought a device that could achieve the same extremely sharp beam focus of the cyclotron but be practical for use within a hospital. Together with Kurt Lidén from the University of Lund and Rune Walstam from the Karolinska Institute in Stockholm, they synthesized the concepts of brain mapping, brain imaging with pneumoencephalograms, stereotactic surgery, SRS, proton therapy, and cobalt-60 teletherapy and came up with the Leksell gamma unit (Leksell 1971). The original device was fabricated by Mottola, a Swedish shipbuilding firm, and contained 179 sealed cobalt-60 sources carefully collimated (through elliptical collimators) to focus with remarkable accuracy on a volume approximately 4, 8, or 14 mm in diameter (Figure 1.11). Leksell deliberately tried to mimic the extremely sharp falloff of the proton therapy beam he had used at Uppsala. The device was first used to treat patients in 1967 at a nuclear physics research center in Studsvik, Sweden, before being moved to Sophiahemmet Hospital in Stockholm later that year. Much of the original collimator design and the dosimetry of the 179 converging beams were attributed to Hans Dahlin and Bert Sarby at the National Institute for Radiation Protection (Dahlin and Sarby 1975). A second device was later fabricated when Leksell moved to Karolinska Hospital, and the first unit was donated to UCLA Medical Center in 1982 (Figure 1.12), where it was used in animal research as well as treating a small number of patients (Rand, Khonsary, and Brown 1987). Both units nicely mimicked the extremely sharp penumbra previously found only with charged-particle therapy devices (90% to 50% in one millimeter for the 4-mm helmet).

Leksell treated primarily functional lesions with the first two gamma units because the imaging limitations at the time precluded visualization of brain tumors. Sweden has had an outstanding tradition

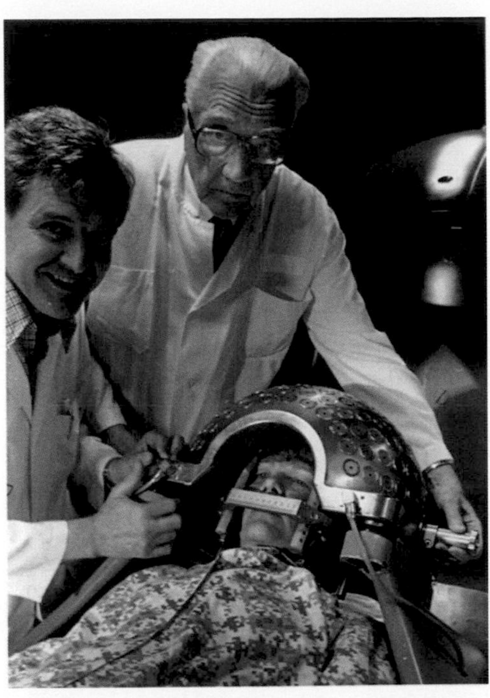

Figure 1.11 First Leksell Gamma Knife with neurosurgeons Christer Lindquist and Lars Leksell.

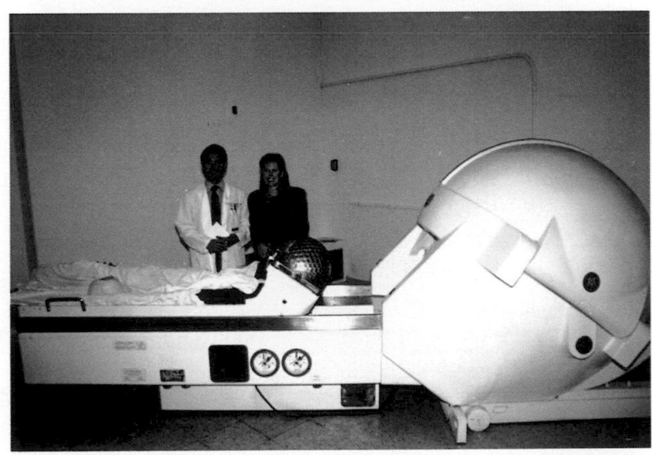

Figure 1.12 First Gamma unit after relocation to UCLA Medical Center with neurosurgeon Dr. Antonio De Salles and Elekta president Catherine Gilmore.

of excellent neurosurgeons since Herbert Olivecrona trained with Harvey Cushing at Johns Hopkins and then began neurosurgery in Stockholm in the 1920s (Ljunggren 1993). He became chair of neurosurgery at Karolinska Hospital, and one of his students was Lars Leksell. Olivecrona pioneered vascular neurosurgery and surgery for acoustic neuromas, two areas of intense interest to early users of the Gamma Knife. Leksell used his new gamma unit to continue the development of stereotactic surgery for movement disorders, which he and others, such as Gerard Guiot, Jean Talairach, Traugott Riechert, and Hirotaro Narabayashi, had pioneered in the late 1940s and early 1950s (Laitinen 1993). Leksell began to use the Gamma Knife to create lesions in the left and right globus pallidus for Parkinson's disease as well as continuing his ongoing work with treatment of trigeminal neuralgia.

Leksell's first collaborating neurosurgical colleague was Ladislau Steiner, who, in 1970, began to capitalize on the development of angiography to visualize complex vasculature in the brain such as AVMs (Steiner et al. 1992). It was discovered that high radiation doses to the congenitally malformed vessels in an AVM cause endothelial cell proliferation, which thereby occludes the abnormal structure that can be any size from a few millimeters to many centimeters in diameter. Steiner made Gamma Knife treatment of AVMs a specialty at the Karolinska Hospital and later at the University of Virginia medical center, reporting on 247 consecutive cases over a 14-year period.

Leksell had the Gamma Knife field entirely to himself until two other neurosurgeons visited his clinic and asked to have duplicate units made for their home countries. Dr. David Forster, a neurosurgeon at the Royal Hallamshire Hospital in Sheffield, and Dr. Hernan Bunge, owner of a private clinic in Buenos Aires, Argentina, both approached Leksell about getting permission to have another Gamma Knife built for their respective clinics (Walton, Bomford, and Ramsden 1987; Bunge, Guevara, and Chinela 1987). Two new units were custom built according to the original blueprints by Nucletec SA of Switzerland (a subsidiary of Scanditronix Medical AB of Sweden). Gamma Knife units 3 and 4 differed from the first two devices built to Dr. Leksell's specifications: They had 201 cobalt-60 sources, and the collimators were circular instead of elliptical, giving flattened spheroid high-dose volumes. Both devices were quite successful for many years and were later replaced by commercial models sold by Elekta Instruments AB, which was founded by Lars Leksell along with his sons Laurent and Daniel in 1972. The new company marketed Lars Leksell's surgical instruments and stereotactic frame and worked on a commercial version of the Gamma Knife.

1.4.4 ELEKTA GAMMA KNIFE INTRODUCED

Elekta Instruments introduced the Leksell Gamma Knife Model U in 1987 (Figure 1.13) with the first unit installed at the University of Pittsburgh Medical Center (Wu et al. 1990). The new Model U had 201 cobalt-60 sources with a nominal activity of 30 Curies each, converging on the unit center point (UCP) with a measured accuracy of 0.25 mm. This new commercial unit, for the first time, offered an 18-mm diameter set of collimator helmets in addition to the 4-, 8-, and 14-mm helmets supplied with the first four custom-designed units. The sources were arranged in a nearly hemispherical pattern about the UCP with a source-to-UCP distance of 40 cm. Andrew Wu determined an absorbed dose rate at the center of a 16-cm diameter spherical acrylonitrile butadiene styrene plastic phantom supplied by the manufacturer. This phantom was chosen to represent a typical human skull (typically 16 cm lateral measurement) with the absorbed dose rate in the center of the phantom, when perfectly aligned with the UCP designated as the calibration point for the Gamma Knife. This definition persists to this day although an appropriate national or international absorbed dose rate for the unique geometry of gamma SRS devices (the official U.S. Nuclear Regulatory Commission designation) has yet to be agreed upon. Wu measured the absorbed dose rate to be just over 3.99 gray per minute using a Capintec PR-05P 0.07 cm^3 volume air ionization chamber, verified by thermoluminescent

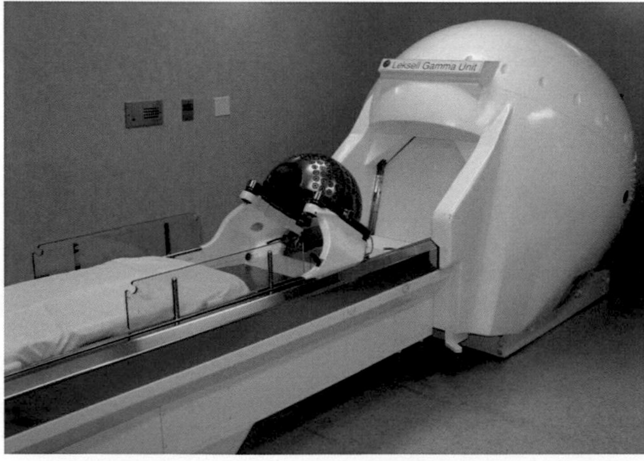

Figure 1.13 Leksell Gamma Knife Model U.

dosimetry (TLD) and diode dosimetry. It was impossible for Wu and subsequent physicists to independently verify the single source profile because, with all 201 sources loaded, the other 200 sources could not be adequately blocked. It was also extremely difficult for clinical users to verify the output of the 4-mm helmet with respect to the "master" 18-mm helmet, a difficulty that persisted for many years.

The original Gamma Knife Model U was a manually operated device. Output of the treatment-planning computer was printed on a list of instructions, and the clinical staff—consisting of a neurosurgeon, radiation oncologist, medical physicist, nurse, and sometimes a radiation therapist—had to manually treat the patient one shot (isocenter) at a time. This made elaborate treatments very lengthy (up to several hours) and led to the possibility of undetected error in setting stereotactic coordinates by eye at the submillimeter level (Flickinger et al. 1993). Elaborate protocols were required to have one person set the Y and Z coordinates on a sliding bar (slider) attached to the Leksell Model G frame while a second team member independently verified the setting. The trunnions, which were used to affix the stereotactic frame to the helmet, were used to set the X (left and right) coordinates and also had to be set manually and then verified on both sides. The 200-kg helmets had to be replaced manually with the aid of a lifting device, which meant that it took 5 to 10 min to change collimator sizes.

Treatment planning in the early days of the gamma unit and the Elekta Model U was quite simplistic. Lars Leksell and Borge Larsson devised a technique with which to calculate the time to give a prescribed dose to a single isocenter or "shot" as they came to be called. They precalculated sample treatment times and created a nomogram used at the time of surgery. Until the introduction of the Kula™ plan computerized system, approximately in 1992, only single shots were ever considered. The earliest computer system required scanning in 14-by-17in. radiographs and working out the Leksell stereotactic coordinates (X, Y, and Z) for the designated target point. A two-dimensional diagram of isodose lines could be printed out on clear film and overlaid on a radiograph with identical magnification with a grease pencil then used to trace the dose lines onto a single axial image. Beam shaping was accomplished primarily by inserting tungsten plugs into as many as 100 of the collimators to create differently shaped treatment volumes (Flickinger et al. 1990). Many patients treated on the Leksell Gamma Knife Model U had to be treated in the prone position because of the angulation of the collimators on the helmet. Later models of the Gamma Knife treated all patients in the supine position.

The advent of CT in 1971 by Hounsfield and Cormack (Isherwood 2005) and its subsequent introduction to stereotaxis (Bergström and Greitz 1976; Brown 1979) revolutionized Gamma Knife radiosurgery. For the first time, both malignant and benign brain tumors could be visualized in any part of the brain and targeted with precisely directed gamma radiation (Lunsford et al. 1989). A CT adapter box with four vertical copper wires and two diagonal wires on the left and right sides was carefully affixed to the Leksell Model G frame (Figure 1.14). The axial images acquired after frame placement then

Figure 1.14 Leksell stereotactic computed tomography fiducial box.

displayed a pattern of six dots (left and right anterior, left and right posterior, and the two diagonal lines). Simple trigonometry allowed the CT coordinates of couch in and out, up and down, and left and right to be converted to the Leksell stereotactic coordinates. This imposed a Cartesian coordinate system on all structures in the patient's brain that was valid until the frame was removed. The original Elekta Kula Plan and later Leksell GammaPlan treatment planning software, at first, required that each image be registered separately, which meant that the patient had to be very precisely aligned with the axis of the CT scanner, and the frame had to be precisely level with the longitudinal couch motion. Later software advancements allowed the entire canonical ensemble of up to several hundred axial (or coronal) slices to be registered simultaneously even if there was a constant offset with respect to one or more of the orthogonal axes. Accuracy of registration of the CT scans compared with known dimensions of the fiducial frame was typically on the order of 0.5 mm or less. The Brown-Robert-Wells (BRW) stereotactic frame, still in widespread use today, was developed at this time (Heilbrun et al. 1983).

One problem with the Leksell Gamma Unit Model U is that Elekta chose not to have the unit certified as a Type B radioactive material–shipping container, which would have required destructive testing. Therefore, the 22-ton central body of the unit was shipped to each hospital site, unloaded, and the radioactive material was installed on site. This led to a very complex, difficult, and expensive procedure. Many tons of shielding material had to be shipped in to create a "hot cell," and a large lead glass window and remote manipulating arms (first developed at Oak Ridge National Laboratory during the Manhattan Project) had to be used to unload the shipping cask and load the Model U unit. The Model U was constructed as a clamshell device so that the upper half of the shielding could be opened up with a motorized crane displaying all 201 source locations. Several hours of work were necessary to load or unload the sources, and the entire operation took 4 to 6 weeks of downtime. Most centers reloaded the unit after about one half-life (5.26 years) although some centers tolerated very low absorbed dose rates by waiting longer to reload their sources. Later manuscripts brought up the question of whether the effectiveness of the treatment was compromised if the treatment times were doubled or quadrupled (Arai et al. 2010).

1.4.5 NEWER GAMMA KNIFE MODELS INTRODUCED

Elekta introduced the new Leksell Gamma Knife Model B (Figure 1.15) in Europe in 1988 (Tlachacova et al. 2005). The major improvement over the previous Model U was that the sources were now rotated into five concentric rings about the UCP with an opening in the back of the removable helmet. This geometry allowed the sources to be loaded with the aid of a specially constructed 11-ton loading device that fit over the source shipping cask and allowed one or two operators to manually remove sources one at a time with the aid of a long tong with a handgrip. No massive "hot cell" needed to be constructed, which saved a great deal of time and expense in reloading. All subsequent Gamma Knife models followed this same procedure. The hydraulic system used in the Model U was replaced with a more reliable electric motor system, which all subsequent models use to this day. The dose configuration of the 4-, 8-, 14-, and 18-mm helmets was

Figure 1.15 Leksell Gamma Knife Model B.

very similar to the previous Model U, but the GammaPlan source profiles were remeasured at the factory by the manufacturer and were specific to each individual Gamma Knife model. The number of installed units grew continually throughout this era with 155 units reported in clinical operation by June 2001 (Leksell Gamma Knife Society 2011).

A new Leksell Gamma Knife Model C (Figure 1.16a) was introduced in 2000 with the optional automatic positioning system (APS) (Horstmann et al. 2000). The GammaPlan treatment-planning system was now connected via an RS232C serial cable with the computer, which interfaced with the motorized APS. If the Leksell stereotactic coordinates were within the space accessible by the APS motors (Figure 1.16b), then as many as 50 different shots could be performed in one sitting. The first set of Leksell coordinates could be verified by the treatment team before leaving the irradiation vault, but subsequent coordinates were changed automatically without the treatment-planning team reentering the room or disturbing the patient. An electronic detection system verified that the planned coordinates were achieved (within 50 μm) much as later multileaf collimator (MLC) positions were verified by a computerized control system (Goetsch 2002). Unfortunately, the APS motors inevitably took up some room between the patient's head and shoulders, and often certain planned shots could not be achieved. The system allowed the removal of the APS and replacement with the original trunnion and Y-Z slider attachments. This was much more time consuming and, again, subject to human error.

The current Gamma Knife Model Perfexion was introduced in 2006 (Lindquist and Paddick 2007). The Gamma Knife Perfexion was dramatically different from its predecessors in appearance as well as in function (Figure 1.17a). The 201 static cobalt-60 sources were replaced with 192 sources affixed to eight movable rods. The rods moved along a well-defined path from a shielded position to one of four possible collimator openings: 4 mm, 8 mm, 16 mm, or blocked (Figure 1.17b). The largest single block of tungsten ever produced was used to collimate these sources (same registered radiation source model as previous

(a)

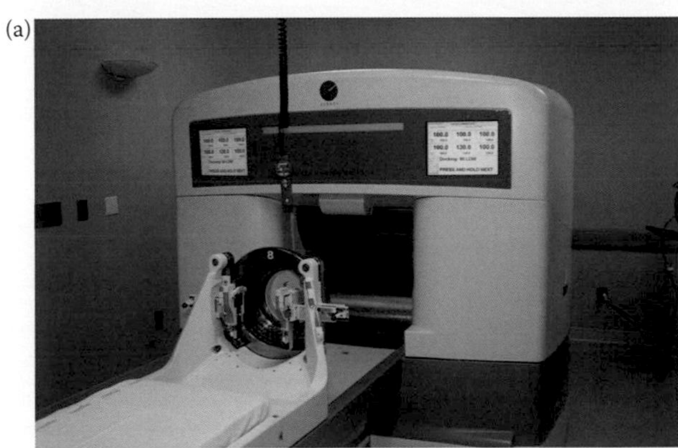

(b)

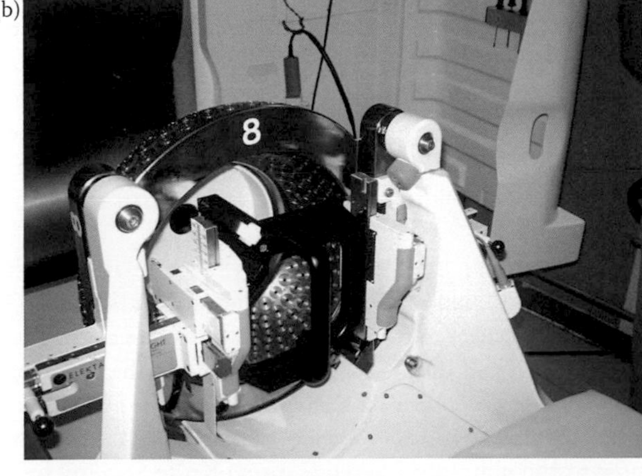

Figure 1.16 (a) Leksell Gamma Knife Model 4C; (b) automatic positioning system (APS).

(a)

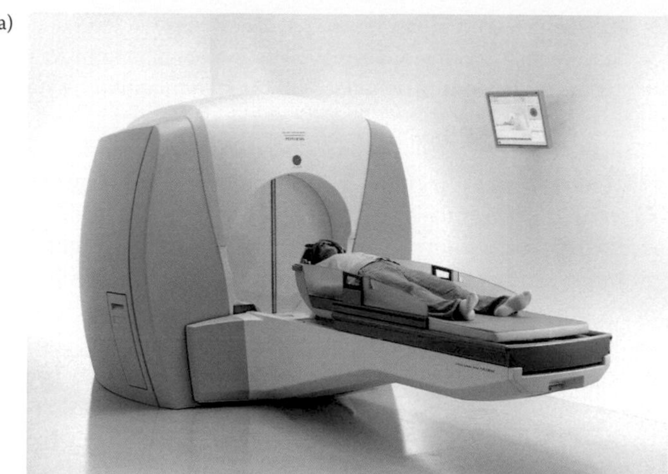

(b)

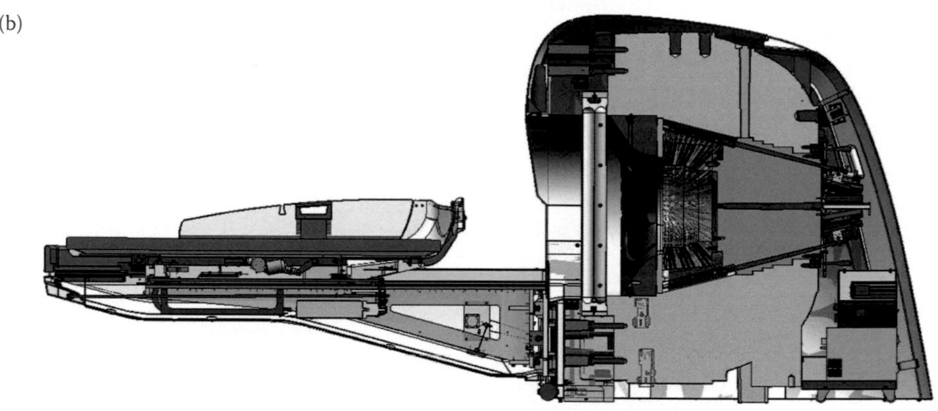

Figure 1.17 (a) Gamma Knife Perfexion with patient; (b) Perfexion cross-section.

units) as the rods independently moved to any of the four possible positions. This permits 4^8, or 65,536, possible source combinations, yielding almost unlimited geometric beam shaping. Extracranial exposure of the patient is also much reduced with respect to previous Gamma Knife units because the sources remain in the blocked position until the patient is in position for the first treatment, whereupon they open up in 50 ms or less. Each shot ends with the sources rapidly closing; the patient is transported to the next position, and the sources rapidly open again. This speeds up overall treatment time substantially compared to the Gamma Knife Model 4C with APS even though the Gamma Knife Perfexion has nine fewer sources and a maximum collimated size of 16 mm instead of 18 mm. Jean Regis at Timone University Hospital in Marseilles, France, reported that during the first 6 months of operation of the first Gamma Knife Perfexion more than five patients per day were treated with this unit with a total time in the treatment room reduced from 65 min per patient with the Model 4C to less than 45 min for the Gamma Knife Perfexion (Regis et al. 2009).

Another innovation of the Gamma Knife Perfexion was the elimination of the interchangeable helmet with all sources collimated within the central body. This innovation allowed an increase in the treatable volume by more than 300% with respect to previous models (Regis et al. 2009). Precise positioning of the intracranial volume to be treated in this unit is achieved by moving the entire patient on a high-precision couch, capable of submillimetric motion with six-sigma reliability for loads of up to 300 kg. By the end of 2009, some 266 Elekta Gamma Knife units were reported to be in use worldwide. The total number of patients treated (as reported to the nonprofit Leksell Gamma Knife Society) by the end of 2011 reached 676,000 (Figure 1.18), which does not include unreported treatments (Leksell Gamma Knife Society 2011).

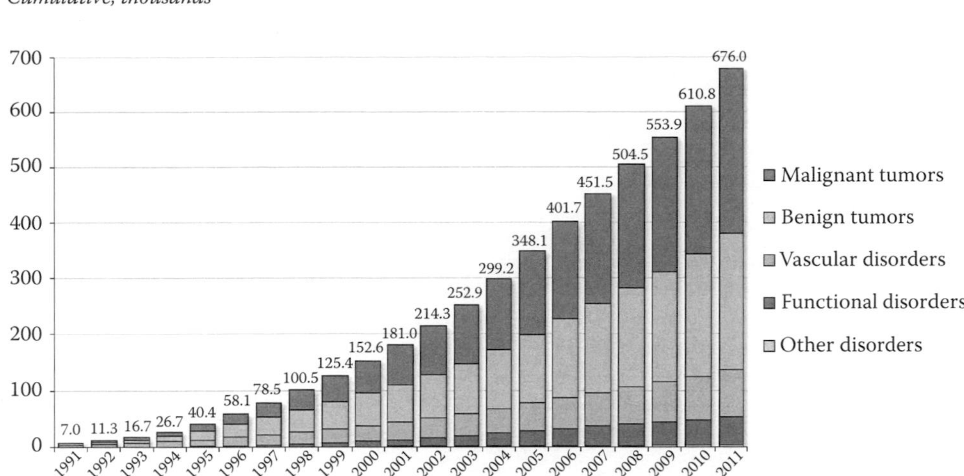

Figure 1.18 Gamma Knife patients treated through calendar year 2011 as reported by the Leksell Society.

1.4.6 ROTATING GAMMA STEREOTACTIC RADIOSURGERY MODELS INTRODUCED

The OUR Company of Shenzhen, China, introduced a revolutionary concept first reported by Steve Goetsch and Brent Murphy (Goetsch et al. 1999). Chinese engineers and scientists realized that a much smaller number of radiation sources rotating with extremely precise alignment could mimic a much larger number of static sources. This potentially reduced manufacturing costs and also eliminated the cumbersome job of manually changing helmets. A clever double system of rotating wheels, one containing 30 cobalt-60 sources of nominal activity 200 Ci each and another containing four sets of precisely aligned openings, produced 4-, 8-, 14-, or 18-mm–sized treatment volumes. Goetsch and Murphy reported measurements made at Auhai Radiosurgery Center in Beijing of absorbed dose rates, treatment volumes, and rapid dose falloff comparable to Elekta Gamma Knife units. The OUR Rotating Gamma Unit was introduced in a number of centers in China in the 1990s with the first one in the United States installed at the University of California, Davis (Kubo and Araki 2002). Kubo also made measurements of the unit at Auhai as well as the modified unit installed in Davis, California. The primary source collimators of the UC Davis device were rebored on site to move the beam focus 3 cm inferior to enable more adequate treatment of posterior fossa tumors. Good consistency was reported for the three sets of measurements.

1.5 LINEAR ACCELERATOR RADIOSURGERY BEGINS: THE EMPIRE STRIKES BACK

1.5.1 THE EARLY LINAC SRS INNOVATORS

After Lars Leksell's work in Stockholm with the Gamma Knife, a number of medical physicists and other innovators around the world began adapting cobalt-60 teletherapy and linear accelerator devices to perform single-dose stereotactic intracranial radiosurgery. One of the earliest efforts took place at the University of Valencia in Spain in 1982 when neurosurgeon Juan Luis Barcia-Salorio adapted a stereotactic head frame and a special collimator for use with a cobalt-60 teletherapy device (Barcia-Salorio et al. 1982). They reported treating a carotid cavernous fistula, which is seldom treated at the present time. This effort represents a bridge between the radionuclide-based Gamma Knife and the linear accelerator–based radiosurgery, which was to come.

One of the earliest adaptations of a linear accelerator for radiosurgery came in that same year, invented by neurosurgeon Osvaldo Betti and an engineer named Victor Derechinsky in Buenos Aires (Betti, Galmarini, and Derechinsky 1982). They combined a Varian Clinac 18 linear accelerator with a custom-built tertiary collimator system and a "rocking chair" with two degrees of freedom in addition to the

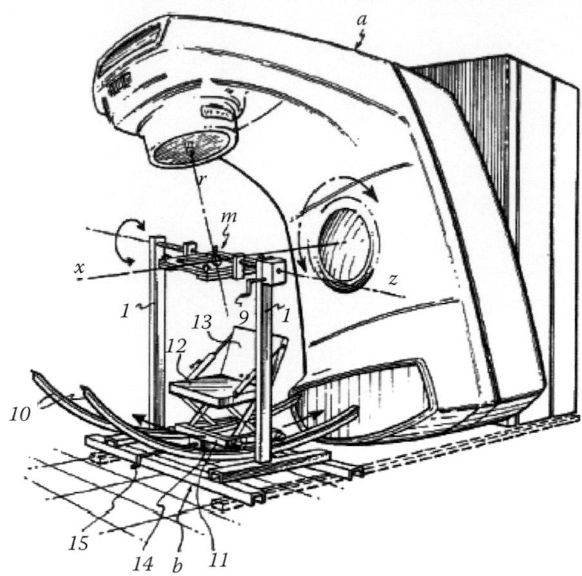

Figure 1.19 Bette and Derechinsky linear accelerator–based radiosurgery system with moveable patient chair.

motion of the linac gantry (Figure 1.19). This was one of the first examples of utilization of multiple noncoplanar arc x-ray beams converging on an intracranial target. Two more systems were later built and installed in French hospitals. A few years later, another group in Vicenza, Italy, lead by neurosurgeon Federico Colombo, developed its own stereotactic frame with a linac and mated it to a Varian Clinac 4 linear accelerator with a 4 MV x-ray beam (Colombo et al. 1985). Colombo sometimes fractionated his intracranial treatment over two fractions 8 to 10 days apart (40 to 50 Gy) and used as many as 11 converging arcs (Benedict et al. 2008). Nearly simultaneously, Gunter Hartman and Wolfgang Schlegel in Heidelberg, Germany, used a Riechert-Mundinger stereotactic frame (Hartmann et al. 1985).

The most historically important of the linear accelerator–based radiosurgery centers came together at the Joint Center for Radiation Therapy (JCRT) in Boston in 1984. Neurosurgeons Ken Winston and Eben Alexander III; medical physicists Wendell Lutz (Figure 1.20), Robert Siddon, Roger Rice, Hanne Kooy, and others; and radiation oncologists William Saunders, Jay Loeffler, and his wife Nancy Tarbell created an extremely important program, which, after the closure of JCRT, continues to this day at Massachusetts General Hospital (Rice et al. 1987; Loeffler et al. 1989). One of a number of small circular collimators 10 to 50 mm in diameter (Figure 1.21) was placed in the accessory tray of the linear accelerator with the beam

Figure 1.20 Physicist Wendell Lutz at Joint Center for Radiation Therapy, Boston, 1985.

Figure 1.21 Circular collimators for linear accelerator radiosurgery. (Courtesy of Aktina Medical, online image.)

stopped down to the outer diameter of the cone (typically about 6 cm in diameter). Lutz then fabricated a floor stand that precisely positioned the stereotactic BRW frame (attached to the patient's skull) with respect to the isocenter of the linear accelerator, which eliminated much of the error due to uncertainty of field lights, in-room laser systems, and gantry sag (Saunders, Winston, and Siddon 1988). Lutz also created a film technique that placed a precisely machined steel ball in the exact center of rotation of the linear accelerator gantry and collimator and exposed a series of films at eight combinations of couch and gantry angles (Lutz, Winston, and Maleki 1988). The "Winston-Lutz" test is still used as the basis for quality assurance of linear-accelerator systems all over the world to this day (Figure 1.22). Wendell Lutz moved on to the University of Arizona, where, together with physicist Bruce Lulu, he produced and distributed many copies of the floor stand and a specialized treatment-planning system on an Apple computer. When commercial radiosurgery linac systems became available, the United States Food and Drug Administration required all "homemade" radiosurgery systems to be retired. This was the end of an era.

Ervin Podgorsak and collaborators at McGill University in Montreal capitalized on their long history of neurosurgical research by combining a stereotactic frame with continuous arc rotation therapy with simultaneous couch and gantry rotation (Pike et al. 1987). They called this technique "dynamic radiosurgery" and had to write their own treatment-planning system to implement it. The technique was never adopted elsewhere. A few years later, Dennis Leavitt at the University of Utah introduced "dynamic

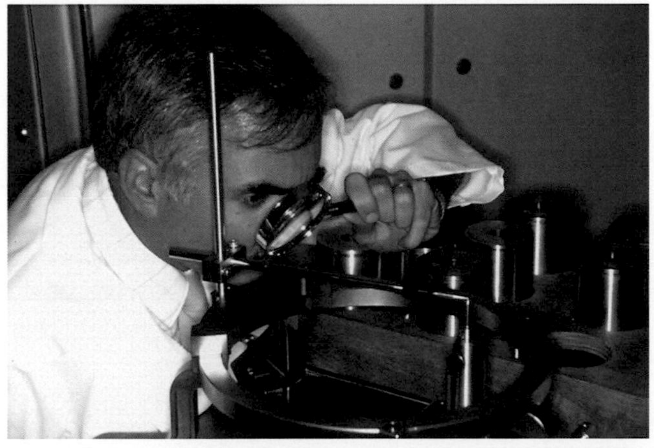

Figure 1.22 Physicist Steve Goetsch analyzing Winston-Lutz Assurance Film, UCLA Medical Center, 1993.

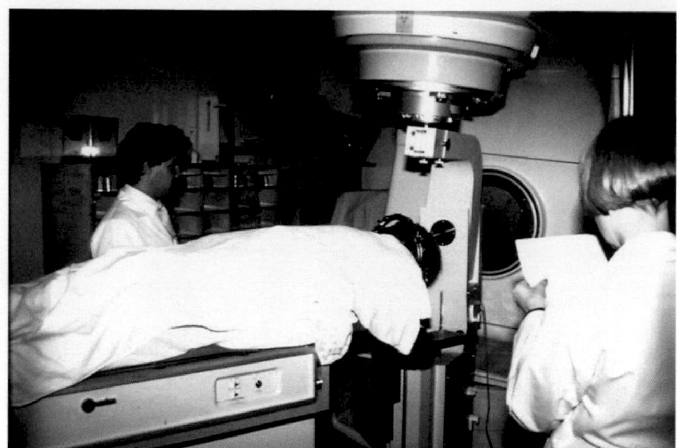

Figure 1.23 A Philips SRS 200 radiosurgery system with floor stand and gimbaled collimator holder with Varian Clinac–18 at UCLA Medical Center in 1992.

field shaping" by adding a custom-made set of shaping vanes upstream from the circular collimators to narrow the field to various polygonal shapes (Leavitt et al. 1991). This was one of the first attempts to reduce a circular field chosen to be large enough to cover a target volume down to a more conformal shape with a reduction in targeting of normal brain tissue.

Two more pioneers of radiosurgery were neurosurgeon Bill Friedman and physicist Frank Bova at the University of Florida. After observing the JCRT floor stand, they took the innovation one step farther and created a precision floor stand with one high-precision ball-bearing system for 360-degree rotation with respect to the floor and an arm holding a circular collimator (Bova and Friedman 1992). This clever system, later commercially marketed by Philips as the SRS-200 radiosurgery system (Figure 1.23), reduced the total overall targeting error of the linear accelerator to 0.5 mm or less, which was remarkable when linear accelerators of that era (1990) typically had trouble achieving pointing accuracy of 2 to 3 mm or less. The Florida system was later sold to the Sofamor Danek division of Medtronic, then to Zmed, Inc., and then sold again to Varian, where it was ultimately discontinued.

Another new device was put forward in 1995 by neurosurgeon Mark Carol. He proposed a system of vanes that could be rapidly opened and closed by air pressure to form a dynamic MLC system capable of delivering intensity-modulated photon beams (Carol 1993). The system was introduced as the Peacock system (Nomos Corporation, Sewickley, PA), which was mounted in the accessory tray of a standard linear accelerator (Woo et al. 1996; Carol et al. 1996). A later innovation with this system was the TALON surgically implanted (but removable) head frame, which allowed the patient to be docked to a table-mounted base frame and precisely repositioned over a period of several days (Salter et al. 2001).

Perhaps the first report of extracranial SRS was by neurosurgeon Andrew Hamilton; physicist Bruce Lulu and radiation oncologist Helen Fosmire, Lulu's wife; and others at the University of Arizona (Hamilton et al. 1995). They created an innovative but very arduous system for attaching immobilization plates to the spinal processes superior and inferior to a spinal tumor, which they then treated with single-dose radiosurgery. The technique required the patient to be placed under general anesthesia, maintained in the prone position, implanted with the immobilization devices, escorted to a CT scanner, kept asleep during treatment planning, and then taken to a linear accelerator and docked in position for multiple noncoplanar arcs. After treatment, the patient was returned to surgery to have the immobilization plates removed. Although attempts were made to market this system thru Leibinger and Fischer (Freiberg, Germany, now owned by Stryker), the system was not commercially successful, and frameless extracranial treatment methods later became popular.

1.5.2 DEDICATED LINEAR ACCELERATOR RADIOSURGERY DEVICES

One of the advantages of the Leksell Gamma Knife system is that it is a dedicated device, useful only for treating intracranial lesions. From the busy neurosurgeon's point of view, this is exactly what they want,

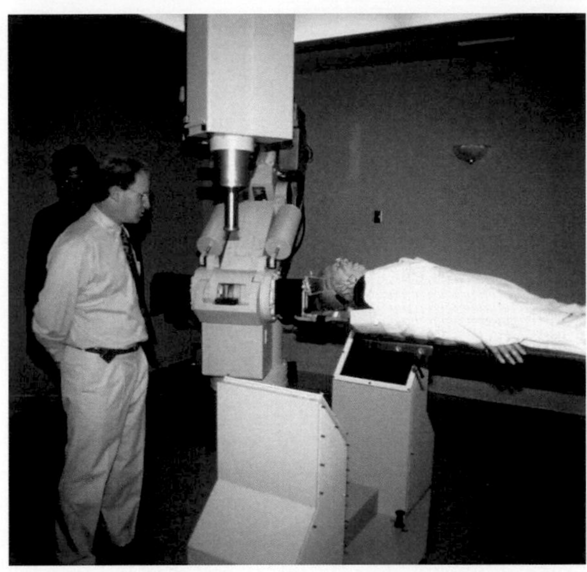

Figure 1.24 John Adler with CyberKnife patient.

rather than treating a patient on a busy linear accelerator, which has to be carefully modified and checked before a radiosurgery patient can be treated over the lunch hour or at the very end of a long clinical day. Sensitive to this concern, a new category of linear accelerators dedicated to radiosurgery began to be introduced in the early 1990s. The first linear-accelerator device that was dedicated to radiosurgery of the brain—or, in fact, the entire body—was the Accuray Neurotron 1000, later renamed the CyberKnife (Adler 1993). This revolutionary device (Figure 1.24) introduced many new concepts at the same time: An x-band very compact 6-MV linac was mounted on a robotic arm and could direct a circular collimated beam of x-rays toward a target from 100 or more positions with six degrees of freedom (Adler et al. 1999). This was also the first image-guided frameless system, utilizing two cross-firing ceiling-mounted x-rays with floor-mounted imaging detectors to localize patient positioning. The advanced imaging system compared real-time patient images against a stored library of digitally reconstructed radiographs (DRR), which was a very new concept at the time (Murphy and Cox 1996). The system was so advanced that it was difficult to implement, and the newly created company (Accuray, Sunnyvale, CA) had to pull back for a time after the first four systems were delivered to do some major redesigns. The CyberKnife later achieved great success, and Accuray became a publically traded company.

UCLA Medical Center was an early adopter of the Philips SRS-200 linac-based radiosurgery system and sponsored early educational conferences and training sessions for those who were new in the field (De Salles and Goetsch 1993). UCLA Medical Center was also the site of the first hybrid Novalis unit installed in 1997 (Solberg et al. 2001). The unit consisted of an upgraded Varian 600SR 6-MV linear accelerator with a 10-cm diameter fixed primary collimator and a new miniature MLC with 2.5-mm leaves, manufactured by Brainlab (Heimstetten, Germany). The linac had a stabilized output in arc mode from 0.3 to 20 cGy/degree and a maximum dose rate of 800 MU per minute in static mode. The unit had specialized treatment-planning software and could utilize either the miniature MLC leaves or circular collimators 5 to 60 mm in diameter. A number of manufacturers began to sell micro-multileaf collimators (mMLCs) as add-ons to existing linear accelerators as linac-based radiosurgery began to spread worldwide (Cosgrove et al. 1999).

1.6 STEREOTACTIC RADIOTHERAPY BEGINS

1.6.1 BIOLOGY OF SINGLE VERSUS MULTIPLE FRACTIONS

Radiation therapy began as an experimental treatment by bold physicians and others just before the dawn of the 20th century. The harmful side effects of diagnostic and therapeutic radiology were evident from the early days, but no one knew quite how to "harness the genie." As with most new medical endeavors, radiation therapy moved ahead mostly by trial and error with the advantage of a large network of dedicated scientific societies in a number of countries within the first decade of this technique. Radiology blossomed in France from the beginning, and French physicians have made enormous contributions to the field. One of the earliest scientific studies on the effects of radiation was published by Jean Bergonié and Louis Tribondeau in 1906 (Bergonié and Tribondeau 2004). These researchers worked at the University of Lyon under the direction of pioneering French radiologist Jean Claude Regaud using a particular strain of white laboratory rats that Regaud had previously developed. Their observations on sterilization of the testis led them to propose the so-called Law of Bergonié and Tribondeau: "The effects of irradiation on the cells are more intense the greater their reproductive activity, the longer their mitotic phases, and the less their morphology and functions are established" (del Regato 1976a). This insight explains many of the adverse effects of both radiation therapy and chemotherapy to this day: The hair follicles, mucosa, and lining of the gastrointestinal tract are very sensitive to either kind of therapy.

The next question to be addressed was the total dose, dose per fraction, and overall length of treatment for patients treated with radiation therapy. The very earliest radiation therapy dose and timing were empirically determined, and treatment was often very lengthy due to the low dose rate from early x-ray tubes. Regaud was named the first director of the Pavillon Pasteur at the University of Paris in 1912, but his work was interrupted by the Great War. Regaud switched to the larger target of the testis of the ram, hypothesizing that sterilization of the animal without unacceptable side effects would serve as a model for sterilization of a cancerous tumor. He found it difficult to sterilize the ram in a single dose without bad injury to healthy tissue. By 1925, he concluded, "The elongation of time of irradiation, without increase in the size of the dose, enhances the effect. Under the circumstances of our experiment, it seems more important to increase the time rather than the dose" (del Regato 1976a). Thus, fractionation in radiation therapy became an accepted principal, almost universally employed. Regaud gave an important invited address to the American Radium Society in Chicago in 1934 where he was hailed as a great innovator. Regaud recruited Henri Coutard in 1918 to provide diagnostic and x-ray therapy for patients at the University of Paris (del Regato 1987). Coutard made a very careful study of the radiosensitivities of the tumors he found in his patient population. Coutard advocated treatment times of 40 days or more with reduced daily fractions to allow time for biological recovery of normal tissue (Coutard 1934).

When Lars Leksell initiated the concept of SRS in a single dose, he was thinking and behaving like a surgeon rather than a traditional radiation oncologist. There has been a creative tension between the surgical approach of neurosurgeons and the fractionated approach of radiation oncology since the field of SRS was invented. Mathematical modeling of radiation biology effects improved substantially with the introduction of the linear-quadratic model (Fowler 1989). Radiation biologist and physicist Jack Fowler once described single-dose SRS as "bad biology saved by good physics." Jay Loeffler, Dennis Schrieve, Nancy Tarbell, and others at Harvard began to introduce the new concept of stereotactic radiotherapy (SRT), which embodied the principles of rigid fixation and extremely conformal delivery of external radiation beams (Schrieve and Loeffler 1994). Using stereotactic principles and conventional fractionation, they sought to combine the best of both traditional radiation therapy and SRS. See Jack Fowler's excellent review of the history of radiation biology as it affects radiation oncology (Fowler 2006).

1.6.2 DAWN OF STEREOTACTIC RADIOTHERAPY

The author of this chapter predicted at a medical physics conference in 1992 that eventually the disciplines of SRS and conventional radiation therapy would merge. This has certainly been true for some time now. The very innovative group at the JCRT in Boston may have been the first to use the term *SRT* (Dunbar et al. 1994). They adapted a relocatable version of the BRW stereotactic head frame, which became known as the Tarbell-Loeffler-Cosman frame for use on both children and adults. Lauri Laitinen, a neurosurgical colleague of Lars Leksell at the Karolinska Hospital, fabricated the "Laitinen stereoadapter" frame, which had two ear bars and a nasion piece held in place by very tight springs (Laitinen et al. 1985). The original CyberKnife units were supplied with a Laitinen frame for use in treating intracranial tumors. Clinicians at the Royal Marsden Hospital in London created a removable stereotactic head frame that could be reattached to a patient's skull for fractionated treatment of intracranial tumors (Thomas et al. 1990). The frame relied on a customized dental appliance that secured to the teeth of the patient's upper jaw during treatment (bite block). The frame was first tested on 10 nontherapy patients with reported repositioning accuracy of 0.5 mm or less (Gill et al. 1991).

Just as other surgical colleagues of Kjellberg used the Harvard cyclotron to treat extracranial targets, innovative physicist Ingmar Lax and radiation oncologist Henrik Blomgren at the Karolinska Hospital created a stereotactic body frame (Lax et al. 1994). Lax and Blomgren created a frame (Figure 1.25) that would both immobilize the patient as well as provide stereotactic fiducial markers visible on a whole body CT scan (Blomgren et al. 1995). They reported using this frame to treat lung and liver tumors. A group led by Minoru Uematsu in Saitama, Japan, developed a device called FOCAL (fusion of CT and linear accelerator) with an in-room CT scanner and a rotating table that immediately repositioned the patient for fractionated treatment after scanning (Uematsu et al. 2001). No stereotactic frame was used in this process, making this one of the first frameless SRT systems.

The prolific University of Florida group lead by physicist Frank Bova and neurosurgeon Bill Friedman adapted a new imaging technique to SRT. The group introduced the concept of the attachment of infrared light-emitting diodes to the patient coupled with a wall-mounted video detector capable of 0.01-mm resolution (Bova et al. 1997). They reported a mean repositioning error of 0.20 mm for the first 11 patients treated. A group at the University of California, Davis, first described the currently used approach when they mounted a ring of six LED emitters around each of three cameras and used the reflection from a group of polished reflecting spheres attached to the patient to determine the 3-D coordinates and orientation of the patient's external anatomy (Rogus, Stern, and Kubo 1999). This type of photogrammetry system (Figure 1.26) is now offered on several commercial linear accelerators and radiosurgery packages under various trade names (Wang et al. 2001).

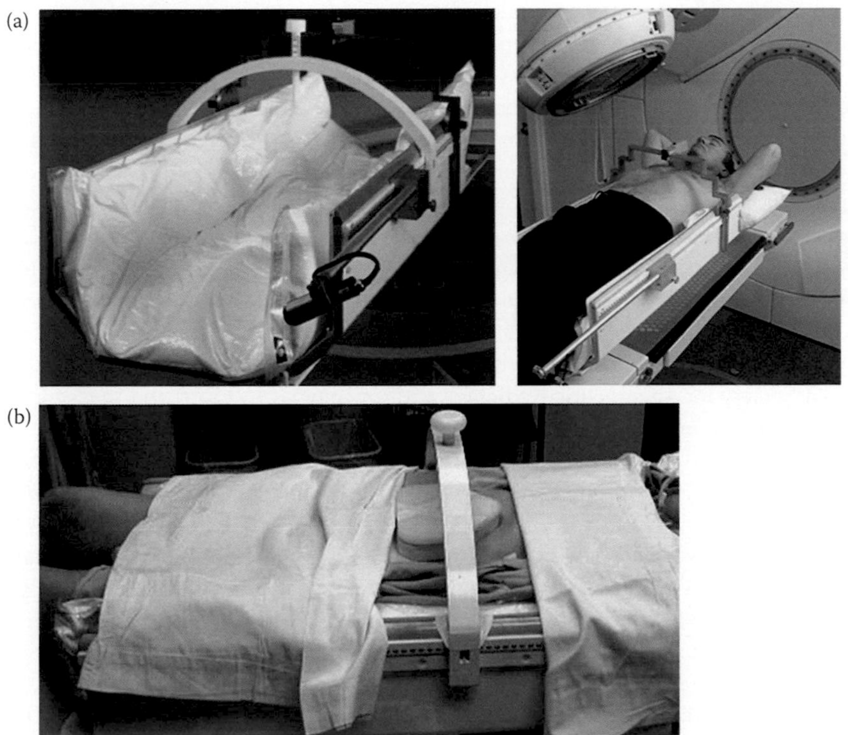

Figure 1.25 (a) Elekta stereotactic body frame based on the work of Lax and Blomgren (with molded insert and with patient on treatment couch). (b) Same frame showing abdominal compression plate.

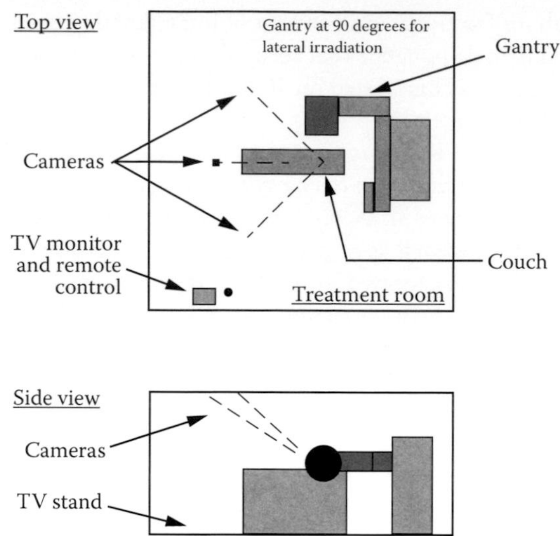

Figure 1.26 In-room photogrammetry system. (From Rogus, R. D. et al., *Med. Phys.* 26, 721–28, 1999.)

1.7 CURRENT STATE OF THE ART IN STEREOTACTIC SURGERY AND RADIOTHERAPY

1.7.1 QUALITY ASSURANCE AND SAFETY

No boldly innovative new field prospers without encountering some growing pains. As radiosurgery groups worldwide sought the best balance between controlling the growth of tumors (some of which would otherwise prove fatal) and being overly aggressive, mistakes were inevitably made and accidents occurred. A report by the United States Nuclear Regulatory Commission called attention to 15 "misadministrations" reported by licensed Gamma Knife users in the United States between 1987 and 1999 (Goetsch 2002). All of these accidents took place at centers using the early Leksell Gamma Knife Models U, B, and B2, which had no automatic checks or record-and-verify systems. Most errors involved interchange of manually set Y and Z coordinates or wrong site, wrong patient errors, which are still troublesome in the present day. Goetsch's analysis of the new Model 4C Gamma Knife with APS demonstrated that virtually all of these human errors would not have been possible with the new system although it could not eliminate all possible errors.

Enhanced quality assurance and quality improvement have been continuing themes throughout the more than 60-year history of radiosurgery, including the first American Association of Physicists in Medicine (AAPM) task group report on SRS (Schell et al. 1995). A "Consensus Statement on Stereotactic Radiosurgery Quality Improvement" was written jointly by the Task Force on Stereotactic Radiosurgery of the American Society for Therapeutic Radiology and Oncology (ASTRO) and the Task Force on Stereotactic Radiosurgery of the American Association of Neurological Surgeons (AANS). The joint report, written by 22 distinguished neurosurgeons, radiation oncologists, and medical physicists, was published simultaneously in January 1994 in both the *International Journal of Radiation Oncology * Biology * Physics* (Red Journal) and *Neurosurgery* (Larson et al. 1993). A special symposium was sponsored jointly by AAPM, ASTRO, and the National Cancer Institute in Dallas, Texas, in February 2007 titled "Quality Assurance of Radiation Therapy: The Challenges of Advanced Technologies." All invited papers were published in a special supplement to the Red Journal in 2008. Four papers presented at that symposium dealt with quality assurance for linear accelerator and Gamma Knife radiosurgery (Goetsch 2008), stereotactic body radiotherapy (Galvin and Bednarz 2008), CyberKnife (Dieterich and Pawlicki 2008), and frameless stereotactic cranial and extracranial hypofractionated radiotherapy (Solberg et al. 2008). Two additional AAPM task groups have issued reports on stereotactic body radiotherapy (Benedict et al. 2010) and quality assurance for robotic radiosurgery (Dieterich et al. 2011). Recommended practice guidelines for the performance of stereotactic body radiotherapy were published jointly in 2010 by ASTRO and the

American College of Radiology (Potters et al. 2010). With recent reports in national news media of tragic radiation therapy accidents and overdoses, patient safety in SRS and radiotherapy remains an active area of investigation and emphasis. It is clear that adequate training of highly skilled staff together with vigorous support of the hospital or clinic are just two of the essential requirements for the successful performance of SRS or SRT.

1.7.2 LATEST INNOVATIONS

The technical challenge of delivery of precisely directed highly conformal beams of photon radiation to deep intracranial or extracranial target volumes was largely solved by the beginning of the 21st century. The remaining problems now deal largely with delivery quality assurance, daily imaging guidance, and better understanding of the biological response of individual tumors versus normal tissue tolerance. The concept of dose volume histograms (DVHs) had been introduced at the Lawrence Berkeley heavy ion facility in the mid-1980s and did not come into general use in conventional radiation therapy until the advent of multislice CT planning and 3-D treatment-planning systems (Austin-Seymour et al. 1986). Plan-evaluation tools like DVHs and the Paddick Conformity Index (Paddick 2000) allow the clinical staff to test their proposed treatment against the published literature and their own in-house records prior to administration of radiosurgery.

Electronic portal imaging, particularly with flat panel electronic devices began to be introduced into radiation therapy about 1991 (Boyer et al. 1992). Enormous advantages of electronic imaging versus film cassettes were immediately apparent: The images were available immediately and could be digitally enhanced to provide improved contrast. Specialized devices, such as the Novalis, the CyberKnife, and TomoTherapy, were designed to be utilized with daily onboard imaging devices (Yang et al. 1997). Clinicians immediately discovered that previous delivery of conventional radiation therapy was much less precise than they had imagined: Early film loops showing the random motion of a patient undergoing 6 to 7 weeks of treatment showed centimeters of motion. Collection of such images has made possible the idea of *adaptive therapy*, in which dose delivery could be changed partway through a planned multifraction treatment to correct for changes in anatomy and failure to deliver dosages accurately to the intended volume (Yan et al. 1995). The quality of such imaging devices improved dramatically when physicist David Jaffray, John Wong, and others at William Beaumont Hospital in Michigan reported on a prototype Elekta SL-20 linear accelerator with the addition of an onboard kilovoltage digital imaging device (Jaffray et al. 1999).

The advent of cone-beam computed tomography (CBCT) on board another Elekta SL-20 linear accelerator was another major breakthrough reported by Jaffray and others at William Beaumont Hospital (Jaffray et al. 2002). This device enabled radiation therapy staff to create the same kind of images (although at lower resolution) as those obtained at the time of CT simulation. Images could be compared side by side or overlaid with a "spyglass" technique. This was a major innovation, especially for stereotactic body radiation therapy (SBRT). Surgical implantation of fiducial seeds now permitted online 3-D position checks of mobile organs, such as the prostate, on a daily basis. Douglas Moseley, David Jaffray, and others at the Princess Margaret Hospital reported submillimetric precision and accuracy for CBCT for determination of localization of spherical gold markers (Moseley et al. 2007).

Stereotactic body radiotherapy, sometimes called stereotactic ablative therapy (SABR), has made rapid progress against previously intractable diseases (Brown, Diehn, and Loo 2010). Bold new treatment schemes, such as 10 Gy times five fractions or 18 Gy times three fractions, are showing remarkable results, such as 90% local control of primary lung tumors at 2 years (Olsen et al. 2011). The textbooks on radiation therapy are being rewritten every year.

At least three new devices are now making their appearance. The Elekta Gamma Knife Perfexion (and Elekta linear accelerators) can now be equipped with the Extend vacuum bite block relocatable frame for fractionated treatment first reported by the Princess Margaret Hospital group (Ruschin et al. 2010). That same group fabricated a CBCT scanner for use with the Gamma Knife, and Elekta has announced a commercial version of the same type of device (Ruschin et al. 2012). Brainlab AG (Feldkirchen, Germany) in collaboration with Mitsubishi Heavy Industries, Ltd. (Tokyo, Japan) has introduced a large ring-shaped device called the Vero containing a 6-MV linear accelerator with a gimbals-mounted collimator and high-precision pointing accuracy for SBRT (Mao et al. 2011). The system has orthogonal built-in dual kVp imaging and receptors as well as CBCT imaging. The ring rotates ±30° with respect to the patient positioning table to permit noncoplanar arc treatments. Lastly, several groups around the world are working on high-energy photon-based treatment devices coupled with real-time magnetic resonance imaging (Raaymakers et al. 2009). The ViewRay, Inc. (Cleveland, OH) at this writing is installing three cobalt-60 rotational units, with the first patient treatments occurring in 2013, at Siteman Cancer Center (Washington University, St. Louis, MO).

The future of SRS and SRT seems very bright indeed. Despite financial challenges in the United States' health care industry, the clear advantages of noninvasive treatment over either inpatient or outpatient surgery are substantial and well recognized.

REFERENCES

Adler, J. R. 1993. Frameless radiosurgery. In *Stereotactic Surgery and Radiosurgery*, ed. A. A. F. De Salles and S. J. Goetsch, 237–48. Madison: Medical Physics Publishing.

Adler, J. R., M. J. Murphy, S. D. Chang, and S. L. Hancock. 1999. Image-guided robotic radiosurgery. *Neurosurgery* 44:1299–306.

Arai, Y., H. Kano, L. D. Lunsford et al. 2010. Does the Gamma Knife dose rate affect outcomes in radiosurgery for trigeminal neuralgia? *J. Neurosurg.* 113:168–71.

Austin-Seymour, M. M., G. T. Y. Chen, J. R. Castro et al. 1986. Dose volume histogram analysis of liver radiation tolerance. *Int. J. Radiat. Oncol. Biol. Phys.* 12:31–5.

Barcia-Salorio, J. L., G. Hernandez, J. Broseta, J. Gonzalez-Darder, and J. Ciudad. 1982. Radiosurgical treatment of carotid-cavernous fistula. *Appl Neurophysiol.* 45:520–2.

Barker, F. G. 1993. The Massachusetts General Hospital: Early history and neurosurgery to 1939. *J. Neurosurg.* 79:948–59.

Benedict, S. H., F. J. Bova, B. Clark et al. 2008. The role of medical physicists in developing stereotactic radiosurgery. *Med. Phys.* 35:4262–77.

Benedict, S. H., K. M. Yenice, D. Followill et al. 2010. Stereotactic body radiation therapy: The report of AAPM Task Group 101. *Med. Phys.* 37:4078–101.

Bergonié, J., and L. Tribondeau. 2004 (original publication 1906). Interpretation of some results from radiotherapy and an attempt to determine a rational treatment technique. *Yale J. Biol. Med.* 76:181–2.

Bergström, M., and T. Greitz. 1976. Stereotaxic computed tomography. *Am. J. Roentgenol.* 127:167–70.

Bertrand, G. 2004. Stereotactic surgery at McGill: The early years. *Neurosurgery* 54:1244–52.

Betti, O. O., D. Galmarini, and V. Derechinsky. 1982. Radiosurgery with a linear accelerator. Methodological aspects. *Stereotact. Funct. Neurosurg.* 57:87–98.

Blomgren, H., I. Lax, I. Näslund, and R. Svanström. 1995. Stereotactic high dose fraction radiation therapy of extracranial tumors using an accelerator. *Acta Oncol.* 34:861–70.

Boone, M. L., J. H. Lawrence, W. G. Connor, R. Morgado, J. A. Hicks, and R. C. Brown. 1977. Introduction to the use of protons and heavy ions in radiation therapy: Historical perspective. *Int. J. Radiat. Oncol. Biol. Phys.* 3:65–9.

Bova, F. J., J. M. Buatti, W. A. Friedman, W. M. Mendenhall, C. Yang, and C. Liu. 1997. The University of Florida frameless high-precision stereotactic radiotherapy system. *Int. J. Radiat. Oncol. Biol. Phys.* 38:875–82.

Bova, F. J., and W. A. Friedman. 1992. The University of Florida radiosurgery system. *Surg. Neurol.* 32:334–42.

Boyer, A. L., L. Antonuk, A. Fenster et al. 1992. A review of electronic portal imaging devices (EPIDs). *Med. Phys.* 19:1–16.

Brodsky, A., R. L. Kathren, and C. A. Willis. 1995. History of the medical uses of radiation: Regulatory and voluntary standards of protection. *Health Phys.* 69:783–823.

Brown, J. M., M. Diehn, and B. W. Loo. 2010. Stereotactic ablative radiotherapy should be combined with a hypoxic cell radiosensitizer. *Int. J. Radiat. Oncol. Biol. Phys.* 78:323–7.

Brown, R. A. 1979. A computerized tomography-computer graphics approach to stereotaxic localization. *J. Neurosurg.* 50:715–20.

Bunge, H. J., J. A. Guevara, and A. B. Chinela. 1987. Stereotactic brain radiosurgery with Gamma Unit III RBS 5000. *Barcelona Proc. 8th Europ. Congr. Neurol. Surg.*

Bushong, S. C. 1995. History of standards, certification and licensure in medical health physics. *Health Phys.* 69:824–36.

Campbell, E. 1951. Walter E. Dandy—Surgeon, 1886–1946. *J. Neurosurg.* 8:249–62.

Carol, M. 1993. Conformal radiosurgery. In *Stereotactic Surgery and Radiosurgery*, ed. A. A. F. De Salles and S. J. Goetsch, 249–66. Madison: Medical Physics Publishing.

Carol, M., W. H. Grant, A. R. Bleier et al. 1996. The field-matching problem as it applies to the Peacock three dimensional conformal system for intensity modulation. *Int. J. Radiat. Oncol. Biol. Phys.* 34:183–7.

Colombo, F., A. Benedetti, F. Pozza et al. 1985. Stereotactic radiosurgery utilizing a linear accelerator. *Appl. Neurophysiol.* 48:133–45.
Cosgrove, V. P., U. Jahn, M. Pfaender, S. Bauer, V. Budach, and R. E. Wurm. 1999. Commissioning of a micro multi-leaf collimator and planning system for stereotactic radiosurgery. *Radiother. Oncol.* 50:325–36.
Coutard, H. 1934. Principles of x-ray therapy of malignant diseases. *Lancet* 2:1–12.
Dahlin, H., and B. Sarby. 1975. Destruction of small intracranial tumours with 60Co gamma radiation. *Acta Radiol. Ther. Phys. Biol.* 14:209–27.
Das, I. J., and K. R. Kase. 1992. Higher energy: Is it necessary, is it worth the cost for radiation oncology? *Med. Phys.* 19:917–25.
del Regato, J. 1975. Wilhelm Conrad Röntgen. *Int. J. Radiation Oncology Biol. Phys.* 1:133–9.
del Regato, J. 1976a. Claudius Regaud. *Int. J. Radiat. Oncol. Biol. Phys.* 1:133–9.
del Regato, J. 1976b. Marie Sklodowska Curie. *Int. J. Radiat. Oncol. Biol. Phys.* 1:345–53.
del Regato, J. 1978. Antoine Beclere. *Int. J. Radiat. Oncol. Biol. Phys.* 4:1069–79.
del Regato, J. 1987. Henri Coutard. *Int. J. Radiat. Oncol. Biol. Phys.* 13:433–43.
del Regato, J. 1989. Magnus Strandqvist. *Int. J. Radiat. Oncol. Biol. Phys.* 17:631–42.
De Salles, A. A. F., and S. J. Goetsch, eds. 1993. *Stereotactic Surgery and Radiosurgery*. Madison: Medical Physics Publishing.
Dieterich, S., C. Cavedon, C. Chuang et al. 2011. Report of AAPM TG 135: Quality assurance for robotic radiosurgery. *Med. Phys.* 38:2914–36.
Dieterich, S., and T. Pawlicki. 2008. Cyberknife image-guided delivery and quality assurance. *Int. J. Radiat. Oncol. Biol. Phys.* 71 (Suppl.):S126–30.
Dunbar, S. F., N. J. Tarbell, H. M. Kooy et al. 1994. Stereotactic radiotherapy for pediatric and adult brain tumors: Preliminary report. *Int. J. Radiat. Oncol. Biol. Phys.* 30:531–9.
Fabrikant, J. I., J. T. Lyman, and Y. Hosobuchi. 1984. Stereotactic heavy-ion Bragg peak radiosurgery: Method for treatment of deep arterio-venous malformations. *Br. J. Radiol.* 57:479–90.
Flickinger, J. C., L. D. Lunsford, D. Kondziolka, and A. Maitz. 1993. Potential human error in setting stereotactic coordinates for radiosurgery: Implications for quality assurance. *Int. J. Radiat. Oncol. Biol. Phys.* 27:397–401.
Flickinger, J. C., A. Maitz, A. Kalend, L. D. Lunsford, and A. Wu. 1990. Treatment volume shaping with selective beam blocking using the Leksell gamma unit. *Int. J. Radiat. Oncol. Biol. Phys.* 19:783–9.
Fodstad, H., M. Hariz, and B. Ljunggren. 1991. History of Clarke's stereotactic instrument. *Stereotact. Funct. Neurosurg.* 57:130–40.
Fowler, J. F. 1989. The linear-quadratic formula and progress in fractionated radiation therapy. *Br. J. Radiol.* 62:679–94.
Fowler, J. F. 2006. Development of radiobiology for oncology—A personal view. *Phys. Med. Biol.* 51:263–86.
Galvin, J., and G. Bednarz. 2008. Quality assurance procedures for stereotactic body radiotherapy. *Int. J. Radiat. Oncol. Biol. Phys.* 71 (Suppl.):S122–5.
Gill, S. S., D. G. T. Thomas, A. P. Warrington, and M. Brada. 1991. Relocatable frame for stereotactic external beam radiotherapy. *Int. J. Radiat. Oncol. Biol. Phys.* 20:599–603.
Goetsch, S. J. 2002. Risk analysis of Leksell Gamma Knife Model C with automatic positioning system. *Int. J. Radiat. Oncol. Biol. Phys.* 52:869–77.
Goetsch, S. J. 2008. Linear accelerator and Gamma Knife–based stereotactic cranial radiosurgery: Challenges and successes of existing quality assurance guidelines and paradigms. *Int. J. Radiat. Oncol. Biol. Phys.* 71 (Suppl.):S118–21.
Goetsch, S. J., B. D. Murphy, R. Schmidt et al. 1999. Physics of rotating gamma systems for stereotactic radiosurgery. *Int. J. Radiat. Oncol. Biol. Phys.* 43:689–96.
Greenblatt, S. H. 1995. Phrenology in the science and culture of the 19th century. *Neurosurgery* 37:790–805.
Grubbé, E. H. 1933. Priority in the therapeutic use of x-rays. *Radiology* 21:156–62.
Hamilton, A. J., B. A. Lulu, H. Fosmire, B. Stea, and J. R. Cassady. 1995. Preliminary clinical experience with linear accelerator-based spinal stereotactic radiosurgery. *Neurosurgery* 54:454–64.
Hartmann, G. H., W. Schlegel, V. Strum et al. 1985. Cerebral radiation surgery using moving field irradiation at a linear accelerator facility. *Int. J. Radiat. Oncol. Biol. Phys.* 11:1185–92.
Heilbrun, M. P., T. S. Roberts, M. L. J. Apuzzo, T. H. Wells, and J. K. Sabshin. 1983. Preliminary experience with Brown-Robert-Wells (BRW) computerized tomography stereotactic guidance system. *J. Neurosurg.* 59:217–22.
Horsley, V., and R. H. Clarke. 1908. The structure and functions of the cerebellum examined by a new method. *Brain* 31:45–124.
Horstmann, G. A., H. Schöpgens, A. T. C. J. van Eck, H. Kreiner, and W. Herz. 2000. First clinical experience with the automatic positioning system and the Leksell Gamma Knife Model C. *J. Neurosurg.* 93 (Suppl.):S193–7.
Ingram, W. R., S. W. Ranson, F. I. Hannett et al. 1932. The direct stimulation of the red nucleus in cats. *J. Neurol. Psychopathol.* 12:219–30.

International Commission on Radiation Units and Measurements. 1959. Report of the International Commission on Radiation Units and Measurements. *National Bureau of Standards Handbook 78*. Washington DC: US Government Printing Office.

Isherwood, I. 2005. Sir Godfrey Hounsfield. *Radiology* 234:975–6.

Jaffray, D. A., D. G. Drake, M. Moreau, A. A. Martinez, and J. W. Wong. 1999. A radiographic and tomographic imaging system integrated into a medical linear accelerator for localization of bone and soft-tissue targets. *Int. J. Radiat. Oncol. Biol. Phys.* 45:773–89.

Jaffray, D. A., J. H. Siewerdsen, J. W. Wong, and A. A. Martinez. 2002. Flat-panel cone-beam computed tomography for image-guided radiation therapy. *Int. J. Radiat. Oncol. Biol. Phys.* 53:1337–49.

Johns, H. E., L. M. Bates, and T. A. Watson. 1952. 1000 Curie cobalt units for radiation therapy. *Br. J. Radiol.* 25:296–302.

Kathren, R. L. 1962. Early x-ray protection in the United States. *Health Phys.* 8:503–11.

Kelly, P. J. 2004. Stereotactic navigation, Jean Talairach, and I. *Neurosurgery* 54:454–64.

Kerst, D. 1975. Betatron-Quastler era at the University of Illinois. *Med. Phys.* 2:297–300.

Kjellberg, R. N., T. Hanamura, K. R. Davis, S. L. Lyons, and R. D. Adams. 1983. Bragg-Peak proton-beam therapy for arteriovenous malformations of the brain. *N. Engl. J. Med.* 309:269–74.

Kjellberg, R. N., A. Shintani, A. G. Frantz, and B. Kliman. 1968. Proton beam therapy in acromegaly. *N. Engl. J. Med.* 278:13.

Kubo, H. D., and F. Araki. 2002. Dosimetry and mechanical accuracy of the first rotating gamma system installed in North America. *Med. Phys.* 29:2497–505.

Laitinen, L. 1993. Functional stereotactic surgery for movement disorders, pain and behavioral disorder in stereotactic surgery and radiosurgery. In *Stereotactic Surgery and Radiosurgery*, ed. A. A. F. De Salles and S. J. Goetsch, 95–106. Madison: Medical Physics Publishing.

Laitinen, L. V., B. Liliequist, M. Fagerlund, and A. T. Eriksson. 1985. An adapter for computed tomography-guided stereotaxis. *Surg. Neurol.* 23:559–66.

Larson, D. A., F. Bova, D. Eisert et al. 1993. Consensus statement on stereotactic radiosurgery quality improvement. *Int. J. Radiat. Oncol. Biol. Phys.* 28:527–30.

Larsson, B., L. Leksell, B. Rexed, P. Sourander, W. Mair, and B. Andersson. 1958. The high-energy proton beam as a neurosurgical tool. *Nature* 182:1222–3.

Lax, I., H. Blomgren, I. Naslund, and R. Svanstrom. 1994. Stereotactic radiotherapy of malignancies of the abdomen. Methodological aspects. *Acta Oncol.* 33:677–83.

Leavitt, D. D., F. A. Gibbs, M. P. Heilbrun, J. H. Moeller, and G. A. Takach. 1991. Dynamic field shaping to optimize stereotactic radiosurgery. *Int. J. Radiat. Oncol. Biol. Phys.* 21:1247–55.

Lederman, M. 1981. The early history of radiotherapy: 1895–1939. *Int. J. Radiat. Oncol. Biol. Phys.* 7:639–48.

Leksell, L. 1950. A stereotaxic apparatus for intracerebral surgery. *Acta Neurochir.* 52:1–7.

Leksell, L. 1951. The stereotactic method and radiosurgery of the brain. *Acta Chir. Scand.* 102:316–9.

Leksell, L. 1971. *Stereotaxis and Radiosurgery: An Operative System*. Springfield: Thomas Publishing.

Leksell Gamma Knife Society. 2011. *Patients treated with the Leksell Gamma Knife, 1968–2011.* Stockholm: Leksell Gamma Knife Society.

Lindquist, C., and I. Paddick. 2007. The Leksell Gamma Knife Perfexion and comparisons with its predecessors. *J. Neurosurg.* 61:130–41.

Ljunggren, B. 1993. Herbert Olivecrona: Founder of Swedish neurosurgery. *J. Neurosurg.* 78:142–9.

Loeffler, J. S., E. Alexander III, R. L. Siddon et al. 1989. Stereotactic radiosurgery for intracranial arteriovenous malformations using a standard linear accelerator: Rationale and technique. *Int. J. Radiat. Oncol. Biol. Phys.* 17:1327–35.

Lunsford, L. D., J. Flickinger, G. Lindner, and A. Maitz. 1989. Stereotactic radiosurgery of the brain using the first United States 201 cobalt-60 source Gamma Knife. *Neurosurgery* 24:151–9.

Lutz, W., K. R. Winston, and N. Maleki. 1988. A system for stereotactic radiosurgery with a linear accelerator. *Int. J. Radiat. Oncol. Biol. Phys.* 14:373–81.

Mao, W., M. Speiser, P. Medin, L. Papiez, and T. Solberg. 2011. Initial application of a geometric QA tool for integrated MV and kV imaging systems on three image-guided radiotherapy systems. *Med. Phys.* 38:2335–41.

Moseley, D. J., E. A. White, K. L. Wiltshire et al. 2007. Comparison of localization performance with implanted fiducial markers and cone-beam computed tomography for on-line image-guided radiotherapy of the prostate. *Int. J. Radiat. Oncol. Biol. Phys.* 67:942–53.

Mould, R. 1993. *A Century of X-Rays and Radioactivity in Medicine*. Bristol, UK, and Philadelphia, PA: Institute of Physics Publishing.

Murphy, M. J., and R. S. Cox. 1996. The accuracy of dose localization for an image-guided frameless radiosurgery system. *Med. Phys.* 23:2043–9.

Mussen, A. T. 1922–1923. A cytoarchitectural atlas of the brain stem of the Macaccus rhesus. *J. Psychol. Neurol.* 29:451–518.

Mutscheller, A. 1925. Physical standards of protection against roentgen ray dangers. *Am. J. Roentgenol.* 13:65.
National Bureau of Standards. 1931. X-ray protection. *National Bureau of Standards Handbook 15*. Washington DC: US Government Printing Office.
Olsen, J. R., C. G. Robinson, I. E. Naqa et al. 2011. Dose-response for stereotactic body radiotherapy in early-stage non-small-cell lung cancer. *Int. J. Radiat. Oncol. Biol. Phys.* 81 (epub):e299–303.
Orton, C. G. 1995. Uses of therapeutic x-rays in medicine. *Health Phys.* 69:662–76.
Paddick, I. 2000. A simple scoring ratio to index the conformity of radiosurgical treatment plans. *J. Neurosurg.* 93 (Suppl.):S219–22.
Piek, J., G. Lidke, T. Terberger, U. von Smekal, and M. R. Gaab. 1999. Stone age skull surgery in Mecklenburg-Vorpommern: A systematic study. *Neurosurgery* 45:147–52.
Pike, B., E. B. Podgorsak, T. M. Peters, and M. Pla. 1987. Dose distributions in dynamic stereotactic radiosurgery. *Med. Phys.* 14:780–9.
Potters, L., B. Kavanaugh, J. M. Galvin et al. 2010. American Society for Therapeutic Radiology and Oncology (ASTRO) and American College of Radiology (ACR) practice guidelines for the performance of stereotactic body radiotherapy. *Int. J. Radiat. Oncol. Biol. Phys.* 76:326–32.
Raaymakers, B. W., J. J. W. Lagendijk, J. Overweg et al. 2009. Integrating a 1.5 T MRI scanner with a 6 MV accelerator: Proof of concept. *Phys. Med. Biol.* 54:N229–37.
Rahman, M., G. J. A. Murad, and J. Mocco. 2009. Early history of the stereotactic apparatus in neurosurgery. *Neurosurg. Focus* 27:1–5.
Rand, R. W., A. Khonsary, and W. J. Brown. 1987. Leksell stereotactic radiosurgery in the treatment of eye melanoma. *Neurol. Res.* 9:142–6.
Regis, J., M. Tamura, C. Guillot et al. 2009. Radiosurgery with the world's first roboticized Leksell Gamma Knife Perfexion in clinical use: A 200-patient prospective, randomized, controlled comparison with the Gamma Knife 4C. *J. Neurosurg.* 64:346–56.
Rice, R. K., J. L. Hansen, G. K. Svensson, and R. L. Siddon. 1987. Measurements of dose distributions in small beams of 6 MV x-rays. *Phys. Med. Biol.* 32:1087–99.
Riechert, T., and F. Mundinger. 1955. Description and use of an aiming device for stereotactic brain surgery. *Acta Neurochir.* 3 (Suppl.):S308–37.
Rogus, R. D., R. L. Stern, and H. D. Kubo. 1999. Accuracy of a photogrammetry-based patient positioning and monitoring system for radiation therapy. *Med. Phys.* 26:721–8.
Ruschin, M., P. T. Komljenovic, S. Ansell et al. 2012. Cone beam computed tomography image guidance system for a dedicated intracranial radiosurgery treatment unit. *Int. J. Radiat. Oncol. Biol. Phys.* 85:243–50.
Ruschin, M., N. Nayebi, P. Carlsson et al. 2010. Performance of a novel repositioning head frame for Gamma Knife Perfexion and image-guided linac-based intracranial stereotactic radiotherapy. *Int. J. Radiat. Oncol. Biol. Phys.* 78:306–13.
Sachs, E. 1958. Victor Horsley. *J. Neurosurg.* 15:240–4.
Salter, B. J., M. Fuss, D. G. Vollmer et al. 2001. The TALON removable head frame system for stereotactic radiosurgery/radiotherapy: Measurement of the repositioning accuracy. *Int. J. Radiat. Oncol. Biol. Phys.* 51:555–62.
Saunders, W. M., K. R. Winston, and R. L. Siddon. 1988. Radiosurgery for arteriovenous malformations of the brain using a standard linear accelerator: Rationale and technique. *Int. J. Radiat. Oncol. Biol. Phys.* 15:441–7.
Schaltenbrand, G., and P. Bailey. 1959. *Introduction to Stereotaxis with an Atlas of the Human Brain*. Stuttgart: Thieme.
Schell, M. C., F. J. Bova, D. A. Larson et al. 1995. *Stereotactic radiosurgery, AAPM report no. 54*. Published for the American Association of Physicists in Medicine. Woodbury, NY: American Institute of Physics.
Schrieve, D. C., and J. S. Loeffler. 1994. Optimal fractionation schedules in small field radiotherapy. *Int. J. Radiat. Oncol. Biol. Phys.* 30:497–9.
Seibert, J. A. 1995. One hundred years of medical diagnostic imaging technology. *Health Phys.* 69:695–720.
Solberg, T. D., K. L. Boedeker, R. Fogg, M. T. Selch, and A. A. F. De Salles. 2001. Dynamic arc radiosurgery field shaping: A comparison with static field conformal and noncoplanar circular arcs. *Int. J. Radiat. Oncol. Biol. Phys.* 49:1481–91.
Solberg, T. D., P. M. Medin, J. Mullins, and S. Li. 2008. Quality assurance of immobilization and target localization systems for frameless stereotactic cranial and extracranial hypofractionated radiotherapy. *Int. J. Radiat. Oncol. Biol. Phys.* 71 (Suppl.):S1131–5.
Spiegel, E. A., H. T. Wycis, E. G. Szekely, M. Marks, and A. S. Lee. 1947. Stereotaxic apparatus for operations on the human brain. *Science* 106:349–50.
Steiner, L., C. Lindquist, J. R. Adler, J. C. Torner, W. Alves, and M. Steiner. 1992. Clinical outcome of radiosurgery for cerebral arteriovenous malformations. *J. Neurosurg.* 77:1–8.
Stone, R. S. 1948. Neutron therapy and specific ionization. *Am. J. Roentgenol. Radium Ther.* 59:771–85.
Suit, H. D., M. Goitein, J. E. Tepper, L. Verhey, A. M. Koehler, and R. Schneider. 1977. Protons and heavy ions-II. Clinical experience and expectation with protons and heavy ions. *Int. J. Radiat. Oncol. Biol. Phys.* 3:115–25.

Talairach, J., M. David, and P. Tournoux. 1957. *Atlas d'Anatomie Stereotaxique*. Vol 1. Paris: Masson.

Tan, T., and P. M. Black. 2002. Sir Victor Horsley (1857–1916): Pioneer of neurological surgery. *Neurosurgery* 50:607–12.

Tlachacova, D., M. Schmitt, J. Novotny Jr., J. Novotny, M. Majali, and R. Liscak. 2005. A comparison of the Gamma Knife Model C and the automatic positioning system with Leksell Model B. *J. Neurosurg.* (Suppl.) 102:25–8.

Thomas, D. G. T., S. S. Gill, C. B. Wilson, J. L. Darling, and C. S. Parkins. 1990. Use of relocatable stereotactic frame to integrate positron emission tomography and computed tomography images: Application in human malignant brain tumors. *Stereotact. Funct. Neurosurg.* 54:388–92.

Thwaites, D. I., and J. B. Tuohy. 2006. Back to the future: The history and development of the clinical linear accelerator. *Phys. Med. Biol.* 51:R343–62.

Tobias, C. A., H. O. Anger, and J. H. Lawrence. 1952. Radiological use of high energy deuteron and alpha particles. *Am. J. Roentgenol.* 67:1–27.

Tubiana, M., J. Dutreiz, and B. Pierquin. 1996. One century of radiotherapy in France, 1896 to 1996. *Int. J. Radiat. Oncol. Biol. Phys.* 35:227–42.

Uematsu, M., A. Shioda, A. Suda et al. 2001. Computed tomography-guided frameless stereotactic radiotherapy for stage I non-small cell lung cancer. *Int. J. Radiat. Oncol. Biol. Phys.* 51:666–70.

Voorhees, J. R., R. S. Tubbs, B. Nahed, and A. A. Cohen-Gadol. 2009. William S. Halsted and Harvey W. Cushing: Reflections on their complex association. *J. Neurosurg.* 110:384–90.

Walton, L., C. K. Bomford, and D. Ramsden. 1987. The Sheffield stereotactic radiosurgery unit: Physical characteristics and principles of operation. *Br. J. Radiol.* 60:897–906.

Wang, L. T., T. D. Solberg, P. M. Medin et al. 2001. Infrared patient positioning for stereotactic radiosurgery of extracranial tumors. *Comput. Biol. Med.* 31:101–11.

Webster, E. W. 1995. X-rays in diagnostic radiology. *Health Phys.* 69:610–35.

Wilson, R. R. 1947. Radiological use of fast protons. *Radiology* 47:487–91.

Woo, S. Y., W. H. Grant, D. Bellezza et al. 1996. A comparison of intensity modulated conformal therapy with a conventional external beam stereotactic radiosurgery system for the treatment of single and multiple intracranial lesions. *Int. J. Radiat. Oncol. Biol. Phys.* 35:593–7.

Wu, A., G. Lindner, A. H. Maitz et al. 1990. Physics of Gamma Knife approach on convergent beams in stereotactic radiosurgery, *Int. J. Radiat. Oncol. Biol. Phys.*, 18:941–9.

Yan, D., J. W. Wong, G. Gustafson, and A. Martinez. 1995. A new model for "accept or reject" strategies in off-line and on-line megavoltage treatment evaluation. *Int. J. Radiat. Oncol. Biol. Phys.* 31:943–52.

Yang, J. N., T. R. Mackie, P. Reckwerdt, J. O. Deasy, and B. R. Thomadsen. 1997. *Med. Phys.* 24:425–36.

Part II

Stereotactic radiation therapy delivery systems

The second part of this book focuses on stereotactic radiosurgery delivery systems. The field of stereotactic radiosurgery, which lead to stereotactic radiotherapy (or SABR, to some authors), advanced clinically as the equipment that made such deliveries possible was first invented and then improved. It is quite true that those of us who have been practicing radiosurgery or stereotactic radiotherapy for 20 years or more are not carrying out these treatments the way we did in the 1990s. Stereotactic frames have disappeared in many cases, fractionated treatments are becoming more common, and extracranial radiosurgery and stereotactic radiation therapy have grown tremendously.

Chapter 2 gives the history of a unique device, the Leksell Gamma Knife, and certain other similar devices described as gamma stereotactic radiosurgery (GSR) units. The original Gamma Knife in 1967 combined elements of stereotactic neurosurgery, single-dose radiosurgery (previously carried out with x-rays and protons), and cobalt-60 teletherapy. From a modest beginning, the Gamma Knife community and the GSR community have spread out worldwide and may have performed as many as 1 million intracranial patient treatments over 46 years. The Gamma Knife Perfexion takes advantage of remarkable engineering achievements in the precision and reliability of patient positioning, automated treatment delivery, and extremely fast dedicated inverse treatment-planning systems.

Linear accelerator–based radiosurgery (Chapters 3 and 4) advanced rapidly beginning in the mid-1980s with the use of add-on circular collimators and ad hoc treatment systems to very sophisticated dedicated or multipurpose devices. The robotic CyberKnife, with its novel treatment philosophy, X-band linear accelerator, and unique inverse planning system, made its first commercial appearance in 1991. After some initial growing pains, it's use spread widely throughout the world, especially in the United States and Asia. At very nearly the same time, Varian Medical partnered with Brainlab to introduce the Novalis dedicated radiosurgery system, with both circular collimators and one of the first commercial micro-multileaf collimator systems. Both the CyberKnife and the Novalis introduced cross-firing x-ray tubes with solid-state imaging devices for 3-D imaging in the room. A new generation of extremely versatile linear accelerators, equipped to "do everything" is now approved for sale, including the Varian Trilogy, TrueBeam, and The EDGE and the Elekta's Versa HD.

Light ion therapy (Chapter 5) began with proton therapy in 1954, and dedicated hospital centers are now being widely installed worldwide, especially in the United States. Ray Kjellberg's pioneering risk assessment for intracranial stereotactic radiosurgery (SRS) (particularly for AVMs) set the groundwork for linac- and Gamma Knife–based radiosurgery. Modern proton therapy centers seem to focus on more conventionally fractionated extracranial radiation therapy.

A new modern generation of very sophisticated animal irradiation devices (Chapter 6), capable of precision stereotactic radiosurgery with ultrasmall collimators (down to 1 mm in diameter) and full cone-beam CT imaging, is now coming on line. Together with such imaging innovations as micro PET, micro CT, and micro MRI, much will be learned from innovative experiments with these remarkable devices.

2 Gamma Knife

Steven J. Goetsch

Contents

2.1 INTRODUCTION

The Leksell Gamma Knife (LGK) and other similar gamma stereotactic radiosurgery (GSR) units are among the most successful medical devices in history. From the initial clinical treatments at a single hospital in Sweden, the devices have spread all across the world and have substantially changed the practice of neurological surgery. Stereotactic radiosurgery began with intracranial surgery and has now been successfully introduced in extracranial sites as well. The single fraction treatment, devised as a "minimally invasive" substitute for open craniotomy, has now been extended to multiple fractions to take advantage of 100 years of fractionated clinical radiation therapy experience.

2.2 FIRST-GENERATION COMMERCIAL DEVICE CHARACTERISTICS

The Elekta Corporation was founded in 1972 by Lars Leksell and his two sons Daniel and Laurent to market the Gamma Knife, the Leksell Model G stereotactic frame, and other surgical devices that the elder Leksell had invented. The first commercial Leksell Gamma Unit (later designated the Model U) was delivered to the University of Pittsburgh Medical Center (UPMC) in 1987 (Wu et al. 1990). The 22-ton unit delivered to UPMC contained 201 cobalt-60 sources of nominal activity 1.11 TBq (30 Ci) each with specially machined interchangeable helmets (approximately 200 kg each) with nominal beam diameters of 4, 8, 14, and 18 mm (Figure 2.1a and b). Unlike the original Gamma Knife, the collimators were circular, not elliptical, and were designed to create a nearly spherical irradiation volume. Wu reported beam alignment within 0.2 mm and an absorbed dose rate (at the center of a 16-cm diameter plastic phantom) of greater than 3 Gy/min. A simple computer program (KulaPlan) was provided to calculate the treatment time and provide two-dimensional isodose contours for a single isocenter or "shot" with a computation time of about 15 min. The X (lateral), Y (anterior-posterior), and Z (superior-inferior) coordinates were set by hand and then verified by the treatment team before each shot. A system of beam plugs was used

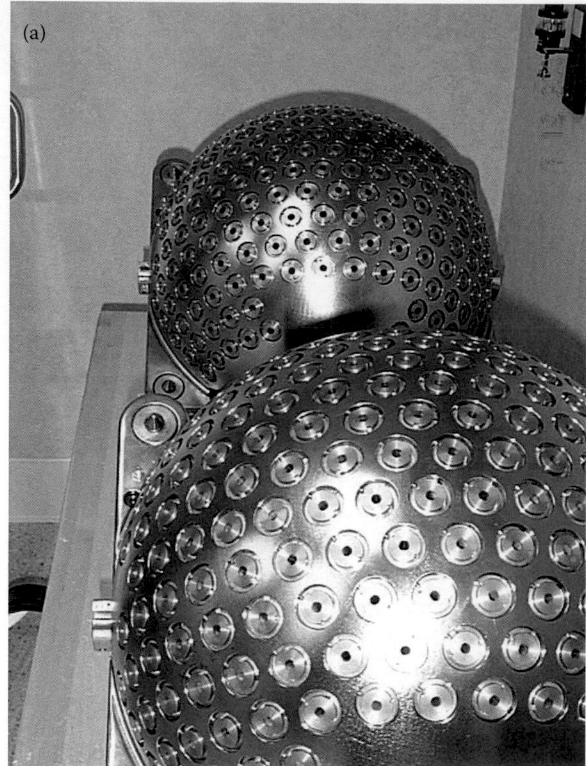

Figure 2.1 (a) LGK Model 4C 14 mm and 18 mm helmets. (b) Color-coded tertiary collimators for LGK Model 4C (4, 8, 14, and 18 mm).

to shape the beam into different three-dimensional (3-D) shapes for individualized patient treatments (Flickinger et al. 1990b).

The advent of computed tomography opened up Gamma Knife radiosurgery to the treatment of malignant and benign intracranial tumors (Lunsford et al. 1990). Although the first diagnostic radiology patient was scanned in London with Hounsfield's pioneering CT scanner in 1972, CT scans were not successfully integrated into stereotactic surgery and radiosurgery until the early 1980s. Most treatments up until 1991 still utilized a single shot tailored by a customized plug pattern. Patient selection rapidly shifted to patients with brain tumors visible on CT, especially CT enhanced with contrast. Benign and malignant brain tumors (both primary and metastatic) were treated in roughly equal numbers by 2001. Treatment of arteriovenous malformations (AVMs) became a much smaller part of the patient load in proportion to the rare incidence of this genetic defect in the general population.

The evolution of the GammaPlan^{TM} treatment-planning system (first introduced by Elekta Instruments in 1991) permitted the rapid calculation of more than one "shot" (X, Y, Z isocenter) in a single patient treatment (Flickinger et al. 1990a and 1990b). Multiple shots with the same- or different-sized collimators could now be used to tailor the treatment precisely to the 3-D shape of the tumor, AVM nidus, or target volume to be treated. Treatment of surgically difficult lesions like acoustic neuromas (also known as vestibular schwannomas) were being treated radiosurgically with some sophistication by 1995 (Foote et al. 1995). Recently, a follow-up study of 440 Gamma Knife patients treated at one center in Japan indicated a 10-year progression-free rate of 93% (Hasegawa et al. 2013). Gamma Knife radiosurgery was also demonstrated to give excellent results for the treatment of pituitary adenomas (Pollock et al. 2008). An excellent meta-analysis of radiosurgery for trigeminal neuralgia was recently published by Jean Regis, who referred to it as a revolution in neurosurgery (Regis and Tuleasca 2011).

The rapid acceptance of the LGK (with 91 units reported worldwide by June 1998 and 267 units reported operational by December 2008) led to the development of similar competing GSR devices (Leksell Society 2008) (Table 2.1). The OUR Rotating Gamma Unit was developed in China and distributed to a number of hospitals in that country (Goetsch et al. 1999). This unit contained 30 cobalt-60 gamma radiation sources of nominal activity 7.4 TBq (200 Ci) each, which rotated about a fixed isocenter (Figure 2.2a and b). The device used an internal rotating collimator that could produce beams of nominal diameters 4, 8, 14, or 18 mm at the isocenter with an absorbed dose rate (at time of initial loading) of 3 Gy/min or greater. The first such OUR Rotating Gamma Unit in North America was installed at the University of California, Davis, in 2000 (Kubo and Araki 2002). Two additional units (distributed by American Radiosurgery, San Diego, CA) are presently operating in Illinois and California. Other gamma stereotactic radiosurgery device manufacturers exist in other parts of the world but have not yet exported their GSR devices to the United States.

The LGK Model B was introduced in Europe in 1988, primarily to make the reloading of the device with new radioactive sources more convenient. The Model U Gamma Knife had to be loaded and unloaded in a "hot cell" with concrete walls up to four feet thick and remote manipulating arms. This was an expensive and time-consuming operation. The nearly hemispherical distribution of sources in the Model U device was changed to a configuration of five concentric rings of collimators beginning at an angle of 90° with respect to the longitudinal axis of the patient's body. This design allowed a loading device to be mated to the central body of the gamma unit without the necessity of constructing remote manipulating arms and a hot cell. Most American Gamma Knife centers adopted a policy of reloading sources after about one half-life (5.26 years) when treatment times would effectively double. The Gamma Knife Model B also relied on manual setting of the coordinate system.

The LGK Model C and later Model 4C (Figure 2.3), first delivered in 1999, introduced the automatic positioning system (APS), which allowed the patient's head to be moved with high-speed precision motors between closely spaced discrete coordinates (within ±2 cm of the starting point). This allowed multiple

Table 2.1 Summary of gamma stereotactic radiosurgery devices with features and year of introduction

GENERATION	NO. OF SOURCES	GEOMETRY	FEATURES	YEAR
Model U	201	Hemispherical	Manual target positioning	1987
Model B	201	Five annular rings	Manual target positioning	1988
Model 4C	201	Five annular rings	Automatic positioning system	1999
Perfexion	192	Eight moveable sectors	Internal collimators (no helmets), couch positioning system	2006
OUR	30	Rotating sources	Internal collimators	1998
American Radiosurgery	30	Rotating sources	Internal collimators, automatic positioning	2007

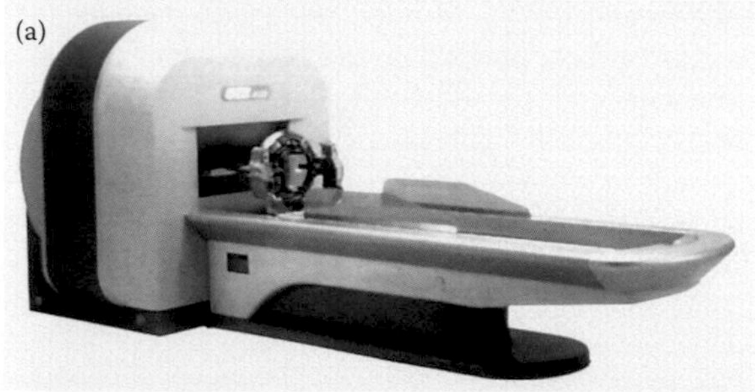

Figure 2.2 (a) OUR Rotating Gamma Unit from front. (b) OUR Rotating Gamma Unit (back side).

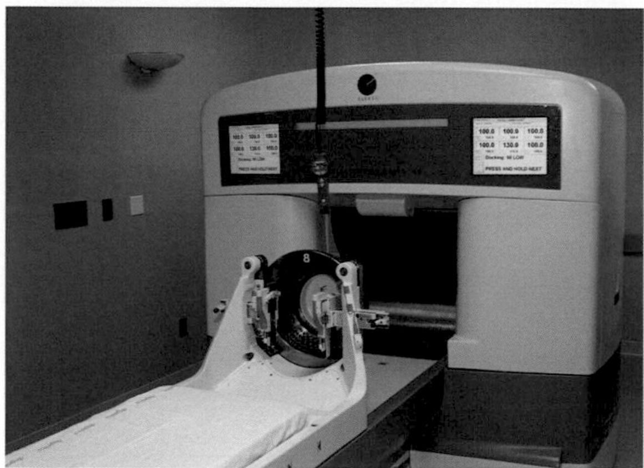

Figure 2.3 LGK Model 4C with 8 mm helmet in trunnion configuration.

shots to be treated without the necessity of the staff entering the treatment vault and setting new X, Y, and Z coordinates manually (Kondziolka et al. 2002; Kuo et al. 2004) (Figure 2.4). This innovation speeded up treatment planning and encouraged the use of more shots, leading to even more conformal plans.

Introduction of the APS and the electronic link between the treatment-planning system and the computer that controls the function of the Gamma Knife Model 4C was credited with making the device safer and more reliable (Goetsch 2002a). Steven Goetsch analyzed 15 patient treatment errors self-reported to the United States Nuclear Regulatory Commission (NRC) over a 10-year period and concluded that all of the reported errors would have been prevented (wrong helmet, wrong coordinates, shot treated twice,

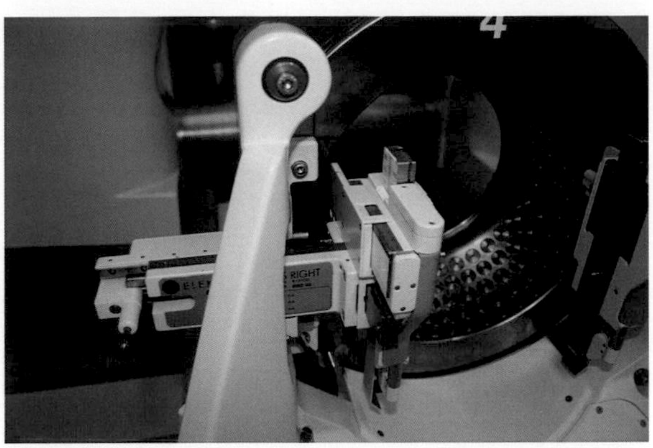

Figure 2.4 APS for LGK Model 4C.

etc.) with the new APS and patient verification system. Radiation therapy treatment errors became a source of much public discussion beginning in calendar year 2010.

2.3 GAMMA KNIFE PERFEXION

2.3.1 DESCRIPTION AND BASIC OPERATION

The LGK Perfexion (2007) introduced two paradigm shifts: First, instead of using stationary sources and changing 200-kg helmets to achieve different beam sizes, a single stationary collimator located inside the Gamma Knife unit is used in conjunction with moving sources on eight sector rods. The highly complex internal tungsten collimator is a tapered cylinder that has 576 collimating channels, which allow three different collimation sizes (4 mm, 8 mm, and 16 mm plus blocked) for each of the 192 sources. The second major change is that the motorized treatment couch provides the required accuracy to position the patient and stereotactic head frame in the treatment volume with submillimetric precision and six sigma reliability.

The ^{60}Co sources are arranged along eight moveable sectors (24 sources per sector) with each source having the same 1.11 TBq (30 Ci) nominal activity as in previous LGK models (Lindquist and Paddick 2007) (Figures 2.5 and 2.6). The sectors slide back and forth over the outer surface of the cylindrical collimator to position the sources over the desired collimating channel. Besides the three beam-on positions (4 mm, 8 mm, and 16 mm), there are two beam-off positions (the home position and sector-blocked position). The order of the positions, from the back to the front of the collimator, is H, 8, B, 4, 16, where

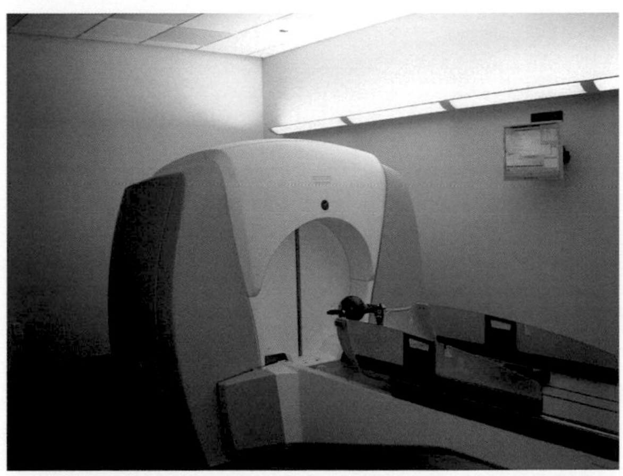

Figure 2.5 LGK Perfexion with dosimetry phantom in place.

Figure 2.6 LGK Perfexion with fiberglass cover off, showing rods with moveable sources.

"H" stands for "Home," and "B" stands for "Blocked." The sources are in the home position when the machine is turned off, and they move immediately to the home position in the case of a system fault or an emergency. The sectors move to the sector-off position (blocked), which physically lies between the 8-mm and 4-mm collimating channels, when the patient position is changed between shots. The mean transit time between isocenters is approximately 3 s, and the sector repositioning time is 0.03 s (Figure 2.7). Each sector can be independently moved to a different position, allowing for hybrid shots (i.e., shots with different collimators in different sectors) or partially blocked shots (i.e., shots with some sectors in the blocked position). It is not possible, however, to block individual sources as it was in previous LGK models. One major advantage of the new system over previous models is that the treatable volume within the LGK frame is over 300% larger than with the Model 4C (Figure 2.8a and b).

Gamma Knife radiosurgery is regulated by the U.S. NRC and designated agreement states. The primary NRC regulations governing Gamma Knife operation are found in Title 10 (Energy) Code of Federal Regulations, Part §35. Motorized movement of the radiation sources caused regulatory concern by the U.S. NRC. NRC staff placed the LGK Perfexion in regulation §35.1000, "Other medical uses of byproduct material or radiation from byproduct material," rather than in §35.600, "Use of a sealed source in … a gamma stereotactic radiosurgery unit." The separate regulatory designation of the Perfexion model Gamma Knife made licensing of this new model very difficult. In response to the NRC, the American Association of Physics in Medicine (AAPM) formed a Task Group 172, "AAPM Recommendations on Regulations for Gamma Stereotactic Radiosurgery Units." TG 172 collaborated with other medical physicists in writing a

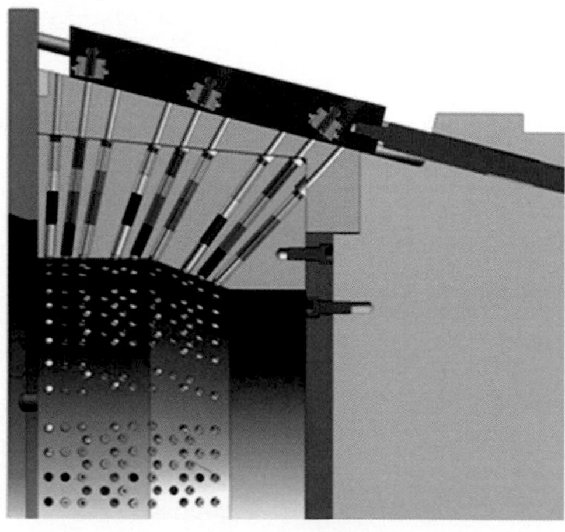

Figure 2.7 Cutaway view of sector positions for Gamma Knife Perfexion.

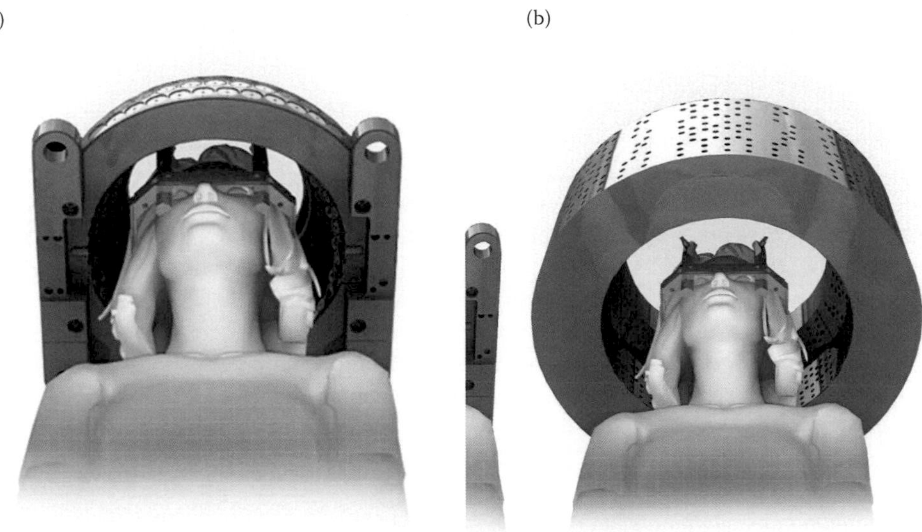

Figure 2.8 (a) Treatment volume for LGK Model 4C. (b) Treatment volume for LGK Perfexion.

new draft of §35.600 that would include the LGK Perfexion and other robotized GSR devices. The final outcome of this effort with regard to the NRC regulations is still pending at the time of this manuscript, but many LGK Perfexion units have been successfully licensed in the United States.

For the Perfexion collimators, the source-axis distances (SADs) for the beam channels are not identical for all source positions. There are five different SADs for each collimator size, resulting in 15 different SADs in total, ranging from 374 to 433 mm. This variation is unique to the Perfexion unit. In earlier LGK designs, the SAD was identical (400 mm) for all beam channels for all collimator sizes, simplifying the dose calculations for the older LGK devices.

Unlike older LGK models for which only the patient's head is moved to place the target at isocenter, the LGK stereotactic frame for the Perfexion, which is attached to the patient's head, is attached to a robotized patient couch via a patient frame adapter. The patient couch is the patient positioner (Figure 2.9) and is officially termed the patient positioning system or PPS for short. The PPS moves in the three orthogonal directions defined by the LGK coordinate system and is capable of submillimeter positioning accuracy. The positioning repeatability is <0.05 mm, and the radiological accuracy is <0.5 mm. Repositioning the patient

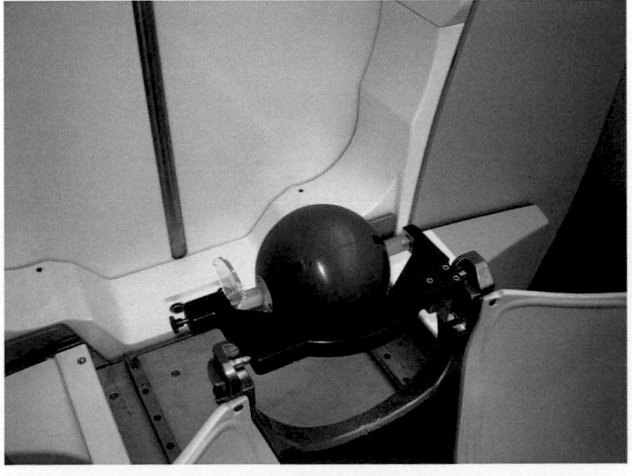

Figure 2.9 Dosimetry phantom in head positioner of LGK Perfexion.

takes 3 s or less. All couch components are manufactured by well-recognized suppliers, and the maximum patient weight specification on the couch is 210 kg (460 lbs).

2.3.2 TREATMENT PLANNING

There are various treatment-planning strategies for the LGK stereotactic radiosurgery (SRS). These strategies will not be discussed in detail here, but some of the parameters and indices used to evaluate an LGK treatment plan will be reviewed. The overall goal of LGK treatment planning is to achieve complete coverage of a target with a prescriptive dose of radiation (the goal is usually >95% coverage of the target with the prescription dose) while limiting the dose to normal brain and central nervous system structures to safe levels. Because of the precision of the LGK systems and very sharp dose falloff, treatment margins are rarely added to the target as defined on contrast-enhanced CT or MR imaging. LGK SRS is often regarded as a form of surgery with the aim of obliterating the target using radiation as opposed to the scalpel. Historically, LGK SRS has been delivered in a single fraction, and highly conformal treatment plans are required to limit the dose to surrounding brain tissue to tolerable levels. Several indices have been defined in the literature to quantify and compare LGK SRS treatment-plan conformity. Two of these indices are

$$\text{(a)} \quad C_S = \frac{V_P}{V_T} \quad \text{(Conformity index proposed by Shaw)} \tag{2.1}$$

$$\text{(b)} \quad C_P = \frac{V_{TP}}{V_T} \times \frac{V_{TP}}{V_P} \quad \text{(Conformity index proposed by Paddick),} \tag{2.2}$$

where V_P is the volume enclosed by the prescription isodose, V_T is the target volume, and V_{TP} is the volume of target enclosed by the prescription isodose volume. The conformity index proposed by Edward Shaw is always greater than or equal to one (Shaw et al. 1993) whereas Ian Paddick's conformity index is necessarily less than or equal to one (Paddick 2000). The first term in the Paddick conformity index, V_{TP}/V_T, is equal to the fractional coverage of the target by the prescription dose. If $V_{TP} = V_T$, then the target is entirely enclosed within the prescription isodose, and $C_P = 1/C_S$.

The second term in the Paddick conformity index, V_{TP}/V_P, is sometimes called the selectivity of the plan and describes the amount of overtreatment of tissue outside of the target. It is thus desirable that this term be as close to one as possible. A selectivity value much less than one indicates that although the prescription isodose volume may cover the target, it does so too generously, potentially irradiating too large a volume of normal brain tissue in the process.

A third commonly used descriptor of LGK plan quality is the so-called gradient index, defined by Paddick and Bodo Lippitz (Paddick and Lippitz 2006) as the ratio of the volume contained within the isodose line whose value equals half of the prescription dose to the volume contained within the prescription isodose, that is,

$$G = \frac{V_{(1/2)P}}{V_P} \tag{2.3}$$

For example, if the dose is prescribed to the 50% isodose volume, then the gradient index is the ratio of the 25% isodose volume to the 50% isodose volume. Paddick and Lippitz conclude that a gradient index less than or equal to about three generally indicates that the shots are reasonably well placed and the prescription isodose level has been reasonably selected. Gradient indices greater than three can occur when prescriptions are made to isodose levels higher than 50%, especially for single-shot treatments, and for plans that use hybrid shots with 16-mm collimators in some of the sectors. More commonly, high-gradient indices are associated with plans in which shots are placed so that most of the volume of larger shots overhangs from the target into the surrounding normal tissue. As always in LGK treatment planning, the advantages of a treatment-planning strategy, such as reduction in treatment time, must be weighed against the disadvantages, such as increased volume of normal brain irradiated to a significant dose.

With eight independent sectors comprising each isocenter, treatment planning for the Perfexion is somewhat more complicated than for earlier GSR units. The number of different static irradiation combinations for any given isocenter is $4^8 = 65{,}536$. Nonetheless, treatment planning for the LGK Perfexion is relatively straightforward, and planning times for typical cases are usually less than 1 hr. When first introduced, treatment planning for the LGK Perfexion was manual, that is, all treatments were forward-planned using strategies such as "sphere packing." However, beginning with Leksell GammaPlan (LGP) version 10.0, released in 2011, a new "inverse" planning function is available that can fill a designated target with isocenters and optimize the resulting dose distribution based on a number of metrics, including specificity, selectivity, and beam time (Ghobadi et al. 2012). The optimization algorithm that was introduced in LGP V10.2 uses three indices—target coverage, plan selectivity, and the gradient index—as tuning parameters. Typically, target-dose conformity (taken as the product of the target coverage and plan selectivity) is improved at the cost of an increased number of isocenters and increased treatment time. However, with the LGK Perfexion, complex treatment plans are delivered much more efficiently than with previous LGK models because there is no need to change collimator helmets or manually set treatment coordinates. A randomized comparison of 100 patients treated with the Model 4C versus 100 patients treated with the Perfexion indicated that treatment using the Perfexion took a significantly shorter period of time (40 min vs. 60 min) (Regis et al. 2009). LGP Version 10.2 also offers dynamic shaping, which permits avoidance of irradiation of defined organs at risk.

2.3.3 RADIATION PROTECTION

The LGK Perfexion weighs approximately 20,000 kg. The outer body of the radiation unit is made mostly of cast iron; the internal collimator is made of tungsten, and the shielding doors are made of steel. According to the manufacturer (Leksell GammaPlan Online Reference Manual 8.3 2008), with the sources in the home position and the shielding doors closed, the average radiation field 5 cm from a newly loaded Perfexion unit (213.1 TBq or 6300 Ci) ranges from about 0.002 mSv/h near the front surface to 0.025 mSv/h in back of the sector drive shafts. The exposure is higher in the back of the unit than it is in the front because the home position is located toward the back of the machine. The manufacturer also states that with the sources in the home position and the shielding doors closed, the average dose equivalent rate near the shielding doors is 0.012 mSv/h. Assuming, as a worst-case scenario, that a radiation worker stands in this position for about 30 min per treatment (10 min to set the patient up, 10 min to change gamma angles if necessary, plus 10 min to remove the patient from the unit) and that this radiation worker treats 300 patients per year on this machine, then he or she will receive an annual dose equivalent on the order of 1.8 mSv (well below the annual limit of 50 mSv for a radiation worker in the United States).

With the shielding doors open (e.g., in treatment position) and all sectors set to the 16-mm collimator for a newly loaded LGK Perfexion, the manufacturer reports the dose rate near the front of the machine is between 2 and 8 mSv/h. This implies that in a worst-case scenario, if a radiation worker took 10 min to respond to an emergency with the newly loaded sources stuck in the 16-mm position, he or she would receive between about 0.3 to 1.3 mSv, which is in the neighborhood of the maximum weekly permissible dose (1 mSv) for a radiation worker in the United States. It should be noted, however, that in an emergency situation, for example, a collision with the collimator cap, the source sectors are programmed to retract to the home position, in which case the expected radiation exposure to the worker responding to an emergency would be at least a factor of 10 less than the figures quoted here. The construction of the LGK Perfexion also leads to a lower scattered dose than the previous Gamma Knife Model 4C, thus reducing the incident dose to the walls of the vault.

2.3.4 EXTRACRANIAL PATIENT DOSE

Peripheral patient doses have been reported in the literature by various investigators for older LGK models and other SRS modalities (De Smedt et al. 2004; Ioffe et al. 2002; Petti et al. 2006; Xu 2008; Yu et al. 2003; Zytkovicz et al. 2007). According to Christer Lindquist and Ian Paddick, typical extracranial doses appear to be on the order of 10 times lower for the Perfexion than they are for these older LGK devices (Lindquist and Paddick 2007). Lindquist and Paddick base their conclusion on measurements they performed themselves as well as on detailed measurements made by the manufacturer. One reason for lower extracranial doses with the Perfexion is that the collimator is composed entirely of tungsten. In older LGK models, although the collimator plugs were made of tungsten, the body of the collimator was composed of steel, which has a lower density and hence lower shielding capacity. Furthermore, in the Perfexion, the sources move to a shielded position while patient coordinates are set whereas, when the automatic patient positioner in the LGK Model C sets the coordinates, the collimator is in a "defocused" position, and the sources are not shielded. Lijun Ma et al. (2011) later computed an integrated dose to normal tissue from radiosurgical treatment of multiple brain metastases and found that the Gamma Knife Perfexion consistently gave lower scattered doses than other types of SRS treatment apparatus.

For the LGK Model U, Vladimir Ioffe et al. (2002) concluded that leakage radiation is the dominant contributor to the peripheral patient dose in Gamma Knife treatments. The data of Francois De Smedt et al. (2004) support this conclusion. They found a strong correlation between the extracranial dose and treatment time but observed no obvious correlation with absolute dose, integral dose, or treatment volume. This is not surprising because leakage radiation is an important consideration in any type of radiotherapy that relies on irradiating a target volume by adding together contributions from small subfields. Data have not yet been published that describe the relative contribution of internally scattered versus leakage radiation to patient peripheral doses for the Perfexion. One might expect the relative contribution from leakage to be lower for the Perfexion due to the improved shielding of this machine. However, the effect of improved shielding may be somewhat mitigated by the fact that because it is relatively simple to do so, typical Perfexion treatment plans may use more isocenters, in particular more 4-mm shots, than treatment plans with older LGK models. The resulting longer treatment times might result in somewhat greater leakage radiation as compared to earlier Gamma Knife models.

Peripheral dose studies for various radiotherapy treatment modalities, including pre–Perfexion Gamma Knife (Chuang et al. 2008), CyberKnife, Tomotherapy, and intensity-modulated radiation therapy (IMRT), are summarized in a review article by X George Xu (2008). Data shown graphically in Figure 1a of this review demonstrate that the patient peripheral dose for 6MV photons at a given distance from the source, expressed in μGy/MU, varies by less than a factor of two for conventional radiotherapy using various linear accelerators with and without a multileaf collimator. Data reported in the AAPM Task Group Report 36 by Marilyn Stovall et al. (1995) are included in this graph. Figure 1b in Xu's review shows that for IMRT, Gamma Knife, CyberKnife, and Tomotherapy—devices that rely on the superposition of many subfields of radiation to treat the target volume—the peripheral dose at a given distance from the source, expressed in μGy/MU, is highly device-dependent and varies by a factor of as much as five or six between the different modalities. The LGK Model C generally delivers the lowest peripheral doses, expressed as in μGy/MU, as compared to the other treatment modalities. Regis et al. (2009) have recently published a study comparing the Perfexion with the Gamma Knife Model 4C. They concluded that the Perfexion reduces the dose to the vertex of the patient's skull by a factor of 8.2, a factor of 10.0 to the thyroid, a factor of 12.9 to the sternum, and by a factor of 15 to the gonads. Thus, by inference, the Perfexion potentially reduces the dose to adjacent body parts by a factor of 50 times less than some of the other radiotherapy and radiosurgery techniques summarized in Xu's review.

2.4 FRACTIONATED TREATMENTS AND RELOCATABLE FRAMES

The LGK was conceived by Lars Leksell as an alternative to an invasive surgical procedure or craniotomy. The Leksell Model G Frame (Figure 2.10) was applied with sharp surgical pins and local anesthetic as an outpatient procedure. It was designed to remain rigidly in place through the stereotactic CT, MRI, and/or angiographic localization studies; the treatment planning; and same-day treatment. The frame was then removed, and the patient was discharged.

Gabriela Simonova et al. (1995) reported on their work using the Leksell Model G frame in one application over a period of 2 to 6 days. They treated a total of 48 patients with two to six fractions over this extended period of time. This technique represents a hybrid of conventional single-dose stereotactic radiosurgery with a "hard" docking frame and use of a frameless or relocatable frame system. Luis Souhami and Ervin Podgorsak had reported on this technique using linear accelerator–based stereotactic

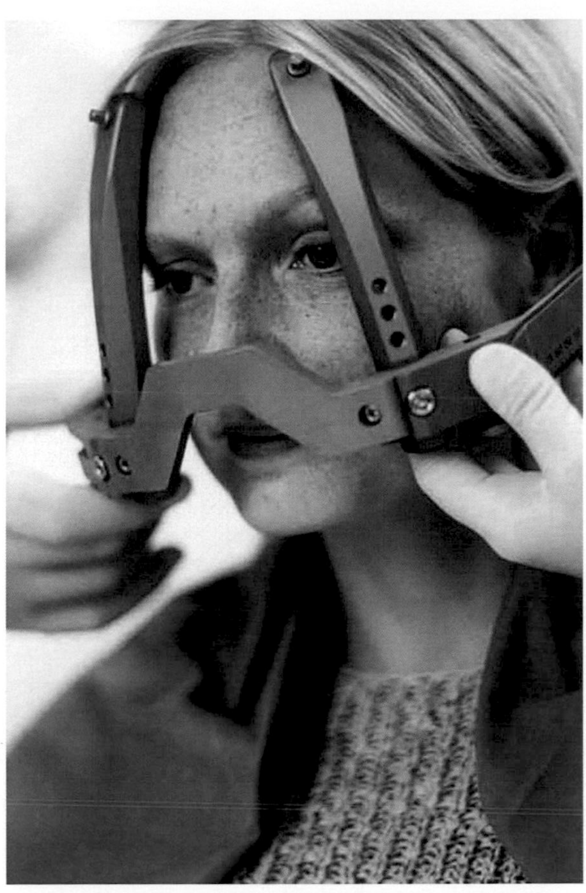

Figure 2.10 Application of Leksell Model G stereotactic frame with model patient.

radiotherapy with a "halo-frame" head ring (49 patients) and later with an in-house stereotactic frame (resembling the Leksell frame) for six fractions over 12 days with a single frame application in eight patients (Souhami et al. 1991; Podgorsak et al. 1993).

Simonova reported treating patients diagnosed with solitary metastatic brain tumors, low-grade gliomas, meningiomas, pinealomas, and a few other malignant tumors. It is believed that some tumors respond better to fractionated treatment than to single doses of radiation while injury to adjacent normal tissue may be reduced (Manning et al. 2000). This response advantage is based on the known coefficients for the linear-quadratic formula for the particular cell pathology to be treated (see chapter on SRT Radiobiology in this volume).

A paper published by Carla D. Bradford et al. (2002) on radiation-induced epilation refers in passing to unpublished fractionated studies done at the New England Gamma Knife Center. This center utilized a protocol administering 3 Gy per fraction for five consecutive days on six patients treated for acoustic neuromas.

A paper by Douglas Kondziolka and L. Dade Lunsford (1991) reported the first known use of the Gamma Knife for extracranial disease. They reported the first single-dose treatment of a recurrent squamous cell carcinoma of the nasopharynx. A similar report was later published by Cynthia Bajada et al. (1994) at UCLA Medical Center using linac-based radiosurgery for three patients.

In 2009 Elekta AB introduced the Extend relocatable bite-block frame for the Gamma Knife Perfexion (Figure 2.11) and also for use with linear accelerators (Ruschin et al. 2010). Ruschin reported on the use of this device for 12 patients treated with a total of 333 fractions using a linear accelerator. Each patient was first fitted with a customized dental mold of the upper mouth that could be used in conjunction with a vacuum device to affix the mold to the patient's upper palate. A specially designed repositioning check tool and digital probe is included with the system for use in assessing patient positioning setup errors.

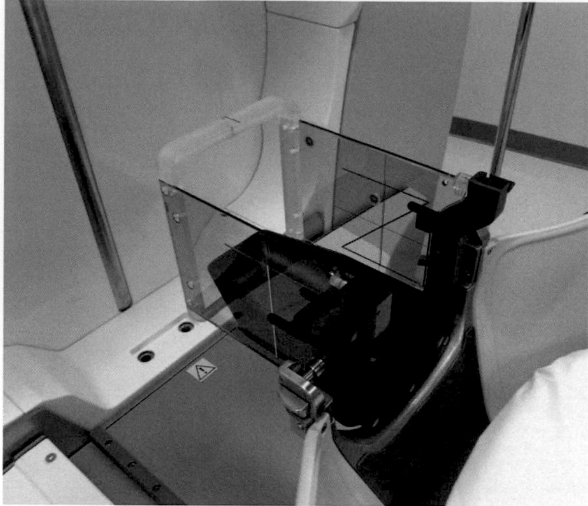

Figure 2.11 Elekta Extend™ relocatable frame system.

The initial, uncorrected mean 3-D positioning displacement was reported to be 2 mm. David Schlesinger et al. (2012) have recently reported on clinical experience with the first 10 patients using this device at the University of Virginia. They reported a mean radial setup distance error of 0.64 mm (standard deviation of 0.24 mm) over the first 36 fractions and concluded that it was satisfactory for hypofractionation treatments of intracranial lesions.

2.5 DOSIMETRY AND QUALITY ASSURANCE

The unique geometry of the original Gamma Knife required creation of a novel dosimetry definition and protocol (Wu et al. 1990). The manufacturer created a 16-cm diameter spherical phantom of acrylonitrile butadiene styrene (ABS) plastic, which was supplied with each Gamma Knife sold. The users provided their own miniature ionization chamber and had a removable 1-cm slice in the center machined to permit the insertion of the chamber so that the center of the sensitive volume was positioned at the exact center of the radiation field of the 18-mm helmet. Because of the hemispherical geometry, 40-cm SAD, 1.8-cm field size, and 8-cm depth of the calibration setup, existing calibration standards were not completely valid for the Gamma Knife. Nevertheless, each user had to choose a protocol (typically the AAPM Task Group 21 protocol in the United States) to follow, resulting in a range of different calibration methods. This unsatisfactory lack of a uniform standard must still be resolved. An AAPM Task Group (number 178) was created in 2008 and given the mission of updating dosimetry and quality assurance for gamma stereotactic radiosurgery devices. A dosimetry round robin measurement series was carried out by that group in 2011 and 2012 to compare and evaluate different calibration techniques (report in press).

The Gamma Knife also requires challenging small field measurements. The largest helmet (18 mm for U, B, C, and 4C models) and largest collimator (16 mm for Perfexion) can be measured adequately by a small-volume ionization chamber, carefully placed in the exact 3-D center of the radiation field. The manufacturer of the Gamma Knife, Elekta AB (and later the OUR Corporation for their alternate model Rotating Gamma Unit), provided the users with suggested helmet output factors (or collimator output factors) based on both physical measurements and Monte Carlo computer simulations. Measurement of small photon fields for radiosurgery (and later for IMRT) has been and continues to be controversial (Kubsad et al. 1990; Nizin 1993; Rice et al. 1987; Serago et al. 1992). Elekta AB revised their estimate for the output factor for the 4-mm helmet in 1997 from 0.80 to 0.87 (9% increase) some 10 years after introduction of the Model U Gamma Knife (Arndt 2002; Goetsch 2002a, 2002b; Nizin 1998). This change was made to accommodate results of both revised Monte Carlo measurements and the invention of a new novel liquid-filled ionization chamber (Wickman and Nystrom 1992). Later investigators utilized new dosimeters, such as microcube thermoluminescent dosimeters, radiochromic film, alanine, diodes, and diamond detectors. A number of informal conferences and users' meetings focused on the output factor for this extremely tiny field until a consensus was finally achieved by a number of investigators in a variety of publications (Ma, Li, and Lu 2000; Mack et al. 2002; Somigliana et al. 1999; Tsai et al. 2003). However, controversy still exists, and Elekta has announced that a second revision of the output factors occurred in the 2011 release of the GammaPlan treatment-planning system, which reflects updated Monte Carlo simulation results. The Gamma Knife Perfexion computes individual collimation factors for each of the 192 sources for each individual shot.

Quality assurance for the Gamma Knife has always been highly specialized due to the unique characteristics of the device. The Radiation Therapy Oncology Group published guidelines for quality assurance for SRS in 1993 (Shaw et al. 1993). Michael Schell and Hanne Kooy (1993) published a manuscript on the interdepartmental collaboration necessary for successful stereotactic radiosurgery. A Consensus Statement of the American Association of Neurological Surgeons and the American Society for Therapeutic Radiology and Oncology (ASTRO) was published simultaneously in two different journals (Larson et al. 1993, 1994). Ann H. Maitz published one of the first review articles on this topic, based on seven years of experience with the original Gamma Knife Model U at the UPMC (Maitz et al. 1995). Goetsch published another review article based on his lecture presented at a Quality Assurance in Radiation Therapy symposium sponsored in 2007 by AAPM, ASTRO, and the National Cancer Institute (Goetsch 2008). Recently, additional publications have offered techniques and measurements proposing quality-assurance and quality-control methods for the unique characteristics of the Gamma Knife Perfexion (Bhatnagar, Novotny, and Huq 2011) and patient-specific quality assurance (Mamalui-Hunter et al. 2013).

REFERENCES

Arndt, J. 2002. In response to Dr. Goetsch (letter). *Int. J. Radiat. Oncol. Biol. Phys.* 54:301.

Bajada, C., M. Selch, A. De Salles et al. 1994. Application of stereotactic radiosurgery to the head and neck region. *Acta Neurochir.* 62 (Suppl.):S114–7.

Bhatnagar, J. P., J. Novotny, and M. S. Huq. 2011. Dosimetric characteristics and quality control tests for the collimator sectors of the Leksell Gamma Knife® Perfexion™. *Med. Phys.* 39:231–6.

Bradford, C. D., B. Morabito, D. R. Shearer, G. Noren, and P. Chougule. 2002. Radiation-induced epilation due to couch transit dose for the Leksell Gamma Knife Model C. *Int. J. Radiat. Oncol. Biol. Phys.* 54:1134–9.

Chuang, C. F., D. A. Larson, A. Zytkovicz, V. Smith, and P. L. Petti. 2008. Peripheral dose measurement for CyberKnife radiosurgery with upgraded linac shielding. *Med. Phys.* 35 (4):1494–6.

De Smedt, F., B. Vanderlinden, S. Simon et al. 2004. Measurements of extracranial doses in patients treated with Leksell Gamma Knife C. In *Radiosurgery*, ed. D. Kondziolka, vol. 5, 197–212. Basel: Karger.

Flickinger, J. C., L. D. Lunsford, A. Wu, A. H. Maitz, and A. Kalend. 1990a. Treatment planning for Gamma Knife radiosurgery with multiple isocenters. *Int. J. Radiat. Oncol. Biol. Phys.* 18:1495–501.

Flickinger, J. C., A. Maitz, A. Kalend, L. D. Lunsford, and A. Wu. 1990b. Treatment volume shaping with selective beam blocking using the Leksell gamma unit. *Int. J. Radiat. Oncol. Biol. Phys.* 19:783–9.

Foote, R. L., R. J. Coffey, J. W. Swanson et al. 1995. Stereotactic surgery using the Gamma Knife for acoustic neuromas. *Int. J. Radiat. Oncol. Biol. Phys.* 132:1153–60.

Ghobadi, K., H. R. Ghaffari, D. M. Aleman, D. A. Jaffray, and M. Ruschin. 2012. Automated treatment planning for a dedicated multi-source intracranial radiosurgery treatment unit using projected gradient and grassfire algorithms. *Med. Phys.* 39:3134–41.

Goetsch, S. J. 2002a. Risk analysis of Leksell Gamma Knife Model C with automatic positioning system. *Int. J. Radiat. Oncol. Biol. Phys.* 52:869–77.

Goetsch, S. J. 2002b. 4-mm Gamma Knife helmet factor (letter). *Int. J. Radiat. Oncol. Biol. Phys.* 54:300.

Goetsch, S. J. 2008. Linear accelerator and Gamma Knife–based stereotactic cranial radiosurgery: Challenges and successes of existing quality assurance guidelines and paradigms. *Int. J. Radiat. Oncol. Biol. Phys.* 71 (Suppl. 1): S118–21.

Goetsch, S. J., B. D. Murphy, R. Schmidt et al. 1999. Physics of rotating gamma systems for stereotactic radiosurgery. *Int. J. Radiat. Oncol. Biol. Phys.* 43:689–96.

Hasegawa, T., Y. Kida, T. Kato, H. Iizuka, S. Kuramitsu, and T. Yamamoto. 2013. Long-term safety and efficacy of stereotactic radiosurgery for vestibular schwannomas: Evaluation of 440 patients more than 10 years after treatment with Gamma Knife surgery. *J. Neurosurg.* 118:557–65.

Ioffe, V., R. S. Hudes, D. Shepard, J. M. Simard, L. S. Chin, and C. Yu. 2002. Fetal and ovarian dose in patients undergoing Gamma Knife radiosurgery. *Surg. Neurol.* 58 (Suppl.):S32–41.

Kondziolka, D., and L. D. Lunsford. 1991. Stereotactic radiosurgery for squamous cell carcinoma of the nasopharynx. *Laryngoscope*, 101:519–22.

Kondziolka, D., A. H. Maitz, A. Niranjan, J. C. Flickinger, and L. D. Lunsford. 2002. An evaluation of the Model C Gamma Knife with automatic patient positioning. *Neurosurgery* 50:429–32.

Kubo, H. D., and F. Araki. 2002. Dosimetry and mechanical accuracy of the first rotating gamma system installed in North America. *Med. Phys.* 29:2497–505.

Kubsad, S., T. R. Mackie, M. A. Gehring et al. 1990. Monte Carlo and convolution dosimetry for stereotactic radiosurgery. *Int. J. Radiat. Oncol. Biol. Phys.* 19:1027–35.

Kuo, J. S., C. Yu, S. L. Giannotta, Z. Petrovich, and M. Apuzzo. 2004. The Leksell Gamma Knife Model U versus Model C: A quantitative comparison of radiosurgical treatment parameters. *Neurosurgery* 55:168–73.

Larson, D., F. Bova, D. Eisert et al. 1993. Consensus statement on stereotactic radiosurgery: Quality improvement. *Int. J. Radiat. Oncol. Biol. Phys.* 28:527–30.

Larson, D., F. Bova, D. Eisert et al. 1994. Consensus statement on stereotactic radiosurgery: Quality improvement. *Neurosurgery* 34:193–5.

Leksell Society. 2008. *Indications treated December 2008.* Stockholm, Sweden.

Leksell Gamma Knife Perfexion. 2007. Instructions for use. Document number 1005337, Rev. 01 (2007/07). ©2007 Elekta Instrument AB. All rights reserved.

Lindquist, C., and I. Paddick. 2007. The Leksell Gamma Knife Perfexion and comparison with its predecessors. *Operat. Neurosurg.* 61:130–41.

Lunsford, L. D., R. J. Coffey, T. Cojocaru, and D. Leksell. 1990. Image-guided stereotactic surgery: A 10-year evolutionary experience. *Stereotact. Funct. Neurosurg.* 54:375–87.

Ma, L., X. A. Li, and C. X. Lu. 2000. An efficient method of measuring the 4 mm height output factor for the Gamma Knife. *Phys. Med. Biol.* 45:729–33.

Ma, L., P. Petti, B. Wang et al. 2011. Apparatus dependence of normal brain tissue dose in stereotactic radiosurgery for multiple brain metastases. *J. Neurosurg.* 114:1580–4.

Mack, A., S. G. Scheib, J. Major, S. Gianolini et al. 2002. Precision dosimetry for narrow photon beams used in radiosurgery—determination of Gamma Knife output factors. *Med. Phys.* 29:2080–9.

Maitz, A., A. Wu, L. D. Lunsford, J. C. Flickinger, D. Kondziolka, and W. D. Bloomer. 1995. Quality assurance for Gamma Knife stereotactic radiosurgery. *Int. J. Radiat. Oncol. Biol. Phys.* 32:1465–71.

Mamalui-Hunter, M., S. Yaddanapudi, T. Zhao, S. Mutic, D. Low, and F. Yin. 2013. Patient-specific independent 3D GammaPlan quality assurance for Gamma Knife Perfexion radiosurgery. *J. Appl. Clin. Med. Phys.* 14:62–70.

Manning, M. A., R. M. Cardinale, S. H. Benedict, B. D. Kavanagh, R. D. Zwicker, and C. Amir. 2000. Hypofractionated stereotactic radiotherapy as an alternative to radiosurgery for the treatment of patients with brain metastases. *Int. J. Radiat. Oncol. Biol. Phys.* 47:603–8.

Nizin, P. S. 1993. Electronic equilibrium and primary dose in collimated photon beams. *Med. Phys.* 20:1721–9.

Nizin, P. S. 1998. On absorbed dose in narrow ^{60}Co gamma-ray beams and dosimetry of the Gamma Knife. *Med. Phys.* 25 (12):2347.

Paddick, I. 2000. A simple scoring ratio to index the conformity of radiosurgical treatment plans. *J. Neurosurg.* 93 (Suppl.):S219–22.

Paddick, I., and B. Lippitz. 2006. A simple dose gradient measurement tool to complement the conformity index. *J. Neurosurg.* 105 (Suppl.):S194–201.

Petti, P. L., C. F. Chuang, V. Smith, and D. A. Larson. 2006. Peripheral doses in CyberKnife radiosurgery. *Med. Phys.* 33 (6):1770–9.

Podgorsak, E. B., L. Souhami, J. Caron et al. 1993. A technique for fractionated stereotactic radiotherapy in the treatment of intracranial tumors. *Int. J. Radiat. Oncol. Biol. Phys.* 27:1225–30.

Pollock, B. E., J. Cochran, N. Natt et al. 2008. Gamma Knife radiosurgery for patients with nonfunctioning pituitary adenomas: Results from a 15-year experience. *Int. J. Radiat. Oncol. Biol. Phys.* 70:1325–9.

Regis, J., M. Tamura, C. Guillot et al. 2009. Radiosurgery with the world's first fully robotized Leksell Gamma Knife Perfexion. *Neurosurgery* 64:346–56.

Regis, J., and M. Tuleasca. 2011. Fifteen years of Gamma Knife surgery for trigeminal neuralgia in the *Journal of Neurosurgery*: History of a revolution in functional neurosurgery. *J. Neurosurg.* 115:207.

Rice, R. K., J. L. Hansen, G. K. Svensson, and R. L. Siddon. 1987. Measurements of dose distributions in small beams of 6 MV x-rays. *Phys. Med. Biol.* 32:1087–99.

Ruschin, M., N. Nayebi, P. Carlsson et al. 2010. Performance of a novel repositioning head frame for Gamma Knife Perfexion and image-guided linac-based intracranial stereotactic radiotherapy. *Int. J. Radiat. Oncol. Biol. Phys.* 76:1–8.

Schell, M., and H. Kooy. 1993. Stereotactic radiosurgery quality improvement: Interdepartmental collaboration. *Int. J. Radiat. Oncol. Biol. Phys.* 28:551–2.

Schlesinger, D., Z. Xu, F. Taylor, C. P. Yen, and J. Sheehan. 2012. Interfraction and intrafraction performance of the Gamma Knife Extend system for patient positioning and immobilization. *J. Neurosurg.* 117:217–24.

Serago, C. F., P. V. Houdek, G. H. Hartmann, D. S. Saini, M. E. Serago, and A. Kaydee. 1992. Tissue maximum ratios (and other parameters) of small circular 4, 6, 10, 15 and 24 MV x-ray beams for radiosurgery. *Phys. Med. Biol.* 37 (10):1943–56.

Shaw, E., R. Kline, M. Gillin et al. 1993. Radiation therapy oncology group: Radiosurgery quality assurance guidelines. *Int. J. Radiat. Oncol. Biol. Phys.* 27:1231–9.

Simonova, G., J. Novotny, J. Novotny Jr., V. Vldyka, and R. Liscák. 1995. Fractionated stereotactic radiotherapy with the Leksell Gamma Knife: Feasibility study. *Radiother. Oncol.* 37:108–16.

Somigliana, A., G. M. Cattaneo, C. Fiorino et al. 1999. Dosimetry of Gamma Knife and linac-based radiosurgery using radiochromic and diode detectors. *Phys. Med. Biol.* 44:887–97.

Souhami, L., A. Olivier, E. B. Podgorsak, J. G. Villemure, M. Pla, and A. F. Sadikot. 1991. Fractionated stereotactic radiation therapy for intracranial tumors. *Cancer* 68:2101–8.

Stovall, M., C. R. Blackwell, J. Cundiff et al. 1995. Fetal dose from radiotherapy with photon beams: Report of AAPM Radiation Committee Task Group no. 36. *Med. Phys.* 22:63–82.

Tsai, J., M. J. Rivard, K. J. Engler, J. E. Mignano, D. E. Wazer, and W. A. Shucart. 2003. Determination of the 4 mm Gamma Knife helmet relative output factor using a variety of detectors. *Med. Phys.* 30 (5):986–92.

Wickman, G., and H. Nystrom. 1992. The use of liquids in ionization chambers for high precision dosimetry in narrow high-energy photon beams. *Phys. Med. Biol.* 37:1789–812.

Wu, A., G. Lindner, A. H. Maitz et al. 1990. Physics of Gamma Knife approach on convergent beams in stereotactic radiosurgery. *Int. J. Radiat. Oncol. Biol. Phys.* 18:941–9.

Xu, X. 2008. A review of dosimetry studies on external-beam radiation treatment with respect to second cancer induction. A topical review. *Phys. Med. Biol.* 53:R193–241.

Yu, C., G. Jozsef, M. L. Apuzzo, D. M. MacPherson, and Z. Petrovich. 2003. Fetal radiation doses for Model C Gamma Knife radiosurgery. *Neurosurgery* 52 (3):687–93.

Zytkovicz, A., I. Daftari, T. L. Phillips, C. F. Chuang, L. Verhey, and P. L. Petti. 2007. Peripheral dose in ocular treatments with Cyberknife and Gamma Knife radiosurgery compared to proton radiotherapy. *Phys. Med. Biol.* 52 (19):5957–71.

3 Conventional medical linear accelerators adapted for stereotactic radiation therapy

William H. Hinson and Timothy D. Solberg

Contents

3.1 TRANSITION FROM GAMMA KNIFE TO LINAC-BASED SRS

The conceptual foundations of stereotactic radiosurgery (SRS) originated in the work of Lars Leksell and his group at the Karolinska Institute on the Gamma Knife® unit. In the wake of the development of the Gamma Knife, several groups of physicians and physicists sought alternative methods of SRS because of the high cost of the Gamma Knife unit and the fact that it could only perform intracranial radiosurgery. The first published report of a stereotactic treatment using a conventional gantry-based system appeared in 1982 from Spain. Luis Barcia-Solorio et al. (1982) published a report on their technique using a special collimator coupled to a Co-60 teletherapy unit to treat a carotid cavernous fistula. The following year, Argentine neurosurgeons Oswaldo Betti and Victor Derechinsky published their work on a multibeam linac SRS technique coupled with a Talairach stereotactic localization system (Betti and Derechinsky 1983; Talairach, de Ajuriaguerra, and David 1952; Talairach et al. 1949). Their system used a 10-MV photon beam from a Varian Clinac 18 accelerator with secondary circular collimators made of heavy alloys, ranging in diameter from 6 to 25 mm. Patients were placed in a moveable chair, shown in Figure 3.1, and attached to a rotating head frame, shown in Figure 3.2. Modified versions of this system were later installed in two other hospitals in France.

Additional programs in Europe began to emerge at that time. Italian neurosurgeon Federico Colombo and his team of physicists, led by R. C. Avanzo, published their SRS technique in an Italian journal (Avanzo et al. 1984; Colombo et al. 1985, 1986). Their technique used a 4-MV beam from a Varian Clinac 4 without the use of secondary collimators. Their treatment deliveries included multiple arcs of 150°–160° obtained by rotations of the gantry and treatment couch. Colombo reported treatment doses of 40–50 Gy to treat tumors of 2–4 cm in size, delivered in two fractions separated by 8–10 days. In Germany, the group at Heidelberg, led by Gűnther H. Hartmann, published their work with a 15-MV beam on a Siemens Mevatron 77 accelerator and using a commercial Riechert-Mundinger stereotactic frame to deliver multiple arc radiosurgery treatments (Hartmann et al. 1985). In their work, the authors described a dosimetry calculation using a pencil beam approximation for circularly collimated beams.

In the United States, John Van Buren and Pavel Houdek and their group at the University of Miami published the first North American radiosurgery work in 1985 (Houdek et al. 1985). Their technique employed their own localization system and used the accelerator jaws to collimate the beam (Van Buren, Houdek, and Ginsberg 1983). The SRS treatments were delivered with arcs of 180° to 240° using a Toshiba

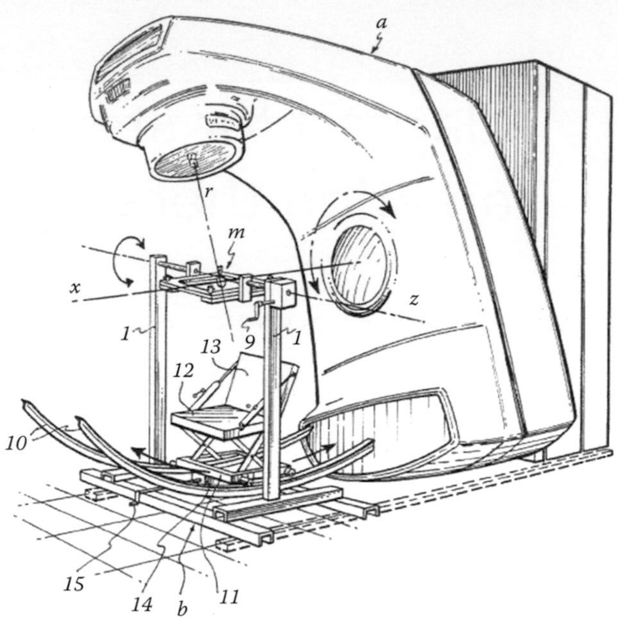

Figure 3.1 Treatment chair drawing from Betti and Derechinsky's original patent; the system is based on the Talairach stereotactic system.

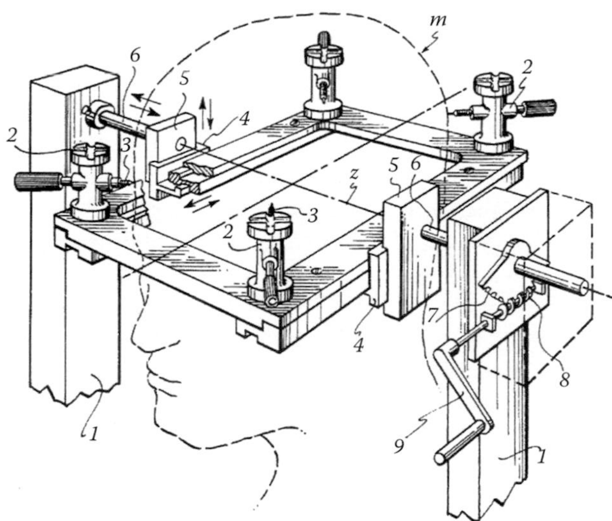

Figure 3.2 Derechinsky's stereotactic fixation device for SRS delivery.

LMR 13 accelerator with a 10-MV photon beam. The authors were concerned about the dose tolerances of the normal tissue in the brain and the accuracy of their targeting systems and therefore designed their treatments to be fractionated.

3.2 INTRODUCTION OF FLOOR STANDS

Beginning in 1985, the groups at the Joint Center for Radiation Therapy and the Harvard Medical School, headed by physicians William Saunders and Jay Loeffler and physicist Wendell Lutz, developed a system

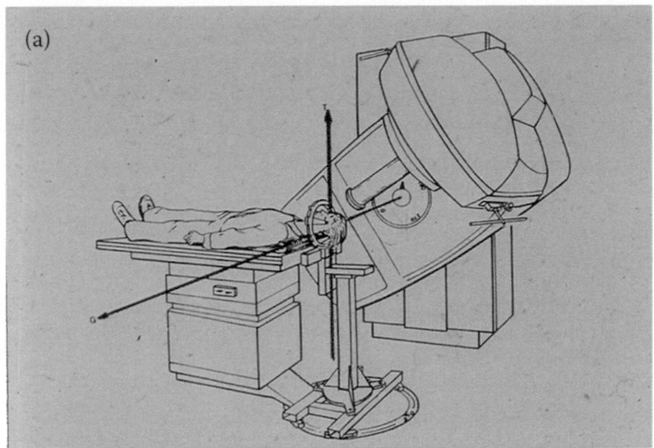

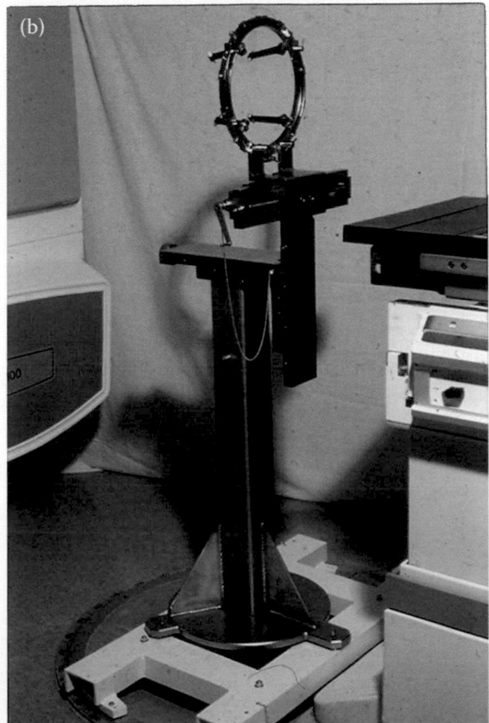

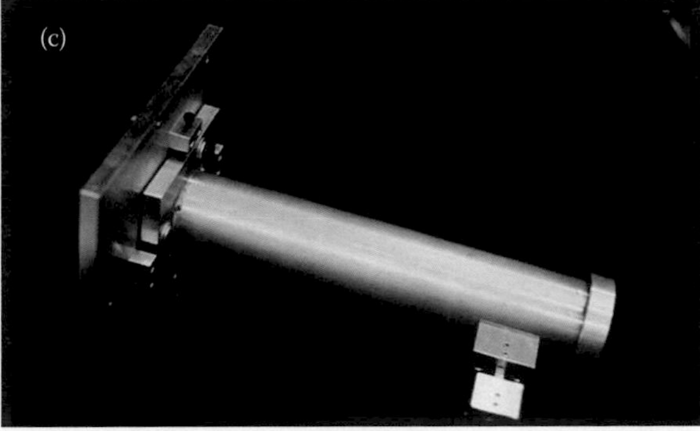

Figure 3.3 Floor stand designed by Winston and Lutz for patient position for SRS: (a) the floor stand with patient in treatment position; (b) the floor stand with screw adjustments for patient positioning; (c) the secondary collimator apparatus, which attached directly to the treatment head.

that would greatly improve the accuracy of linac radiosurgery (Loeffler et al. 1989; Saunders et al. 1988). At that time, the lack of mechanical precision of the commercially available accelerators, particularly the treatment couches, limited the use of linear accelerators in radiosurgery treatments. To address these limitations, Lutz and physician Ken Winston designed and built a floor-mounted immobilization stand, shown in Figure 3.3, to more precisely position the patient's head for treatment (Lutz, Winston, and Maleki 1984, 1986, 1988). The floor stand removed the dependency of the treatment on the mechanical accuracy of the treatment couch. The stand was attached to a conventional 6-MV accelerator. Fundamentally, it was Lutz's quality-assurance (QA) procedure, later deemed the "Winston-Lutz" test, that established a standard for patient-specific QA for linac radiosurgery. The Winston-Lutz test uses a radio-opaque ball mounted to a Brown-Robert-Wells ring on which the patient-specific coordinates are set using a phantom base, shown in Figure 3.4. Once the coordinates were established on the floor stand and the radio-opaque ball was positioned, a series of films were taken using the treatment collimators at various gantry and couch positions to verify that the ball was placed at the proper isocenter coordinates.

At McGill University, Ervin Podgorsak, shown in Figure 3.5, modified a Clinac 18 with a 10-MV photon beam to deliver multiple dynamic arc treatments (Podgorsak et al. 1987, 1988). His design included secondary collimators, which defined small circular fields delivered with simultaneous gantry and couch

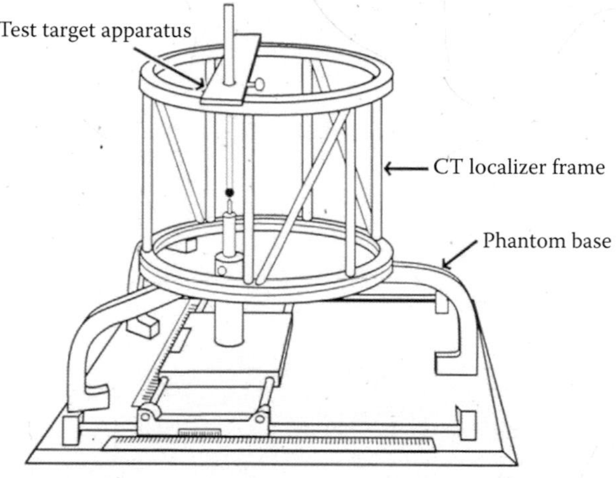

Figure 3.4 Lutz's phantom base for defining the patient-specific coordinates on a QA frame. A radio-opaque ball was positioned at the isocenter coordinates and filmed on the accelerator to verify that the treatment isocenter was properly positioned.

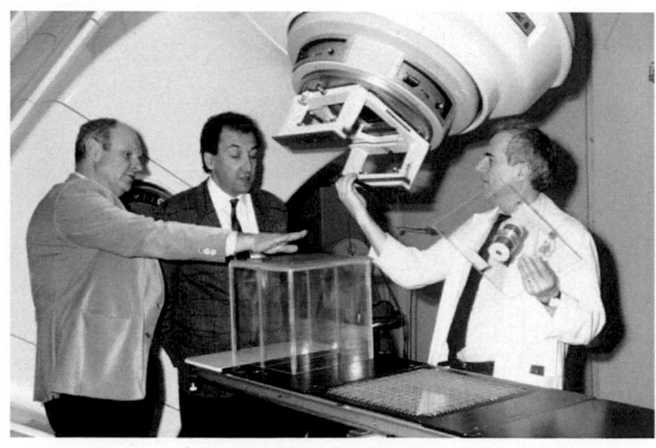

Figure 3.5 Ervin Podgorsak (right) in 1987, showing the setup of the McGill radiosurgery system to Dan Leksell (center), son of Lars Leksell, and Juergen Arndt (left), clinical director of Elekta.

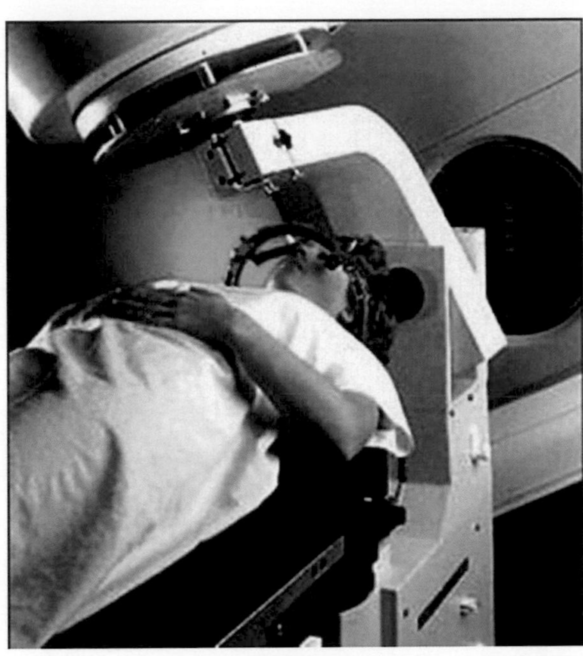

Figure 3.6 The Gainesville "Linac Scalpel" system; the rotational arm attached to the floor stand held the tertiary collimator and was attached to the linac gantry with a gimbal bearing that avoided torque on the linac head.

rotations that eliminated the need to enter the room during treatment. The McGill group created a three-dimensional treatment-planning system based on the Milan and Bentley algorithm that used measured central axis depth doses and profiles to calculate tissue maximum ratio distributions (Milan and Bentley 1974).

While the floor-stand additions made by several groups overcame much of the mechanical concerns of the treatment couches at the time, the gantry rotational characteristics remained a source of inaccuracy for radiosurgery treatments. Physicist Frank Bova and neurosurgeon William Friedman and their group at the University of Florida improved the floor-stand design by adding a rotational arm to the stand that supported the tertiary circular collimators, shown in Figure 3.6 (Friedman and Bova 1989). The new design coupled the beam collimation to the positioning device and improved the mechanical accuracy of the linac system to within 0.3 mm, which was similar to the reported accuracy of the Gamma Knife. This design led to the first commercial linac radiosurgery system: the SRS 200 from Philips Medical Systems. The system later became known as the Gainesville "Linac Scalpel" system and was quite common in the early to mid-1990s (Bova et al. 1997).

3.3 FIELD SHAPING USING MICRO-MLCs

The advances in SRS techniques followed the path from the early days using small square/rectangular fields, defined by the linac jaws, to circular fields, defined by secondary collimators. Early dose-delivery techniques included both static and dynamic arc fields. The next logical step in the advancement of SRS was toward shaped field apertures. In 1991, Dennis Leavitt et al. first presented the idea of field shaping by adding a dynamic field-shaping collimator to the auxiliary circular collimators (Leavitt et al. 1991). The upstream trimmers were motor controlled and could rotate about the beam axis. The ability to shape the field apertures increased the conformality of the treatment dose. Dan Bourland and Edwin McCollough at the Mayo Clinic (Rochester, MN) investigated the use of static, shaped fields for conformal SRS in 1994, prior to the general availability of the multileaf collimator (MLC) (Bourland and McCollough 1994). The growing interest in field shaping led to the development of the first micro-MLC for SRS by the German Cancer Research Center group in Heidelberg (Schlegel et al. 1992). The German micro-MLC had

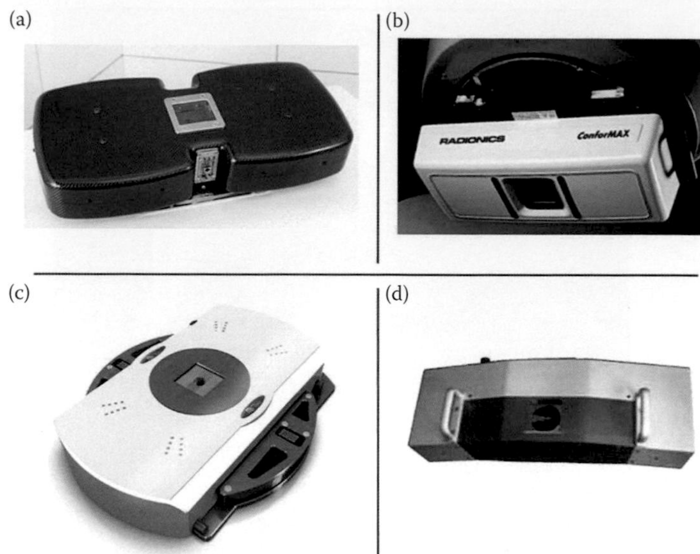

Figure 3.7 Micro-MLCs: (a) the ModuLeaf, developed at Heidelberg and subsequently sold to Siemens; (b) ConforMAX, originally developed at MD Anderson in Houston and later marketed by Radionics; (c) the m3, developed through a collaboration between Varian and Brainlab; (d) micro-MLC, developed at the University of Florida in collaboration with Wellhofer Dosimetrie.

3-mm-wide leaves that mounted to the auxiliary device holders on most conventional accelerators. The device, shown in Figure 3.7a, later became the ModuLeaf™ MLC that was marketed by MRC Systems and was subsequently sold to Siemens Medical Solutions.

Other MLC designs soon followed. Almon Shiu et al. (1997) developed a micro-MLC for cranial SRS with 15 pairs of 4-mm leaves that provided a maximum field size of 6×6 cm^2. Treatment planning for the micro-MLC was done using the XKnife® RTP systems (Radionics, Burlington, MA). That connection led to the development of a 27-leaf pair MLC with a maximum field size of 13.4×10.8 cm^2 that became commercially available from Radionics under the name ConforMAX, shown in Figure 3.7b. Varian and Brainlab (Heimstetten, Germany) collaborated on the development of the m3, a 52-leaf micro-MLC with 14 pairs of 3-mm leaves centered in the middle of the field and six pairs of 4.5-mm leaves at the periphery of the field, shown in Figure 3.7c. The m3 had a maximum field size of 10.2×10.0 cm^2. Vivian Cosgrove et al. (1999) and Xia et al. (1999) described the physical and dosimetric characteristics of the m3. Sanford Meeks et al. (1999) reported the development of a double-focused miniature MLC, built in conjunction with Wellhofer Dosimetrie (Schwarzenbruck, Germany). This miniature MLC is shown in Figure 3.7d.

Further developments and advances in SRS led to dedicated SRS linear accelerator systems, which are discussed in Chapter 4. However, conventional accelerators played a significant role in the development and growth of another important treatment paradigm: stereotactic body radiation therapy (SBRT).

3.4 STEREOTACTIC BODY RADIATION THERAPY

SBRT grew from the same Swedish roots as cranial SRS. Lars Leksell designed and built the Gamma Knife at the Karolinska Hospital in Stockholm, Sweden. Similarly, physician Henrik Blomgren and physicist Ingmar Lax, also from the Karolinska Hospital, first proposed the idea of extending the concepts of SRS to other body sites (Blomgren et al. 1995; Lax et al. 1994). Patient immobilization and localization in the absence of a rigidly attached frame was the real challenge in the transition from cranial lesions to extracranial treatment sites. Lax designed and built an immobilization frame with embedded computed tomography (CT) fiducials that would allow targeting of thoracic and abdominal lesions from CT simulation images. The device included a mechanism for applying abdominal compression to limit the motion of abdominal and thoracic tumors due to respiration. The device, shown in Figure 3.8, was later bought by Elekta AB (Stockholm, Sweden) and marketed as the Stereotactic Body Frame®. This frame was used worldwide for most of the early SBRT treatments and has only recently been discontinued (Hansen, Petersen, and Hoyer 2006; McGarry et al. 2005; Nagata et al. 2002; Sinha and McGarry 2006; Timmerman et al. 2003, 2006; Wulf et al. 2000, 2001).

The Elekta Stereotactic Body Frame was used extensively in most of the largest SBRT services at the time. In Europe, Joern Wulf and his colleagues at the University of Wurzburg (Germany) treated primary and metastatic patients with lung and liver lesions using the Elekta frame (Wulf et al. 2000, 2001, 2004a,b, 2006). In Japan, Nagata and his colleagues at Kyoto University reported their SBRT experience at 13 different Japanese institutions (Hiraoka and Nagata 2004; Nagata et al. 2001, 2002). In the United States, the University of Indiana group, led by Robert Timmerman and Ronald McGarry, treated medically inoperable patients with this emerging treatment schema in several Phase I and Phase II trials (McGarry et al. 2005; Papiez et al. 2003; Timmerman, Papiez, and Suntharalingam 2003; Timmerman et al. 2003, 2006). The work of the Indiana group led to the first large cooperative group trial on SBRT, Radiation Therapy Oncology Group protocol 0236. This protocol studied the local control rates of SBRT treatments in medically inoperable, early-stage. non-small-cell lung cancer patients. Dr. Timmerman served as the principal investigator for this clinical trial, which accrued 59 patients and closed in 2009. The data showed that the local control for SBRT was significantly higher than traditional dose fractionations.

Neurosurgeon Andrew J. Hamilton and physicist Bruce Lulu first described SBRT treatments of spinal and paraspinal lesions at the University of Arizona Medical Center in 1995 (Hamilton and Lulu 1995; Hamilton et al. 1995, 1996). Their treatment technique used an immobilization frame that was rigidly attached to the spinal processes, as shown in Figure 3.9. The stereotactic coordinates used for the localization were defined relative to a small radio-opaque sphere that was visible in the CT simulation images. Because the immobilization and localization frame was surgically attached, the simulation, treatment planning, and subsequent treatment was done in a single visit under general anesthesia. Hamilton reported targeting accuracies within 2 mm. The immobilization frame was never successfully marketed commercially.

The group at the German Cancer Research Center in Heidelberg developed another stereotactic body frame for SBRT (Herfarth et al. 2000, 2001a,b, 2003; Hof, Herfarth, and Debus 2004; Hof et al. 2003; Lohr et al. 1999). The frame provided targeting coordinates through V-shaped fiducials that rigidly mounted to a carbon fiber board. Patients were immobilized in the frame using a vacuum pillow.

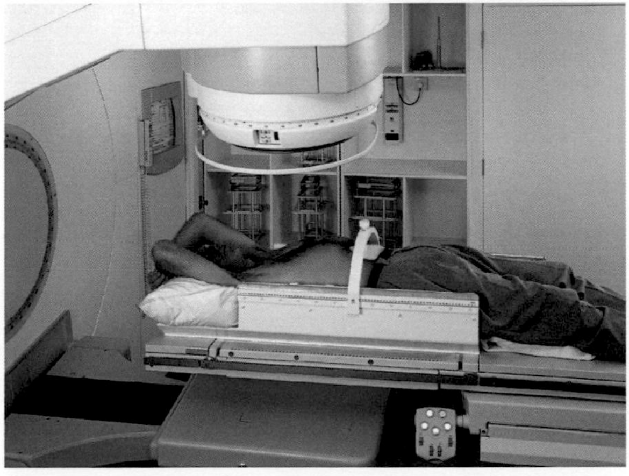

Figure 3.8 The Elekta stereotactic body frame, developed by Lax and Blomgren at the Karolinska Hospital in Stockholm, Sweden.

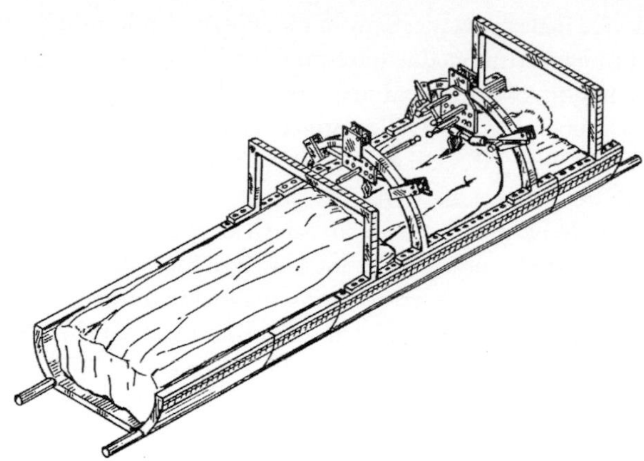

Figure 3.9 Spinal immobilization frame for spinal radiosurgery, developed at the University of Arizona by Hamilton and Lulu.

An abdominal compression device was added to later versions of the frame. The frame, shown in Figure 3.10, was purchased and marketed by Leibinger (Freiburg, Germany) but was later sold and eventually discontinued by Stryker (Kalamazoo, MI).

Physician Martin Fuss and his physics colleague, Bill Salter, at the University of Texas San Antonio employed a technique called serial tomotherapy to treat SBRT patients using a binary MLC called MIMiC (Rassiah-Szegedi et al. 2007; Salter 2001). The serial tomotherapy approach used a relatively unique combination of intensity modulation and linac-based arc therapy. The MIMiC collimator consists of two rows of 20 8-cm-tall and 10-mm-wide tungsten vanes, shown in Figure 3.11. Serial tomotherapy treatments are analogous to axial CT, wherein a slice is treated and then the patient is translated by a fixed amount for the next slice treatment. The linac gantry is arced back and forth, repeating the process until the entire lesion is treated. The precise translations of the treatment couch are performed by a vendor-supplied, retrofitted couch attachment known as Crane II. The translations are done with 0.1-mm accuracy. Planning for serial tomotherapy treatments was done using a Monte Carlo planning system (PEREGRINE, NOMOS, Cranberry Township, PA) (Rassiah-Szegedi et al. 2007).

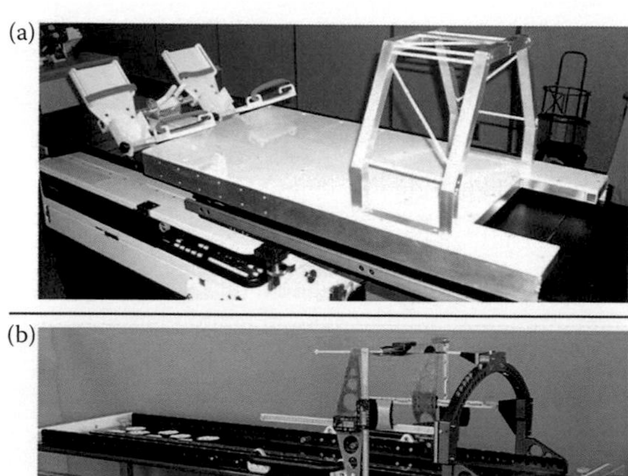

Figure 3.10 The commercial Leibinger frame, originally developed at Heidelberg. (a) A prototype immobilization frame developed at UCLA in 1996. (b) The Leibinger frame, commercially marketed in Germany and later sold by Stryker (Kalamazoo, MI).

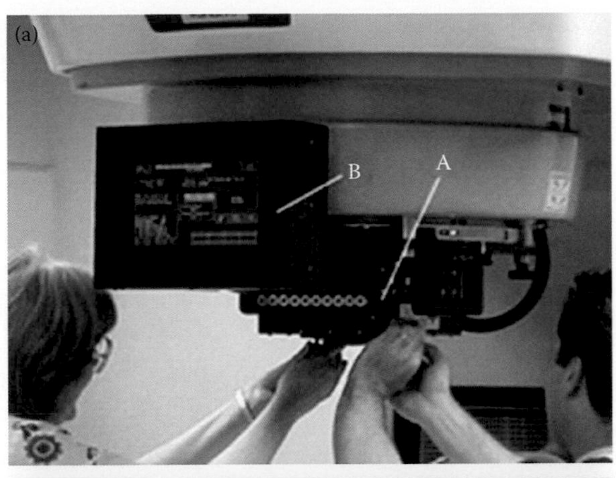

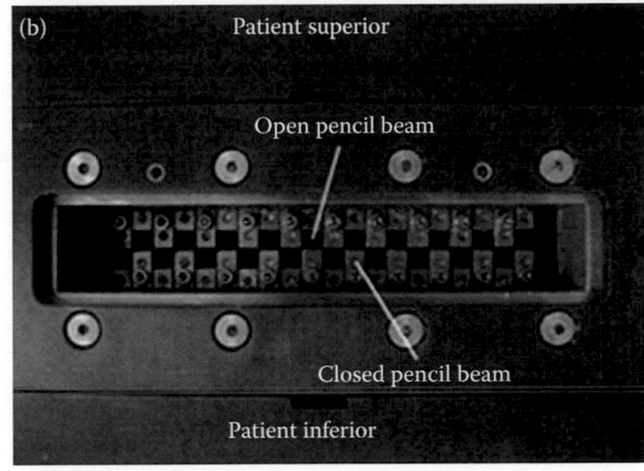

Figure 3.11 (a) MIMiC (indicated by letter A) and the computer controller (indicated by letter B), attached to conventional accelerator; (b) view into MIMiC opening, depicting 20 cm by 3.5 cm "slit" aperture. Note two rows of 20 pencil beams.

3.5 SUMMARY

Both linac-based SRS and SBRT originated from adaptations made to conventional accelerators. Accelerators from several different manufacturers were adapted and studied in the early work for both cranial and extracranial SRT. However, over time, the standards that define a "conventional" clinical accelerator have greatly improved. The mechanical tolerances of the modern-day accelerators and treatment couches are significantly tighter than those linacs used in much of the early SRS and SBRT work. The additions of onboard imaging systems and robotic couches have reduced the uncertainties in patient positioning to the submillimeter range. Routine image-guided treatments have removed the need for external fiducials and stereotactic frames for target localization. Most of these advances have been incorporated into accelerators specifically designed to deliver SRT treatments in both the head and body. These dedicated accelerators are discussed in the following chapter. Later chapters will specifically discuss the principles of image-guided radiation therapy treatments.

REFERENCES

Avanzo, R. C., G. Chierego, C. Marchetti et al. 1984. Stereotaxic irradiation with a linear accelerator. *Radiol. Med.* (Torino) 70:124–9.

Barcia-Solorio, J. L., G. Hernandez, J. Broseta, J. Gonzalez-Darder, and J. Ciudad. 1982. Radio surgical treatment of carotid cavernous fistula. *Appl. Neurophysiol.* 45:520–2.

Betti, O., and V. Derechinsky. 1983. Multiple-beam stereotaxic irradiation [in French]. *Neurochirurgie* 29:295–8.

Blomgren, H., I. Lax, I. Naslund, and R. Svanstrom. 1995. Stereotactic high dose fraction radiation therapy of extracranial tumors using an accelerator. Clinical experience of the first thirty-one patients. *Acta Oncol.* 34:861–70.

Bourland, J. D., and K. P. McCollough. 1994. Static field conformal stereotactic radiosurgery: Physical techniques. *Int. J. Radiat. Oncol. Biol. Phys.* 28:471–9.

Bova, F. J., J. M. Buatti, W. A. Friedman, W. M. Mendenhall, C. C. Yang, and C. Liu. 1997. The University of Florida frameless high-precision stereotactic radiotherapy system. *Int. J. Radiat. Oncol. Biol. Phys.* 38:875–82.

Colombo, F., A. Benedetti, F. Pozza et al. 1985. External stereotactic irradiation by linear accelerator. *Neurosurgery* 16:154–60.

Colombo, F., A. Benedetti, F. Pozza et al. 1986. Radiosurgery using a 4MV linear accelerator. Technique and radiobiologic implications. *Acta Radiol.* 369 (Suppl.):S603–7.

Cosgrove, V. P., U. Jahn, M. Pfaender, S. Bauer, V. Budach, and R. E. Wurm. 1999. Commissioning of a micro multi-leaf collimator and planning system for stereotactic radiosurgery. *Radiother. Oncol.* 50:325–36.

Friedman, W. A., and F. J. Bova. 1989. The University of Florida radiosurgery system. *Surg. Neurol.* 32:334–42.

Hamilton, A. J., and B. A. Lulu. 1995. A prototype device for linear accelerator-based extracranial radiosurgery. *Acta Neurochir.* 63 (Suppl.):S40–3.

Hamilton, A. J., B. A. Lulu, H. Fosmire, and L. Gossett. 1996. LINAC-based spinal stereotactic radiosurgery. *Stereotact. Funct. Neurosurg.* 66:1–9.

Hamilton, A. J., B. A. Lulu, H. Fosmire, B. Stea, and J. R. Cassady. 1995. Preliminary clinical experience with linear accelerator-based spinal stereotactic radiosurgery. *Neurosurgery* 36:311–9.

Hansen, A. T., J. B. Petersen, and M. Hoyer. 2006. Internal movement, set-up accuracy and margins for stereotactic body radiotherapy using a stereotactic body frame. *Acta Oncol.* 45:948–52.

Hartmann, G. H., W. Schlegel, V. Sturm, B. Kober, O. Pastyr, and W. J. Lorenz. 1985. Cerebral radiation surgery using moving field irradiation at a linear accelerator facility. *Int. J. Radiat. Oncol. Biol. Phys.* 11:1185–92.

Herfarth, K. K., J. Debus, F. Lohr et al. 2000. Extracranial stereotactic radiation therapy: Set-up accuracy of patients treated for liver metastases. *Int. J. Radiat. Oncol. Biol. Phys.* 46:329–35.

Herfarth, K. K., J. Debus, F. Lohr et al. 2001a. Stereotactic single-dose radiation therapy of liver tumors: Results of a phase I/II trial. *J. Clin. Oncol.* 19:164–70.

Herfarth, K. K., J. Debus, F. Lohr, M. L. Bahner, and M. Wannenmacher. 2001b. Stereotactic irradiation of liver metastases [in German]. *Radiologe* 41:64–8.

Herfarth, K. K., H. Hof, M. L. Bahner et al. 2003. Assessment of focal liver reaction by multiphasic CT after stereotactic single-dose radiotherapy of liver tumors. *Int. J. Radiat. Oncol. Biol. Phys.* 57:444–51.

Hiraoka, M., and Y. Nagata. 2004. Stereotactic body radiation therapy for early-stage non-small-cell lung cancer: The Japanese experience. *Int. J. Clin. Oncol.* 9:352–5.

Hof, H., K. Herfarth, and J. Debus. 2004. Stereotactic irradiation of lung tumors [in German]. *Radiologe* 44:484–90.

Hof, H., K. K. Herfarth, M. Munter et al. 2003. Stereotactic single-dose radiotherapy of stage I non-small-cell lung cancer (NSCLC). *Int. J. Radiat. Oncol. Biol. Phys.* 56:335–41.

Houdek, P. V., J. V. Fayos, J. M. Van Buren, and M. S. Ginsberg. 1985. Stereotaxic radiotherapy technique for small intracranial lesions. *Med. Phys.* 12:469–72.

Lax, I., H. Blomgren, I. Naslund, and R. Svanstrom. 1994. Stereotactic radiotherapy of malignancies in the abdomen. Methodological aspects. *Acta Oncol.* 33:677–83.

Leavitt, D. D., F. A. Gibbs Jr., M. P. Heilbrun, J. H. Moeller, and G. A. Takach Jr. 1991. Dynamic field shaping to optimize stereotactic radiosurgery. *Int. J. Radiat. Oncol. Biol. Phys.* 21:1247–55.

Loeffler, J. S., E. Alexander III, R. L. Siddon, W. M. Saunders, C. N. Coleman, and K. R. Winston. 1989. Stereotactic radiosurgery for intracranial arteriovenous malformations using a standard linear accelerator. *Int. J. Radiat. Oncol. Biol. Phys.* 17:673–7.

Lohr, F., J. Debus, C. Frank et al. 1999. Noninvasive patient fixation for extracranial stereotactic radiotherapy. *Int. J. Radiat. Oncol. Biol. Phys.* 45:521–7.

Lutz, W., K. R. Winston, and N. Maleki. 1984. Stereotactic radiosurgery in the brain using a 6 MV linear accelerator. *Int. J. Radiat. Oncol. Biol. Phys.* 10(Suppl. 1):189.

Lutz, W., K. R. Winston, and N. Maleki. 1986. A system for stereotactic radiosurgery with a linear accelerator and its performance evaluation. *Int. J. Radiat. Oncol. Biol. Phys.* 12(Suppl. 1):100.

Lutz, W., K. R. Winston, and N. Maleki. 1988. A system for stereotactic radiosurgery with a linear accelerator. *Int. J. Radiat. Oncol. Biol. Phys.* 14:373–81.

McGarry, R. C., L. Papiez, M. Williams, T. Whitford, and R. D. Timmerman. 2005. Stereotactic body radiation therapy of early-stage non-small-cell lung carcinoma: Phase I study. *Int. J. Radiat. Oncol. Biol. Phys.* 63:1010–5.

Meeks, S. L., F. J. Bova, S. Kim, W. A. Tome, J. M. Buatti, and W. A. Friedman. 1999. Dosimetric characteristics of a double-focused miniature multileaf collimator. *Med. Phys.* 26:729–33.

Milan, J., and R. E. Bentley. 1974. The storage and manipulation of radiation dose data in a small digital computer. *Br. J. Radiol.* 47:115–21.

Nagata, Y., Y. Negoro, T. Aoki et al. 2001. Three-dimensional conformal radiotherapy for extracranial tumors using a stereotactic body frame [in Japanese]. *Igaku Butsuri* 21:28–34.

Nagata, Y., Y. Negoro, T. Aoki et al. 2002. Clinical outcomes of 3D conformal hypofractionated single high-dose radiotherapy for one or two lung tumors using a stereotactic body frame. *Int. J. Radiat. Oncol. Biol. Phys.* 52:1041–6.

Papiez, L., R. Timmerman, C. DesRosiers, and M. Randall. 2003. Extracranial stereotactic radioablation: Physical principles. *Acta Oncol.* 42:882–94.

Podgorsak, E. B., A. Olivier, M. Pla, J. Hazel, A. de Lotbiniere, and B. Pike. 1987. Physical aspects of dynamic stereotactic radiosurgery. *Appl. Neurophysiol.* 50:263–8.

Podgorsak, E. B., A. Olivier, M. Pla, P. Y. Lefebvre, and J. Hazel. 1988. Dynamic stereotactic radiosurgery. *Int. J. Radiat. Oncol. Biol. Phys.* 14:115–26.

Rassiah-Szegedi, P., M. Fuss, D. Sheikh-Bagheri et al. 2007. Dosimetric evaluation of a Monte Carlo IMRT treatment planning system incorporating the MIMiC. *Phys. Med. Biol.* 52:6931–41.

Salter, B. J. 2001. NOMOS Peacock IMRT utilizing the Beak post collimation device. *Med. Dosim.* 26:37–45.

Saunders, W. M., K. R. Winston, R. L. Siddon et al. 1988. Radiosurgery for arteriovenous malformations of the brain using a standard linear accelerator: Rationale and technique. *Int. J. Radiat. Oncol. Biol. Phys.* 15:441–7.

Schlegel, W., O. Pastyr, T. Bortfeld et al. 1992. Computer systems and mechanical tools for stereotactically guided conformation therapy with linear accelerators. *Int. J. Radiat. Oncol. Biol. Phys.* 24:781–7.

Shiu, A. S., H. M. Kooy, J. R. Ewton et al. 1997. Comparison of miniature multileaf collimation (MMLC) with circular collimation for stereotactic treatment. *Int. J. Radiat. Oncol. Biol. Phys.* 37:679–88.

Sinha, B., and R. C. McGarry. 2006. Stereotactic body radiotherapy for bilateral primary lung cancers: The Indiana University experience. *Int. J. Radiat. Oncol. Biol. Phys.* 66:1120–4.

Talairach, J., J. de Ajuriaguerra, and M. David. 1952. Etudes stereotaxiques des structures encephaliques profondes chez l'homme technique, interet physiologique et therapeutique. *Presse Med.* 28:605–9.

Talairach, J., M. He'caen, M. David, M. Monnier, and J. Ajuriaguerra. 1949. Recherches sur la coagulation therapeutique des structures sous-corticales chez l'homme. *Rev. Neurol.* 81:4–24.

Timmerman, R., R. McGarry, C. Yiannoutsos et al. 2006. Excessive toxicity when treating central tumors in a phase II study of stereotactic body radiation therapy for medically inoperable early-stage lung cancer. *J. Clin. Oncol.* 24:4833–9.

Timmerman, R., L. Papiez, R. McGarry et al. 2003. Extracranial stereotactic radioablation: Results of a phase I study in medically inoperable stage I non-small cell lung cancer. *Chest* 124:1946–55.

Timmerman, R., L. Papiez, and M. Suntharalingam. 2003. Extracranial stereotactic radiation delivery: Expansion of technology beyond the brain. *Technol. Cancer Res. Treat.* 2:153–60.

Van Buren, J. M., P. Houdek, and M. Ginsberg. 1983. A multipurpose CT-guided stereotactic instrument of simple design. *Appl. Neurophysiol.* 46:211–6.

Wulf, J., M. Guckenberger, U. Haedinger et al. 2006. Stereotactic radiotherapy of primary liver cancer and hepatic metastases. *Acta Oncol.* 45:838–47.

Wulf, J., U. Hadinger, U. Oppitz, B. Olshausen, and M. Flentje. 2000. Stereotactic radiotherapy of extracranial targets: CT-simulation and accuracy of treatment in the stereotactic body frame. *Radiother. Oncol.* 57:225–36.

Wulf, J., U. Hadinger, U. Oppitz, W. Thiele, and M. Flentje. 2004a. Stereotactic boost irradiation for targets in the abdomen and pelvis. *Radiother. Oncol.* 70:31–6.

Wulf, J., U. Haedinger, U. Oppitz, W. Thiele, G. Mueller, and M. Flentje. 2004b. Stereotactic radiotherapy for primary lung cancer and pulmonary metastases: A noninvasive treatment approach in medically inoperable patients. *Int. J. Radiat. Oncol. Biol. Phys.* 60:186–96.

Wulf, J., U. Hadinger, U. Oppitz, W. Thiele, R. Ness-Dourdoumas, and M. Flentje. 2001. Stereotactic radiotherapy of targets in the lung and liver. *Strahlenther. Onkol.* 177:645–55.

Xia, P., P. Geis, L. Xing et al. 1999. Physical characteristics of a miniature multileaf collimator. *Med. Phys.* 26:65–70.

4 Dedicated linear accelerators for stereotactic radiation therapy

Malika Ouzidane, Joshua Evans, and Toufik Djemil

Contents

4.1 INTRODUCTION

Following the technological developments that allowed for the adaptation of conventional medical linear accelerators to stereotactic radiotherapy (SRT), a new generation of radiation therapy machines designed to perform stereotactic radiosurgery (SRS) and radiotherapy was born. This new class of machines is designed from the ground up to integrate high-precision radiation dose delivery at a high dose rate and precise tumor localization using image-guided radiotherapy (IGRT). These systems deliver high-precision conformal doses to irregularly shaped tumors while sparing adjacent organs at risk, which makes them suitable for dose escalation protocols, potentially resulting in higher tumor control rates and fewer side effects (Jaffray et al. 2007). Many of these new devices come with integrated treatment planning and delivery hardware for radiation therapy treatments.

Different manufacturers have adopted different technologies and incorporated different hardware and software tools into their designs. Each modality is distinguished by the beam produced and its modeling, treatment delivery, multileaf collimators (MLCs), applicators, dose rate, specialized treatment planning, and IGRT techniques used. Some of the commercially available technologies are different from the classical C-arm type accelerator (CyberKnife, Hi-ART, ViewRay, and Vero). This chapter describes different technologies and devices created specifically for SRT. The main machines used are presented along with some future technologies (Table 4.1).

Table 4.1 Summary of the main characteristics of linear accelerators dedicated for radiosurgery available on the market

	NOVALIS	NOVALIS TX	CYBERKNIFE	TRILOGY	SYNERGY-S	AXESSE	ARTISTE	TOMOTHERAPY	TRUEBEAM	EDGE	VIEWRAY	VERO
ENERGY	6 MV	6–20 MV, 6 SRS	6 MV	6–25 MV, 6 SRS	6–18 MV	6–18 MV	6–18 MV	6 MV	6–20 MV, 6× HI, 10× HI (high dose rate 6 and 10 MV)	6 MV, 6× HI, 10× HI (high dose rate 6 and 10 MV)	Co-60: Equivalent to 4 MV	6 MV
MAX DOSE RATE	800 MU/ min	600 MU/ min 1000 MU/ min SRS mode	1000 MU/min	600 MU/ min 1000 MU/ min SRS mode	600 MU/min	600 MU/ min	500 MU/ min	850 cGy/min	600 MU/min 6× HI: 1400 MU/min 10× HI: 2400 MU/min	600 MU/ min 6× HI: 1400 MU/ min 10× HI: 2400 MU/ min	600 cGy/ min at installation	500 cGy/ min
FIELD SIZE	10×10 cm 9.8×9.8 cm clinical	40×40 cm 15×15 cm SRS mode 22×40 cm MLC	6-cm diameter 12×10 cm with InCise MLC	40×40 cm 15×15 cm SRS mode 34×40 cm MLC	16×21 cm	16×21 cm	40×40 cm	5×40 cm	40×40 cm	40×22 cm	30×30 cm 27.3×27.3 cm clinical	15×15 cm
JAWS	Variable	Variable	N/A	Variable	Fixed	Fixed	Variable (no X-jaw)	N/A	Variable	Variable	N/A	N/A
BEAM SHAPING	m3 microMLC 52 leaf	HD120 MLC 120 Leaf	InCise MLC 82 leaf	Millenium 120 MLC 120 leaf	MLC 80 leaf	MLC 80 leaf	MLC 160 leaf	MLC (binary) 64 leaf	Millenium 120 MLC (HD120 Optional) 120 leaf	HD120 MLC 120 leaf	MLC 60 leaf	MLC 60 leaf
MLC LEAF SIZE	3 mm for central 4.2 cm	2.5 mm for the central 8 cm	2.5 mm	5 mm for central 20 cm	4 mm	4 mm	5 mm	6.25 mm	5 mm for central 20 cm	2.5 mm for the central 8 cm	1 cm	5 mm
APPLICATORS	Conical 4–50 mm	Conical 4–50 mm	Circular 5–60 mm Iris variable aperture collimator dodecagonal 5–60 mm	Conical 5–30 mm	Conical 5–50 mm	Conical 5–50 mm	N/A	N/A	Conical 4–50 mm	Conical 4–50 mm	N/A	

IGRT	Room-based ExacTrac (2 planar kV x-rays)	Room and gantry-based ExacTrac (2 planar kV x-rays) kV CBCT MV portal imaging	Room-based 6D Skull Xsight Spine Xsight Lung (2 planar kV x-rays)	Gantry-based kV CBCT MV PortalVision	Gantry-based XVI kV CBCT MV portal imaging	Gantry-based XVI kV CBCT MV portal imaging	Gantry-based MV CBCT MV portal imaging Optional CT on-rail	Gantry-based MV CT	Gantry-based kV CBCT MV PortalVision	Room and gantry-based kV CBCT MV PortalVision Calypso® 4D VisionRT	Gantry-based On-board MRI	Gantry-based ExacTrac (2 planar kV x-rays) kV CBCT MV portal imaging
GATING/ TRACKING	ExacTrac x-rays Supports gating	ExacTrac x-rays Supports gating	Synchrony respiratory system tracking (synchronizes beam delivery to tumor motion)	RPM respiratory gating	N/A	N/A	4-D gating system Respiratory sensor and breathing belt	N/A	RPM respiratory gating	RPM respiratory gating	Image-based gating (continous imaging 4 frames/s)	ExacTrac x-rays Supports gating
COUCH	Standard	6-D Robotic	6-D RoboCouch with optional seated load table top	Standard	Standard	HexaPOD Robotic	Standard	Standard	Standard	PerfectPitch 6D radiosurgery	Robotic couch	6-D Robotic
TREATMENT PLANNING SYSTEMS	iPlan Pencil Beam or Monte Carlo	iPlan Pencil Beam or Monte Carlo Other commercial systems	MultiPlan Ray Tracing or Monte Carlo	Eclipse Other commercial systems	CMS XiO Other commercial systems	CMS XiO Other commercial systems	Commercial systems	Hi-ART II Convolution Superposition	Eclipse Other commercial systems	Eclipse Other commercial systems	ViewRay TPS (Monte Carlo)	iPlan Pencil Beam or Monte Carlo

As with all new technologies, these dedicated machines and techniques require their own specialized quality assurance (QA) and quality control (QC) procedures. The original American Association of Physicists in Medicine (AAPM) task group report on radiosurgery is now being superseded by newer reports (Schell et al. 1995). The AAPM has published a task group report on SBRT (Benedict et al. 2010). Task group reports have also been published by AAPM for robotic radiosurgery (Dieterich et al. 2011) and for helical tomotherapy (Langen et al. 2010).

4.2 BRAINLAB NOVALIS AND THE NOVALIS Tx

The Novalis® radiosurgery platform, featuring Novalis and Novalis Tx systems, brings together technologies from Varian Medical Systems, Inc. (Palo Alto, California) and Brainlab AG (Feldkirchen, Germany) to create a fully integrated SRS/SBRT solution (Solberg et al. 2001). The linac is a Varian accelerator, whereas the beam-shaping device, the localization and tracking system, and the treatment planning system (TPS) are all Brainlab components for the Novalis (Figure 4.1). For the Novalis Tx model, the MLC is also a Varian device, whereas the localization has components from both manufacturers (Figure 4.2). The TPS is available from both Varian and Brainlab.

4.2.1 KEY FEATURES AND COMPONENTS

The Novalis radiosurgery family delivers noninvasive, frameless, shaped-beam treatments for cancerous and noncancerous conditions of the entire body. The high-definition shaped beam technology dynamically conforms the treatment beam to the lesion while protecting surrounding healthy tissue. Although these machines do come with a stereotactic frame for SRS, the intracranial radiosurgery treatments are usually achieved without the invasive frame. With the use of robust IGRT methods and immobilization techniques, the quality of the SRS treatment is not compromised. Frameless linac-based SRS in many institutions is replacing frame-based treatments as the standard of care. A head-to-shoulder mask is used improving flexibility for imaging, planning, and treatment, as well as overcoming the restrictions of frame-based radiosurgery. The treatment of a variety of intracranial targets is monitored by specific software tools. X-ray images are acquired and verified with the ExacTrac® x-ray 6D IGRT imaging system, during patient setup and treatment delivery. They provide detection and visualization of displacements. The setup, including rotational corrections, is thereby adjusted.

Some of the features of the Novalis system include an energy of 6 MV; a 52-leaf micro-MLC (Brainlab m3® high-resolution MLC), allowing for the delivery of highly conformal dose, improved margins, and better protection of surrounding tissue; Brainlab iPlan® Radiation Therapy (RT) Treatment Plan for treatment planning; and ExacTrac for accurate and automated patient positioning (Brainlab). The system includes a full range of stereotactic hardware for immobilization, setup, and QA, such as head rings, conical collimators, and target positioning hardware (Figure 4.3).

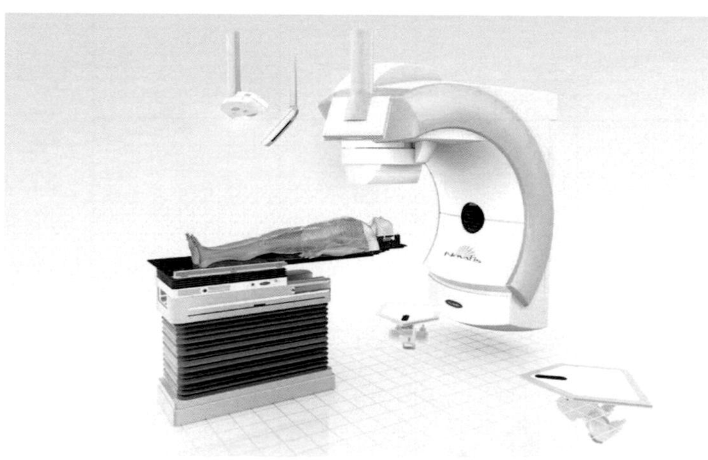

Figure 4.1 Brainlab Novalis® radiosurgery linear accelerator. (Image courtesy of Brainlab AG.)

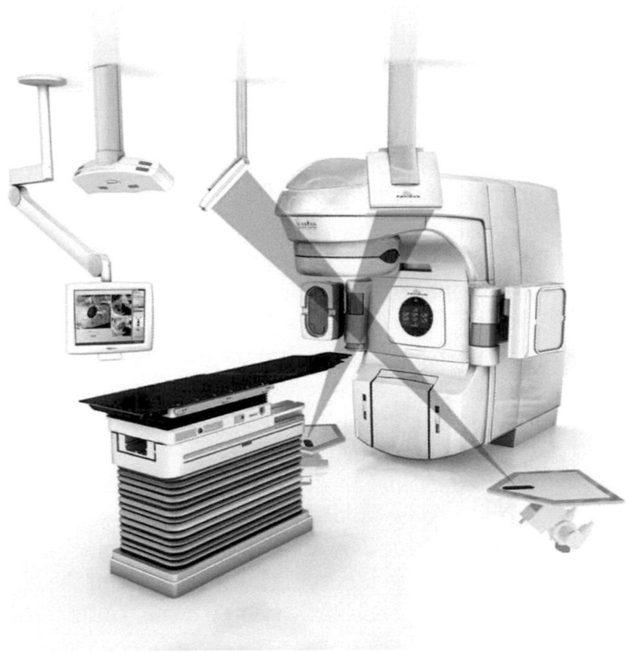

Figure 4.2 Varian/Brainlab Novalis® Tx radiosurgery system with On-Board Imager® and ExacTrac® x-rays. (Image courtesy of Brainlab AG.)

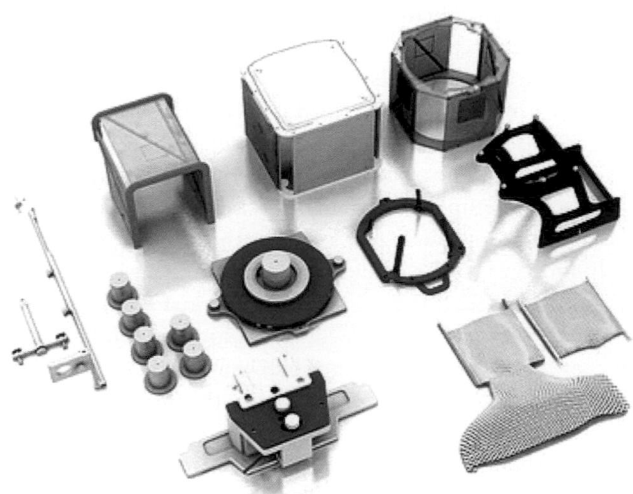

Figure 4.3 Novalis® radiosurgery accessories. (Image courtesy of Brainlab AG.)

The Novalis Tx system includes a dose delivery with a range of high energies from 6 to 20 MV; a 120-leaf high-definition (HD120®) MLC; iPlan RT for treatment planning; ExacTrac and 6D robotic couch for image-guided patient positioning, correcting directional inaccuracies and rotational misalignments; and PortalVision™, which offers treatment verification during dose delivery, and dosimetry QA for verifying system performance. Additionally, the Novalis Tx also has the On-Board Imager® (OBI), providing cone beam computed tomography (CBCT) for volumetric soft tissue discrimination and kilovoltage (kV) fluoroscopic imaging for respiratory motion verification (Brainlab). The system combines both room- and gantry-based imaging solutions, delivering precise positioning for intracranial and extracranial targets. The conical collimators originally offered with Novalis are offered as an option with the Novalis Tx model.

These systems provide a wide variety of fractionation schemes. The spectrum includes single stage and fractionated treatments of functional disorders, and benign and malignant tumors. Supported modalities include dynamic conformal arcs, circular arcs, intensity-modulated radiosurgery (IMRS), and intensity-modulated radiation therapy (IMRT). The system also provides real-time detection and compensation of tumor motion including daily internal adaptation of the patient breathing cycle. Novalis is suitable for patients with intracranial, spinal, and lung lesions, where the bony anatomy of the skull and the spinal column, and the large density contrast of the lungs, can be easily localized with x-rays from the ExacTrac system. For lesions requiring soft tissue localization, such as liver, pancreas, and kidneys, kV-CBCT imaging in the Novalis Tx model offers superior three-dimensional (3-D) anatomical information and soft-tissue contrast for more accurate tumor localization and patient positioning. The Novalis Tx comes equipped with three different imaging devices: the ExacTrac x-ray 6D system, kV-CBCT, and MV portal imaging system. These imaging technologies help localize the tumor before and during treatment with submillimeter accuracy (Ackerly et al. 2011). With continual imaging guidance during the treatment to detect movement, the Novalis Tx robotic treatment couch provides automatic correction in patient positioning.

The Novalis Tx is also characterized by an extended collimator size (increased from 10 × 10 cm in the classic Novalis to 40 × 40 cm) and a finer resolution of the beam shaper as the leaves are 2.5 mm wide (3 mm for Novalis) for a more conformal painting of the dose to the tumor and sparing of the normal tissues around it. Larger lesions not treatable on Novalis can be treated on the Novalis Tx. In SRS mode (only available on the Novalis Tx), a special SRS 6 MV photon beam, going through a thinner flattening filter, is used, and the dose rate is 1000 monitor units (MU) per minute. This is in a higher dose rate for SRT on the Novalis Tx compared to the regular Novalis (up to 800 MU/min).

4.2.2 MLCs AND CONICAL APPLICATORS

Brainlab supports the use of different micro and standard MLCs. The m3 micro-MLC, which is a Brainlab component, is fully integrated with Varian accelerators. It communicates with the Varian C-Series linacs as an internal part of the accelerator, providing safety interlocks and additional capabilities, like dynamic conformal arcs and dynamic IMRT. The original micro-MLC was detachable, and in recent years, it has been permanently attached to the head of the linac, such as in the case of the Novalis. This mount design occludes the optical distance indicator (ODI) projection, so the Novalis lacks a distance indicator. Patient setup does not rely on source-to-skin distance (SSD) information; it uses external body markers and the ExacTrac system. In the Novalis Tx model, the MLC, Varian HD120®, is not externally added to the linac, but it is built into it. It is a high-definition (HD) tungsten 120-leaf collimator. It has 32 leaf pairs with 2.5 mm leaf width (projection at the isocenter), surrounded by 28 leaf pairs of 5-mm projection width. The total length is 22 cm across the leaves at the isocentric plane (Chang et al. 2008). The ODI is not blocked and SSD could be used in patient setup for conventional cases. For the purpose of radiosurgery though, this feature is not used, and setups rely on more robust systems such as OBI and ExacTrac.

MLCs with 10-mm leaf width are commonly used for standard radiotherapy. For stereotactic treatments, a higher resolution and a steep dose falloff are required to protect normal tissue and critical structures from high doses. MLCs with 5-mm leaves or less are recommended by the AAPM. Five-millimeter leaves allow the delivery of fractionated SRT but are not generally acceptable for single fraction radiosurgery, especially if the lesion is small (less than 3 cm) or irregularly shaped (Benedict et al. 2010). For these situations, special requirements have been defined by the AAPM in the report of Task Group 42 (Schell et al. 1995). To increase the steepness of the dose, the penumbra of the collimator system must be minimized. For radiosurgery, the recommended limit for dose gradient in the beam penumbra (from 80% to 20%) is at least 60%/3 mm. Most commercially available MLCs have a penumbra specification from 4 to 8 mm. The m3 micro-MLC has an effective penumbra of less than 3 mm for all SRS field sizes and meets the SRS requirements (Schell et al. 1995). The dose can then be tailored to the shape of the lesion while sparing healthy tissue.

For even smaller circular targets, 1 cm or less, especially if they are close to sensitive structures, such as the brainstem for example, a steeper dose falloff and a sharper penumbra become necessary. In such radiosurgery procedures, stereotactic cones are used. The divergence of the beam is taken into account by the conical aperture of the collimator, the penumbra is reduced, and the dose gradient is steep. One typical case where a cone would be used in linac-based SRS would be the treatment of trigeminal neuralgia. As part of Brainlab components, these cones integrate seamlessly with Brainlab software; but as hardware components, they are also compatible with most linear accelerators and major stereotactic frames. The brass shelled lead collimators are available in sizes from 4 to 50 mm. They are securely mounted to the gantry head through a collimator mount with a bayonet locking mechanism (Novalis Radiosurgery). Newer circular collimators now have both a hardware and software interlock with Varian accelerators, which previous models including the Novalis lacked. Lack of interlock between the add-on cones and the accelerator could lead to radiosurgery overdosing. An Urgent Field Safety Notice issued by Brainlab in 2009 warned of an incident in which the collimator jaw settings defined a radiation field that was not completely within the shielded area of the conical collimator accessory. When this occurs, radiation may be delivered outside the area shielded by the conical collimator, potentially resulting in an unintended radiation dose to patients, which could result in patient injury or even death. Varian Medical Systems also published in 2010 a similar Urgent Field Safety Notice, limiting primary collimator settings for Brainlab conical collimator accessories.

4.2.3 ExacTrac

The ExacTrac x-ray 6D imaging system was released by Brainlab in 2008 (Jin et al. 2008). It allows for patient setup with high-resolution stereo x-ray imaging. ExacTrac relies mostly on bony anatomy registration. For soft tissue targets, implanted fiducial markers are needed for localization. ExacTrac corrects patient positioning with submillimeter precision (Verellen et al. 2003). It is also suitable to the growing need of detection and compensation for intrafraction tumor motion, whether induced by patient movement, breathing, or other. This targeting applies not only to tumors that move but also to tumors that change size and shape during and over the course of treatment.

The ExacTrac stereotactic system consists of two kV x-ray units in the treatment room floor and two ceiling-mounted amorphous silicon flat panel detectors. It also comes with an integrated optical infrared tracking system for continuous monitoring of patient position throughout treatment. The system uses a pair of infrared video cameras to triangulate on an external set of fiducial markers, the Body Marker Array, which is affixed to the anterior thorax and/or abdomen of the patient (Brainlab). It is calibrated in the treatment reference space of the linear accelerator, and it controls the treatment couch motions, guiding the lesion to the linac isocenter. The system quantifies the x, y, and z displacements from the treatment position as well as the three angles of body rotation. This room-based design supports continuous target tracking and IGRT verification during treatment.

4.2.4 iPlan RT TREATMENT PLANNING SOFTWARE

The Novalis family systems come with an integrated TPS (BrainSCAN® and iPlan RT). The iPlan RT TPS offers image fusion capabilities, automatic coregistration of anatomic, metabolic, and functional data often used for stereotactic treatments, and automatic segmentation for the definition of relevant critical structures for brain, spine, head and neck, and prostate treatments. Full DICOM RT connectivity allows the export of generated objects to existing dose planning systems. The automation, displays, and templates focused on SRS/SBRT support the use of different treatment options including conformal arcs, conformal beams, dynamic conformal arcs, cones, and IMRT, or a composite plan combining several treatment modalities, resulting in an optimal individualized treatment for each patient. iPlan RT does not support volumetric modulated arc therapy (VMAT), but it uses HybridArc™, a combination of static IMRT beams and dynamic conformal arcs as an alternative solution. HybridArc is suitable for shaping dose for large structures and concave regions.

The iPlan RT TPS offers a choice of either Pencil Beam (PB) or Monte Carlo (MC) algorithms for dose calculation. For the purpose of stereotactic treatments, especially single fractions and escalated dose protocols, with small targets and inhomogeneous media, the use of MC dose engine is superior to PB algorithms. The software delivers both forward and inverse planning capabilities for photon-beam treatment. It considers linac head geometry, secondary electron dose effects, and tissue inhomogeneities, commonly found in extracranial treatments, providing an accurate dose calculation for SBRT. The iPlan

RT TPS also features an adaptive dose calculation grid in function of the object volume. It automatically adjusts the dose grid for the calculation of dose distributions in the case of very small structures (Brainlab).

MC algorithm is based on the x-ray voxel MC algorithm, XVMC. The inverse planning (IP) algorithm is based on the dynamically penalized likelihood (DPL) method (Llacer et al. 2001). It was developed in collaboration with physicists from UCLA (Solberg et al. 2001). DPL is as precise in generating treatment plans as a stochastic (simulated annealing) algorithm and two well-recognized gradient methods (conjugated gradients and Newton gradients), but it is faster (Llacer et al. 2001).

When changes in tumor shape during the course of treatment occur, iPlan RT TPS allows adaptive RT. Periodic plan adaptations are used when substantial change occurs. Intuitive plan updates are based on periodic CT scans prompted by ExacTrac. Adapted objects are exported with DICOM RT to the dose engine.

Novalis Tx can deliver treatments planned with other TPS commonly used with Varian accelerators, such as Varian Eclipse™ and Philips Pinnacle. They all provide integrated treatment planning technologies for radiosurgery. With Eclipse, the use and operation of RapidArc® volumetric arc therapy is enabled. It sculpts a 3-D dose distribution by modulating gantry speed, MLC leaf position, and linac output. In addition, it supports PortalVision, CBCT, 4-D planning, and portal dosimetry.

4.2.5 CLINICAL APPLICATIONS

Brainlab IMRS/IMRT is suitable for targets next to or wrapped around critical organs, and it allows for compensation of tissue inhomogeneities and surface curvature. Treatments can be stereotactic or nonstereotactic. After automatic CT and magnetic resonance (MR) image fusion, the lesion and critical organs are outlined. Noncoplanar conformal beams are directed to the isocenter, and the beam intensities are optimized by the IP algorithm. The desired dose and the minimum dose allowed to the target and also the maximum dose per volume allowed to each organ at risk (OAR) are prescribed graphically using "forbidden areas" in dose–volume histograms. These constraints are used by the IP algorithm to compute an optimized plan. Alternative plans are simultaneously calculated with different OAR weightings, allowing a selection from several optimized plans. The leaf sequencing is incorporated into the optimization process, taking leakage and tongue-and-groove effect into account. Delivery may be accomplished in either step-and-shoot or dynamic MLC modes.

In prostate treatments, real-time IGRT and CBCT soft tissue localization improve setup time and quality. Prostate patients are imaged during treatment delivery to ensure that the target area is covered and OARs are avoided.

For small spherical lesions, Brainlab stereotactic conical collimators can be used with either one or multiple isocenter circular arc treatments. Circular arc radiosurgery is the classical method to deliver linac-based radiosurgery. The dose is delivered in several arcs, all intersecting at the isocenter. For irregularly shaped lesions, several isocenters are combined to generate a plan with acceptable dose coverage and protection of normal tissue and critical organs. During treatment delivery, the patient is set up to the different isocenters in sequence.

Another important application is spine SBRT. Spinal radiosurgery offers a treatment alternative to spinal surgery for a variety of indications. From spinal column metastases and epidural disease, to arteriovenous malformations (AVMs) and inoperable tumors, spine SBRT is becoming the standard of care (Haley et al. 2011). Multimodality image fusion is used for target delineation. Two-dimensional stereo x-ray and CBCT imaging provide high-precision patient positioning, and the localization does not require implanted fiducials.

Brainlab is also used for treatments sensitive to patient breathing. The Novalis system supports adaptive gating, delivering consistent treatments to areas affected by respiratory motion. ExacTrac addresses changes in daily respiration and intrafraction variations. It uses optical infrared patient tracking, thereby reducing imaging dose compared to fluoroscopic approaches, and IGRT x-ray verification of internal tumor position during gated treatment. This feature is well suited for the treatment of lungs, kidneys, liver, and pancreas. Varian's Real Time Position Management® system (RPM) is available for both the CT and positron emission tomography (PET) imaging systems and for the Novalis Tx radiosurgery platform, offering both prospective and retrospective gating and 3-D monitoring (Novalis Radiosurgery). SBRT for early-stage lung cancer is therefore becoming the standard of care. The lung gating technique pauses radiation delivery until the tumor is in the optimal position. Tumor coverage is assured and normal tissue is spared. A user-defined gating window allows adjustment for specific treatment margins and dose escalation protocols beyond those possible with standard positioning. Some of the key features for adaptive lung gating include 4-D CT data integration, elastic morphing for automatic internal target volume (ITV) contouring, MC dose calculations, daily internal adaptation of the patient breathing cycle, and real-time detection and compensation of tumor motion.

4.3 ACCURAY ROBOTIC RADIOSURGERY: THE CYBERKNIFE

The CyberKnife® M4™ is a frameless robotic system for radiosurgery developed by Accuray, Inc. (Sunnyvale, California). It combines a lightweight linear accelerator, a robotic delivery system, and noninvasive image-guided localization (Adler et al. 1997).

The robotic nature of the CyberKnife (Figure 4.4) requires smaller dimensions and weight than conventional linacs. The CyberKnife uses a 9.3-GHz, X-band 6-MV linear accelerator that runs at a dose rate of 1000 MU/min in the current version and weighs just 285 lbs. The target is fixed and is made of a tungsten alloy. The linac is mounted on a highly maneuverable robotic manipulator arm. The modern version of the system uses the robot KUKA KR 240 (KUKA Roboter GmbH, Augsburg, Germany). This manipulator can position the linac with 6 degrees of freedom and 0.12-mm precision. The system has a straight-through waveguide, steering coils, no flattening filter, sealed air-filled ionization chambers, a primary collimator, and a secondary set having circular apertures with diameters ranging from 5 to 60 mm at a source-to-axis distance (SAD) of 80 cm. Since 2008, a collimator (Figure 4.5) with moving segments (Iris™ Variable Aperture Collimator) is available (Pantelis et al. 2012). It consists of two stacked hexagonal banks of tungsten blades forming a dodecagonal opening. It can be set to 12 circular field sizes (from 5 to 60 mm, nominally equal to the system's 12 fixed collimator sizes). This allows the use of several different aperture sizes in a single treatment plan, optimizing dosimetry and treatment time for SRS. The beam flatness is within 14% when measured at a depth of 5 cm, a SAD of 80 cm, and a secondary collimator of 4 cm. With the same field size and SAD, the penumbra is within 4.5 mm at a depth of 5 cm.

Most treatments on the CyberKnife are non-isocentric, but a reference point serving as the origin of coordinate systems is defined in the room. Both the robotic arm and the imaging calibration use this room or geometric isocenter. It is defined by an isocrystal, a light-sensitive detector, mechanically mounted at the tip of a rigid post, the isopost. In the non-isocentric technique, which constitutes the majority of treatments, the beams originate from arbitrary points in the workspace and are delivered into the lesion. They are pointing away from the geometric isocenter. Their directions are optimized to conform to

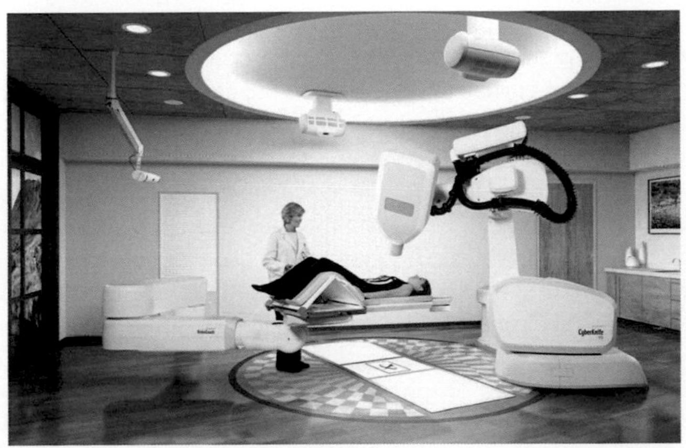

Figure 4.4 CyberKnife® VSI robotic radiosurgery system. (Image courtesy of Accuray, Inc.)

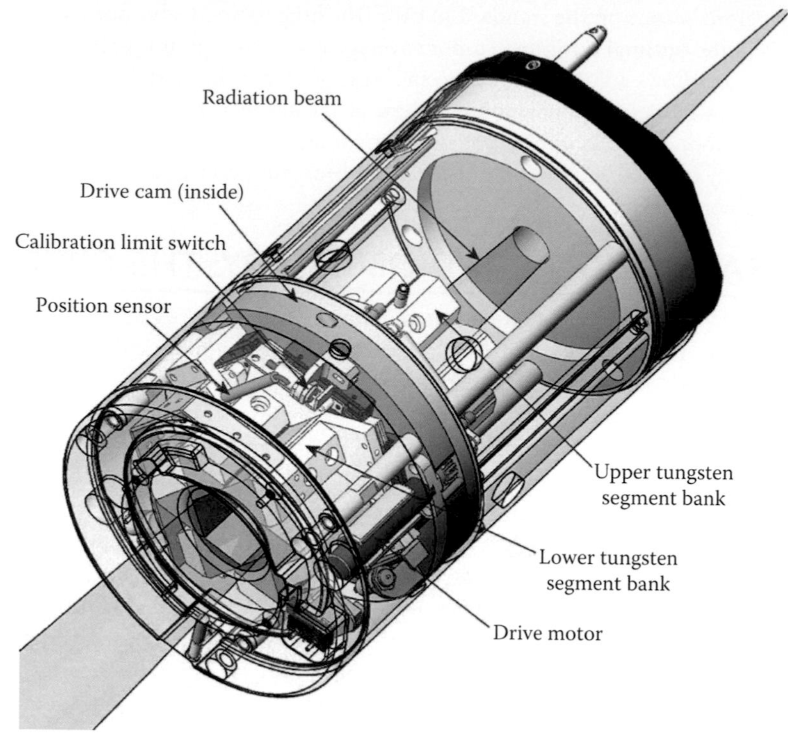

Figure 4.5 CyberKnife® Iris™ variable aperture collimator design. (Courtesy of Accuray, Inc. All rights reserved.)

highly irregular target shapes that may contain surface concavities and complex volumes (Dieterich et al. 2011). This geometric isocenter is different from the treatment isocenter, in the case of an isocentric or an overlay of isocentric shots of different collimator sizes. In this technique, only used in a small fraction of CyberKnife treatments, the treatment isocenter is a point defined in the planning CT during treatment plan creation, usually within or near the treatment target, and is used to define how the patient will be placed with respect to the robotic arm during initial alignment. This point is, in general, within 10 cm of the geometric isocenter.

The imaging hardware consists of two diagnostic flat-panel detectors illuminated by x-ray sources. Each source is 3.6 m from its corresponding detector. The sources are orthogonal with respect to the patient. The patient is positioned on an automated robotic couch controlled by a computer system. The centerline from each x-ray focal spot to its image receptor is at 45° from the floor. Both x-ray sources and detectors are rigidly attached to the room. The sources are conventional rotating anode tubes with 2.5 mm or more of aluminum-added filtration. In the current system version, the detectors are amorphous silicon, 41 × 41 cm, and mounted on the floor.

The CyberKnife image guidance system uses several targeting methods depending on the area being treated. Fiducial tracking is based on locating radio-opaque implanted markers rigidly associated with a soft tissue target, for example, in liver, pancreas, and prostate treatments. Bony structure tracking, 6D Skull Tracking, and the XSight® Spine Tracking systems are used for cranial and spinal lesions where the bony structures are characterized by a very high radiographic contrast. Soft tissue tracking (XSight Lung and 1-View Tracking) is used for lungs and relies on density differences between tumor and surrounding lung tissue, avoiding invasive implantation of fiducials whenever the target is not obscured by a radiographically dense structure (heart or spine).

The system is based on CT scans for anatomical spatial relationships in planning and targeting. Up to 512 axial slices can be used, generally with 1–1.5 mm slice thickness. Multimodality image fusion of CT, MR, PET, and 3-D angiography, advanced contouring, automatic segmentation, and dose optimization features are available. The TPS is DICOM RT compatible and can import images and contour data sets from other commercial TPS. The planning system offers forward or inverse planning, isocentric planning with single or multiple isocenters, IP by means of sequential or simplex optimization, fractionation or single-treatment planning, and the possibility of planning for simultaneous treatment of multiple tumors.

The TPS (MultiPlan®) is specific to the device (integrated) and starts by defining an initial set of beam configurations based on the tumor geometry. Optimization techniques are used to determine dose weighting of the beams to satisfy the constraints. It is an iterative process of beam selection and optimization. The planner selects the preconfigured treatment path, collimator size(s), and dose calculation algorithm (ray-tracing or MC) and sets the dose and MU constraints and dose objectives. The IP system uses linear optimization for MUs (Dieterich et al. 2011). The range of motion of the robotic arm allows a wide choice of beam directions; the optimization engine will generally start with several thousands of candidate beams as a solution set. A typical treatment will consist of approximately 100 different beams per fraction. Patients are usually set up head-first supine, but for some anatomies being treated, it is permissible to treat prone and/or feet first. For extracranial treatments, a vacuum fixation cushion and conventional RT positioning aids are employed to hold the patient in position. For cranial and C1–2 spinal lesions, a thermoplastic head mask is used. C3–7 setups may include a cervical collar.

During treatment, when treating targets that do not move with respiration, patient position is tracked frequently using either cranial anatomy, spinal bony anatomy (Xsight Spine), or implanted fiducials. The tracking is accomplished by correlating live radiographic images with digitally reconstructed radiographs (DRRs) generated from the treatment planning CT. At chosen time intervals, images are taken to detect any shifts in the bony landmarks or fiducials in 6 degrees of freedom, which, in turn, are used to extrapolate changes in the target location based on their defined spatial relationship. Initial patient positioning is aided by an automated couch (AXUM™ or RoboCouch®) that is integrated with the image guidance system to access patient positioning information and automatically correct for translations, roll and pitch (AXUM), or all 6 degrees of motion (RoboCouch). Once treatment has begun, changes in target position are automatically communicated to and compensated for by the robotic arm, which repositions the linac to retarget each radiation beam.

The clinically relevant accuracy of the CyberKnife is determined by conducting end-to-end (E2E) tests. The E2E test shows that the CyberKnife system can deliver radiation so that the centroids of the planned and actual delivered dose distributions are the same within the desired tolerance. It includes all elements in the treatment process: CT scan, treatment planning, image-guidance system, robot, and accelerator beam delivery. The test can be done using an anthropomorphic head phantom, anthropomorphic spinal phantom, and anthropomorphic lung phantom, with or without implanted fiducials.

Targets that move with respiration are treated using the real-time, dynamic capabilities of the CyberKnife by means of the Synchrony system. Three externally placed markers emitting pulsed, visible red light are positioned on the patient to track surface respiratory motion. The position and time stamp of the marker data captured by an optical camera are correlated to either internal fiducials or the treatment target itself (Xsight Lung and 1-View treatments) taken by the x-ray cameras. From multiple samples, a correlation model between the internal and external positions is established. The software determines the model (which can be either linear or nonlinear) based on the best fit to the data, updating it with each new x-ray image. During treatment, the system continually corrects for tumor motion using the marker data and correlation model, dynamically moving the linac to follow the internal target location in real time. The Synchrony® System has a targeting accuracy of better than 1.5 mm for fiducial and Xsight Lung treatments. Treatment time per session is generally less than 30 min for non-Synchrony treatments and less than 45 min for Synchrony treatments, with the actual time for each case depending on the dose, fractionation, and complexity and location of the tumor.

The CyberKnife M6™ FI, M6 FM, and M6 FIM models received Food and Drug Administration (FDA) 510(k) premarket approval in 2012. These units feature the robotic couch and the models FM and FIM include the InCise™ MLC with 41 tungsten 2.5-mm leaf pairs and a maximum field size of 10 × 12 cm at 80-cm SAD (Fahimian et al. 2013). The first patient was treated with this unit in February 2013 at the European CyberKnife Center Munich–Grosshadern (Figure 4.6).

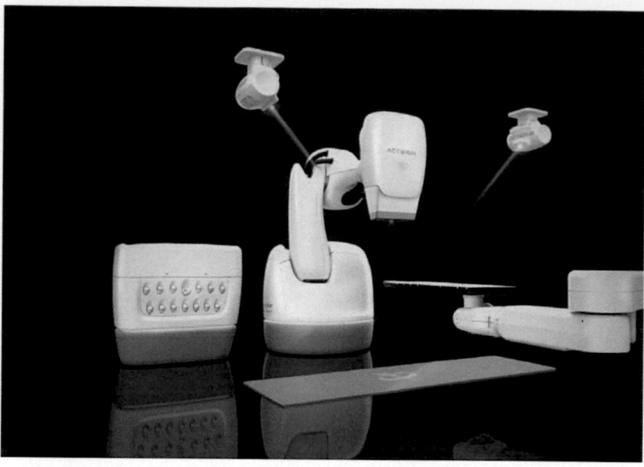

Figure 4.6 CyberKnife® M6™ radiosurgery suite with the InCise™ MLC. (Image courtesy of Accuray, Inc.)

4.4 VARIAN, ELEKTA, AND SIEMENS DEDICATED LINEAR ACCELERATORS

In this section, conventional C-arm gantry S-band linacs from the main manufacturers are presented. These machines are SRS/SBRT systems in addition to their more conventional use in radiotherapy.

4.4.1 THE TRILOGY SYSTEM

The Varian Trilogy® is a comprehensive delivery system. It can be used for 3-D conformal radiation therapy (CRT), IMRT, IGRT, dynamic adaptive radiotherapy (DART), and both intracranial and extracranial stereotactic techniques. In 2004, the Trilogy Tx model optimized for SRS, with improved amorphous silicon (a-Si) flat panel image guidance, HD120 MLC, and a higher dose rate, was released (Varian Medical Systems). The Trilogy platform combines a multimodality linac with MLC, advanced imaging capabilities, patient immobilization, target tracking and motion management, and treatment planning (Eclipse), although other commercially available TPSs may be used (Figure 4.7).

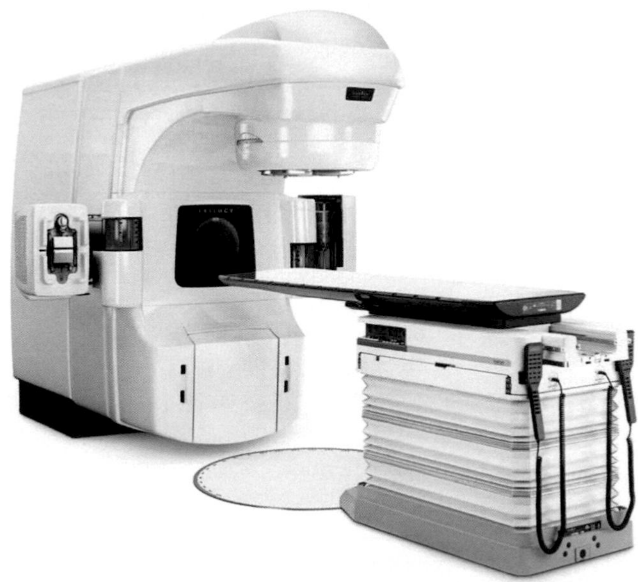

Figure 4.7 Varian Trilogy® linear accelerator. (Courtesy of Varian Medical Systems, Inc. All rights reserved.)

Varian's linac is characterized by a gridded electron gun that controls the beam from full on to full off in 2 to 3 ms, a waveguide, and an achromatic three-field bending magnet coupled with a 3-D servo-system and a solenoid (Varian Medical Systems). Real-time beam steering and dual sealed ion chambers working as a feedback loop to keep this steering accurate result in a tightly focused beam with optimal flatness, symmetry, and dosimetry. The 2-mm circular focal spot allows for steeper dose gradients and sharper portal images.

Varian SmartBeam™ IMRT paints tumors with 2.5 × 5 mm resolution. It can deliver more than 250 radiation intensity levels. Painting the dose with the small beamlets achieves a sculpted and sharply defined field that highly conforms to the target, sparing radiation to normal and critical structures.

Another feature of the new Varian linacs is an energy switch that optimizes the guide length to deliver the highest dose rates for all energies up to 1000 MU/min for stereotactic beams at maximum dose depth (dmax) and 100-cm SAD. The Trilogy has a separate small filter optimized for small field treatments in the 6-MV high dose rate mode (HDRM) of 1000 MU/min. This feature is useful for SRT treatments considering the large dose per fraction delivered.

The Millennium™ MLC is a two-bank, 120-leaf collimator with 5-mm leaf widths at the isocenter for the central 20 cm of the 40 × 40 cm field. The small leaf resolution makes it suitable for small lesions as in stereotactic treatments. Fast leaf speed and interdigitation capabilities result in shorter treatment times. The MLC operates in static, dynamic (step-and-shoot and sliding window delivery), and conformal arc modes. For stereotactic techniques, the Trilogy Tx model comes with the HD120 MLC and conical collimators. The maximum field size for SRS is 15 × 15 cm in HDRM, and the maximum dose per field is 6000 MUs (Varian Medical Systems).

The system's dynamic targeting IGRT process consists of several tools: optical, radiographic, CBCT, fluoroscopic, and respiratory gating. The MV, kV, and optical image guidance systems are used to set up patients and pinpoint targets and verify treatment delivery. The PortalVision™ MV imager is used for verification of patient setups, treatment portals, and pretreatment QA. It is an a-Si detector with an active area of 40 × 30 cm. Image acquisition is possible before, during, and after treatment. It also records the intensity patterns of IMRT fields for pretreatment QA of IMRT planning and delivery. This portal dosimetry capability comes with an integrated image viewing and analysis software. CBCT images are acquired using the On-Board Imager®. It is a kV imaging system used for target localization, patient positioning, and motion management. For gated treatment portals, this system provides kV fluoroscopy for pretreatment verification. The RPM™ respiratory gating system keeps track of intrafraction movement. It provides a passive, real-time monitoring of patient respiration. The gating system includes an infrared tracking camera, external marker block, and RPM workstation.

4.4.2 THE TRUEBEAM SYSTEM

The TrueBeam® is a new line of accelerators manufactured by Varian. This platform for image-guided radiotherapy and radiosurgery is a fully integrated system (imaging, beam delivery, and motion management) designed from the ground up to treat moving targets and to improve operation, precision, and speed. The product line includes TrueBeam STx released in 2010, specially configured for advanced radiosurgery. The system uses a completely reengineered control system, the Maestro, with sophisticated architecture and a multitude of technical innovations to dynamically synchronize imaging, patient positioning, motion management, beam shaping, and treatment delivery. The system also delivers Varian's new gated RapidArc® radiotherapy, which compensates for tumor motion by synchronizing imaging with dose delivery during a continuous rotation around the patient.

In this new generation of linear accelerators, many key elements including the waveguide system, carousel assembly, beam generation, and monitoring control system differ significantly from those found in previous models (Hrbacek et al. 2011). One of the key features is the availability of two types of photon beams: standard flattened filtered beams and flattening filter-free (FFF) beams. The TrueBeam linac has a slightly different design of the head and related components from its predecessors. The carousel system has been modified to permit the use of several photon energies (flattened and FFF modes). It has an integrated bending magnet with an in-air target instead of the vacuum-sealed target found in the standard Varian models. The TrueBeam also contains a thicker primary collimator of slightly different design to permit

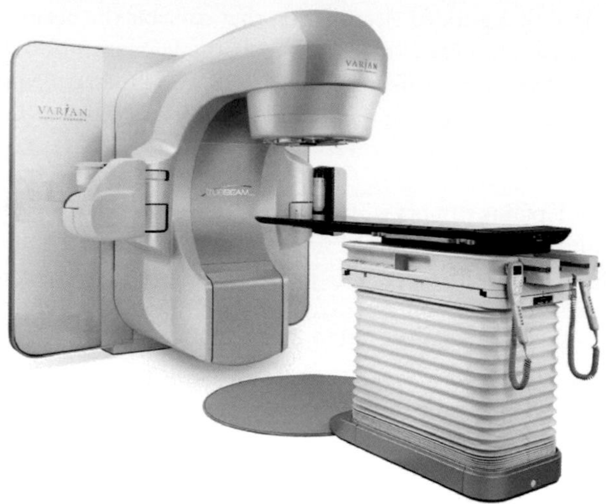

Figure 4.8 Varian TrueBeam® STx system. (Image courtesy of Varian Medical Systems, Inc.)

sharper beam falloff and uses an antibackscatter filter, which can reduce the dose dependency on field size (Beyer 2013).

The TrueBeam STx system for SRS (Figure 4.8) comes equipped with the new Varian 120-leaf MLC (HD120) and has an increased dose rate (1400 MU/min for 6-MV FFF and 2400 MU/min for 10-MV FFF) due to the use of an unflattened beam (Kielar et al. 2012).

4.4.3 THE EDGE RADIOSURGERY SUITE

The Edge™ radiosurgery suite is the newest Varian dedicated linear accelerator to SRT. It is a TrueBeam platform constituting an end-to-end solution for planning and delivering radiosurgery treatments. It was developed with the assistance of clinicians from different medical centers, such as the Champalimaud Foundation in Lisbon, Portugal. Varian Medical Corporation received FDA 510(k) premarketing approval in January 2013 to begin marketing the Edge radiosurgery suite (Figure 4.9). This new linear accelerator system is designed to fit in smaller vaults (smaller profile than the TrueBeam system) for use in dedicated radiosurgery or fractionated stereotactic radiation therapy programs. It is a stand-alone product, which does not rely on components from any other vendors.

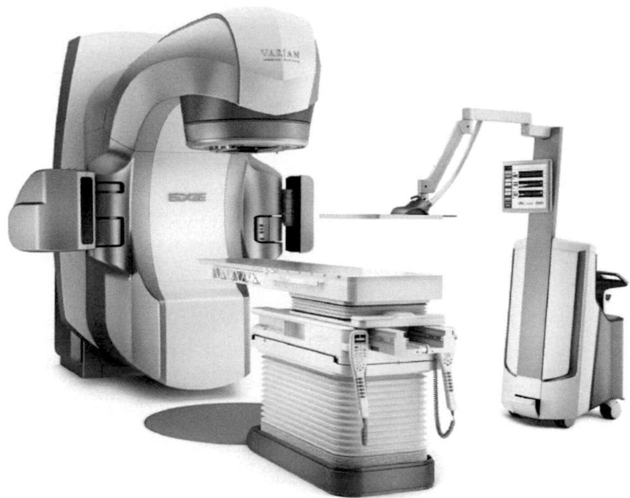

Figure 4.9 Varian's Edge™ radiosurgery suite with the Calypso® system. (Courtesy of Varian Medical Systems, Inc.)

The Edge intracranial SRS package provides real-time tracking and motion management for radiosurgical lesions by using an Optical Surface Monitoring System and a 3-D surface mapping of the patient's external surface (Vision RT motion management solution). The extracranial SBRT package comes with the Calypso® 4-D Localization and Tracking System™ (Varian Medical Systems) to keep the tumor in the path of the radiation beam at all times. It is an electromagnetic system, and no additional ionizing radiation is delivered to the patient for tracking. Calypso tracks the target using radiofrequency waves, an innovation referred to as GPS for the Body® Technology. Three electromagnetic Beacon® Transponders implanted during an outpatient procedure transmit location information about the target. The tracking system locks onto the signal during patient setup and tracks it throughout the treatment. The system comes with electromagnetic transponders, a 4-D electromagnetic array (antenna that transmits and also receives the radiofrequency signals), an optical system with three infrared cameras, and a 4-D tracking station. The internal transponders are FDA-cleared only for implantation into the prostate so far. The Calypso Surface Beacon Transponders expand the utility of the Calypso System for real-time motion tracking anywhere on the body where intrafraction motion may be a concern. They are cleared by the FDA for general use.

The Edge suite includes the new PerfectPitch™ 6 degrees of freedom radiosurgery couch, which makes it possible to adjust the patient's position along six axes of motion versus four that were in earlier generations of technology; high-intensity flattening filter free mode for radiosurgery (1400 MU/min for 6 MV and 2400 MU/min for 10 MV); and 2-D, 3-D, and 4-D imaging systems (Varian Medical Systems). Beam shaping options include the HD120 MLC collimator and circular radiosurgery cones, interlocked to the device operating system. The Edge supports both frame-based and frameless approaches to patient immobilization.

4.4.4 THE SYNERGY S SYSTEM

The Elekta Synergy® S solution (Elekta AB, Stockholm, Sweden) combines integrated imaging with high-resolution radiation delivery (Figure 4.10). It is optimized for extracranial SRT and SRS applications. Synergy S is an image-guided robotic linac that combines high-conformal beam shaping with 4-D Adaptive™ IGRT technology. It provides planar, fluoroscopic quality and 3-D x-ray volume imaging (XVI) of the patient in the treatment position to help minimize setup errors and identify critical structures during treatment, allowing clinicians to compensate for motion. Synergy S features a small treatment head, 62 cm in diameter, increasing clearance around the patient and providing a wide variety of treatment approaches, including noncoplanar beams that are usually common in stereotactic treatments.

The iBEAM® evo Couchtop is of homogenous carbon fiber design containing no metal in the treatment area, improving radiotranslucency, and reducing attenuation for IMRT, intensity modulated arc therapy (IMAT), and IGRT.

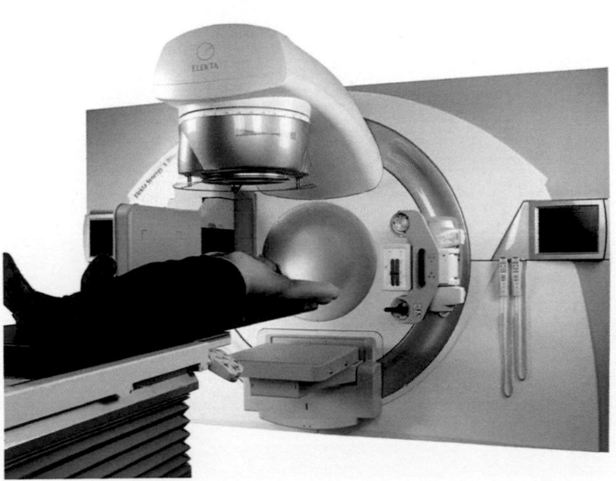

Figure 4.10 Elekta Synergy® S Linear Accelerator. (Image courtesy of Elekta AB.)

The 80-leaf MLC provides a field size of 16 × 21 cm with interdigitation capabilities of the 4-mm-thick leaves. It allows conformality to the target and avoidance of critical structures. The newly introduced Agility™ 160-leaf MLC covers a field size up to 40 × 40 cm with 5-mm width at the isocenter with a leaf speed of up to 3.5 cm/s. Interleaf leakage was reported to be 0.2% (Bedford et al. 2013). It was designed to be used in VMAT.

For IGRT, Synergy S is equipped with 3-D imaging and the VolumeView™ software. The system enables volume imaging of soft tissue and critical structures at the time of treatment and in the treatment position. Imaging can be repeated as often as necessary throughout the course of treatment. This allows margins to be reduced and uncertainties in target dimensions, location, and movement to be accommodated. Complete capture of the area of interest is done in a single acquisition. Reconstruction occurs simultaneously, enabling rapid registration against the CT treatment plan image.

Imaging capabilities also include 2-D monitoring with MotionView™ and PlanarView™ software. MotionView is a sequence mode that enables real-time viewing of anatomy while the patient is in the treatment position. PlanarView offers fast low-dose snapshots.

The iViewGT™ megavoltage portal imaging system provides a beam's eye view of the treatment delivery.

4.4.5 THE AXESSE SYSTEM

Elekta Axesse™ is a system capable of delivering SRS, SRT, VMAT, IGRT, IMRT, and CRT treatments (Figure 4.11). It provides a comprehensive and integrated solution to apply SRS and SRT. The system combines 3-D image guidance at the time of treatment with highly conformal beam shaping and robotic 6-D submillimeter patient positioning. All treatment processes from planning to delivery are controlled from a single workstation supported by an electronic medical record-centered workflow. Another feature of the system is the high degree of automation and the integrated QA tools, which help reduce the physicist's workload.

High-resolution beam shaping is supported by planning and delivery techniques to improve target conformance. Beam shaping options include circular cones and MLC-based delivery, as in IMRT, IMRS, and dynamic conformal delivery utilizing VMAT. Multiple targets can be treated within one plan. One characteristic of Axesse is the large stereotactic field size, enabling more beam configurations and treatment options, combined with the flexibility of a 360°, 90-cm clearance around the patient and a wide range of noncoplanar angles, very suitable for SRS/SBRT. Fusion with CT, MR, PET, and angiograms is available and so is the automatic localizer and determination of stereotactic coordinates when used. Field-fluence optimization, based on targets and critical structures, is achieved with fast simulated annealing algorithms. Radiobiological optimization is also possible. Three-dimensional dose clouds and dose volume histograms (DVHs) are displayed live during iterations.

To improve setup efficiency, Axesse incorporates integrated indexing accessories designed to be compatible with CT, MR, PET, single photon emission computed tomography (SPECT), and ultrasound for larger field size requirements. Patient immobilization incorporates noninvasive dual vacuum activated

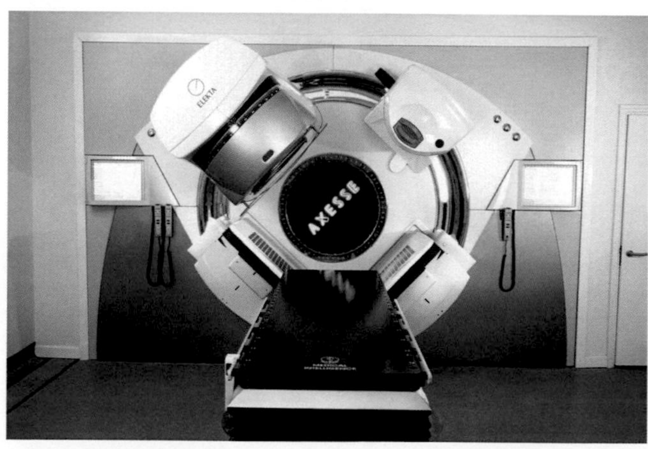

Figure 4.11 Elekta Axesse™ system. (Image courtesy of Elekta AB.)

fixation to minimize patient movement and maintain accurate targeting during delivery. The noninvasive head frame, which comes with a vacuum-sealed bite-block fixation, is constructed from carbon fiber and is both CT and MR compatible.

The Axesse system uses XVI combined with 6-D robotic table adjustments. With kV imaging, patient movement can be monitored in real time eliminating the need for tumor surrogates. Axesse integrates remote, automatic positional correction in 6-D (*x*, *y*, *z*, roll, pitch, and yaw) with submillimeter positional accuracy. The 6-D correction provides realignment in 3-D space within the patient at the time of treatment.

4.4.6 THE ARTISTE SYSTEM

The Siemens Artiste™ solution (Siemens AG, Munich, Germany) is an integrated imaging and therapy workflow solution designed specifically for adaptive radiation therapy (ART). It offers a comprehensive portfolio of 2-D and 3-D imaging modalities and advanced treatment delivery tools, enabling clinicians to choose the appropriate treatment technique and IGRT application and make critical adjustments instantly (Figure 4.12).

Artiste supports CRT, IMRT, IGRT, gated treatments, SRS, and SBRT. For IMRT treatments, the MLC is a 160-leaf, high-resolution and fast field-shaping collimator, with a 5-mm leaf thickness over the full field of view (FOV). Leaf-positioning accuracy is of 0.5 mm, and leaf motion is at 4 cm/s.

To manage respiratory motion, a special hardware and software package is available. The 4-D gating system is commonly used for CT and megavoltage (MV) imaging. The system uses a respiratory sensor

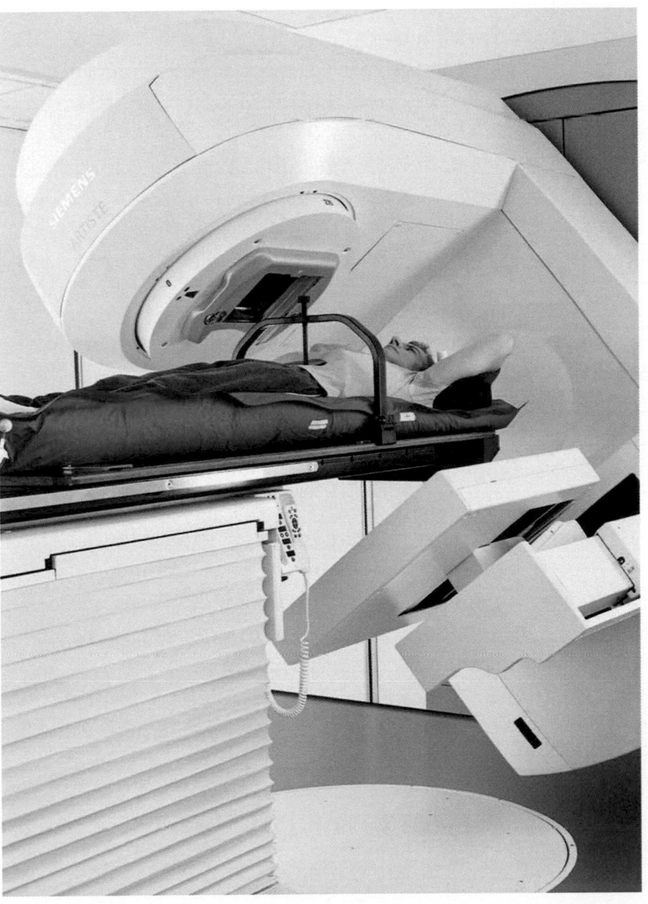

Figure 4.12 Siemens Artiste linear accelerator with an abdominal compression device. (Image courtesy of Siemens AG.)

secured on the patient to detect changes in respiratory motion and a breathing belt used as a trigger to define the breathing phase.

The treatment couch is a robotic flat-panel positioner accommodating a patient weight capacity of up to 550 lb. (250 kg). It is a robust, stable, and accurate table. It enables flexible off-isocenter imaging, supporting verification of patient positioning even for larger patients.

The imaging modalities include OPTIVUE™ for portal imaging, MVision™ for 3-D megavoltage cone beam (MVCB) imaging, and CTVision™ for an in-room diagnostic CT imaging (Gayou 2012). For daily clinical routines, the OPTIVUE robotic positioning is used to verify accurate treatment delivery. It is a 2-D a-Si, MV-based electronic portal imaging device (EPID). Alternatively, when additional imaging information is required during the course of treatment, clinicians can choose 3-D MVCB imaging.

MVision minimizes artifacts and provides the soft-tissue resolution required in challenging cases, such as imaging prostheses, orthopedic hardware, and large patients. Doses applied from the MV imaging can be calculated into the prescription to ensure that OARs do not receive more than permissible doses. It offers variable field sizes, extended FOV (useful for large patients and tumors located off-axis), and selectable start and stop angles for sparing structures at risk and avoiding patient collisions with the gantry during imaging. This 3-D adaptive targeting is a fully integrated technology. Software includes automated registration and visualization tools for rapid, user-friendly patient setup verification.

CTVision brings the same fast, high-contrast diagnostic imaging standard used for TPS into the treatment room. For clinical applications requiring in-room, high-contrast diagnostic image quality, ARTISTE offers a Siemens' high-precision CT-on-Rails gantry. It is a modified SOMATOM® diagnostic CT. The SOMATOM Sensation Open combines the advantages of a large-bore CT with fast, multislice CT technology. The 82-cm gantry bore and FOV accommodate large patients and patient positionings requiring greater clearance. Patients with breast cancer, lung cancer, or Hodgkin's lymphoma, and those weighing up to 550 lb., can be scanned in the correct treatment position. This approach allows direct comparison of daily patient anatomy to the original planning data and implementation of benchmark concepts such as daily replanning.

Siemens Medical Systems announced in November 2011 that it was leaving the radiation oncology business. Their linear accelerators will continue to be used and supported for a number of years.

4.5 ACCURAY HELICAL TOMOTHERAPY: THE Hi-ART SYSTEM

Helical tomotherapy is an IMRT delivery technique developed at the University of Wisconsin-Madison later commercialized as the TomoTherapy® Hi-ART® System by TomoTherapy, Inc. of Madison, Wisconsin (Mackie et al. 1993; Mackie 2006). Now marketed by Accuray Inc. (Sunnyvale, California), it is still the only treatment unit in the market that uses the helical slice-based delivery process since it became available routinely for clinical practice in 2003 (Langen et al. 2010). The TomoTherapy HD unit combines IMRT treatment delivery, megavoltage computed tomography (MVCT) imaging capabilities, and an integrated treatment planning (Figure 4.13).

The Hi-ART system is characterized by a unique geometry similar to that of a helical CT scanner. The beam is generated by a 3-GHz S-band, 40-cm-long 6-MV linac mounted on a slip ring gantry. The SAD is 85 cm, and there is no flattening filter, which allows for a high dose rate. The beam is collimated by a jaw into a fan beam shape. At the isocenter, the fan beam has an extension of 40 cm in the lateral or x direction and 5 cm in the longitudinal or y direction. In the y direction, the jaw is adjustable, allowing the beam to be collimated to a smaller size (typically slice widths are 1, 2.5, and 5 cm at the isocenter).

For further beam shaping, a binary MLC is used. It is a two-bank, 64-leaf collimator that divides the fan beam in the x direction. The MLC is pneumatically driven so that each leaf can be rapidly opened or closed. When the leaves are closed, they move across the entire field and stop outside of it. Intensity modulation is achieved by leaf-specific opening times. The leaves are made of tungsten and are 10 cm thick and 0.625 cm wide at the isocenter (40 cm divided by 64).

Another component of the machine is its imaging capability. The ring gantry contains a detector system. This is an arc-shaped xenon CT detector that is a standard array from a third-generation CT scanner (Keller et al. 2002; Judy et al. 1977; Ruchala et al. 1999), mounted opposite to the accelerator and used to

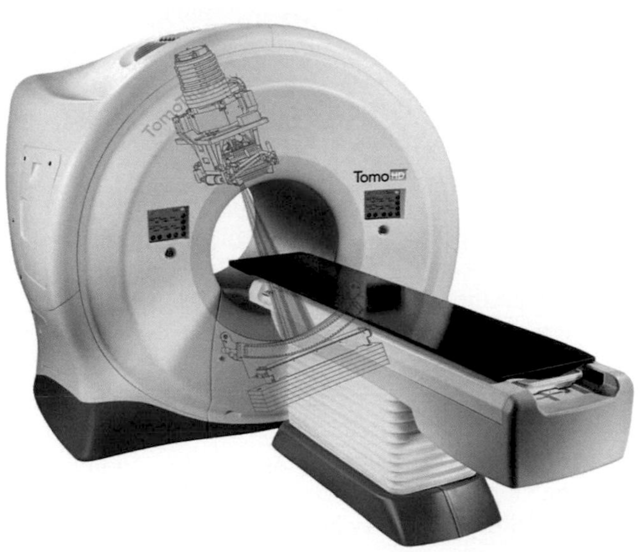

Figure 4.13 TomoTherapy Hi-ART System. (Courtesy of Accuray Inc. All rights reserved.)

collect data for MVCT acquisition. For imaging, the accelerator is adjusted such that the nominal peak energy of the x-ray beam is approximately 3.5 MeV and average energy is approximately 1 MeV (Jeraj et al. 2004). The standard image matrix size is 512 × 512 pixels, and the FOV has a diameter of 40 cm. A filtered back-projection algorithm is used for image reconstruction (Ruchala et al. 2002).

The acquisition rate is 1 slice per 5 s (10 s per 360° gantry rotation with half-scan reconstruction technique), and reconstruction is concurrent with acquisition. The imaging dose depends on the selected pitch and the thickness of the imaged anatomy, but it is typically in the range of 1–3 cGy (Shah et al. 2008). This is a sufficiently low dose for daily imaging to be standard practice. The total scan time depends on the number of selected slices. The unit also comes equipped with a lead beam stopper to absorb primary radiation and therefore reduce shielding requirements. The TomoTherapy operator station includes image registration tools for manual or automatic rigid-body registration.

During treatment, the gantry continuously rotates while the patient is moved through the rotating beam plane. The gantry, treatment couch, and MLC leaves are all in motion simultaneously, which makes the Hi-ART machine a fully dynamic delivery unit. The treatment plane is inside the bore, so for patient setup, a virtual isocenter is defined outside the bore and two distinct laser systems are installed in the room. The distance in the *y* direction from the virtual to the treatment isocenter is 70 cm.

Some of the characteristics of this system are the machine output defined in terms of absorbed dose per unit time and not per MU and the Hounsfield units (HU) calibrated against mass density rather than relative electron density. Plan parameters are all time based. A constant, but machine-specific, dose rate (approximately 850 cGy/min at the isocenter, at 1.5-cm depth in water, for a 40 × 5 cm field) is assumed for treatment planning purposes, and delivery is finished after all projections are complete (Langen et al. 2010). Two parallel-plate sealed ion chambers are located upstream of the *y* jaw to monitor the dose rate. The MLC leaves are closed for the initial 10 s of every planned delivery to ensure beam stabilization. It should be noted that the slowest permissible gantry period is 60 s. Large fraction doses, such as for SBRT treatments, which would require longer gantry periods, need to be broken into two or more subfractions.

TomoTherapy's Hi-ART II is the only TPS available to generate treatment plans for TomoTherapy machines. For each plan, the treatment slice width, pitch, and modulation factor (MF) need to be selected. Each rotation is divided into 51 projections. For each projection, each MLC leaf has a unique opening time.

The slice width is the fan-beam width defined by the *y* jaws in the longitudinal direction at the isocenter. The pitch is the ratio of the couch travel per gantry rotation to the treatment slice width. Pitch is usually set to less than 0.5 for better dose homogeneity (Olivera et al. 1999). The MF is the longest leaf opening time divided by the average of all nonzero leaf opening times. The longest opening time determines the gantry rotation speed. The MF selected is the maximum value available to the optimization software. The actual MF, from the final plan, is often smaller and is typically in the range 1.5–2.5. A higher MF (up to approximately 3.5) may improve the plan quality and is typically used for more complex target volumes that require dose conformality to highly concave shapes.

Once the plan parameters are selected, the dose distribution for each beamlet is calculated. The number of beamlets depends on the field size, pitch, target volume, and shape. The dose calculation engine uses the convolution/superposition method (Mackie et al. 2003). The next step is the optimization process. A least squares method is used to optimize the objective function (Olivera et al. 1999). Measured MLC leaf latency data are used in the final calculation of leaf-opening times. Times shorter than the leaf transition time (18 ms) are deleted from the control sinogram.

As of yet, no commercial software for independent calculation of helical TomoTherapy dose distribution exists, so for each individual plan, a calculation in a phantom geometry is done and verified by measurements. TomoTherapy, Inc., was acquired by Accuray, Inc., in June 2011. A newly revised TomoTherapy system featuring TomoEDGE™ dynamic jaws technology was announced in October 2012 and first treated patients at Heidelberg University Hospital in March 2013.

4.6 NEW NOVEL MULTIMODALITY MACHINES

The role of radiation therapy in the treatment of cancers and some nonmalignant conditions has been well established. Relatively recent technological advancements in imaging and computers have allowed radiotherapy to make considerable advancements. Motion management still represents a challenge for clinicians. Different approaches are being used in the clinic.

4.6.1 THE VIEWRAY SYSTEM

A new hybrid technology has been introduced by ViewRay™, Inc. (Cleveland, Ohio). It consists of a unique and patented combination of radiotherapy delivery and simultaneous MR imaging (MRI). The system is designed to provide continuous soft-tissue imaging during treatment so that clinicians can see the tumor, adapt to changes in the patient's anatomy, and deliver a precise treatment, all in real time. The ViewRay MRI-guided radiation therapy system is expected to provide more sophisticated treatments of tumors that move significantly during treatment. The motion of tumors in lung, prostate, liver, head and neck, and other sites is under study at the company (ViewRay). MRI guidance will give a clearer view of the patient's internal organs without the added risk of ionizing radiation, such as in the case of CT guidance or the use of invasive markers or procedures. Subsecond continuous tracking can be performed in one to three planes simultaneously.

ViewRay's purpose-built MRI scanner utilizes a split 0.35-T superconducting actively shielded magnet and is radiation compatible. Siemens AG Healthcare Sector, which is collaborating with ViewRay, supplies the electronics. Volumetric positioning scans are acquired prior to treatment in 15 s, and a planning scan is achieved in less than 3 min. The RT delivery system, which is fully compatible with simultaneous MRI operation, has three equally spaced Co-60 therapy sources with 600-cGy/min output at installation (Figure 4.14). The multihead, rotating gantry configuration is capable of treating a full 360°. The MLC is double focused, which results in sharper penumbra, and the maximum field size is 30 × 30 cm. Treatment planning is integrated with the system and uses PB and MC dose computation (ViewRay). It allows for deformable image registration and the transfer of CT density to MRI for dose calculation. The ViewRay System (Figure 4.15) is presently in its final stages of development and is being installed at its first site, the Siteman Cancer Center at Barnes-Jewish Hospital and Washington University School of Medicine in St. Louis, Missouri. It has received FDA 510(k) clearance for its MRI-guided radiation therapy system in 2012.

4.6.2 THE VERO SBRT SYSTEM

Another novel IGRT device under assessment, developed by Brainlab (Brainlab AG, Munich, Germany), is the Vero™ SBRT system, also known as MHI-TM2000 in Japan (Speiser et al. 2009). It is a joint product

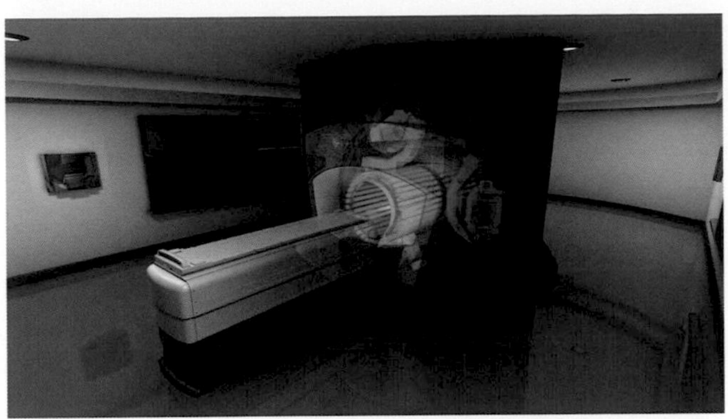

Figure 4.14 The ViewRay™ design: skinless illustration of the system. (Courtesy of ViewRay Inc. All rights reserved.)

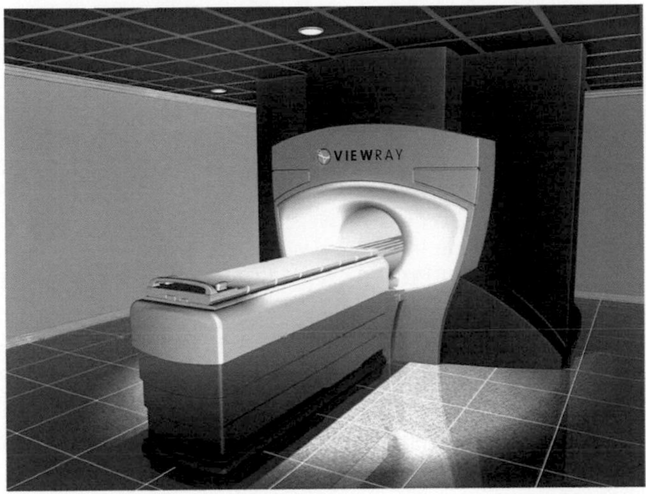

Figure 4.15 The ViewRay™ System. (Image courtesy of ViewRay Inc.)

of Mitsubishi Heavy Industries Ltd. (Tokyo, Japan) and Brainlab AG (Feldkirchen, Germany). The Linac is an in-line 6-MV standing wave, C-band 38-cm-long accelerator with a beam stopper (152 mm of lead with an 18-mm steel cover) to reduce shielding requirements. It is mounted with a 60-leaf MLC on two orthogonal gimbals integrated into an O-ring gantry (Figure 4.16). This design provides a high structural stability. Instead of a couch rotation, the ring can rotate by ±60° around the vertical axis (skew direction) to eliminate patient movement during noncoplanar treatment approaches. The maximum dose rate is 500 cGy/min, and the minimum and maximum field sizes are 0.25 × 0.5 and 15 × 15 cm, respectively. The tungsten alloy leaves are single-focused and have a width of 5 mm at the isocenter (physical 2.5 mm) and a thickness of 11 cm.

The most novel innovation of this system is the gimbaled linac head mechanism. The gimbal mechanism enables the entire linac head to be rotated up to 2.4° in the pan and tilt directions, allowing the MV beam to be displaced up to ±4.2 cm in each direction in the isocenter plane with a maximum speed of up to 6 cm/s (Depuydt et al. 2011). The gimbals can be utilized in both static and dynamic delivery modes. In the static mode, the gimbal mechanism can be used to compensate for any residual geometric distortion in the O-ring gantry, allowing the mechanical isocenter to be reproduced to within 0.1 mm root mean square error over all 360° of rotation (Kamino et al. 2006). In the dynamic mode, the gimbaled gantry allows dynamic tracking of moving tumors, for example, tracking lung tumors moving under respiration. The tracking system has a lag time in the order of 50 ms. A prediction algorithm, incorporating this lag time, can be used to accurately track motion. Studies showed that the E90% (the 90% percentile of the residual

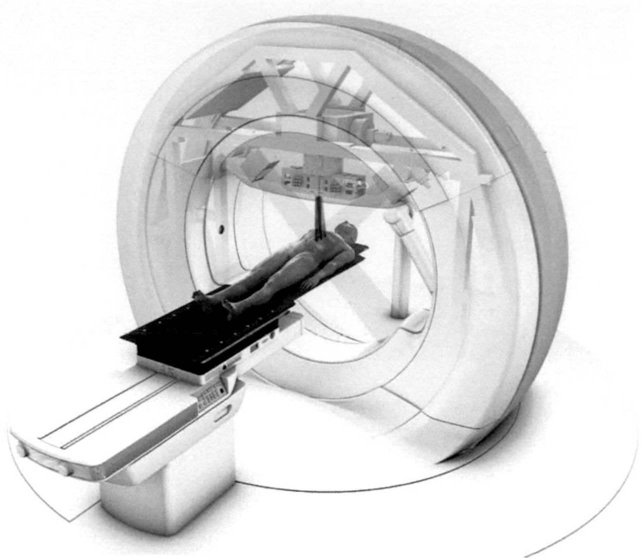

Figure 4.16 The Brainlab Vero™ System. (Courtesy of Brainlab AG. All rights reserved.)

error distribution) was less than 1 mm for both simple sinusoidal and real-patient breathing trajectories (Depuydt et al. 2011).

Several options are available for patient imaging, positioning, and tracking, such as conventional laser-based, IR, stereo x-ray, CBCT, and EPID. The Vero SBRT system incorporates Brainlab ExacTrac x-ray imaging system directly into the ring gantry. In the ring are two kV imaging devices with flat-panel detectors, and an EPID for MV portal imaging. The kV x-ray sources are mounted at 90° to one another. A stereoscopic IR camera system is used for real-time patient position monitoring. The orthogonal kV system can also be utilized for cone-beam imaging to provide 3-D volumetric imaging for soft-tissue-based setup verification. Both kV x-ray systems can be utilized simultaneously in dual cone-beam mode to increase the efficiency of data acquisition. The Vero SBRT system includes a robotic couch to allow for 6-D translational and rotational setup corrections, with an automatic 5-D couch motion to correct for translational, roll, and pitch inaccuracies. The couch itself can tilt and roll to correct for detected pitch and roll rotational errors, while yaw rotation corrections can be achieved by rotating the O-ring gantry around the couch. The integration of a gimbaled treatment beam, 2-D and 3-D imaging, tracking capabilities, and real-time feedback constitute unique features of the dedicated SBRT device and provide unique opportunities for advanced delivery techniques. The system also comes with a modern suite of Brainlab applications, including automatic image fusion, autosegmentation, planning tools, and MC dose calculation, making it capable of real-time adaptive radiotherapy.

The final accuracy of the Vero SBRT system's IGRT functionality relies on accurate geometric calibration of the imaging subsystems, the linear accelerator, and the robotic treatment couch. An initial E2E performance analysis performed by University of Texas Southwestern Medical Center in 2009 utilized the stereotactic Lucy® 3-D QA Phantom (Standard Imaging, Middleton, Wisconsin) to perform a hidden target test. Their results showed the imaging system capable of correcting setup errors to within a maximum residual displacement of 0.4 mm (Speiser et al. 2009). A separate study later reported on the E2E positioning accuracy of the Vero SBRT system showing that the imaging (both stereoscopic and cone beam), autoregistration, and robotic couch systems can align a target to within 0.5 mm of the radiation beam isocenter (Miyabe et al. 2011).

Studies of the Vero SBRT's dynamic tracking capabilities, using a 3-D motion phantom with simple motion patterns including a simulated respiration-like waveform, with an amplitude of 2 cm and a frequency of 0.25 Hz, were conducted in 2009. Under motion, the dose distribution is known to become blurred. Historically, this motion blurring has been dealt with by increasing the target margins, leading to larger irradiated volumes of surrounding normal tissue. Using the stereoscopic kV imaging at a rate of 7.5 frames per second and a linear autoregressive predictive protocol to compensate for system lag, the results showed that gimbal-tracking for a moving target can very nearly match the intended static dose distribution (Takayama et al. 2009). While more work is needed to better define the optimal trade-off between imaging frequency, imaging dose, and tracking accuracy, the initial results of using the gimbal mechanism for pursuing irradiation seem to be a promising approach.

REFERENCES

Ackerly, T., C. M. Lancaster, M. Geso, and K. J. Roxby. 2011. Clinical accuracy of ExacTrac intracranial frameless stereotactic system. *Med. Phys.* 38:5040–8.

Adler, J. R., S. D. Chang, M. J. Murphy et al. 1997. The CyberKnife: A frameless robotic system for radiosurgery. *Stereotact. Funct. Neurosurg.* 69:124–8.

Bedford, J. L., M. D. R. Thomas, and G. Smyth. 2013. Beam modeling and VMAT performance with the Agility 160- leaf multileaf collimator. *J. Appl. Clin. Med. Phys.* 14:172–85.

Benedict, S. H., K. M. Yenice, D. Followill et al. 2010. Stereotactic body radiation therapy: The report of AAPM Task Group 101. *Med. Phys.* 37 (8):4078–101.

Beyer, G. P. 2013. Commissioning measurements for photon beam data on three TrueBeam linear accelerators, and comparison with Trilogy and Clinac 2100 linear accelerators. *J. Appl. Clin. Med. Phys.* 14 (1):273–88.

Brainlab. Radiation Oncology. www.Brainlab.com.

Chang, Z., Z. Wang, Q. J. Wu et al. 2008. Dosimetric characteristics of Novalis Tx system with high definition multileaf collimator. *Med. Phys.* 35:4460–3.

Depuydt, T., D. Verellen, O. Haas et al. 2011. Geometric accuracy of a novel gimbals based radiation therapy tumor tracking system. *Radiother. Oncol.* 98 (3):365–72.

Dieterich, S., C. Cavedon, C. F. Chuang et al. 2011. Quality assurance for robotic radiosurgery: Report of the AAPM Task Group 135. *Med. Phys.* 38 (6):2914–36.

Fahimian B., S. Soltys, L. Xing et al. 2013. Evaluation of MLC-based robotic radiotherapy. *Med. Phys.* 40:344.

Gayou, O. 2012. Influence of acquisition parameters on MV-CBCT image quality. *J. Appl. Clin. Med. Phys.* 13:14–26.

Haley, M. L., P. C. Gerszten, D. E. Heron et al. 2011. Efficacy and cost effectiveness analysis of external beam and stereotactic body radiation therapy in the treatment of spine metastases: A matched-pair analysis. *J. Neurosurg. Spine* 14:537–42.

Hrbacek, J., S. Lang, and S. Klock. 2011. Commissioning of photon beams of a flattening filter-free linear accelerator and the accuracy of beam modeling using an anisotropic analytical algorithm. *Int. J. Radiat. Oncol. Biol. Phys.* 80 (4):1228–37.

Jaffray, D., P. Kupelian, T. Djemil, and R. M. Macklis. 2007. Review of image-guided radiation therapy (IGRT). *Expert Rev. Anticancer Ther.* 7 (1):89–103.

Jeraj, R., T. R. Mackie, J. Balog et al. 2004. Radiation characteristics of helical tomotherapy. *Med. Phys.* 31 (2):396–404.

Jin, J., F. Yin, S. Tenn, P. Medin, and T. Solberg. 2008. Use of the Brainlab ExacTrac X-ray 6D system in image-guided radiotherapy. *Med. Dosim.* 33:124–34.

Judy, P. F., S. Balter, D. Bassano et al. 1977. Phantoms for performance evaluation and quality assurance of CT scanners: The Report of Diagnostic Radiology Committee Task Force on CT Scanner Phantoms. *AAPM Report No. 1.*

Kamino, Y., K. Takayama, M. Kokubo et al. 2006. Development of a four-dimensional image-guided radiotherapy system with a gimbaled X-ray head. *Int. J. Radiat. Oncol. Biol. Phys.* 66 (1):271–8.

Keller, H., M. Glass, R. Hinderer et al. 2002. Monte Carlo study of a highly efficient gas ionization detector for megavoltage imaging and image-guided radiotherapy. *Med. Phys.* 29 (2):165–75.

Kielar, K., E. Mok, A. Hsu, L. Wang, and G. Luxton. 2012. Verification of dosimetric accuracy on the TrueBeam STx: Rounded leaf effect of the high definition MLC. *Med. Phys.* 39:6360–71.

Langen, K. M., N. Papanikolaou, J. Balog et al. 2010. QA for helical tomotherapy: Report of the AAPM Task Group 148. *Med. Phys.* 37 (9):4817–53.

Llacer, J., T. D. Solberg, and C. Promberger. 2001. Comparative behavior of the dynamically penalized likelihood algorithm in inverse radiation therapy planning. *Phys. Med. Biol.* 46 (10):2637–63.

Mackie, T. R. 2006. History of tomotherapy. *Phys. Med. Biol.* 51:R427–53.

Mackie, T. R., T. Holmes, S. Swerdloff et al. 1993. Tomotherapy: A new concept for the delivery of dynamic conformal radiotherapy. *Med. Phys.* 20 (6):1709–19.

Miyabe, Y., A. Sawada, K. Takayama et al. 2011. Positioning accuracy of a new image-guided radiotherapy system. *Med. Phys.* 38 (5):2535–41.

Novalis Radiosurgery, Technology, Delivery System, www.novalis-radiosurgery.com.

Olivera, G. H., D. M. Shepard, K. Ruchala et al. 1999. *The Modern Technology of Radiation Oncology*, edited by J. Van Dyk. Medical Physics Pub., Madison, 521–87.

Pantelis, E., A. Moutsatsos, K. Zourari et al. 2012. On the output factor measurements of the CyberKnife iris collimator small fields: Experimental determination of the k correction factors for microchamber and diode detectors. *Med. Phys.* 39:4875–85.

Ruchala, K. J., G. H. Olivera, E. A. Schloesser, and T. R. Mackie. 1999. Megavoltage CT on a tomotherapy system. *Phys. Med. Biol.* 44 (10):2597–621.

Ruchala, K. J., G. H. Olivera, J. M. Kapatoes et al. 2002. Multi-margin optimization with daily selection (mmods) for image-guided radiotherapy. *Int. J. Radiat. Oncol. Biol. Phys.* 54 (2):318.

Schell, M. C., F. Bova, D. Larson et al. 1995. *Stereotactic Radiosurgery: The Report of AAPM Task Group 42. AAPM Report No. 54*, American Institute of Physics.

Shah, A. P., K. M. Langen, K. J. Ruchala et al. 2008. Patient dose from megavoltage computed tomography imaging. *Int. J. Radiat. Oncol. Biol. Phys.* 70 (5):1579–87.

Solberg, T. D., K. L. Boedeker, R. Fogg, M. T. Selch, and A. A. F. De Salles. 2001. Dynamic arc radiosurgery field shaping: A comparison with static field conformal and noncoplanar circular arcs. *Int. J. Radiat. Oncol. Biol. Phys.* 49:1481–91.

Speiser, M., P. Medin, W. Mao et al. 2009. First assessment of a novel IGRT device for stereotactic body radiation therapy. In *World Congress on Medical Physics and Biomedical Engineering,* edited by O. Dössel and W. C. Schlegel, 25/1:266–9. Munich, Germany: IFMBE Proceedings. http://www.springerlink.com.

Takayama, K., T. Mizowaki, M. Kokubo et al. 2009. Initial validations for pursuing irradiation using a gimbals tracking system. *Radiother. Oncol.* 93 (1):45–9.

Varian Medical Systems, Oncology Systems, Varian Treatment Delivery Systems, Radiotherapy Machines, www.varian.com.

Verellen D., G. Soete, N. Linthout et al. 2003. Quality assurance of a system for improved target localization and patient set-up that combines real-time infrared tracking and stereoscopic X-ray imaging. *Radiotherapy and Oncology* 67:129–41.

ViewRay, The Future of Radiotherapy, www.viewray.com.

Light ion beam programs for stereotactic radiosurgery and stereotactic radiation therapy

Michael F. Moyers, Marc R. Bussière, and Richard P. Levy

Contents

5.1 CHARACTERISTICS OF LIGHT ION BEAMS FOR SRS AND SRT

According to Chu et al. (1993) and later agreed upon by the International Commission on Radiation Units and Measurements (ICRU) (Wambersie et al. 2004) light ions are ion species with an atomic number less than or equal to 10 (neon). Most patients treated with light ions for stereotactic radiosurgery (SRS) or stereotactic radiotherapy (SRT) to date have been irradiated with either hydrogen-1 ions (protons) or helium-4 ions. Recently, there have been several clinical trials to study reducing the number of treatment fractions to one to five for some large-field radiotherapy treatments using carbon-12 ions; these programs will not be specifically discussed in this chapter, but the techniques that are described here are applicable to all light ions. There are two primary dosimetric characteristics that suggest the use of light ion beams for SRS/SRT. The first is that the large mass of the ions minimizes scattering as they penetrate a patient, resulting in a narrow lateral penumbra width. The second is that, unlike x-ray beams with which the dose decreases with depth because the x-rays are attenuated exponentially as they pass through tissue, the dose deposited by light ions increases with depth because the ions lose energy and slow down, thereby transferring energy to tissue at an even greater rate. This energy deposition peaks just before the ions lose all their energy and stop, thereby creating a peak in the depth dose distribution.

The first patients treated with light ion beams occurred in the 1950s before the availability of reliable depth information, such as afforded by CT scanning. Without knowing exactly at what depth to stop the ions, these first treatments used the so-called "plateau cross-fire" technique in which multiple beams passed completely through the patient (Larsson et al. 1958; Lawrence 1957; Tobias et al. 1958). This technique is similar to that currently used with megavoltage x-rays, but the narrower penumbra of the light ions can be used advantageously to reduce the dose delivered to normal tissues adjacent laterally to the target. Figure 5.1 shows a beam arrangement used for some early SRS/SRT treatments. This technique did not, however, take advantage of the ability of light ion beams to reduce the dose to normal tissues beyond the distal aspect of the target. In fact, the exit dose of each individual beam was slightly higher than the entrance dose.

Methods were later developed to estimate the water equivalent depth of the target from different beam directions, and the so-called "stopping" technique was implemented. In this technique, the energy of the beam was varied from each direction such that the ions stopped immediately after traversing the target. Figure 5.2 shows several depth dose distributions measured along the central beam axis of small square proton fields of different sizes. The depth dose distribution is a strong function of field size, at least for small sizes. This is primarily due to an accumulation of lateral scattering with depth resulting in protons leaving the primary beam; that is, the lateral penumbra from all sides of the field encroach upon the central axis. Figure 5.3 is a plot of field size dependence factors (FSDF) for several different ranges and two different applicator sizes. With increasing range, the sharp drop in the FSDF with decreasing field size is seen to occur at larger field sizes. One advantage of helium-4 ions over protons is that the scattering in water is only about one half as much, resulting in a smaller lateral penumbra and less dependence of the dose/MU on field size. Scattering of carbon-12 ions is only about one-third that of protons.

Except for a few cases, most targets are wider in the depth direction of the beam than the width of the peak of the depth dose distribution. If the stopping technique is used for treatment, the penetration of the beam must be modulated to place multiple peaks within the target and cover the target with a uniform

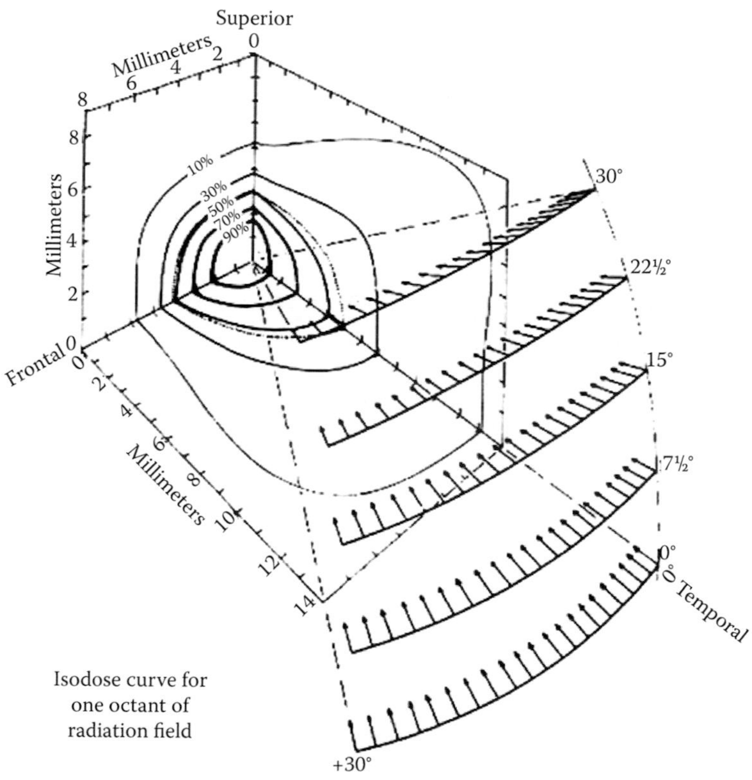

Figure 5.1 Arrangement of beams for plateau cross-fire technique at LBL. The figure represents only one octant of the total treatment beam entry space. Each curved line with arrows pointing toward the center represents one piece of an arc performed by rotating the head at one positioner angle. Typically the positioner was moved 7.5° between head rotations. (Reprinted from Tobias, C. A., Pituitary irradiation: Radiation physics and biology. In *Recent Advances in the Diagnosis and Treatment of Pituitary Tumors*, ed. J. A. Linfoot, 221–43, Raven Press, New York, 1979 with permission from Lippincott-Williams-Wilkins.)

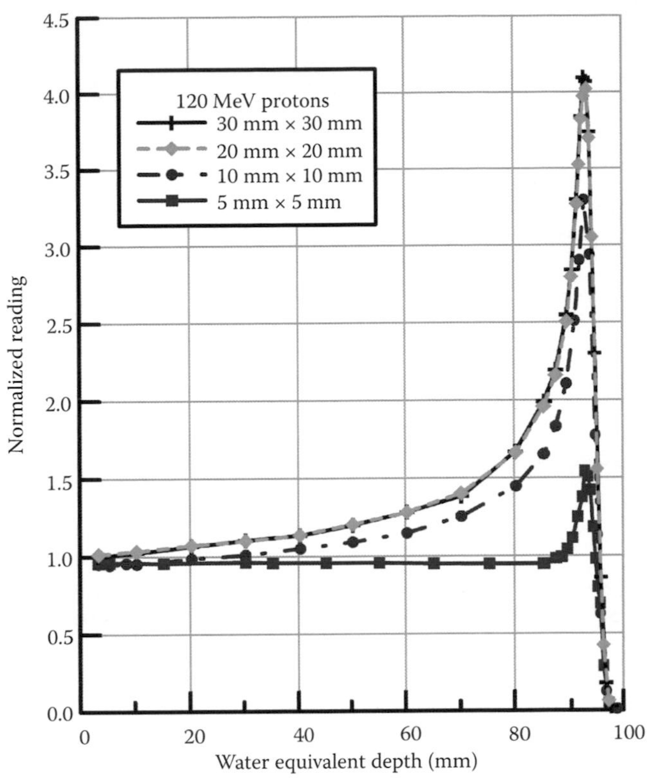

Figure 5.2 Nonmodulated depth dose distributions for different field sizes. (From Moyers, M. Vatnitsky, S. Miller, D. "Small proton beams for stereotactic radiosurgery" *Acta Neurochirurgica 122*, p. 173, Congress of the International Stereotactic Radiosurgery Society, Stockholm, Sweden, June 16–19, 1993.)

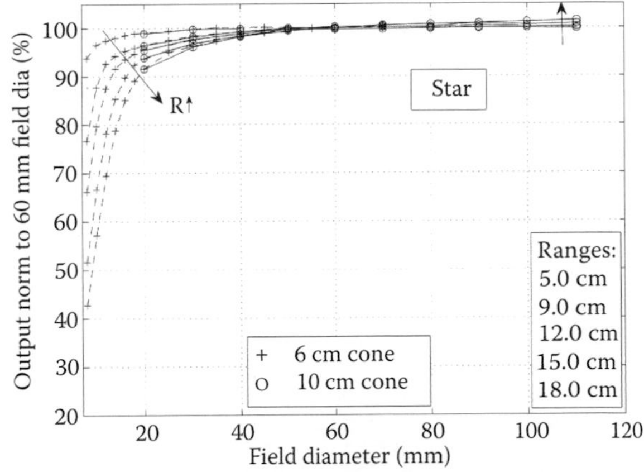

Figure 5.3 Field-size dependence factor as a function of beam range for the FHBPTC/STAR beamline. Data are shown for two different cone (applicator) sizes. All fields have a 90%–90% modulation width of 20 mm. (Reprinted from Daartz, J. et al., *Med. Phys.*, 36, 5, 1886–94, 2009. With permission from American Association of Physicists in Medicine.)

dose. In light ion beam treatments, tumors have traditionally been covered with a dose that was at least 90% of the dose delivered to the center of the tumor. This meant that the width for range modulation was typically specified as the difference in water equivalent depths between the location where the dose was 90% of the prescribed dose proximal to the target and the location where the dose was 90% of the prescribed dose distal to the target. Figure 5.4 shows a depth dose distribution created by delivering several

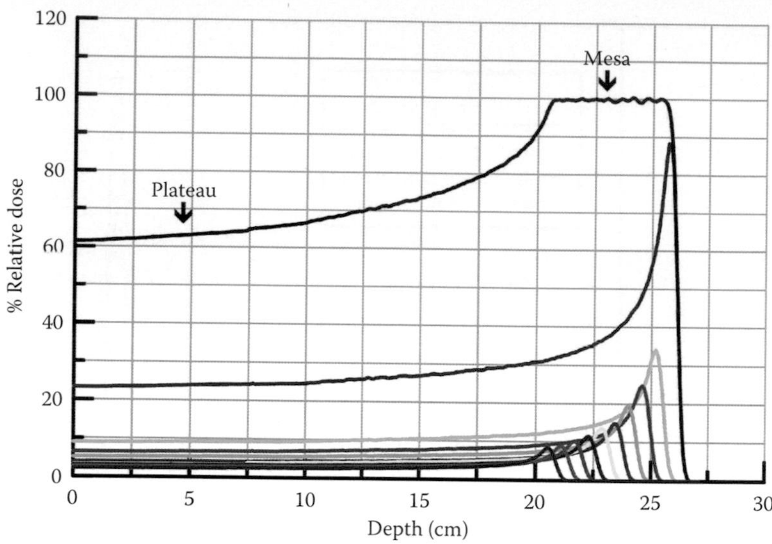

Figure 5.4 Range-modulated depth dose distribution. (Reprinted from Moyers, M. F., and Vatnitsky, S. M., *Practical Implementation of Light Ion Beam Treatments*, Medical Physics Publishing, Madison, WI, 2012. With permission.)

beams with different energies. Different facilities have devised different techniques to modulate the depth of penetration, and some facilities have used more than one technique.

Normal tissues present just distal to the target receive no benefit from being irradiated, and often it is critical that they receive as little dose as possible. For irregularly shaped targets, a three-dimensionally shaped bolus can be placed into the path of the beam to differentially modify the penetration of the beam at different lateral positions. These boluses are typically made of machinable wax (MW) or polymethylmethacrylate (PMMA). Unlike their use with electron beams when they are usually placed directly on the skin, boluses for light ion beams are usually installed onto the radiation head to facilitate accurate alignment. Figure 5.5 shows a MW bolus used for SRS and SRT that was manufactured using multiple vertical plunges from a milling machine tool. Figure 5.6 is a close-up view of a PMMA bolus highlighting the hexagonal pattern of the plunge points.

Figure 5.5 Picture of shaped bolus manufactured from MW. The diameter of the individual plunges used for machining is smaller than those used in large-field radiotherapy.

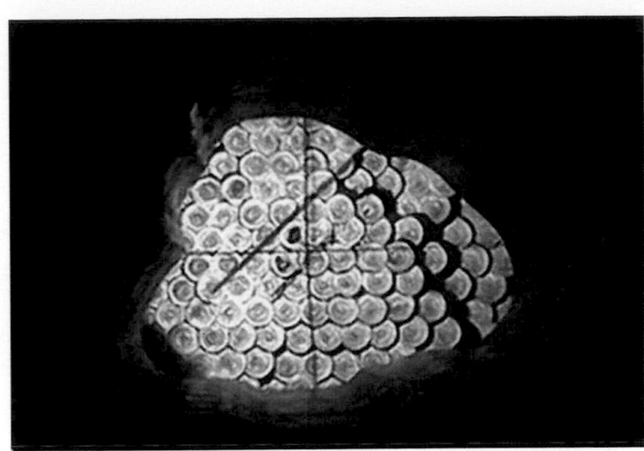

Figure 5.6 Close-up image of a PMMA bolus showing the hexagonal pattern of the vertical plunges used for milling. The image was taken through a conformal aperture and also shows the shadow of the alignment crosshairs.

5.2 BERKELEY PROGRAM (LBL)

The first treatments of humans with proton beams occurred in 1954 in Berkeley, California (Lawrence 1957) at the Lawrence Berkeley Laboratory (LBL). Early treatments concentrated upon treating the pituitary gland for hormone suppression for metastatic breast cancer (Lawrence and Tobias 1956; Tobias et al. 1958). Over the years, many different conditions were treated with SRS and SRT at LBL. A review of the treatment techniques and patients was given by Levy et al. (1990). In 1980, a special program to treat arteriovenous malformations (AVMs) and angiographically occult vascular malformations (Chang et al. 1998) was started. Reviews of the technique and patients for this type of treatment were given by Levy et al. (1989), Steinberg et al. (1990), and Fabrikant et al. (1992). Following Levy, Table 5.1 lists several of the conditions that have been treated with light ions using SRS/SRT.

Many patients receiving SRS/SRT at LBL were treated with the plateau cross-fire technique (see Figure 5.1), but after CT scanners became available in 1973 to provide accurate depth information, the stopping technique was implemented. Starting in 1954, the first 30 patients were treated with 340 MeV protons accelerated by the 184″ diameter synchrocyclotron (SC) that was built in 1947 (Brobeck et al. 1947). In 1957, the beam for SRS patient treatments was changed to 934 MeV (234 MeV/n) helium-4 ions (Lawrence, Tobias, and Born 1962; Castro et al. 1980). When the 184″ SC was closed in 1987, the SRS/SRT programs were moved to the Bevalac synchrotron that had been treating large field patients since 1975 (Alonso et al. 1989). Both the SRS/SRT and large-field programs continued through 1993 when the Bevalac was closed. AVM patients from Stanford University (SU) that had previously been treated at LBL were then entered into a joint program with the Loma Linda University Medical Center (LLUMC). Until 2004, these patients were worked up and had their planning CT and MRI scans and cerebral angiograms performed at SU, treatment plans prepared at LBL and LLUMC, patient-specific devices made at LLUMC, and patients treated at LLUMC.

The beamline used at LBL for SRS/SRT during the 1980s was described by Lyman et al. (1986). This stationary horizontal beamline used a range shifter placed 3.8 m upstream of the final collimator to reduce the beam range from 315 mm to 145 mm of water. Immediately after the range shifter, a 22-mm-diameter brass collimator was inserted to minimize the effective source size and reduce the lateral penumbra at the patient. The range shifter was partly composed of polyethylene and partly of copper to scatter the beam sufficiently to cover uniformly with dose a 40-mm-diameter field at the isocenter. Patient-specific range shifting was performed by insertion of additional polyethylene sheets just upstream of the primary range

Table 5.1 Example conditions treated with light ion SRS/SRT

Hormone suppression	metastatic breast cancer
	metastatic prostate cancer
	diabetic retinopathy
	Nelson's syndrome
Benign tumors	acromegaly
	Cushing's
	prolactinoma
	nonsecreting pituitary tumors
	craniopharyngioma
	acoustic neuroma
	meningioma
	hemangioblastoma
	chordoma
Malignant tumors	low-grade astrocytoma
	anaplastic astrocytoma
	glioblastoma multiforme (boost only)
	oligodendroglioma
	pineal
	optic neuroma
	malignant meningioma
	chondrosarcoma
	metastases
Vascular disorders	arteriovenous malformation
	angiographically occult malformation
	wet macular degeneration*
Functional disorders	Parkinson's disease
	intractable epilepsy
	trigeminal neuralgia

*Single fraction, not stereotactic.

shifter. A rotating propellor, similar to what would be used later at LLUMC (see Figure 5.25), provided modulation of the penetration depth. The dose rate at the center of the target was typically 4 Gy per min.

Patients were typically immobilized and registered to the patient positioner using a face mask. Originally a two-piece fiberglass and polyester resin mask was used with thumbscrews tightened until less than 0.5 mm motion was obtained (Tobias et al. 1958). Later a transparent polystyrene vacuum-formed mask was used, but this mask took 2 days for completion (Lyman et al. 1989). Eventually a thermoplastic head mask and stereotactic frame with embedded fiducial markers was developed (Lyman et al. 1989). Figure 5.7 is a picture of the latter mask and frame, and Figure 5.8 is a diagram of its various components. Accurate alignment of the patient was performed through localization of the embedded fiducial markers with orthogonal x-ray imaging systems.

After the target was localized, the patient was repositioned with a five degrees of freedom (DOF) patient positioner. Lyman and Chong (1974) described the ISAH (irradiation stereotaxic apparatus for humans)

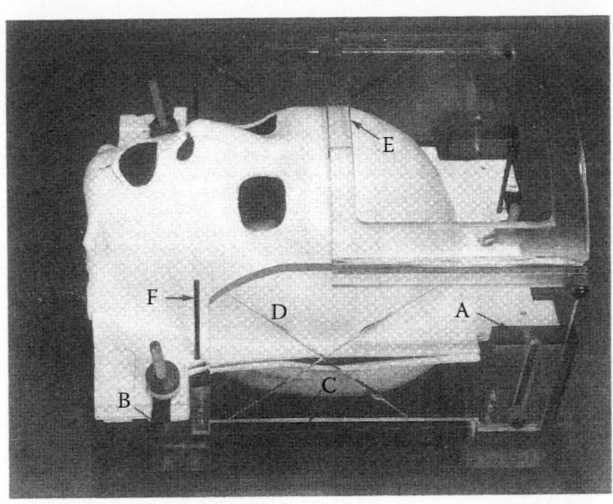

Figure 5.7 Picture of stereotactic frame and thermoplastic head mask. (Reprinted from *Int. J. Radiat. Oncol. Biol. Phys.*, 16, Lyman, J. T., Phillips, M. H., Frankel, K. A., and Fabrikant, J. I., Stereotactic frame for neuroradiology and charged particle Bragg peak radiosurgery of intracranial disorders, 1615–21, Copyright 1989, with permission from Elsevier.)

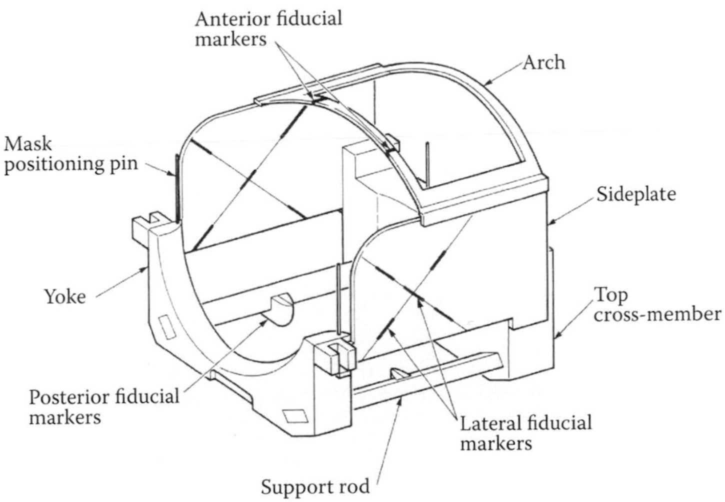

Figure 5.8 Detailed diagram of stereotactic frame. (Reprinted from *Int. J. Radiat. Oncol. Biol. Phys.*, 16, Lyman, J. T., Phillips, M. H., Frankel, K. A., and Fabrikant, J. I., Stereotactic frame for neuroradiology and charged particle Bragg peak radiosurgery of intracranial disorders, 1615–21, Copyright 1989, with permission from Elsevier.)

patient positioner. The base of the unit was a granite slab of 0.5-m thickness with its surface polished to a flatness of ±5 μm. Moving on top of the granite slab via four air bearings was a 5 DOF (three translations and two rotations) support assembly. A sixth DOF was available via a restricted angle head rotator. Figure 5.9a is a diagram of the slab and support assembly. The range of motions was 400 mm in two of the translational motions and 200 mm in the other. The readout accuracy of the positioner in each of the translational directions was 0.1 mm and in each of the rotational directions 0.1°. Figure 5.9b is a picture of the device installed at the treatment beamline. In the picture, a tabletop is installed to allow patients to be treated lying down. Alternatively, a chair could be attached for seated patients. During treatment, the positioner was continuously monitored and equipped with a beam termination interlock in case any positions changed during treatment.

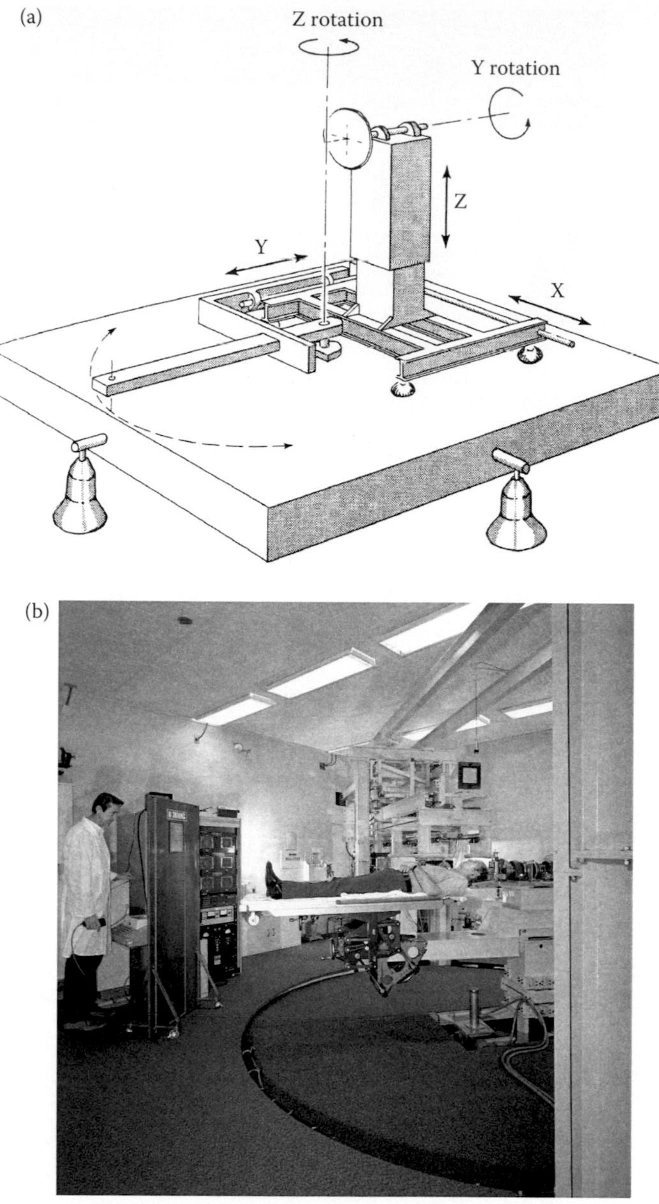

Figure 5.9 (a) Diagram of ISAH positioner with five of the motions indicated. The yaw and roll motions can be seen in Figure 5.9b. (Lyman, J. T., and Chong, C. Y.: ISAH: A versatile treatment positioner for external radiation therapy, *Cancer.* 1974. 34. 12–16. Copyright Wiley-VCH Verlag GmbH & Co. KGaA. Reproduced with permission.) (b) Picture of ISAH positioner and beamline. The pitch mechanism is seen immediately below the patient, and the roll mechanism is seen just superior to the patient's head. The flat tabletop may be removed and replaced with a chair for posterior and anterior beam access to the head.

At shallow depths within the entrance (plateau) region of the beam, the relative biological effectiveness (RBE) of the helium-4 ion beam was about the same as the proton beam, that is, 1.1 compared to cobalt-60 radiation. For the stopping beam delivery technique, however, the RBE was observed to increase slightly with depth. For these treatments, the range modulator propellor was manufactured to give a nonuniform physical dose distribution across the target such that the distal part of the target received a lower dose than the proximal part. The average RBE over the target was typically 1.3.

5.3 UPPSALA PROGRAM (GWI/TSL)

In 1957, the Gustav Werner Institute (GWI) in Uppsala, Sweden (Falkmer et al. 1962) first treated a tumor in a human patient with a proton beam. This palliative treatment for a relapsing gynecologic cancer delivered 19 Gy in a single fraction with a perineal beam entry direction (Blomquist 2010). By 1976, when the accelerator complex was shut down for upgrades, the program had treated 73 patients (Sisterson 1990) having a variety of conditions, including SRS for Parkinson's disease and chronic pain (Larsson, Leksell, and Rexed 1963). The design and construction of the "Gamma Knife" was a significant spin-off and achievement as part of the activities at GWI.

The proton beam for this program was produced by a 230-cm-diameter SC that was constructed in 1952 and had an energy of 185 MeV (Graffman, Brahme, and Larsson 1985). The SC produced protons in frequency-modulation mode, but could accelerate ions as heavy as neon in continuous-wave mode (Larsson and Graffman 1985). The SC was less than 1 km from the University Hospital making the collaboration by the technical and medical teams easier than at Berkeley and other research programs to follow. The temporal structure of the beam consisted of 10 μs pulses every 3 ms. The radiosurgery beam was broadened laterally using a pair of electromagnets to scan the beam across the target volume and produce a uniform lateral dose profile (Larsson et al. 1959). A thin aluminum foil was placed into the beam path to smooth the resulting dose distribution. Figure 5.10 shows the Lissajous scanning pattern used for these treatments. Depending upon the size of the field, the dose rate varied from 1 to 100 Gy/min (Graffman, Brahme, and Larsson 1985). For tumor therapy, a water balloon was placed between the beamline and the patient to add thickness to the patient, thereby setting the range of the beam just beyond the distal depth of the target (Larsson 1961, 1967). For SRS/SRT however, the plateau cross-fire beam technique was used without modulating the penetration. Alignment of the target with the proton beam was not done with daily orthogonal imaging as was done at Berkeley but rather with a stereotactic head frame with coordinates in three directions as commonly used for neurosurgery. This precise alignment was an enabling factor for performing the functional SRS mentioned earlier.

In 1989, treatments resumed at the renamed The Svedberg Laboratory (TSL) and have continued ever since. In 1991, SRS treatments of AVMs began. This new treatment program uses the stopping technique with a custom bolus and rotating modulator propellor. The majority of proton beam treatments are delivered using a stationary horizontal beam with the patient in a seated position. Patients are immobilized with a bite block and supporting helmet attached to a movable chair as seen in Figure 5.11. Orthogonal

Figure 5.10 Scanning pattern used. (After Larsson, B. L. et al., *Acta Radiol. Stock.*, 51, 52–64, 1959.)

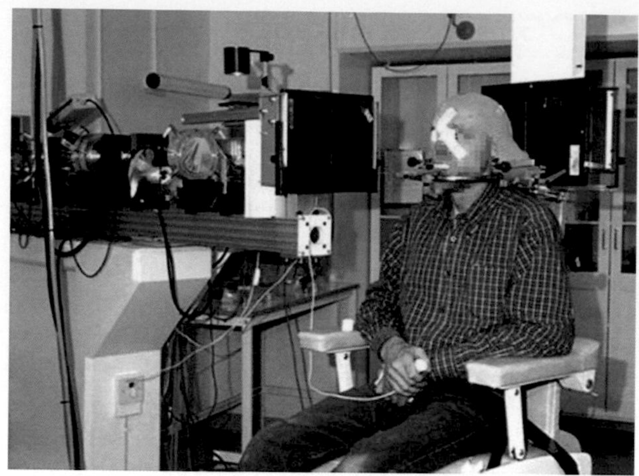

Figure 5.11 Treatment in sitting position with partial mask and bite block. Beam enters from the left, passing through a rotating modulator propellor and custom bolus. (Courtesy of Teddy Thörnlund.)

x-rays are taken before each treatment to align the patient. During the second half of 2005, the availability of the SC for medical treatments increased, thereby opening up the possibility of fractionated treatments at smaller doses per fraction. More than 1300 patients have now been irradiated with protons or combined techniques with protons and photons.

5.4 BOSTON PROGRAM (HCL AND FHBPTC)

In 1961, a proton treatment program was established in Boston, Massachusetts, for the treatment of pituitary disorders (Kjellberg et al. 1962). This was a collaborative effort between the Massachusetts General Hospital (MGH) and the Harvard Cyclotron Laboratory (HCL). This program utilized a horizontal beam generated by a 700-ton proton SC that was originally installed at the HCL. The first beam was obtained from the SC in 1949 and subsequently upgraded from 95 MeV to 160 MeV in 1955 (Wilson 2004). The driving RF had a frequency of 23 to 30 MHz with a repetition rate of 300 Hz. Pituitary dysfunctions were ideal candidates for stereotactic targeting in the pre-CT imaging era because of the location of the sella within the patient skull. This small target required a small field size, so it was sufficient to broaden the beam laterally using a single scatterer placed at a distance 130 cm upstream of the isocenter. Due to the low energy available from the SC, the beam could not penetrate through the entire head, and thus a stopping technique was developed. As was used at Uppsala for early tumor treatments, the beam passed through a water column and a balloon that allowed protons to be extracted from the accelerator at a single energy. This approach provided a simple yet robust control over the delivered range. Figure 5.12 and its caption describe the range modulation procedure. Figure 5.13 shows the equipment used in the early years of the treatment program.

In 1965, the MGH/HCL group treated the first AVM with radiosurgery (Kjellberg 1986); however, routine AVM treatments did not start until 1972. Alignment of noncentralized targets, such as AVMs, were achieved using x-ray and angiography overlay techniques. Between 1961 and 1991, 2927 proton radiosurgery cases were treated for pituitary dysfunctions (1441) and AVMs (1486). During the late 1980s, equipment and technique upgrades enabled full 3-D imaging–based alignment and dosimetry to be integrated into the proton radiosurgery program. This involved the development of CT-based treatment planning software (Goitein and Abrams 1983; Goitein et al. 1983), fabrication of a new isocentric patient positioner named STAR for "STereotactic Alignment Radiosurgery" (Chapman, Ogilvy, and Butler 1993), and the implementation of a fiducial-based alignment system (Gall, Verhey, and Wagner 1993). The water balloon technique continued to be used, but the scattering system was upgraded to a double scattering system in order to irradiate larger lesions with uniform doses. As with

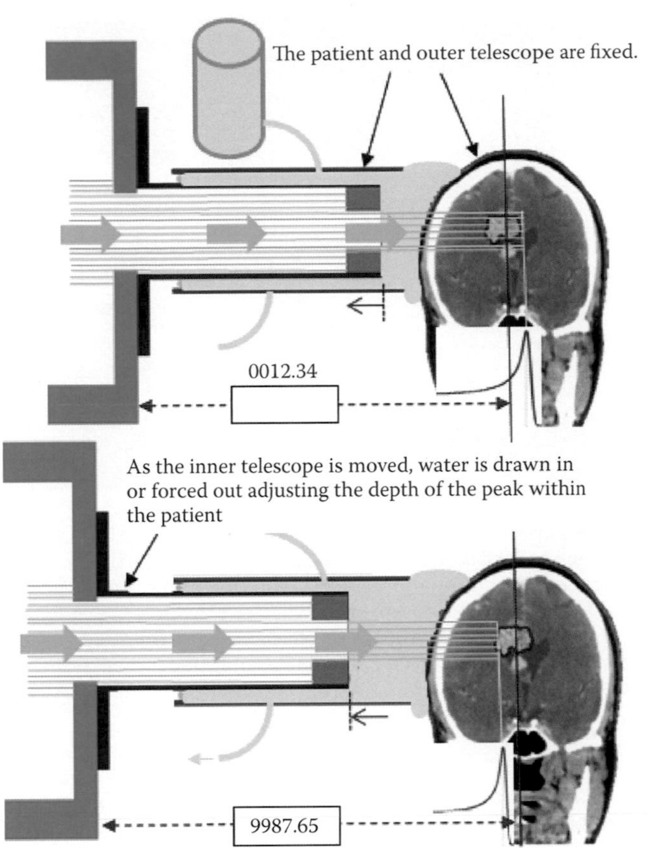

Figure 5.12 Water balloon collimator/modulation system. The applicator consisted of a sealed inner cone, which held the field-specific conformal aperture; an open outer cone, which slid over the inner cone; and a water balloon, which created a flexible membrane to contain the water between the two cones. A single proton energy was used that provided a maximum range equal to 14.3 cm of water. The initial position of the inner cone/aperture was determined by calculating the necessary water thickness to be held between the cones to irradiate the deepest part of the target. This thickness was calculated by subtracting the water equivalent thickness of the patient from the surface to the distal side of the target from the range of the beam (14.3 cm). Once the dose from the deepest range component had been delivered, the inner cone was retracted to increase the water content between the aperture and the patient and create a second range component. The process was then repeated until the target was covered with a uniform dose. The numbers on the diagram represent the distance in mm by which the peak was shifted relative to the mechanical isocenter with the counter looping from 9999.9 to 0. Zero represents the peak centered at isocenter.

the initial equipment configuration, orthogonal x-ray imaging enabled isocentric alignment. Figure 5.14a–c shows the STAR installed for use at the HCL horizontal beamline. This upgraded program expanded on the original diagnoses to include metastatic disease, acoustic neuromas, meningiomas, glioblastomas, and other malignant diseases. In the period from 1991 to 2002, this system delivered 754 treatment courses. Figure 5.15 summarizes the patients treated at the HCL from the start of the treatment program to its closing in 2002.

Radiosurgery treatments continued uninterrupted as the MGH started treating at the new Frances H. Burr Proton Therapy Center (FHBPTC, formerly known as the NPTC) in 2002. The facility, located on the main campus of the hospital, relies on a fixed-energy 230 MeV isochronous cyclotron (IC) to generate the necessary protons for the treatment rooms, which include two beamlines mounted on rotating gantries (see Figure 5.16), one horizontal beamline dedicated to eyes, and one horizontal beamline used for SRS/SRT. Several techniques were modified to accommodate the modernized program, including use of double scattering for broadening the beam laterally in the rotating gantry rooms, custom boluses for protecting

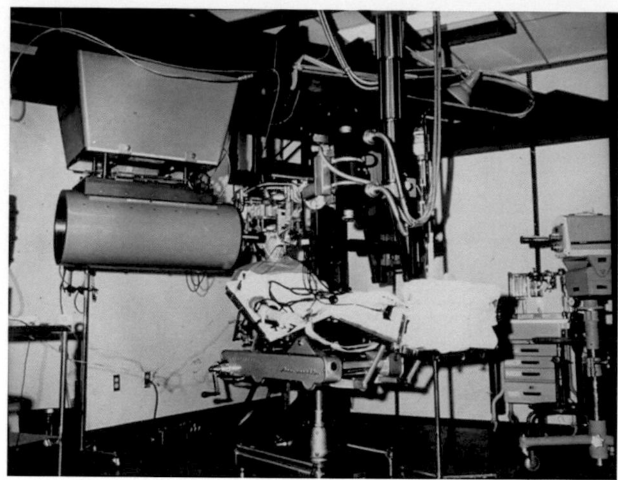

Figure 5.13 HCL treatment equipment used during the 1960s and 1970s. The large structure at the left of the image shows a 1-inch-thick steel cylinder suspended from a ceiling I-beam. This cylinder contained the scatterer and beam monitors and could be moved parallel to the beam axis to accommodate adjustment of the water balloon cones. The downstream face of the cylinder provided precollimation as well as a mounting surface for the cones. A stereotactic head-ring assembly was also suspended from the I-beam and could be rotated about the isocenter to provide flexibility when selecting beam directions. A surgical table had to be adjusted independently.

tissues distal to the target, and a commercial planning system that uses a pencil beam dose calculation algorithm (Hong et al. 1996). Different techniques are used in each of the beamline types to achieve the desired modulation. In the rotating gantries, instead of delivering different energies one step at a time, a series of range shifter steps are arranged around an axle like a propellor and spun through the beam at 600 rpm (Gottschalk 1987). To support different energy ranges, several of these propellors are located in an exchanging assembly placed at the entrance of the radiation head where the beam diameter is narrow. To achieve different widths of modulation for different portals and patients, the beam flux is gated and modulated over different thicknesses of the propellor.

Further program improvements have included the reintegration of the refurbished STAR device in 2006 for use with one of the horizontal beamlines. This beamline was designed specifically for small fields using a binary range shifter and single scattering system to achieve precise range and range modulation control while achieving the necessary lateral beam spreading. The combination of a 4.6-m source-to-isocenter distance with scattering and shielding optimized for small diameter beams reduces the neutron exposure to the patient by a factor of 10 compared to if the patient was treated in a rotating gantry at the same facility (Daartz, Bangert et al. 2009). Figure 5.17 shows the refurbished STAR system in place at the new facility. The proton SRS/SRT program currently treats up to six single fraction radiosurgery cases per week with the bulk of cases being AVM, pituitary disorders, acoustic neuromas, meningiomas, and metastatic lesions. An additional 12–14 fractionated intracranial stereotactic radiotherapy cases are treated daily in this beamline. The use of rotating gantries has also enabled select single high-dose treatments of extracranial lesions. In addition to those patients treated at HCL shown in Figure 5.15, as of 2012 6550 patients have been treated at FHBPTC.

As seen in Figure 5.14, patients can be fixated to the STAR positioner with a stereotactic frame and four cranial screws. The STAR positioner also accommodates noninvasive fixation for image-based treatments using a GTC stereotactic frame from Integra-Radionics (Burlington, MA). Upon relocation of the SRS/SRT program from the HCL to the FHBPTC, the frame was modified to use a thin carbon fiber composite occipital cup, a low-density Mold Care cushion from CIVCO (Orange City, IA), and an altered bite mold and tray as seen in Figure 5.18. Alignment is confirmed by obtaining orthogonal x-ray images and triangulation of fiducials implanted into the skull prior to obtaining the planning CT.

(a)

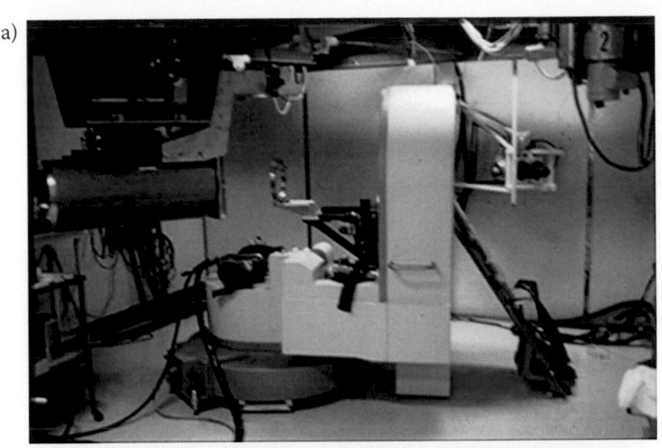

(b)

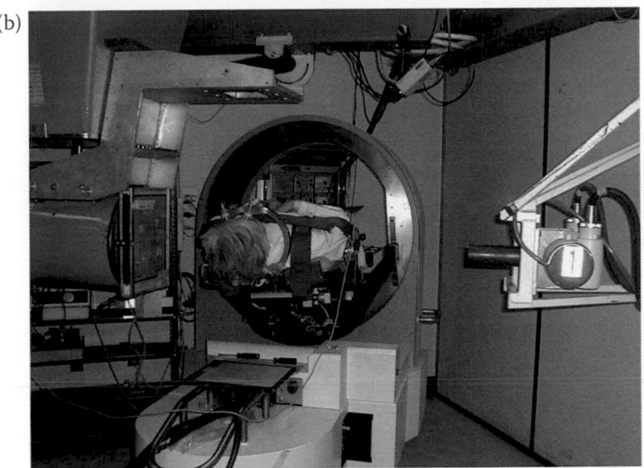

(c)

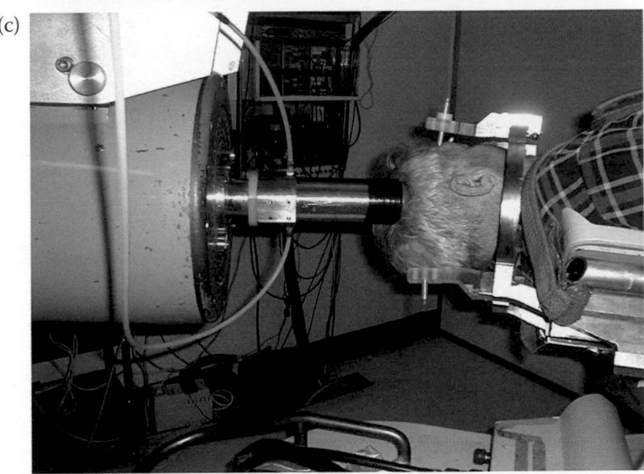

Figure 5.14 (a) HCL horizontal radiosurgery beamline with water column and STAR positioner (circa 1992). Patients would initially perch themselves on a cushioned platform with one end of the platform positioned at an angle of 50° to the floor. The platform was hinged to the patient positioner so that staff could lift the floor end and glide the patient horizontally through the head fixation ring. The patient's head would then be rigidly fixated to the ring, and the body was secured with ergonomic side bolsters. (b) To accommodate the horizontal-only beamline, patients could be rotated isocentrically in the coronal plane up to ±115° and rolled ±95° from the supine position as necessary to direct the horizontal beam to any point within the cranium superior to the isocenter. Isocentric alignment was performed through the use of orthogonal x-rays. The x-ray tube on the right of the picture is installed to provide a back projection along the treatment beam axis. (c) The beam applicator cone and balloon positioned against the head in readiness for treatment.

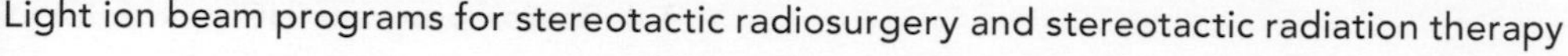

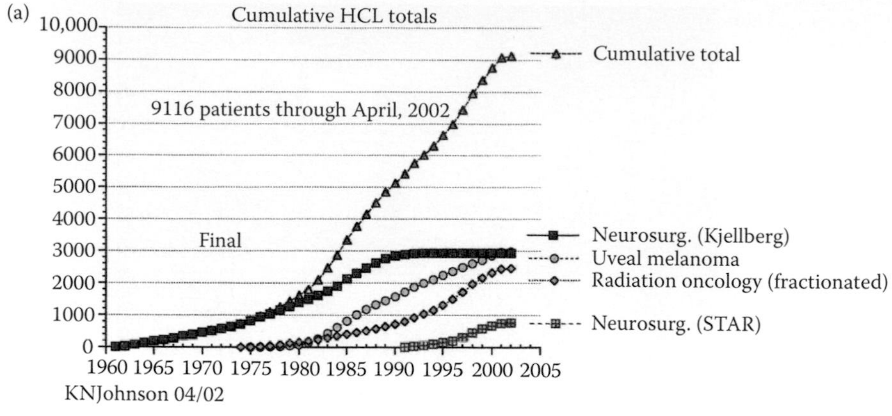

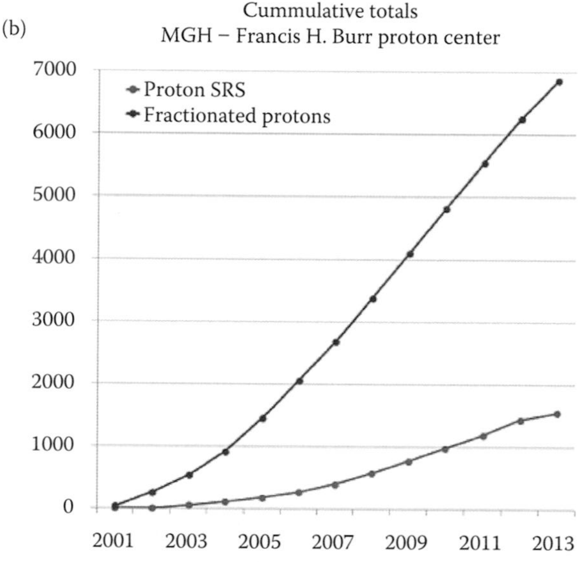

Figure 5.15 (a) MGH patient treatment numbers for all programs at the Harvard Cyclotron Laboratory between 1961 and 2002 with the STAR program inception in 1991. (b) MGH hospital-based treatment numbers at the Francis H. Burr Proton Center since inception in 2001.

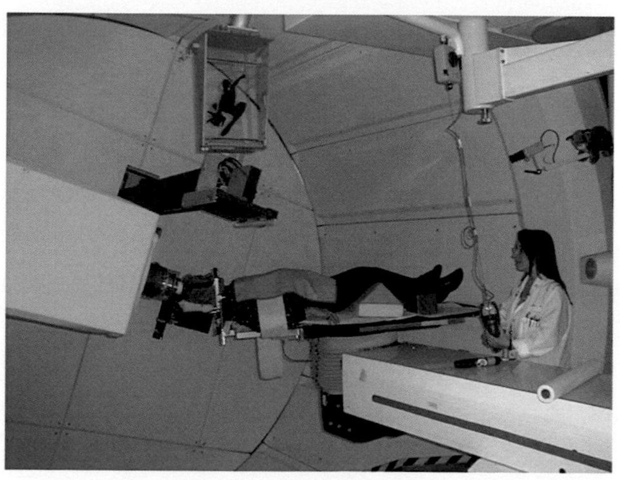

Figure 5.16 Patient positioned for an anterior vertex field in one of the two gantries at the FHBPTC.

(a)

(b)

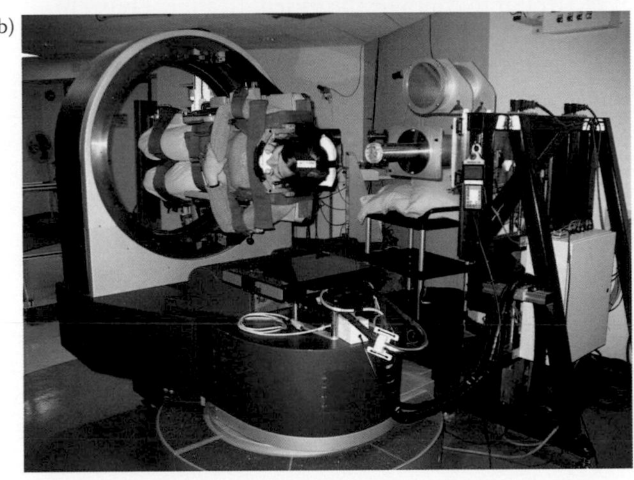

Figure 5.17 Dedicated radiosurgery beamline at the FHBPTC which incorporates the refurbished STAR patient positioner and uses the noninvasive frame shown in Figure 5.18. (a) Patients can climb a few stairs to get to the elevated bed and no longer have to be lifted into the device by the staff. (b) Patient in rotated position for delivery of posterior beam.

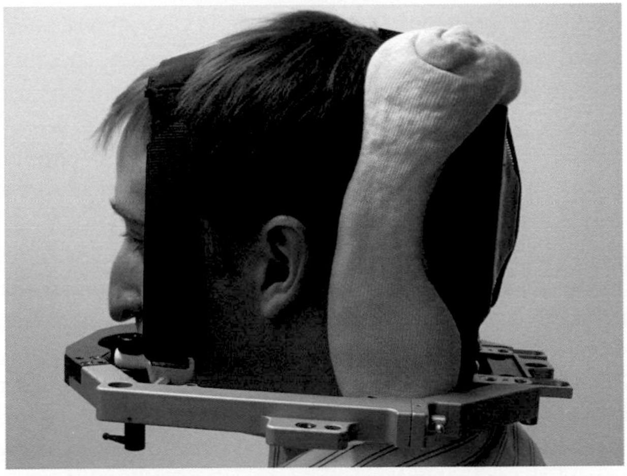

Figure 5.18 Modified GTC head immobilization using bite block, low-density headrest, and cranium straps.

5.5 DUBNA PROGRAM (JINR)

A proton treatment program began at the Joint Institute for Nuclear Research (JINR) in Dubna, Russia (125 km north of Moscow) in 1967 (Goldin et al. 1989) and treated patients during three periods; 1967 to 1975, 1989 to 1996, and 2000 to 2010 (Mytsin 2003). Although JINR has treated more than 900 patients with proton beams, the number of patients treated with SRS or SRT over the years has been small. In the most recent series, 60 patients were treated for AVMs with a smaller number of patients being treated for an assortment of other brain targets (Kamanin 2010).

The early program used proton beams of 680 MeV produced by a 270-cm-diameter SC. Protons were extracted at a pulse rate of about 250 Hz. For treatment of large tumors, the energy was range-shifted down to between 70 and 200 MeV, but for SRS/SRT the plateau cross-fire technique was used at an energy of 660 MeV (Mytsin 2003). The facility has recently switched to using the stopping technique for SRS/SRT using custom boluses and ridge filters for range modulation (Tseytlina et al. 2009). Alignment of the patient is now accomplished using orthogonal digital radiography (Shipulin and Mytsin 2008). Treatment planning is currently performed using a system originally developed at LLUMC. The facility plans on installing a commercial accelerator to continue the patient treatment program.

5.6 MOSCOW PROGRAM (ITEP)

A proton treatment program started in Moscow in 1969 at the Institute for Theoretical and Experimental Physics (ITEP) and has continued to this day (Chuvilo et al. 1984; Minakova 1987; Goldin et al. 1989). Clinical trials with narrow beams began in 1972 (Minakova et al. 1990). The accelerator for this program is a synchrotron originally built in 1961 to accelerate protons to 7 GeV, but the energy was later increased to 10 GeV, and the capability to accelerate any ion species up to uranium was later added. For medical purposes, proton beams were extracted at five different energies between 70 and 200 MeV. Pulses of protons were extracted every 2.7 s and were 150 ns long (Chuvilo et al. 1984). The resulting high dose rate prevented ion chambers from being used to monitor the beam, so current transformers were used instead. Calibration of the beam monitors in terms of dose/MU was performed by measuring the induced activity in carbon targets through the $^{12}C(p, n)^{11}C$ reaction. Accuracy of this method of calibration was originally estimated at 7%–9%, but later improved to better than 5% (Nichiporov 2003).

Three rooms have been available at the facility for patient treatments, all of which have only horizontal beamlines (Goldin et al. 1987). In two of the rooms, patients can be treated lying down, and in the third room, the patient can be treated sitting upright in a chair. A summary of the early results of the program were given by Minakova (1987), Minakova et al. (1990), and Minakova and Burdenko (1990). For SRS/SRT, proton beams of 5 to 30 mm diameter are produced using thin scatterers installed in the beamline several meters upstream of the isocenter. Many intracranial targets were treated with the plateau cross-fire technique, including pituitaries and trigeminal nerve ganglion. Most AVMs were treated with the stopping technique, which used a fixed energy of 180 MeV, a variable thickness water column placed between the aperture and the patient's head (see Figure 5.19 and description in Section 5.4), and range modulation via spiral ridge filters.

Figure 5.20 shows a specialized patient positioner used in the early program for delivering SRS/SRT to patients lying down. The patient positioner provided translational movements of ±70 mm in two directions and ±85 mm in the other direction, all with an accuracy of 0.1 mm. The positioner also provided rotations of ±126° about the vertical axis and ±27° and ±36° about two other axes (see Figure 5.20). Figure 5.21 shows a patient in position for pituitary irradiation. This system was later replaced with the system shown in Figure 5.22, which had similar ranges and accuracy of motion but provided greater access for the treatment personnel and more automation.

During treatment, the patient is immobilized to the positioner using a custom thermoplastic mask, which, over the years, has consisted of various materials and designs, some of which included a bite block. Alignment of the patient is performed using two orthogonal x-ray tubes, one located downstream of the isocenter and pointing toward the radiation head while the other is placed above the patient pointing downward.

Figure 5.19 Beam-delivery applicator cone and variable-thickness water column placed against a patient face mask as used for treatment. (Courtesy of Valeri Kostjuchenko.)

Figure 5.20 Picture of patient positioner mechanism for radiosurgery. (Reprinted from *Int. J. Rad. Oncol. Biol. Phys.*, 10, Chuvilo, I. V., Goldin, L. L., Khoroshkov, V. S. et al., ITEP synchrotron proton beam in radiotherapy, 185–195, Copyright 1984, with permission from Elsevier.)

Figure 5.21 Picture of pituitary setup. (Reprinted from *Int. J. Rad. Oncol. Biol. Phys.*, 10, Chuvilo, I. V., Goldin, L. L., Khoroshkov, V. S. et al., ITEP synchrotron proton beam in radiotherapy, 185–195, Copyright 1984, with permission from Elsevier.)

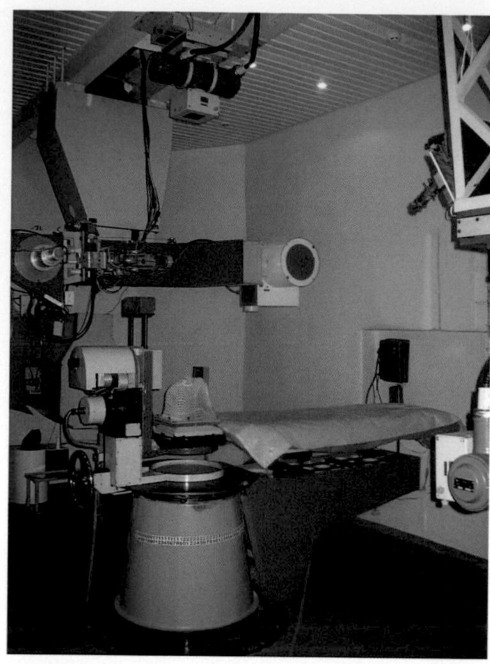

Figure 5.22 New ITEP positioner. Orthogonal x-ray tubes are seen at top and right of the picture. The orange c-arm in the center of the picture rotates ±90° to position, alternatively, the beam applicator cone and water column in line with the beam and adjacent to the patient or the x-ray sensor panel during patient setup localization. (Courtesy of Valeri Kostjuchenko.)

5.7 GATCHINA PROGRAM (LINP/PNPI)

A proton treatment program was started in 1975 in Gatchina, Russia, at the Leningrad Institute for Nuclear Physics (LINP). The facility was later renamed the Petersburg Nuclear Physics Institute (PNPI). The proton beam was produced with a SC that was built in 1967 (Abrossimov, Vorobyov, and Riabov 2001). The magnet poles had a diameter of 6.85 m and weighed 7800 tons. The energizing coils weighed an additional 120 tons. Protons were extracted with a repetition rate of 30 to 50 Hz at an energy of 1 GeV with a 30% extraction efficiency. Because of this high energy, all patients were treated with the cross-fire technique only (Abrossimov, Ivanov et al. 2001). Unique among all SRS/SRT programs is that the proton fields were magnetically focused to a 5 to 10 mm FWHM diameter without using collimators prior to entering the patient (Konnov 1987). Typically the head was rotated ±30° about the cranial-caudal axis and the positioner was rotated ±40° about a vertical axis (Chervjakov et al. 1987; Ermakov et al. 1987). The patient was aligned using two orthogonal x-ray tubes and imagers (Konnov 1987). Figure 5.23 shows the beamline, patient positioner, and patient alignment system. As of 2006, 1281 patients had been treated with SRS/SRT for all varieties of conditions as given in Table 5.1, including five patients for epilepsy (Abrossimov et al. 2006).

5.8 LOMA LINDA PROGRAM (LLUMC)

In 1990, the first hospital-based proton facility was installed at the LLUMC in Loma Linda, California, and, as of 2012, had treated more than 16,800 patients. In 1993, the Berkeley SRS/SRT program was transferred to Loma Linda. These patients were primarily treated for AVMs. Subsequently, LLUMC began treating their own patients and expanded the conditions treated.

LLUMC was the first facility to treat patients with light ions using a rotating gantry. Although the main aspect of this feature was better optimization of treatment plans than afforded by facilities having only stationary beamlines, it also increased the efficiency of treatment. Figure 5.24 shows a model of the double

Figure 5.23 Patient setup for plateau cross-fire irradiation using 1000 MeV protons at Gatchina. Beam enters from right. (Reprinted from Abrossimov, N. K. et al., *J Phys. Conference Series*, 41, 424–32, 2006. With permission from IOP Publishing, Ltd.)

ring isocentric gantry that is approximately 10 m in diameter and weighs approximately 90 Mg. Moyers and Lesyna (2004) showed that the beam could be delivered to a point in space within 1 mm for all gantry angles without specialized correction mechanisms.

This facility produces protons using a 250 MeV (maximum achieved energy of 304 MeV) weak focusing synchrotron. In research mode, the cycle time of the accelerator can vary from 2.2 to 8.8 s with the beam being extracted in a pulse from 0.25 to 7.1 s long, but typical clinical use is 2.2 s and 0.25 s, respectively (Moyers 2002). The proton beam energy is checked each cycle before extraction, and if it deviates from the prescribed energy by more than 0.1 MeV (approximately 0.1 mm of range at SRS/SRT energies) the extraction process is aborted (Moyers and Ghebremedhin 2008). Standard beams used for tumor therapy

Figure 5.24 Front view of model of LLUMC isocentric rotating gantry. The gantry rotates on a pair of rings, one located in the front and one in the rear. The beam enters from the rear and transits several dipole bending magnets (shown in yellow) and focusing quadrupole magnets (shown in red). (Reprinted from *Int. J. Radiat. Oncol. Biol. Phys.*, 60/5, Moyers, M. F., and Lesyna, W., Isocenter characteristics of an external ring proton gantry, 1622–1630, Copyright 2004, with permission from Elsevier.)

are broadened laterally by a double scatterering system and collimated by Lipowitz metal apertures. Distal blocking of normal tissues is performed using custom boluses made of MW. Range modulation is generated by inserting a rotating modulator propellor about halfway between the first scatterer and the isocenter. Each modulation propellor is designed to produce only one modulation width but works over a large range of energies (140–250 MeV). Figure 5.25 shows one of the propellors used at LLUMC. For SRS/SRT patients, a single scatterer system is used, which reduces the size of the effective source, thereby reducing the width of the lateral penumbra. The effective source to isocenter distance is about 280 cm. Custom apertures and boluses are used, similar to those used in large-field radiotherapy, but are placed into a special long, small-diameter applicator cone built to permit them to be placed as close as possible to the patient even when the target is near the neck and/or the beam is entering from beneath. Figure 5.26 shows this SRS/SRT applicator cone installed onto the radiation head. The 30-cm length usually allows passage of the distal end of the cone past the mechanical components of the patient positioner for most cases, but occasionally the cone must still be retracted. Figure 5.27 shows the components associated with the applicator cone, including a precollimator, spacers to prevent the aperture and bolus from sliding inside the applicator when the gantry is rotated, and a removable crosswire to indicate the central axis of the

Figure 5.25 Range modulator propellor to produce 120 mm of modulation. The modulation for radiosurgery cases is typically limited to less than 60 mm.

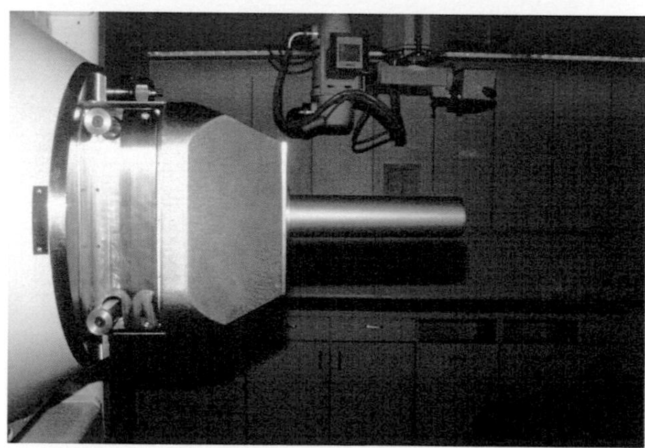

Figure 5.26 Picture of Loma Linda radiosurgery applicator cone installed on beamline. (Reprinted from Moyers, M. F., and Vatnitsky, S. M., *Practical Implementation of Light Ion Beam Treatments*, Medical Physics Publishing, Madison, WI, 2012. With permission.)

Figure 5.27 Picture of Loma Linda radiosurgery applicator cone components. (Reprinted from Moyers, M. F., and Vatnitsky, S. M., *Practical Implementation of Light Ion Beam Treatments*, Medical Physics Publishing, Madison, WI, 2012. With permission.)

beam during radiographic imaging. The applicator can be configured with the aperture either upstream or downstream of the bolus. Boluses, as seen in Figure 5.5, are produced with a smaller milling tool bit than normally used in large field proton therapy.

Typically, targets in the cranium require a beam with a range less than 10 cm of water, so a scatterer was optimized for an energy of 126 MeV. The width of the peak portion of the depth dose distribution is very narrow for this energy, so modulator propellors designed for higher energies with wider peaks do not work well at this energy. Rather than designing a separate set of range modulator propellors, a ripple filter was built that is placed just upstream of the precollimator. The ripple filter, seen in Figure 5.28, consists of small aluminum strips that extend completely across the field, shifting the range of a few protons to fill in the valleys produced by the high-energy modulator propellor. Figure 5.29 shows the resulting depth dose distribution for 30 mm of range modulation. Although the dose outside the target decreases very rapidly, inside the target the dose distribution is quite uniform, so the dose at the isocenter is usually equal to the prescribed dose.

Patients in the Loma Linda program were immobilized with either a perforated thermoplastic face mask or registered to the table top and immobilized with the vacuum-assisted maxilla immobilization device (VAMID) seen in Figure 5.30 (Schulte et al. 1998). The VAMID is secured to the patient using vacuum against the maxilla (roof of the mouth). The frame holding the VAMID is secured to the tabletop also using vacuum. During treatment, the patient holds an enable switch that they can depress if an emergency occurs thereby releasing both the VAMID from the mouth and the frame from the tabletop.

Figure 5.28 Picture of ripple filter.

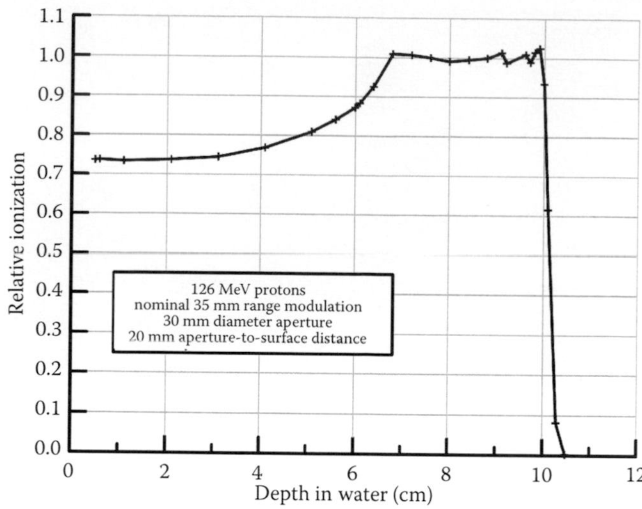

Figure 5.29 Depth dose distribution generated by combination of rotating modulator propellor and ripple filter.

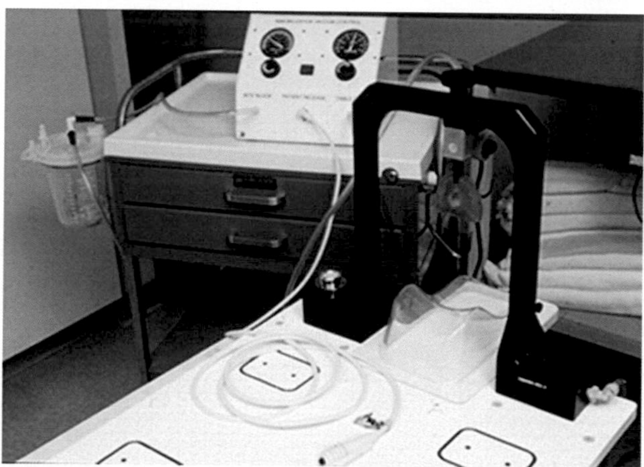

Figure 5.30 Picture of VAMID attached to tabletop interface board. Vacuum is applied to both the mouthpiece against the roof of the mouth and the black frame attaching to the white positioner interface board. The white cable lying on the table is a patient-activated emergency release. The box with dials in the background is the vacuum control box.

5.9 FAURE PROGRAM (NAC/iThemba)

Proton treatments began at iThemba LABS (formerly National Accelerator Centre) in Faure, South Africa, in 1993. A variety of conditions have been treated with SRS/SRT, including AVMs (Vernimmen et al. 2005), acoustic neuromas, meningiomas (Vernimmen et al. 2001), pituitary adenomas, brain metastases, and gliomas. SRS/SRT treatments are given in one to four fractions. The dose is prescribed to the ICRU reference point, which usually coincides with the isocenter. As of 2011, more than 500 patients had been treated at iThemba.

The proton beam is provided by a separated-sector cyclotron that is capable of accelerating and extracting multiple ion species to variable energies. Switching the ions and energies does, however, take a couple of hours, so for SRS/SRT, protons are extracted only at 200 MeV with reduced energies being provided through the use of a carbon double-wedge range shifter. Only a horizontal beam direction is available for SRS/SRT. Lateral broadening of the beam is performed by double scatterers with an occluding ring and plug placed near the second scatterer. The first scatterer-to-isocenter distance is approximately

7 m. The maximum circular field size is 10 cm in diameter with custom apertures constructed of Lipowitz metal. Range modulation is provided by a rotating propellor. The dose rate is typically 3 Gy/min.

SRS/SRT treatments are performed using both plateau cross-fire and stopping techniques. For treatments using the stopping technique, it is important to deliver the beam with correct penetration. The range of the proton beam is checked immediately before each treatment using a multilayer Faraday cup and, if the range deviates from the standard range by more than 0.4 mm, thin range shifter plates are inserted or removed (Jones et al. 2004; Schreuder et al. 1996).

The patient is immobilized and registered to a 5 DOF patient positioner (chair) using a double-sided partial face mask shown in Figure 5.31. Alignment of the patient to the beamline is facilitated through the use of a thin carbon-fiber marker carrier attached via a bite block to the patient (see Figure 5.32). This allows the marker carrier to move with the patient because the bite block is not coupled to the immobilization system or the positioner. The marker carrier includes several 0.8-mm-diameter radio-opaque markers and a combination of 10-mm and 12-mm-diameter retro reflective optical markers. The patient is CT scanned with the marker carrier and bite block in place, thus allowing the target coordinates to be extracted relative

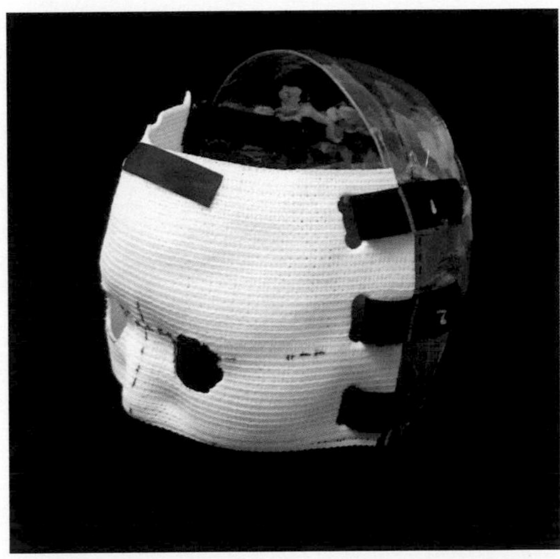

Figure 5.31 Double-sided partial face mask used to immobilize patients. The rear section, made of PMMA, is attached to the patient positioner.

Figure 5.32 Markers and bite block attached to marker carrier. The bite block and marker carrier are free to move with the patient independent of the double-sided face mask.

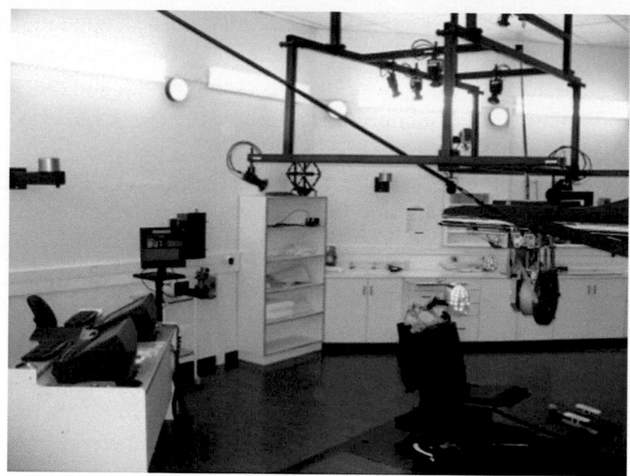

Figure 5.33 The horizontal beamline and patient positioner chair at iThemba Labs. Beam enters from the right of the picture. The chair may be lowered below the floor level to allow equipment such as water phantoms to be placed near the isocenter. Above the chair is a frame with the SPG cameras for monitoring the patient's position.

to the markers. A photograph of the patient end of the beamline and positioner is shown in Figure 5.33. For treatment, the patient is aligned to the beam using a real-time multi-camera stereophotogrammetry (SPG) system (Jones 1995; Schreuder et al. 1997). Several of these cameras can be seen in Figure 5.33. After the images are analyzed, the patient is repositioned using the chair positioner, which has a translational and rotational accuracy of 0.02 mm and 0.03°, respectively. Typically, two to three iterations are required to obtain both the correct position and orientation. During beam delivery, the markers are continuously monitored by the SPG system, which terminates the delivery if a significant change in position is detected. Overall accuracy, including patient motion during treatment, is estimated at about 1 mm (1 s.d.).

5.10 SPECIAL CONSIDERATIONS FOR LIGHT ION BEAM PLANNING

When the stopping technique is used, the planners must pay careful attention to several items that are often casually dismissed when doing megavoltage x-ray planning. Most of these are related to CT scanning.

The absolute values of CT numbers are sensitive to many different factors, including the incident x-ray energy spectrum, applied filters, volume and shape of object being scanned, contrast material being used for image enhancement, presence of embolic materials, metal objects, and others. Increased accuracy can be achieved by individually scaling the CT numbers for each patient. A method for scaling was provided by Wrightstone et al. (1989) and has been adopted by several treatment planning systems (Moyers et al. 2010).

After the CT numbers are scaled, the CT numbers must be converted to values useful for designing boluses, selecting range modulators, and calculating the dose distributions. For planning megavoltage x-ray and electron treatments, most planning systems convert CT numbers to relative electron density, which is approximately proportional to the relative linear attenuation coefficients or relative linear stopping powers (RLSPs), respectively, but this approximation is not sufficient for light ion planning. It is imperative that each institution determine a conversion function (such as a table of paired numbers) for each CT scanning protocol in use. The use of CT number and stopping powers for light ion treatments has been reviewed by Moyers et al. (2010), including the presentation of several conversion functions.

Because light ion beams have sharp distal dose gradients, it is critical to minimize variables that may affect the beam penetration. Slight lateral misalignments of the patient with respect to the beam and shaped bolus can result in substantial differences in the range. To minimize possible penetration errors, tabletops and immobilization devices through which the beam passes should be constructed with shallow thickness gradients. Light ion penetration can even be affected by seemingly negligible things, such as wet hair and gel in the hair.

Similar to electron beams, the penumbra of light ion beams is very sensitive to the aperture-to-patient distance. To minimize the width of the penumbra, it is preferable to bring the aperture (or multileaf collimator [MLC]) as close as possible to the patient (Urie et al. 1986). Applicator cones with reduced outer diameters (see Figures 5.25 and 5.26) allow these devices to be brought close to the patient for tangential fields and reduce shoulder interference for accessing inferior base of skull lesions. A review of 750 cases treated with proton SRS/SRT at the HCL between 1991 and 2002 showed that limiting the cone size to provide a maximum field size of 6 cm diameter would allow 95% of cases to be treated with very little collision risk. The other 5% would require a larger 10-cm cone and thus be more likely to encounter a collision with the stereotactic frame, shoulders, etc. Fractionated cases require even larger cones to hold the apertures and boluses. Typically, the applicator cone is mounted to a translatable applicator carriage that can extend or retract the applicator toward the isocenter to adjust for different isocenter-to-surface distances, which also helps in avoiding collisions. Despite having smaller diameter applicators, there is still a large risk of collision, especially when rotating the gantry. X-ray tubes and imagers that move into and out of position are other devices with high risk of collision with the patient or patient positioner. For accurate dose calculations and collision avoidance planning, it is imperative that the true applicator carriage extension be used in treatment planning.

The injection of contrast during CT scanning for the purpose of delineating tumors results in those materials that uptake the contrast having artificially high CT numbers compared to what they would have if scanned at the time of treatment. If the contrast scan were to be used blindly during treatment planning, the conversion of CT numbers to RLSPs would be in error, boluses would be designed incorrectly, and the beam would penetrate too deeply during treatment. To reduce these errors, back-to-back noncontrast and contrast CT planning studies are routinely performed with the patient in the same position and orientation. The contrast scan is used for target and normal structure delineation, and the noncontrast scan is used for designing boluses and calculating dose distributions. This procedure is especially important for vascular regions where the contrast material is not absorbed uniformly.

Fiducial markers can be any small, easily discernible objects whose position relative to the target and/or normal tissues remains constant. Fiducials can be external, such as a Z box; internal, such as a bony prominence or surgical hardware from a prior surgical procedure; or implanted specifically for the radiosurgical procedure. A minimum of three fiducials are necessary for stereotactic triangulation. Because the coordinates for each fiducial are obtained from the CT study, spherical markers provide better accuracy than elongated screws or rods. As described earlier, localization at the treatment unit for light ion SRS/SRT is usually done using kV imaging; therefore, commercially available 2-mm gold seed fiducials that are necessary for MV imaging are unnecessary for light ion treatments. In fact, because of the unnecessarily large CT artifacts they generate, smaller surgical steel fiducials, such as those available from Cosman Medical (Burlington, MA) are preferred. Generally, planners try to avoid aiming beams that pass through the fiducials. If this is impossible, the planner should contour the fiducials and change their RLSPs to realistic values because the CT numbers of the metal fiducials are generally above the maximum CT number produced by most scanners and thus the converted RLSPs would be in error.

There are many other materials that are often present in CT images that do not lie on the standard CT number-to-RLSP conversion curve and therefore are not converted properly. Implanted materials may include drainage tubes, bone cement, prostheses, surgical clips, stabilization hardware, and injected "glue-like" materials, such as Onyx® (EV3 Neurovascular, Irvine, CA) for embolization of AVMs. In some instances, surgical records may provide enough information to determine the elemental composition of the implanted material, and thus an RLSP may be calculated. Typically, however, this information is not available. It is best not to direct a beam through implanted materials, but if there is no other reasonable choice of beam directions, an estimate of the RLSP can be made, the material contoured, and the estimated RLSP assigned to the contoured structure. Moyers et al. (2010) have provided typical CT numbers and RLSPs for a number of frequently encountered materials. When this procedure is used, larger uncertainties must be applied to the target and normal tissue margins. For high–atomic number, high-density materials, artifacts can be generated in the image surrounding the material. A limited amount of work has been done in an attempt to reduce these artifacts by using megavoltage CT scanning (Moyers et al. 2007; Moyers et al. 2010; Newhauser et al. 2008). Generally, however, the artifacts are contoured, an estimate made of the RLSP for the actual material contained within the artifact, and the RLSP assigned to the contoured artifact region. As before, larger uncertainties must be applied to the margins.

5.11 DOSE MEASUREMENTS IN LIGHT ION BEAMS

Measuring dose in light ion beams is very similar to that performed in x-ray, electron, or neutron beams. In fact, the old American Association of Physicists in Medicine Task Group 21 report (AAPM 1983) could be used with ion chambers to measure the dose if appropriate w-values and stopping powers were used. On the other hand, several proton and other light ion–specific protocols have been developed over the years to address the special needs of these beams (AAPM 1986; ICRU 1998). The most recent protocol is Report 78 by the ICRU (2007). A detailed discussion of these protocols and their applications is provided by Moyers and Vatnitsky (2012).

ICRU Report 78 is based upon the use of ionization chambers for reference dosimetry. Many of the chambers listed in the report may not be appropriate for small SRS/SRT fields, so alternate methods must be used. Vatnitsky et al. (1999) have described the use of silver halide film, radiochromic film, silicon diodes, thermoluminescent detectors, and diamond detectors for narrow fields for proton beam SRS/SRT. For field sizes less than 5 mm, only the film techniques can be used. One problem with film, however, is its dose response as a function of ion energy spectrum. Moyers (1993) and Moyers, Reder, and Lau (2007) have devised a technique for using film to measure the lateral beam profiles and an ion chamber with a diameter larger than the beam to provide a nearly energy-independent dose response. This technique has been used to measure a 1-mm-diameter experimental beam and should be useful for functional SRS beams less than 5 mm in diameter. Scatter effects from the bolus and aperture can be significant in small fields. These effects are particularly important when the cone must be retracted away from the skin. As an example, with a 75-mm-thick bolus retracted 200 mm from the skin, the dose/MU can differ by 65% compared to if the bolus was adjacent to the skin. Because these effects are so strongly dependent upon the bolus thickness, field size, and bolus-to-skin distance, the dose/MU for many small SRS/SRT fields are verified with physical measurements.

5.12 FUTURE DIRECTIONS IN LIGHT ION BEAM DELIVERY

Although SRS and SRT with light ion beams have been performed for more than 55 years, there are still some improvements that can be made, and new features are continuously being investigated. The following paragraphs provide some speculation about potential future enhancements in light ion beam SRS/SRT.

To achieve a sharp lateral penumbra, many beamlines have used a single scatterer to broaden the beam profile. Unfortunately, less than 4% of the ions reach the patient with this technique while the remaining ions impact the beam delivery equipment producing neutrons and other stray radiation (Koehler, Schneider, and Sisterson 1977). Proper shielding design of the equipment can minimize the effects of this stray radiation; however, a more efficient method of laterally broadening the beam is scanning the beam across the field with magnets. There are several scan patterns that can be used, but a particularly attractive one is a circle with a small radius no larger than needed to cover the aperture (Moyers and Vatnitsky 2012). This technique should dramatically reduce the neutron dose received by patients. Another advantage afforded by this technique is that it can, if desired, greatly increase the dose rate to the patient decreasing the treatment time and reducing the chance of patient motion. This increased dose rate is particularly attractive for functional radiosurgery that may require a dose as high as 120 Gy.

MLCs for light ion beams were proposed before 1983 (Kanai, Kawachi, and Matsuzawa 1983), and one was built for the Berkeley facility in 1991 (Ludewigt et al. 1995). Several Japanese facilities have implemented MLCs into their large-field radiotherapy clinical programs. Typically, the leaves of these devices have a width of 3.75 mm (Mitsubishi Electric 2010). For SRS/SRT, an advantage of MLCs over manually installed apertures is that the field shape can be changed quickly as the gantry is moved from one treatment angle to another, thus allowing more beam entry angles to be practically delivered during a given treatment fraction. For SRS/SRT, it is desirable for these devices to have smaller leaf widths than standard for large-field tumor therapy. Such devices are often referred to as mini-MLCs. Some mini-MLCs have been tested with proton beams (Torikoshi et al. 2007; Daartz et al. 2009), but thus far, none have been implemented into the clinic.

The use of rotating gantries has increased the solid angle through which beams may enter the patient. The size and weight of these devices inevitably, however, result in slight geometrical distortions of the beam aiming and patient imaging systems upon rotation. For general radiotherapy, these distortions have minimal effect, but for functional SRS/SRT with critical structures immediately adjacent to targets, submillimeter accuracy is desired. One method to correct for these distortions is to premeasure the distortions and apply compensating moves of a scanning beam, collimator, or patient positioner (Moyers 2004). Another method is to align the applicator containing the aperture and bolus to the patient using an array of sensors. One such method was described by Schulte et al. (2005) wherein a trio of optical cameras monitors two sets of reflective markers, the first set mounted on a stereotactic coordinate head frame secured to the patient and the second set mounted on the applicator. Software can then analyze the positions of all markers and move the patient positioner to bring the beam aiming point in alignment with the target.

The possibility of using ions for CT was first discussed by Cormack (1963). Over the years, a number of investigators have reconstructed tomographic images from ion beams (Goitein 1972; Hanson et al. 1981; Zygmanski et al. 2000; and Ryu et al. 2007). Using the same radiation to acquire the patient information as used for treatment reduces some of the uncertainty in converting the CT number to relative linear stopping power. There is also a theoretical dose advantage in favor of ions. Recently, Bruzzi et al. (2007) have investigated using silicon strip detectors to measure the spatial distribution of protons entering and exiting the patient to predict the most probable path through the patient, thereby increasing the effective spatial resolution. During 2010, a prototype system sized for a head was under construction at LLUMC using this technology.

Designs for several new accelerators are being pursued. The most mature of these is a superconducting synchrocyclotron (SSC), its smaller size and weight allowing its placement directly onto an isocentric rotating gantry. Based upon Nb_3Sn, an advanced superconducting material, the SSC itself weighs only about 19 tons as opposed to a room temperature IC that weighs about 230 tons and a superconducting IC that weighs about 90 tons. This design avoids the need to have separate rooms for the accelerator and the switchyard that delivers the beam from the accelerator to the treatment room. Figure 5.34 is a conceptual illustration of an installation. Beam testing of this equipment at the first clinical installation site is underway during the year 2013. Most clinical facilities that have implemented light ion beam treatments have had medium to large patient loads and used only one accelerator to deliver beam to three to five treatment rooms. This new SSC technology allows smaller institutions without a large enough number of patients to justify a multiroom facility to implement proton therapy. Whether this new technology can decrease the cost per patient remains to be seen.

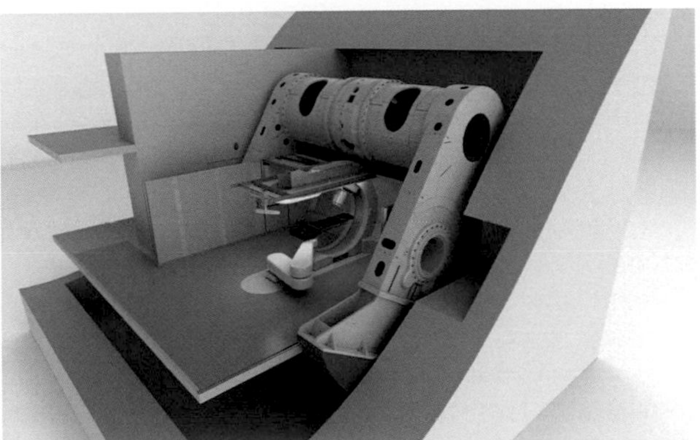

Figure 5.34 Digital rendering of Mevion single-room design with SSC mounted directly onto the gantry. (Courtesy of Mevion, Inc., Littleton, MA.)

Another new accelerator being pursued is the dielectric wall linear accelerator (DWA). This technology is based upon the ability to produce high-voltage gradients and fast optical switching. Preliminary designs expect accelerating gradients of up to 100 MV per meter, suggesting that the entire accelerator could be placed onto a rotating gantry (Caporaso et al. 2009). The DWA is expected to provide pulses at a frequency of 10 to 50 Hz with a different light ion flux and energy extracted for each pulse. For SRS/SRT, an energy of 120 MeV is probably sufficient for most intracranial targets. With the expected accelerating gradient, this might mean an accelerator length of only 1.2 m plus another 0.4 m for beam monitoring and shaping equipment. It might be feasible for such equipment to be installed onto the arm of a very large robot or other gimbaling device and moved about the patient similar to what is done with the CyberKnife® linac (see Chapter 4).

Another new accelerator being pursued is based upon experiments that have shown that high-flux, short-duration laser pulses on thin targets can accelerate protons to energies as high as 68 MeV. Theoretical studies have shown that energies above 200 MeV can be reached, and several groups around the world are currently building prototype devices. If high enough energies and flux rates can be achieved, then this technology could eliminate the heavy magnets and beamlines used with current accelerators. Currently, the spectrum of protons produced by laser experiments is quite wide, and an energy selection system may be required to deselect those protons not wanted (Fourkal et al. 2003). Collimation and beam monitoring systems will, of course, still be required for treatment systems.

5.13 COMPARISON OF MODALITIES

The difference between the dose distributions of plateau cross-fire light ion beam techniques and most megavoltage x-ray beam techniques should be small except in the region just outside of the target. In that region, the lateral penumbra for helium and heavier ions should be a little smaller than for megavoltage x-ray beams and, thus, the volume of normal tissue just outside of the target receiving a high dose should be reduced. There may also be a treatment time advantage for functional SRS because very high dose rates can be achieved with nonscattered light ion beams. On the other hand, the small number of available facilities is currently a disadvantage for light ion beams.

Much bigger differences can be seen between delivery of light ion beams using the stopping technique and megavoltage x-ray beam techniques. Phillips et al. (1990) compared calculated dose distributions for AVMs from a Gamma Knife, arced megavoltage x-ray beams from a linear accelerator, protons, helium-4 ions, carbon-12 ions, and neon ions. Two dosimetric analyses are useful for discussion; the ratio of integral doses to the whole brain outside of the target and the ratio of volumes irradiated to high doses in the region just outside of the target. Essentially no difference between the different x-ray techniques was found using these analyses. There was also little difference seen between the various ions although helium and carbon did exhibit a slight integral dose advantage for large targets. Figure 5.35 plots the integral dose to the nontarget brain as a function of target volume for the different beams. When the target was centered within the cranium, the integral dose from x-rays compared to the dose from any of the ion beams was about a factor of 1.6 higher. When the target was near the periphery, the integral dose of x-rays compared to the integral dose from any of the ion beams was about a factor of two higher for all volumes tested. The normal tissue volume irradiated to a high dose (>80% of the prescribed dose) for different modalities is plotted in Figure 5.36 as a function of target volume. The difference in high dose normal tissue volume irradiated by x-rays versus light ion beams is seen to increase as the target volume increases. For the largest target treated, the high-dose volume irradiated by x-rays was about 2.5 times greater.

Serago et al. (1995) performed a comparison study of several different x-ray and proton beam configurations for four different target sizes and shapes. They found little differences in the volumes of normal tissues irradiated to doses above the 50% level, that is, near the target, from the different modalities and configurations, but large differences in volumes at lower dose levels. Verhey, Smith, and Serago (1998) and Smith, Verhey, and Serago (1998) compared several configurations of Gamma Knife, linac-based x-ray beams, and proton beams for five different target sizes and shapes. Based upon an analysis using normal tissue complication probabilities, they found the biggest advantage for proton beams to be for peripheral targets and/or large targets.

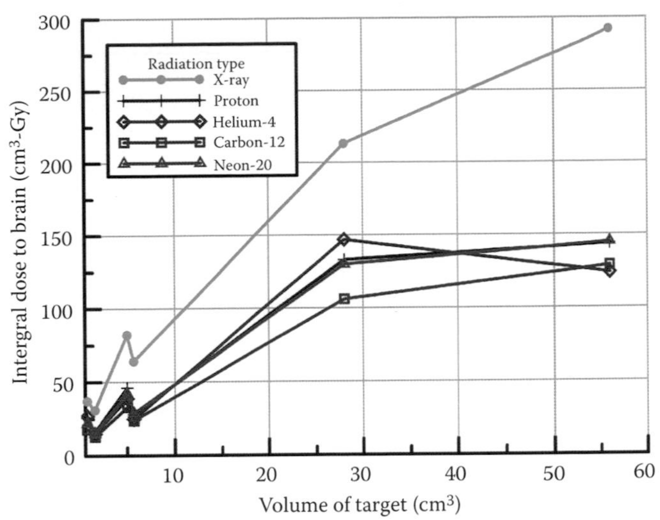

Figure 5.35 Integral dose to nontarget brain versus volume of target for various radiation modalities. Points at 1.6, 5.7, and 56 cm^3 were for targets near the periphery while all other points were for targets centered in the cranium. Lines were drawn connecting data from the same beam to aid viewing. (Data was taken from *Int. J. Radiat. Oncol. Biol. Phys.*, 18, Phillips, M. H., Frankel, K. A., Lyman, J. T., Fabrikant, J. I. and Levy R. P., Comparison of different radiation types and irradiation geometries in stereotactic radiosurgery, 211–220, Copyright 1990, with permission from Elsevier.)

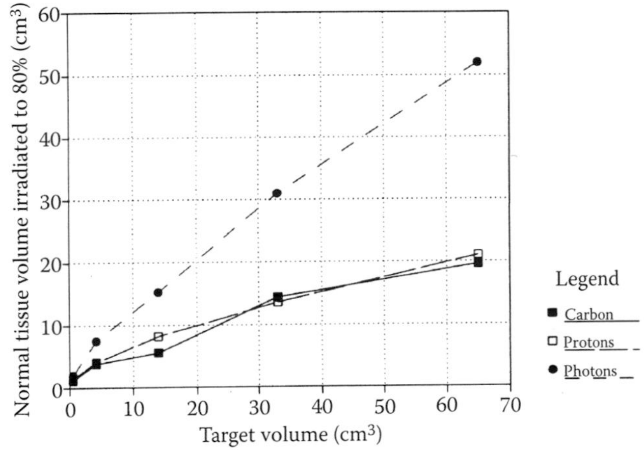

Figure 5.36 Volume of normal tissue irradiated to greater than 80% of the prescribed dose for megavoltage x-rays, protons, and carbon ions. (Reprinted from *Int. J. Radiat. Oncol. Biol. Phys.*, 18, Phillips, M. H., Frankel, K. A., Lyman, J. T., Fabrikant, J. I. and Levy R. P., Comparison of different radiation types and irradiation geometries in stereotactic radiosurgery, 211–220, Copyright 1990, with permission from Elsevier.)

ACKNOWLEDGMENTS

The authors thank Ken Frankel, Dan Jones, Evan de Kock, Erik Blomquist, and Dmitri Nichiporov for providing assistance in obtaining pictures and information for this chapter.

REFERENCES

AAPM 1986. *Protocol for Heavy Charged-Particle Therapy Beam Dosimetry.* Report 16 of the American Association of Physicists in Medicine (AAPM) Task Group 20. New York: American Institute of Physics.

AAPM Task Group 21. 1983. A protocol for the determination of absorbed dose from high-energy photon and electron beams. *Med. Phys.* 10:741–71.

Abrossimov, N. K., Y. A. Gavrikov, E. M. Ivanov et al. 2006. 1000 MeV proton beam therapy facility at Petersburg Nuclear Physics Institute synchrocyclotron. *J. Phys. Conference Series* 41:424–32.

Abrossimov, N. K., E. M. Ivanov, V. I. Lazarev et al. 2001. Stereotaxic proton therapy at PNPI synchrocyclotron. In *Main Scientific Activities PNPI XXX (1997–2001)*, eds. A. A. Vorobyov, V. A. Elisev, I. V. Lopatin, and G. E. Solyakin. Russian Academy of Sciences, Petersburg Nuclear Physics Institute, High Energy Physics Division PNPI, 312–3.

Abrossimov, N. K., A. A. Vorobyov, and G. A. Riabov. 2001. Status of PNPI synchrocyclotron. In *Main Scientific Activities PNPI XXX (1997–2001)*, eds. A. A. Vorobyov, V. A. Elisev, I. V. Lopatin, and G. E. Solyakin. Russian Academy of Sciences, Petersburg Nuclear Physics Institute, High Energy Physics Division PNPI.

Alonso, J. R., J. Bercovitz, W. T. Chu et al. 1989. Relocation of the helium ion radiotherapy program from the 184″ synchrocyclotron to the BEVALAC. In *Proceedings of the 1989 IEEE Particle Accelerator Conference*, 669–71.

Blomquist, E. 2010. Personal communication.

Brobeck, W. M., E. O. Lawrence, K. R. McKenzie et al. 1947. Initial performance of the 184-inch cyclotron of the University of California. *Phys. Rev.* 2:449.

Bruzzi, M., N. Blumenkrantz, J. Feldt et al. 2007. Prototype tracking studies for proton CT. *IEEE Transactions on Nuclear Science* 54 (1):140–145.

Caporaso, G. J., Y. J. Chen, S. Sampayan et al. 2009. Status of the dielectric wall accelerator. *Proceedings of the 2009 U.S. Part. Acc. Conf.:* TH3GAI02. Vancouver, BC, Canada.

Castro, J. R., J. M. Quivey, W. M. Saunders et al. 1980. Clinical results in heavy particle radiotherapy. In *Biological and Medical Research with Accelerated Heavy Ions at the Bevalac, 1977–1980 LBL-11220* eds. Pirruccello, M. C. and Tobias, C. A. (Regents of the University of California, Berkeley) 305–318.

Chang, S. D., R. P. Levy, J. R. Adler, D. Martin, P. R. Krakovitz, and G. K. Steinberg. 1998. Stereotactic radiosurgery of angiographically occult vascular malformations: 14-year experience. *J. Neurosurg.* 43:213–20.

Chapman, P. H., C. S. Ogilvy, and W. E. Butler. 1993. A new stereotactic alignment system for charged particle radiosurgery at the Harvard cyclotron laboratory, Boston. In *Stereotactic Radiosurgery*, eds. E. Alexander III, J. S. Loeffler, and L. E. Lunsford, 105–8. New York: McGraw-Hill.

Chervjakov, A., I. Ermakov, B. Konnov, L. Saltykova, and S. Vatnitsky. 1987. Perspective and some experience on photon narrow beam in radiation therapy. Papers presented at a joint meeting of the Proton Therapy Cooperative Group and the Second International Charged Particle Therapy Workshop. Loma Linda, California, October 12–14, 1987.

Chu, W. T., B. A. Ludewigt, and T. R. Renner. 1993. Instrumentation for treatment of cancer using proton and light-ion beams. *Rev. Sci. Instruments* 64:2055–122.

Chuvilo, I. V., L. L. Goldin, V. S. Khoroshkov et al. 1984. ITEP synchrotron proton beam in radiotherapy. *Int. J. Rad. Oncol. Biol. Phys.* 10:185–95.

Cormack, A. M. 1963. Representation of a function by its line integrals, with some radiological applications. *J. Appl. Phys.* 34 (9):2722–7.

Daartz, J., M. Bangert, M. R. Bussière, M. Engelsman, and H. M. Kooy. 2009. Characterization of a mini-multileaf collimator in a proton beamline. *Med. Phys.* 36 (5):1886–94.

Daartz, J., M. Engelsman, M. Bangert, and M. R. Bussière. 2009. Field size dependence of the output factor in passively scattered proton therapy: Influence of range, modulation, air gap and machine settings. *Med. Phys.* 36 (7):3205–10.

Ermakov, I. A., A. M. Chervijakov, L. M. Saltykova, and M. L. Razymnaja. 1987. Computer program for two axis rotation 1000 MeV proton beam therapy treatment planning. Papers presented at a joint meeting of the Proton Therapy Cooperative Group and the Second International Charged Particle Therapy Workshop. Loma Linda, California, October 12–14, 1987.

Fabrikant, J. I., R. P. Levy, G. K. Steinberg et al. 1992. Charged-particle radiosurgery for intracranial vascular malformations. *Neurosurg. Clin. N. Am.* 3 (1):99–139.

Falkmer, S., B. Fors, B. Larsson, A. Lindell, J. Naeslund, and S. Stenson. 1962. Pilot study on proton irradiation of human carcinoma. *Acta Radiol.* 58:33–51.

Fourkal, E., J. S. Li, M. Ding, T. Tajima, and C. Ma. 2003. Particle selection for laser-accelerated proton therapy feasibility study. *Med. Phys.* 30 (7):1660–70.

Gall, K. P., L. J. Verhey, and M. Wagner. 1993. Computer-assisted positioning of radiotherapy patients using implanted radiopaque fiducials. *Med. Phys.* 20 (4):1153–9.

Goitein, M. 1972. Three-dimensional density reconstruction from a series of two-dimensional projections. *Nuclear Instruments and Methods* 101:509–518.

Goitein, M., and M. Abrams. 1983. Multi-dimensional treatment planning: I. Delineation of anatomy. *Int. J. Rad. Onc. Biol. Phys.* 9 (6):777–87.

Goitein, M., M. Abrams, D. Rowell, H. Pollari, and J. Wiles. 1983. Multi-dimensional treatment planning: II. Beam's eye view, back projection, and projection through CT Sections. *Int. J. Rad. Onc. Biol. Phys.* 9 (6):789–97.

Goldin, L. L., V. S. Khoroshkov, S. I. Blokhin et al. 1987. Physical and technical aspects of proton therapy at the ITEP synchrotron. Papers presented at a joint meeting of the Proton Therapy Cooperative Group and the Second International Charged Particle Therapy Workshop. Loma Linda, California, October 12–14, 1987.

Goldin, L. L., V. S. Khoroshkov, E. I. Minakova, and I. A. Voronzov. 1989. Proton therapy in USSR. *Strahlenther. Onkol.* 165:885–90.

Gottschalk, B. 1987. Proton radiotherapy nozzle for combined scatterer/modulator. *Harvard Cyclotron Laboratory (HCL) internal report.*

Graffman, S., A. Brahme, and B. Larsson. 1985. Proton radiotherapy with the Uppsala cyclotron. Experience and plans. *Strahlentherapie* 161 (12):764–70.

Hanson, K. M., J. N. Bradbury, T. M. Cannon et al. 1981. Computed tomography using proton energy loss. *Physics in Medicine and Biology* 26 (6):965–983.

Hong, L., M. Goitein, M. Bucciolini et al. 1996. A pencil beam algorithm for proton dose calculations. *Phys. Med. Biol.* 41 (8):1305–30.

International Commission on Radiation Units and Measurements (ICRU). 1998. Clinical proton dosimetry part 1: Beam production, beam delivery and measurement of absorbed dose. *ICRU Report 59*. Bethesda, MD.

International Commission on Radiation Units and Measurements (ICRU). 2007. Prescribing, recording, and reporting proton beam therapy. *ICRU Report 78, J. ICRU.* 7(2).

Jones, D. T. L. 1995. NAC—the only proton therapy facility in the southern hemisphere. In *Ion Beams in Tumor Therapy*, ed. U. Linz, 350–7. Weinheim, Germany: Chapman and Hall.

Jones, D. T. L., A. N. Schreuder, E. A. de Kock et al. 2004. Proton therapy at iThemba LABS. *Radiat. Phys. Chem.* 71:983–4.

Kamanin, D. V. 2010. Radiation oncology at JINR in brief. Presented at the International Center for Scientific and Technical Information Conference, Baku, Republic of Azerbaijan, May 20, 2010.

Kanai, T., K. Kawachi, and H. Matsuzawa. 1983. Broad beam three-dimensional irradiation for proton radiotherapy. *Med. Phys.* 10 (3):344–6.

Kjellberg, R. N. 1986. Stereotactic Bragg peak proton beam radiosurgery for cerebral arteriovenous malformations. *Ann. Clin. Res.* 18(Suppl. 47):17–9.

Kjellberg, R. N., W. H. Sweet, W. M. Preston, and A. M. Koehler. 1962. The Bragg peak of a proton beam in intracranial therapy of tumors. *Trans. Amer. Neurol. Assoc.* 87:216–8.

Koehler, A. M., R. J. Schneider, and J. M. Sisterson. 1977. Flattening of proton dose distributions for large-field radiotherapy. *Med. Phys.* 4 (4):297–301.

Konnov, B. A. 1987. Proton therapy at Leningrad synchrocyclotron. Papers presented at a joint meeting of the Proton Therapy Co-operative Group and the Second International Charged Particle Therapy Workshop. Loma Linda, California, October 12–14, 1987.

Larsson, B. 1961. Pre-therapeutic physical experiments with high energy protons. *Br. J. Radiol.* 34:143–51.

Larsson, B. 1967. Radiological properties of beams of high-energy protons. *Radiation Res.* S7:304–11.

Larsson, B., and S. Graffman. 1985. Experience with the Uppsala 230 cm cyclotron and preparations for future use in radiotherapy. *Proceedings of a Medical Workshop on Accelerators for Charged-Particle Beam Therapy:* 7–39, Batavia, IL, January 24–25, 1985.

Larsson, B. L., L. Leksell, and B. Rexed. 1963. The use of high energy protons for cerebral surgery in man. *Acta Chir. Scand.* 125:1.

Larsson, B. L., L. Leksell, B. Rexed, and P. Sourander. 1959. Effect of high energy protons on the spinal cord. *Acta Radiol. Stock.* 51:52–64.

Larsson, B., L. Leksell, B. Rexed, P. Sourander, W. Mair, and B. Andersson. 1958. The high-energy proton beam as a neurosurgical tool. *Nature* 182:1222–3.

Lawrence, J. H. 1957. Proton irradiation of the pituitary. *Cancer* 10:795–98.

Lawrence, J. H., and C. A. Tobias. 1956. Radioactive isotopes and nuclear radiations in the treatment of cancer. *Cancer Res.* 16 (3):185–193.

Lawrence, J. H., C. A. Tobias, and J. L. Born. 1962. Acromegaly. *Trans. Am. Clin. Climatol. Assoc.* 73:176–85.

Levy, R. P., J. I. Fabrikant, K. A. Frankel, M. H. Phillips, and J. T. Lyman. 1989. Stereotactic heavy-charged-particle Bragg peak radiosurgery for the treatment of intracranial arteriovenous malformations in childhood and adolescence. *Neurosurgery* 24 (6):841–52.

Levy, R. P., J. I. Fabrikant, K. A. Frankel, M. H. Phillips, and J. T. Lyman. 1990. Charged-particle radiosurgery of the brain. *Neurosurg. Clin. N. Am.* 1 (4):955–90.

Ludewigt, B., J. Bercovitz, M. Nyman, and W. Chu. 1995. Method for selecting minimum width of leaf in multileaf adjustable collimator while inhibiting passage of particle beams of radiation through sawtooth joints between collimator leaves. *U.S. Letters* Patent No. 5,438,454. Issued August 1, 1995.

Lyman, J. T., and C. Y. Chong. 1974. ISAH: A versatile treatment positioner for external radiation therapy. *Cancer* 34:12–6.

Lyman, J. T., L. Kanstein, F. Yeater, J. I. Fabrikant, and K. A. Frankel. 1986. A helium-ion beam for stereotactic radiosurgery of central nervous system disorders. *Med. Phys.* 13 (5):695–9.

Lyman, J. T., M. H. Phillips, K. A. Frankel, and J. I. Fabrikant. 1989. Stereotactic frame for neuroradiology and charged particle Bragg peak radiosurgery of intracranial disorders. *Int. J. Radiat. Oncol. Biol. Phys.* 16:1615–21.

Minakova, E. I. 1987. Proton therapy on synchrotron of the Institute for Theoretical and Experimental Physics in Moscow (a review of twenty-year clinical experience). Papers presented at a joint meeting of the Proton Therapy Co-operative Group and the Second International Charged Particle Therapy Workshop. Loma Linda, California, October 12–14, 1987.

Minakova, Y. I., and N. Burdenko. 1990. Twenty years clinical experience of narrow proton beam therapy in Moscow. In *Proceedings of the International Heavy Particle Therapy Workshop,* ed. H. Blattmann, 158–62. Villigen, Switzerland: Paul Scherrer Institut.

Minakova, Y. Y., L. L. Goldin et al. 1990. Proton therapy at ITEP. *Proceedings of the International Heavy Particle Therapy Workshop,* ed. H. Blattmann, 154–7. Villigen, Switzerland: Paul Scherrer Institute.

Mitsubishi Electric 2010. Information available at http://global.mitsubishielectric.com/bu/particlebeam/technology/treatment.html#c01.

Moyers, M. F. 1993. Generation and characterization of a proton microbeam. *Med. Phys.* 20 (3):867.

Moyers, M. F. 2002. LLUPTF: Eleven years and beyond. *Nuclear Physics in the 21st Century*, 305–9. New York: American Institute of Physics.

Moyers, M. F. 2004. Method and device for delivering radiotherapy. *International Patent Cooperation Treaty Publication Number WO 0131039212 A1, PCT/US02/34556,* Issued June 7, 2004. *U.S. Letters* Patent No. 6,769,806. Issued August 3, 2004.

Moyers, M. F., and A. Ghebremedhin. 2008. Spill-to-spill and daily proton energy consistency with a new accelerator control system. *Med. Phys.* 35 (5):1901–5.

Moyers, M. F., and W. Lesyna. 2004. Isocenter characteristics of an external ring proton gantry. *Int. J. Radiat. Oncol. Biol. Phys.* 60 (5):1622–30.

Moyers, M. F., C. S. Reder, and D. C. Lau. 2007. Generation and characterization of a proton microbeam for experimental radiosurgery. *Technol. Cancer Res. Treat.* 6 (3):205–11.

Moyers, M. F., M. Sardesai, S. Sun, and D. W. Miller. 2010. Ion stopping powers and CT numbers. *Med. Dosim.* 35 (3):179–94.

Moyers, M. F., S. Sun, M. Sardesai et al. 2007. Tomotherapy MVXCT numbers versus RLSP for various ions and materials. *Med. Phys.* 34 (6):2383.

Moyers, M. F., and S. M. Vatnitsky. 2012. *Practical Implementation of Light Ion Beam Treatments*. Madison, WI: Medical Physics Publishing.

Moyers, M., S. Vatnitsky, and D. Miller. 1993. Small proton beams for stereotactic radiosurgery. *Acta Neurochir.* 122:173.

Mytsin, G. V. 2003. Development of the hadrontherapy complex at the phasotron in the Dzhelepov Laboratory of Nuclear Problems. Presented at the Second International Summer Student School, Poznan, Poland, June 19–30, 2003.

Newhauser, W. D., A. Giebeler, K. M. Langen, D. Mirkovic, and R. Mohan. 2008. Can megavoltage computed tomography reduce proton range uncertainties in treatment plans for patients with large metal implants? *Phys. Med. Biol.* 53 (9):2327–44.

Nichiporov, D. 2003. Verification of absolute ionization chamber dosimetry in a proton beam using carbon activation measurements. *Med. Phys.* 30 (5):972–8.

Phillips, M. H., K. A. Frankel, J. T. Lyman, J. I. Fabrikant, and R. P. Levy. 1990. Comparison of different radiation types and irradiation geometries in stereotactic radiosurgery. *Int. J. Radiat. Oncol. Biol. Phys.* 18:211–20.

Ryu, H., B. Choi, J. Lee, and J. Kim. 2007. Design of the beam line elements for proton computed tomography. *Journal of the Korean Physical Society* 50 (5):1489–1493.

Schreuder, A. N., D. T. L. Jones, J. E. Symons, T. Fulcher, and A. Kiefer. 1996. The NAC proton therapy beam delivery system. *Proc. 14th Int. Conf. on Cyclotrons and their Applications*, ed. J. Cornell, 523–6. Singapore: World Scientific.

Schreuder, A. N., D. T. L. Jones, J. E. Symons et al. 1997. Three years' experience with the NAC proton therapy patient positioning system. In *Advances in Hadrontherapy,* eds. U. Amaldi, B. Larsson, and Y. Lemoigne, 251–8. Amsterdam: Elsevier BV.

Schulte, R. W., R. P. Levy, M. F. Moyers, M. Neupane, and K. E. Schunbert. 2005. Image-guided alignment verification with submillimeter precision for functional proton radiosurgery. *Int. J. Radiat. Oncol. Biol. Phys.* 63 (S):516–7.

Schulte, R. W., W. J. Wicks, H. J. Meinass, and W. J. Nethery. 1998. Vacuum-assisted fixation apparatus. *U.S. Letters* Patent No. 5,730,745. Issued March 24, 1998.

Serago, C. F., A. F. Thornton, M. M. Urie et al. 1995. Comparison of proton and x-ray conformal dose distributions for radiosurgery applications. *Med. Phys.* 22 (12):2111–6.

Shipulin, K., and G. Mytsin. 2008. Verification of patient positioning in proton therapy based upon digital x-ray images. *Radioprotection* 43 (5):231.

Sisterson, J. M. 1990. Overview of proton beam applications in therapy. *Nuclear Instruments and Methods in Physics Research Section B: Beam Interactions with Materials and Atoms* 45 (1–4):718–23.

Smith, V., L. Verhey, and C. F. Serago. 1998. Comparison of radiosurgery treatment modalities based on complications and control probabilities. *Int. J. Radiat. Oncol. Biol. Phys.* 40 (2):507–13.

Steinberg, G. K., J. I. Fabrikant, M. P. Marks et al. 1990. Stereotactic heavy-charged-particle Bragg-peak radiation for intracranial ateriovenous malformations. *N. Engl. J. Med.* 323:96–101.

Tobias, C. A. 1979. Pituitary irradiation: Radiation physics and biology. In *Recent Advances in the Diagnosis and Treatment of Pituitary Tumors*, ed. J. A. Linfoot, 221–43. New York: Raven Press.

Tobias, C. A., J. H. Lawrence, J. L. Born et al. 1958. Pituitary irradiation with high-energy proton beams—a preliminary report. *Cancer Res.* 18 (2):121–38.

Torikoshi, M., S. Minohara, N. Kanematsu et al. 2007. Irradiation system for HIMAC. *J. Radiat. Res. Tokyo* 48:A15–25.

Tseytlina, M., Y. I. Luchin, A. V. Agapov et al. 2009. Proton radiosurgery of intracranial arteriovenous malformations: Dubna experience. Presented at the 48th meeting of the Particle Therapy Co-operative Group, Heidelberg, Germany, September 28 to October 3, 2009.

Urie, M. M., J. M. Sisterson, A. M. Koehler, M. Goitein, and J. Zoesman. 1986. Proton beam penumbra: Effects of separation between patient and beam modifying devices. *Med. Phys.* 13 (5):734–41.

Vatnitsky, S. M., D. W. Miller, M. F. Moyers et al. 1999. Dosimetry techniques for narrow proton beam radiosurgery. *Phys. Med. Biol.* 44:2789–801.

Verhey, L. J., V. Smith, and C. F. Serago. 1998. Comparison of radiosurgery treatment modalities based on physical dose distributions. *Int. J. Radiat. Oncol. Biol. Phys.* 40 (2):497–505.

Vernimmen, F. J. A. I., J. K. Harris, J. A. Wilson, R. Melvill, B. J. Smit, and J. P. Slabbert. 2001. Stereotactic proton beam therapy for skull base meningiomas. *Int. J. Radiat. Oncol. Bio. Phys.* 49 (1):99–105.

Vernimmen, F. J. A. I., J. P. Slabbert, J. A. Wilson, S. Fredericks, and R. Melvill. 2005. Stereotactic proton beam therapy for intracranial arteriovenous malformations. *Int. J. Radiat. Oncol. Bio. Phys.* 62 (1):44–52.

Wambersie, A., P. M. Deluca, P. Andreo, and J. H. Hendry. 2004. Light or heavy ions: A debate of terminology. *Radiother. Oncol.* 73(S2):iiii.

Wilson, R. 2004. *A Brief History of the Harvard University Cyclotron.* Cambridge, MA: Harvard University Press.

Wrightstone, T., C. Crowell, M. Urie, and M. Goitein. 1989. The Treatment Planning Program (RX): The User's Manual. (Massachusetts General Hospital: Boston, Mass.) 8–1.

Zygmanski, P., K. P. Gall, M. S. Z. Rabin, and S. J. Rosenthal. 2000. The measurement of proton stopping power using proton-cone-beam computed tomography. *Physics in Medicine and Biology* 45 (2):511–528.

6 Small animal irradiators for stereotactic radiation therapy

Ke Sheng, Patricia Lindsay, and John Wong

Contents

6.1 BACKGROUND

With the adoption of image-guided stereotactic radiotherapy (SRT) in clinical practice, the study of radiobiological responses to high fractional dose treatments has intensified. Radiobiology of conventionally fractionated radiotherapy is relatively well understood, but the biological response to a high dose from a single fraction of treatment is less clear because of relatively few cases, the heterogeneity of the patient population, and short follow-up time. Animal experiments are an important way to mechanistically understand radiobiology under a more controlled environment, but traditional animal irradiation is drastically different from human treatments in the accuracy of both dosimetry and geometrical targeting. The disparity can render results from animal studies less useful in indicating human radiobiology. For example, compared with conventionally fractionated radiotherapy, the microenvironment may play a more important role in the SRT treatment. It was demonstrated that the tumor cell death from a single radiation dose of 20 Gy was mediated by the apoptosis of surrounding endothelial cells (Garcia-Barros et al. 2003). Further investigation revealed a threshold dose for such transition to the new tumor cell death pathway (Ch'ang et al. 2005). Possibly because of these new tumor cell death pathways, a large error is present when the classical linear quadratic model is use to predict the outcome from treatments with a high dose fraction (Guerrero and Li 2004; Liu et al. 2003; Park et al. 2008). To further correlate these findings with human treatments, more rigorous animal experiments are needed, but less targeted animal irradiation would result in wider normal tissue damage that inevitably contaminates the final results.

Characterization of the radiobiology to high fractional dose treatments remains highly controversial because of the difficulty of studying it using an animal model strictly adhering to human treatment protocols, which typically include four important elements: 3-D CT simulation, tumor and organs-at-risk delineation, treatment planning, and accurate dose delivery. Conventional small animal irradiation techniques are generally not targeted and inaccurate in dosimetry. These techniques are applicable under the assumption that tumor responses and normal tissue toxicities can be studied separately, and approximate point doses suffice to characterize the treatment. On the other hand, they are inadequate when both the exact dose and volume of irradiation to the tumor and surrounding normal tissue are important to the outcome. For instance, to study the response of an orthotopic tumor to radiation, it was not possible to focally irradiate the tumor to the expected dose without an animal irradiation system compatible with the aforementioned four basic elements in human treatments. To treat the inoculated tumor using conventional techniques, extensively irradiating surrounding normal tissue to a very high dose produces a disproportional amount of injury to normal tissue compared with human treatment, which, in turn, generates a disproportional amount of injury responses that can potentially alter the response of tumor cells (Barcellos-Hoff, Park, and Wright 2005). Moreover, the tissue attenuation to kV x-rays results in substantial dose heterogeneity from the beam entrance to the beam exit of the animal. Subsequently, evaluation of radiation injury can only be performed qualitatively on a traditional small animal irradiator without the knowledge of the dose volume histogram of the normal tissue. Further limitation from conventional small animal irradiation stems from the lack of reproducibility to reposition the animals for treatments. Multiple fractions of treatments are often needed in human SRT, but it is nearly impossible to irradiate a small target reproducibly without x-ray imaging guidance for animal setup (Armour et al. 2010).

To address these problems, two separate paths were taken by the investigators: utilization of existing infrastructures for human treatment and development of new dedicated small animal irradiators.

6.2 ANIMAL STUDIES ON EXISTING CLINICAL TREATMENT MACHINES

Although unintended for the animal experiments, human treatment machines with image guidance and capability to deliver fine collimated radiation therapy fields have been used to focally target and treat a well-defined volume of larger mammals. Pigs were frequently used to study the skin injury from radiotherapy, but with a human clinical linear accelerator, a fraction of the lung can be treated reproducibly in a multiple fractional experiment (Hopewell, Rezvani, and Moustafa 2000), in which setup of the pigs was assisted by the x-ray portal images on the linear accelerator. More recently, the pig spinal cord tolerance to radiosurgery was studied (Medin et al. 2011). Dogs were used to test the radiation effects of non-uniform irradiation of the intestine on a Cobalt 60 machine (Vigneulle et al. 1990; Zeman et al. 1990). In a more elaborate experiment to study the volume and dose response effects of spinal cord injury, 4 and 20 cm lengths of the spinal cord of 89 beagle dogs were irradiated by a clinical linac in 4-Gy fractions for up to 21 fractions (Powers et al. 1998). Using the same experimental platform, beagle dogs were irradiated by 1.5 Gy daily to 33%, 67%, and 100% of the lung for up to 6 weeks for severe symptomatic pneumonitis (Poulson et al. 2000). In principle, animals close to human sizes can always be treated on the regular clinical machines, but it is prohibitively expensive to perform large-scale experiments with these animals. It is also more difficult to obtain animal care and use committee approval to conduct radiobiological studies on them.

The geometrical precision of the conventional human treatment machine is on the order of 2 mm as reported in AAPM TG 40 (Kutcher et al. 1994). Additional errors are likely when the setup uncertainty is included. This means that to treat an animal one order of magnitude smaller than a human submillimeter mechanical accuracy is needed.

With the technological development of modern radiotherapy, partially for the purpose of SRT of very small lesions, treatment platforms specialized in delivering radiation doses to a very small volume with high precision were invented. These developments enable focal treatment of smaller animals, particularly rodents for radiobiological assessments. CyberKnife (Quinn 2002) is a frameless non-isocentric therapeutic machine suited for both intracranial and extracranial SRT. A 6-MV compact linear accelerator is mounted on a robotic arm that can direct the x-ray with varying positions and angles. Guided by stereotactic x-ray images, the volume to be treated can be localized in the 3-D space; the radiation beam can be directed to the target with a geometrical precision of 1 mm (Adler et al. 1999). The smallest size of the treatment field on CyberKnife is defined by a 5-mm cone. CyberKnife was used to treat rat glioma (Psarros et al. 2004).

Another system dedicated to intracranial SRT is Gamma Knife. Two hundred and one cobalt sources are concentrically placed in a heavily shielded hemisphere. A highly conformal plan can be generated using varying sizes of cones that produce radiation fields as small as 4 mm at the isocenter. Accuracy of 0.5 mm was achievable from the fixed geometry. Gamma Knife was used to treat rats with epilepsy (Mori et al. 2000) in one study and to introduce a cataract in the lens of a rat by a single dose of 15 Gy (DesRosiers et al. 2003). Yet these experiments were performed without computer-based treatment planning and consequently did not provide dose volume histogram (DVH) data. The first animal experiments strictly followed a human image-guided radiation therapy (IGRT) protocol and were performed on the helical tomotherapy (HT) platform. HT integrates image guidance, inverse optimization, and intensity modulation in a single package (Beavis 2004; Mackie 2006). In the HT machine, a compact 6-MV linac is mounted on a ring gantry that rotates continuously around the patient. The beam is modulated by 6.25-mm-wide binary MLC leaves. A CT detector array is located at the opposite direction of the x-ray source for fan-beam megavoltage CT (MVCT) acquisition. Prior to receiving FDA clearance in 2002, HT was used to treat dogs with head-and-neck cancer (Lawrence and Forrest 2007). In the process, a planning CT was obtained, tumor and organs at risk were delineated, and a treatment plan was created using the tomotherapy planning software. Before treatment, each dog was sedated and positioned based on the MVCT scan. Excellent treatment outcome was observed from treating these dogs with tumor remission and discoloration of fur that matched the treatment field.

The feasibility of delivering SRT treatment on rodents by HT was not demonstrated until 2009 (Cai et al. 2009). In the study, stereotactic lung radiosurgery to the right rabbit lung was prescribed a dose of 60 Gy (20 Gy × three fractions in one week), similar to human protocols (Timmerman et al. 2003; Timmerman, Park, and Kavanagh 2007). A cylindrical treatment volume of 1.6 cm^3 in the lower right lung was contoured. Inverse treatment planning was performed to minimize dose to the heart and large pulmonary and mediastinal vessels using a 1-cm field width. A MVCT scan was performed prior to each dose delivery and registered to prior kVCT treatment-planning imaging to position the rabbit correctly. Large discrepancies caused by yaw, pitch, or body flex, if any, were corrected manually by adjusting the rabbit position and rescanning with MVCT. Isodose distributions and DVHs of the dosimetry with low-dose spillage to adjacent lung volumes are shown in Figure 6.1. The dose was verified by film and ion chamber measurements prior to treatment.

Contrast enhanced perfusion MRI was used to measure pulmonary parenchyma perfusion.

Rabbits were sacrificed 4 months post radiation, and the lungs were removed and fixed. The lungs were inflated with air at a constant pressure of 30 cm H_2O for 2 days. Once the tissue was fully dried, samples from the irradiated right lung regions and control left lung were excised for microscopic study. The size of the visibly injured lung volume was measured and compared with the radiation isodoses to determine the injury threshold.

Both lungs enhanced homogenously at baseline in all rabbits (Figure 6.2a). After radiation, the irradiated lung showed substantial progressive perfusion deficits (Figure 6.2b). The location of the perfusion

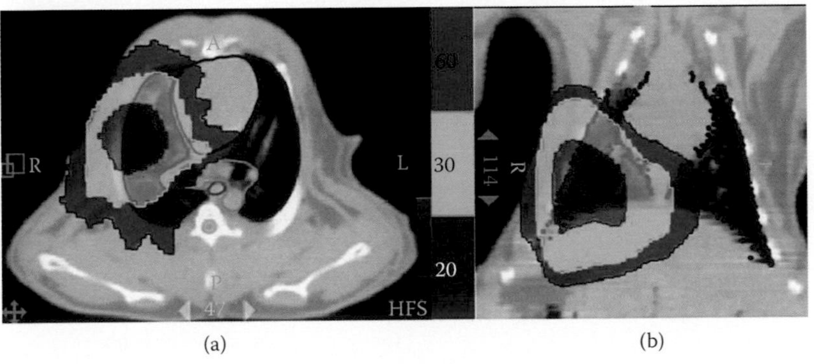

Figure 6.1 Treatment planning on New Zealand rabbits using tomotherapy. Images show the color isodose distribution overlaid on the (a) axial and (b) coronal CT images.

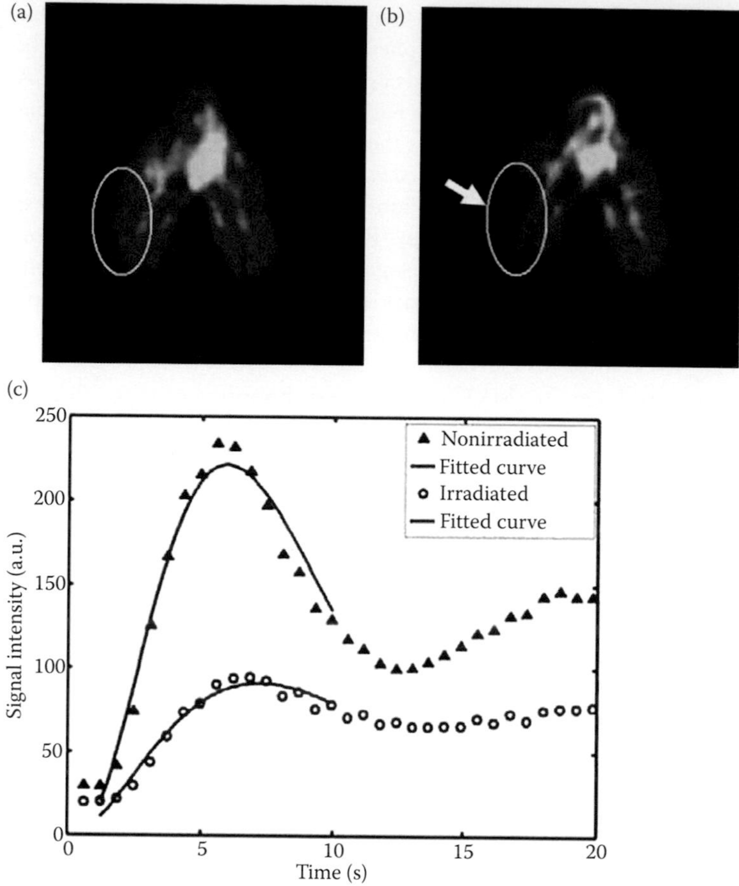

Figure 6.2 MRI perfusion images show a defect at the irradiated volume in (b) compared with the pre-irradiation level in (a). The time resolved perfusion is shown in (c).

deficits correlated well with the radiation target volume. Time-intensity-curve (TIC) analysis verified the large reduction in maximal enhancement ratio (MER) and slope of enhancement (SLE) in the irradiated lung region (Figure 6.2c). The radiosurgery-induced perfusion deficits emerged approximately 1–2 months after radiation and progressively increased with time.

Injury to the rabbit lungs were also manifested in the longitudinal CT images. At 2 months post radiation, the CT images showed apparent low density in the irradiated lung regions (Figure 6.3b). At 4 months post radiation, the lung density in the irradiated lung regions was further decreased (Figure 6.3c). The regions with reduced lung density were well defined by a clear boundary as indicated by white arrows in Figure 6.3.

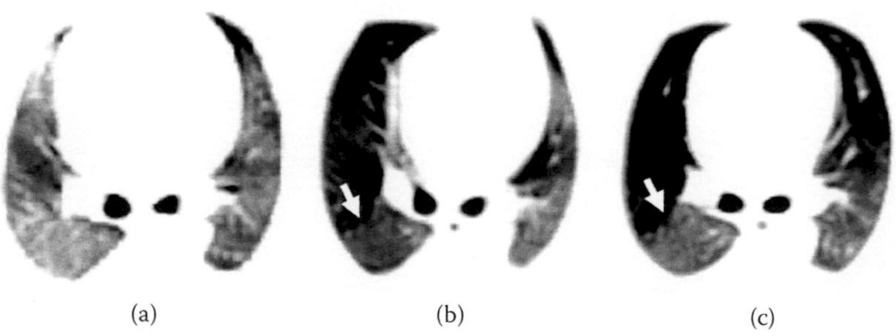

Figure 6.3 Longitudinal CT images of the irradiated rabbits. A hypodense area was observed at the irradiated volume 2–4 months post radiation. (a) Baseline. (b) 2 months. (c) 4 months.

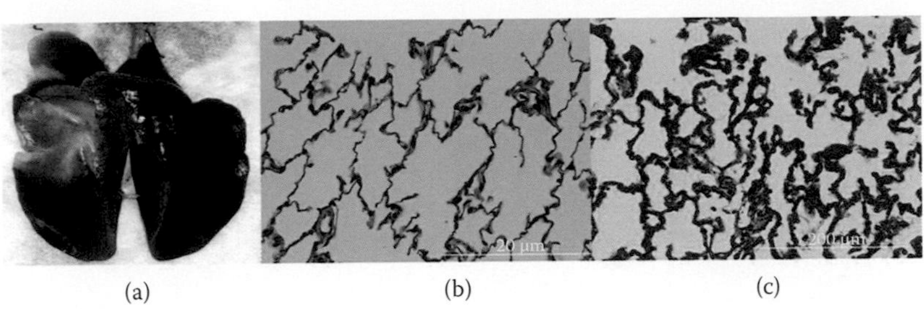

Figure 6.4 Autopsy and histopathology images of the rabbits. (a) A poorly perfused area congruent to the 30 Gy isodose line. (b) Significantly reduced red blood cell and partial destruction of the alveolar structure at the irradiated volume, (c) compared with the contralateral lung that was not irradiated.

Evident gross lung radiosurgery-induced injury was observed as shown in Figure 6.4. The discoloration was caused by significantly reduced red blood cell concentration as confirmed by the microscopic images. The diameter of the discolored area is 2.6 cm, which coincides approximately with the 36-Gy isodose line. There was no sign of inflammation or fibrosis at the time of dissection.

The same platform based on HT was further used to evaluate the efficacy of Amifostine, a radiation protector. Amifostine is a thiol-containing prodrug that metabolizes to its active moiety, WR1065. Amifostine was approved to protect the salivary glands of head-and-neck cancer patients from the toxicity of radiotherapy. It was used off label on lung cancer patients receiving conventionally fractionated radiotherapy (Antonadou et al. 2003; Castiglione, Porcile, and Gridelli 2000; Choi 2003). A significant protective effect was demonstrated, but daily administration of Amifostine can lead to severe side effects (Vardy et al. 2002), and the timing of drug delivery relative to the radiotherapy was not clear (Komaki 2005; Werner-Wasik et al. 2002). A potentially improved application of Amifostine is in SRT treatment when the number of the treatment fraction is significantly reduced to 3–5. Fewer doses of Amifostine can be more tolerable and manageable if the radiation protection effect is observed. For this purpose, an animal study adhering to the human treatment protocol was necessary.

In this study, 13 rabbits were included; four received pilot radiation doses in three equal fractions of 7, 9, 11, or 13 Gy to the right lower lobe of the lung within a week to determine the threshold dose that induced significant lung injury. The dose was determined to be 3 × 11 Gy, which was the dose used for the remaining rabbits. Six received 50 mg/kg of Amifostine before each radiation dose; the remaining were used as controls and received stereotactic body radiation therapy (SBRT) only. Contrast-enhanced magnetic resonance angiography (Ce-MRA), computed tomography, and He-3 MRI were performed at baseline, 4, 8, 12, and 16 weeks post-SBRT. Ce-MRA perfusion patterns for the radioprotector group remained essentially unchanged, and those for control animals showed perfusion deficits in the irradiated area as early as 4 weeks post-SBRT. The difference in perfusion at 16 weeks between the control and radioprotector groups for the irradiated lungs was statistically significant ($p = 0.03$) and that for the non-irradiated lungs was not ($p = 0.32$). CT and histopathology suggest that changes were primarily related to perfusion, but structural changes were also present. No changes in He-3 MRI lung ventilation were observed in either group. Radiation-induced pulmonary perfusion injury was detected at 4 weeks post-SBRT by Ce-MRA and CT and was significantly reduced by Amifostine (Figure 6.5).

Another investigational radiation protector, X-82, a platelet derived growth factor receptor (PDGFR) inhibitor used for radioprotection (Abdollahi et al. 2005), was also tested using the same methodology. Although X-82 was effective in protecting animals against lethal full thoracic irradiation, it is not effective to protect the lung against higher dose to a smaller volume.

These experiments showed that larger rodents can be accurately irradiated by state-of-the-art human treatment machines. Moreover, the ability to focally irradiate a small volume of the animal lends the ability to conduct radiobiological experiments not feasible on a conventional platform and subsequently reveal results previously unavailable. SRT dose to the lower right lung of rabbits up to 60 Gy was well tolerated, but full thoracic irradiation of the same dose would be lethal. The differences in dose-volume effect allow one to study the efficacy of radioprotectors with higher clinical relevancy.

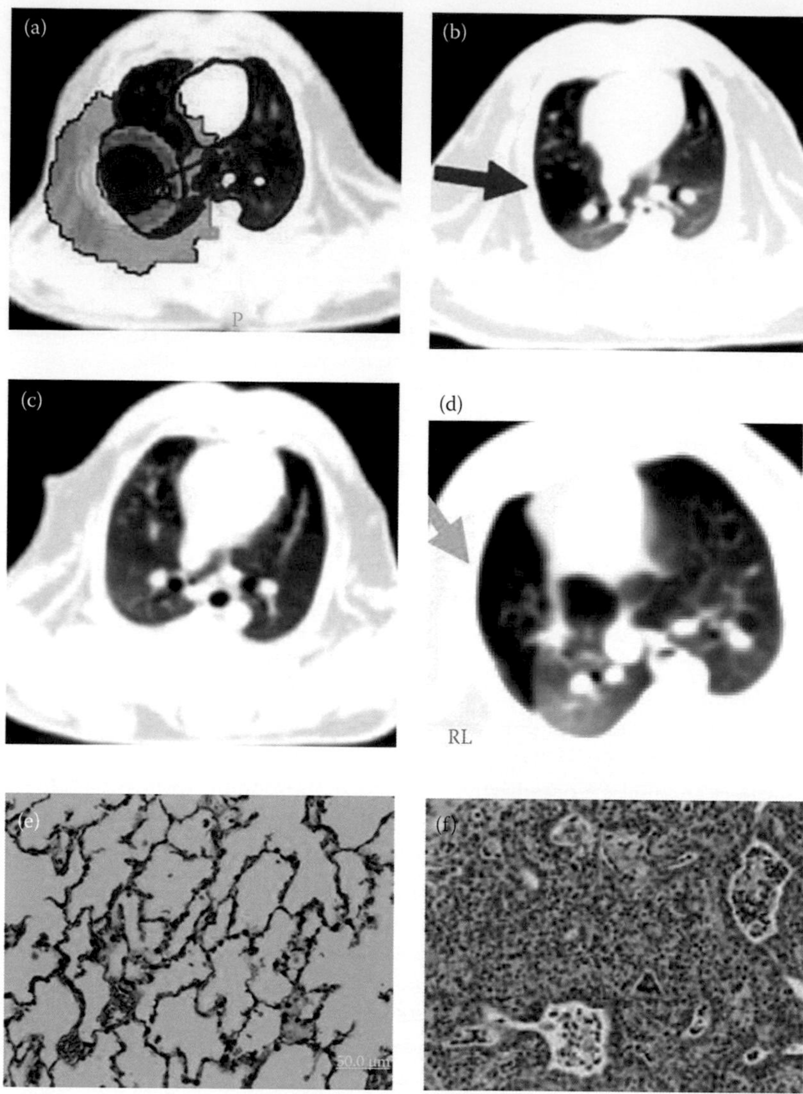

Figure 6.5 (a) Conformal dose plan to deliver 33 Gy in three fractions to the rabbit lung. (b) Hypodense area 2 months after radiation as a consequence of radiation-induced endothelial cell damage in the control group. (c) No radiological sign of injury was observed in the group receiving radiation + Amifostine. (d) Hypodense area appeared in the group receiving radiation + X-82. (e) Histopathology image of the irradiated lung 6 months post treatment by radiation + Amifostine. (f) Histopathology image of the irradiated lung from rabbit receiving radiation + X-82. Severe fibrosis was observed.

Another advantage of this approach is that the dosimetry of human treatment machines are very well characterized and rigorously checked. The difference between actual delivered dose and the planned dose is typically less than 3% (Tailor, Hanson, and Ibbott 2003). The accuracy is significantly superior to that of traditional small animal irradiators that are less frequently and rigorously calibrated. This is in addition to the fact that the accurate treatment-planning system of the human treatment machines is not part of the traditional small animal irradiation process. However, there are drawbacks to performing animal studies on patient treatment machines. Logistics can be complicated as the majority of available treatment time is, by definition, devoted to treating human patients. Additionally, treating animals on clinical machines may be prohibited in certain institutions due to infectious control rules.

6.3 DEDICATED SMALL ANIMAL IRRADIATOR FOR PARTIAL VOLUME IRRADIATION

With the progress of animal radiobiological studies, particularly the recognition of the volume effect of SRT radiation to the lung and spinal cord, animal irradiation protocols were established at the institutional level to deliver partial volume irradiation to larger rodents.

Novakova-Jiresova et al. (2005) fabricated custom collimators using 3-mm lead blocks with shapes based on the CT images to irradiate six different positions of the rat lung (Figure 6.6). The radiation was delivered by an orthovoltage machine operated at 200 kVp. The accuracy of treatment was verified by portal films as in human treatment. Using this experimental setup, the authors were able to associate the sensitivity of varying stages of radiation injury, including the pneumonitic phase (6–12 weeks), intermediate phase (16–28 weeks), and fibrotic phase (34–38 weeks) to the different locations of radiation.

Using the ^{60}Co source and 10-cm-thick lead blocks, Khan et al. (2003) irradiated the upper, lower, or whole lung of rats up to 20 Gy. X-ray portal images were used to verify the radiation field. Sporadic radiation injury was observed in the shielded upper lung when the lower lung was irradiated. Semenenko et al. (2008) further quantified the dose-volume effect by irradiating the rat lung using a 300-kVp x-ray machine to the same mean dose but different total volumes. To achieve the goal, more elaborate dose

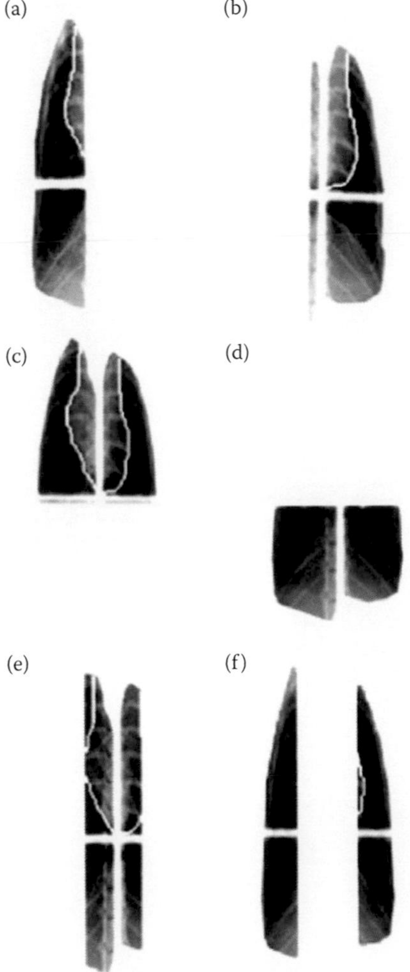

Figure 6.6 Radiation field to cover different 50% of the rat lung. (a) Left, (b) right, (c) apical, (d) basal, (e) mediastinal, and (f) lateral lung regions. (From Novakova-Jiresova, A. et al., *Cancer Res.*, 65, 9, 3568–76, 2005.)

modeling was achieved by measurement of the exit and entrance doses using Gafchromic films. DVHs were calculated by interpolating the exit and entrance doses and superimposing the result on the cone beam CT (CBCT) images of the rat. It was then observed that mortality was clearly limited to rats receiving a dose to the whole lung and was dependent on the mean lung dose. Elevation of the breathing rate increased with the volume of lung receiving 5 Gy or higher. It was also clear that a higher dose to a smaller volume resulted in a lower degree of damage even when the mean lung dose remained the same as the treatment with a lower dose to a larger volume (Figure 6.7).

Animal models were also valuable to assess the dose-volume effect of spinal cord radiation. Most studies showed a marginal volume effect for cord lengths longer than 1 cm but a steep increase in tolerance doses for irradiated lengths of less than 1 cm (van der Kogel 1993).

While it is relatively straightforward to collimate photon radiation fields for spinal treatments due to the position of the target, it is more difficult to account for uncertainty of the delivered dose and the size of the photon penumbra. For this reason, proton beams, which have more predictable dosimetry (particularly in the plateau region before the Bragg peak) and sharper penumbra, were also used to study the dose-volume effect. To study the spinal cord injury from spinal SRT, the plateau region of the depth-dose profile of 150-MeV proton beams generated by a cyclotron was used to irradiate the spinal cord of rats (Bijl et al. 2002). The broad proton beam was collimated down to a narrow strip of a 2-mm-wide field and variable

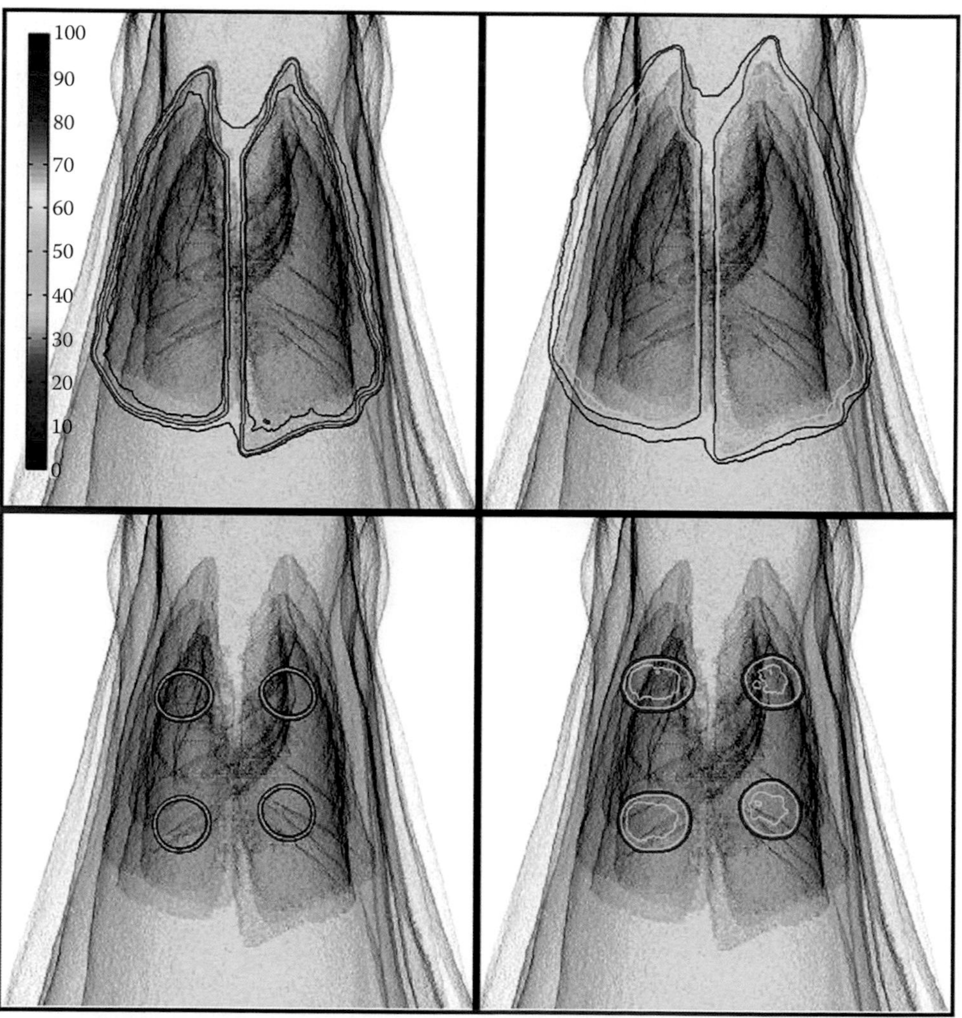

Figure 6.7 To study the dose-volume effect in lung radiotherapy, the whole lung, half the lung, and four small regions of the lung.

lengths. The dose during treatment was monitored by a parallel plate ion chamber mounted in front of the collimator. A sharp increase in the dose required to cause 50% paralysis (ED_{50}) was observed for radiation fields smaller than 4 mm, similar to the previous experiments using photon beams.

The above results show that traditional radiation sources, modified with custom collimation and x-ray imaging guidance, can be used to deliver partial volume radiation. These techniques provided invaluable data for radiobiological study. However, the improvement of existing systems is limited to treating larger animals, such as pigs, dogs, and rabbits. Exclusion of the most widely available, cost effective, and engineerable species, the mouse, can hamper experiments that require a large number of test samples, genetically modified, or immune-deficient animals. There is also a much wider selection of mouse reagents for studies tightly related to radiobiology (e.g., antibodies for the inflammatory cytokines). Furthermore, results from custom systems are hard to reproduce and are inefficient to acquire. Dosimetry, such as the one by Semenenko et al. (2008) is very time consuming, preventing large-scale experiments from being performed. There is also a substantial dosimetric uncertainty in the small field dose. Therefore, completely redesigned small animal irradiators featuring the accuracy of proportionally scaled state-of-the-art human treatment are highly desired and have been investigated. Among them there are two relatively mature systems developed independently. The details of the two systems are described in the following sections.

6.4 SARRP MACHINE

6.4.1 GENERAL INFORMATION

Although devices dedicated for small animal irradiation can be backtracked to more than 3 decades ago (Hranitzky et al. 1973), modern image-guided small animal therapy machines have not emerged until recently, following the development of human IGRT systems. These animal systems are, in many ways, miniaturized human systems, but unique approaches were taken to manage challenges from the difference in the form factor.

The small animal radiation research platform (SARRP) irradiator (Wong et al. 2008) (developed at Johns Hopkins University and marketed by the Gulmay Medical Corporation) integrates fine collimation, accurate dose modeling, and 3-D volumetric imaging guidance into a single platform, thereby allowing highly conformal dose distributions to be achieved in animals. The SARRP system is shown in Figure 6.8. It consists of a 225-kV x-ray tube (Varian, Palo Alto, CA or GE ISOVOLT 225 M2, Lewistown, PA), a charge-coupled device (CCD), an electronic portal imaging device (EPID), a robotic stage, and a flat panel detector. Compared with a c-arm human system, one outstanding difference is in the dynamics between

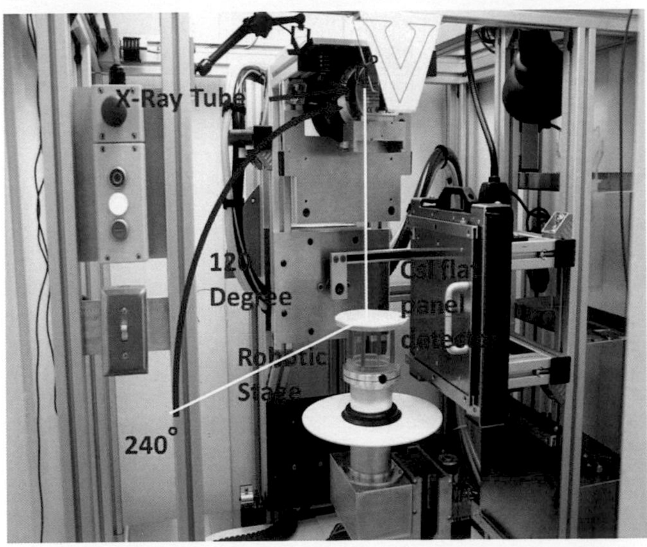

Figure 6.8 A SARRP machine installed at the University of Virginia. Newer SARRP machines allow full posterior-anterior irradiation at 180°.

the gantry and the robotic stage (RTR-6, Danaher Motion, Salem, MA). The stage offers 4° of freedom in x, y, z, and yaw. The accuracy of motion in the XY direction and in the Z direction is 65 μm and 125 μm, respectively. The accuracy of rotation is 0.05° (Matinfar et al. 2009). The gantry rotation is limited between 240° and 0° under International Electrotechnical Commission convention. (The limitation has been recently increased to between 180° and 0°.) In order to acquire a CT image or to deliver arc beams instead of rotating the gantry, the stage rotates a full 360° with the gantry at 270°.

The open field size is 20 × 20 cm at a 35-cm nominal source to surface distance (SSD) with the potential of larger field sizes by increasing the SSD. The field can be collimated down to smaller sizes with a set of collimators as shown in Figure 6.9. The smallest field that can be treated is 0.5 mm. Nominal output for selected field sizes is shown in Table 6.1. The SARRP system uses a 225-kV, dual-focus tube (Varian, Palo Alto, CA). The tube assembly consists of a mono polar x-ray tube with a cooled anode at ground potential and a high-voltage receptacle socket. The tube housing has fittings for water hose connections. The robotic specimen stage is used for imaging the specimen (359°) by rotation to acquire the full CT data set. It is also used to position the specimen in relation to the system's isocenter. Position accuracy is within 0.5 mm. The SARRP system uses a PerkinElmer digital x-ray detector, based on a Gd_2O_2S or CsI scintillator operating a 2-D photodiode array. There are two options for the flat panel detector: 512 × 512 or 1024 × 1024, providing resolution of 400 and 200 μm, respectively. The imaging system operates with the gantry at 270°, using horizontally opposed stationary x-rays while the specimen is rotated via the robotic stage. The EPID is located under the robotic stage and allows the operator to take an anterior/posterior portal image of the specimen prior to irradiation. The image can be captured and remained for setup confirmation if the experiment requires a fractionation sequence of exposures. An example of the CBCT image is shown in Figure 6.10.

Compared with a traditional micro CT using the fan-beam geometry, there are more imaging artifacts and bone saturation. This has several reasons. First, the animal is rotated around the dorsal-ventral axis, resulting in substantial variation in the radiological path length. The flat panel detector has a limited dynamic range so that it is not possible to cover the entire range of the signal intensity, resulting in streak artifacts and bone saturation. Adding lateral compensators to the animal can alleviate the problem by providing a uniform attenuation profile. Second, CBCT geometry is more susceptible to the scatter photons from the animal, resulting in scatter noise that is difficult to remove. Nevertheless, CBCT imaging

Figure 6.9 Collimators of different sizes.

Table 6.1 Collimation filter output*

COLLIMATOR SIZES	FILTER	DOSE RATE
0.5 mm diameter	0.5 mm Cu	90 cGy/min
1.0 mm diameter	0.5 mm Cu	92 cGy/min
5 mm × 5 mm sq	0.5 mm Cu	146 cGy/min
6 cm × 6 cm sq	4 mm Al	378 cGy/min

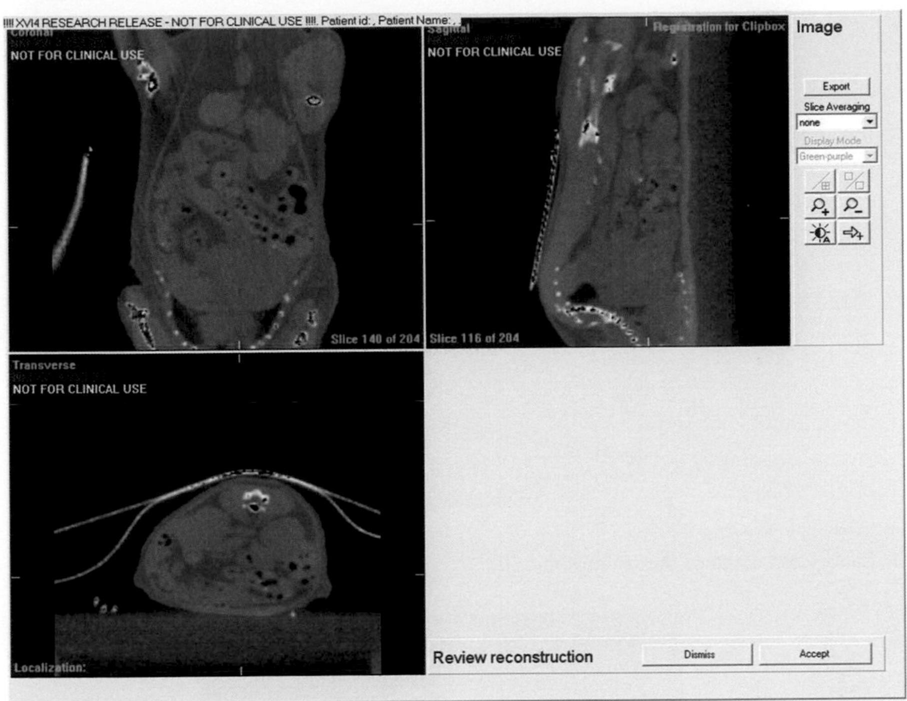

Figure 6.10 CBCT images from SARRP. Imaging parameters: kVp 50, 1.5 mA, and 1 mm Al filtration. The window/leveling were adjusted to show soft tissue contrast.

with good soft tissue contrast was achieved. The imaging dose from a typical CBCT procedure is less than 1 cGy (Wong et al. 2008).

For 2-D portal images and simpler animal positioning, orthogonal EPI images can be acquired using the detector panel and the CCD camera under the robotic stage.

6.4.2 CALIBRATION AND DOSIMETRIC PLANNING

Calibration and commissioning of the SARRP machine was, in general, guided by the AAPM Task Group 61. Yet, because of the CBCT imaging guidance and extremely fine collimation, special procedures were involved as outlined by Tryggestad et al. (2009). Energy-independent Gafchromic EBT2 film (International Specialty Products, Wayne, NJ) was used as the primary dosimeter after calibration by known doses as shown in Figure 6.11. To avoid variation between batches of films, EBT2 films from the same box are recommended for a single commissioning task. EBT2 films are scanned by a flatbed scanner. (Epson Expression 10000 XL was recommended by the authors, but a less expensive model Epson Precision V700 was also acceptable.) Only transmission data from the red channel was evaluated as per standard EBT2 protocol. To measure the percentage of depth dose, a customer calibration jig (Figure 6.12) was used to set up the film reproducibly. Films were cut to 5 cm × 5 cm squares and sandwiched between water equivalent spacers and then fixed by the calibration jig. The transmission of the red channel was converted to radiation doses and plotted in Figure 6.13 for a 5-mm field size.

Once the depth dose curve is established for each cone size, treatment planning based on table lookup can be performed by polynomial fitting of the curve. In order to obtain a 3-D dose distribution and DVH, Monte Carlo simulation can be used, but it can be overly time-consuming for actual animal experiments. A more practical treatment-planning system is therefore needed based on convolution/superposition of the poly-energetic kernels computed using the Monte Carlo simulation. Still, due to the sensitivity of x-ray energies in the kV range to the atomic composition of animal tissue, a great amount of research will be required to obtain a better understanding the dosimetric accuracy (Wong et al. 2008).

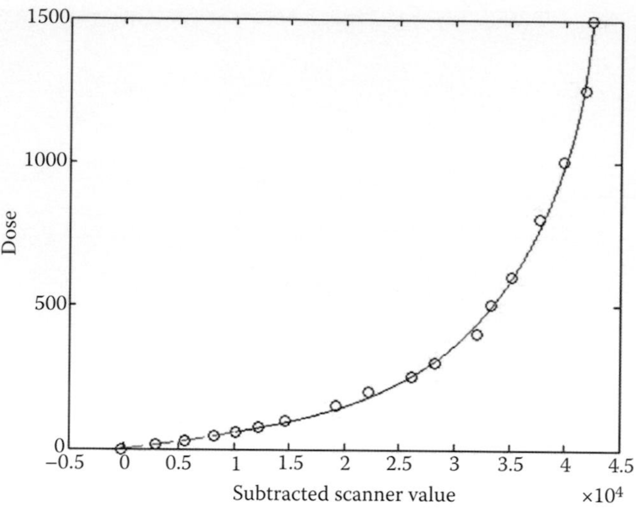

Figure 6.11 Calibration curve of the Gafchromic film.

Figure 6.12 Calibration jig designed for SARRP. 5 cm × 5 cm EBT2 films were placed in between water equivalent phantoms for both PDD and profile measurement.

Compared with the existing uncertainty of dosimetric accuracy, a very high geometrical accuracy of targeting is achievable on SARRP after meticulous calibration of the machine. The process was described in detail by Matinfar et al. (2008, 2009). Although the robotic stage is capable of moving at 0.1 mm accuracy, to achieve targeting accuracy at this level, transformation matrices between the three coordinate systems, the gantry rotation, robotic stage, and CBCT, must be established. A visual servo method was used to automatically calculate the transformation matrix for varying gantry positions. To do so, an x-ray camera is 45° mounted on the robotic stage. X-rays are collimated by a 1-mm collimator and roughly aligned to the center of the camera. When the robotic stage rotates, a pattern of elliptical dots are recorded. By measuring the center of gravity at two different heights, the rotation axis is then determined. The second step is to

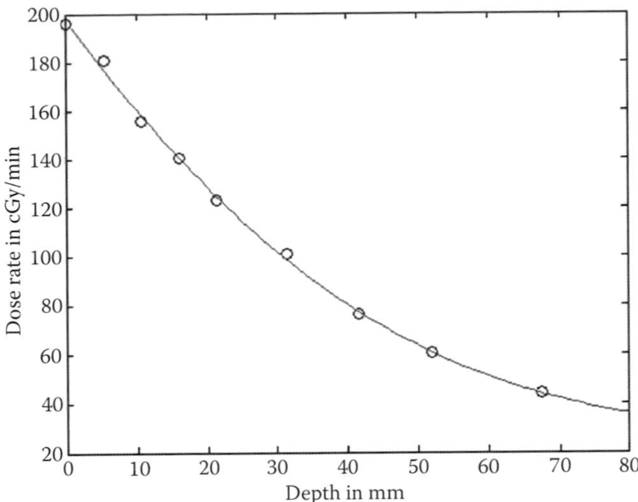

Figure 6.13 Percent depth dose of 5 mm square field.

determine the beam axis. To measure this, the robotic stage is moved to a different location until the new center of gravity is the same as the center of rotation. A transformation matrix can then be derived from the least square optimization of Matinfar et al. (2009):

$$\min \sum_{\varphi} \left\| P(\varphi) + D(\varphi)s(\varphi) - C \right\|^2 \quad \varphi = 0, 15, \ldots$$

where $P(\varphi)$ and $D(\varphi)$ are the point and direction vectors that define the beam axis at the gantry angle φ, C is the unknown "best fit" isocenter, and $s(\varphi)$ is the unknown parameter that defines the point on the line that is closest to C.

In order to align the CBCT isocenter and the beam isocenter, the CBCT of a piece of foam with a ball bearing attached to it is acquired. The distance between the ball bearing and the current MVCT isocenter is calculated. The robotic stage is commanded to move for the distance, so the ball bearing is expected to be at the CBCT isocenter. With the gantry rotated to 270°, portal images of the ball bearing are acquired with the 5 × 5 mm cone from two orthogonal directions from the rotary stage as shown in Figure 6.14. The

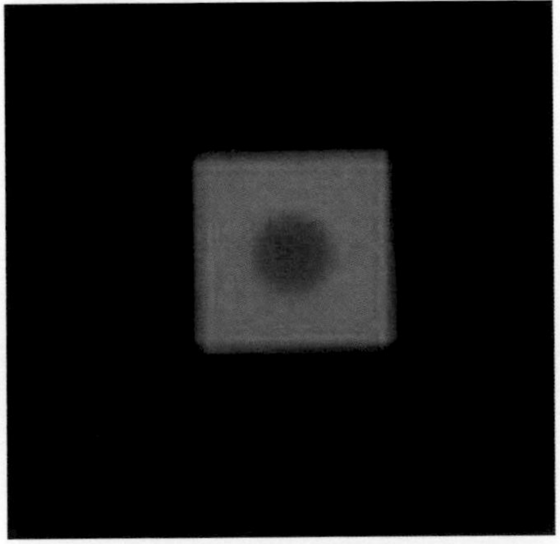

Figure 6.14 A radio opaque ball bearing shows at the center of a 5 mm field when the robotic stage and the gantry coordinates are precisely calibrated.

distances between the center of the ball bearing and the center of the square field are then determined. A translation matrix is then calculated from the offset.

6.4.3 VERIFICATION OF THE GEOMETRICAL ACCURACY

There are a number of ways to examine the accuracy of image-guided focal radiation on SARRP. A very sensitive yet simple method is given as follows: Three pieces of EBT2 film are sandwiched between phantoms slabs with the identical thickness and imaged by CBCT. The center one is punched so that the pinhole is visible in the CBCT. The robotic stage is commanded to reposition, so the pinhole is expected at the beam center. The gantry is set at 315°, and the 0.5-mm collimator is used to deliver an arc treatment.

Figure 6.15 shows the resultant films. The procedure is expected to generate two rings with identical diameter on the upper and lower films. On the film at the center, it is expected to create a dot on the punched film. XY localization error can be calculated from the distance between the pinhole and the dots. Z localization error can be calculated from the difference in diameter between the two rings. Localization accuracy of 200 μm is reported on SARRP (Matinfar et al. 2009).

6.4.4 EXAMPLES OF ANIMAL IRRADIATION

An example to show the biological outcome for a narrow field is demonstrated by Figure 6.16, in which a 1-mm, 10-Gy strip is created in a mouse brain. γH2Ax was used to stain the cells with DNA strand breaks caused by radiation. The result is shown in Figure 6.16b that the irradiated region correlates with the radiation pattern well. A sharp gradient can be observed at the edge of the radiation path that drops off in ~10 cells. The method to deliver focal radiation to the mouse brain using arc beams is similar to that of the aforementioned localization test that a MVCT is performed on the mouse, and the region to be irradiated is selected from the graphic user interface. The distance between the region and the current radiation isocenter is calculated, and the robotic stage is moved according to the distance, so the radiation isocenter overlaps with the focal spot. A convolution/superposition calculation is performed using a customized treatment-planning system (modified Pinnacle software, Philips, Madison, WI) for the treatment time and dose. The robotic stage is then rotated at speed to complete 359° in the calculated time while radiation is turned on. Figure 6.17 shows the arrangement of radiation beams and the dosimetry of arc dose delivery to the mouse brain.

6.5 PXI SYSTEM

Similar to the SARRP system, the X-Rad225Cx was a joint venture between academia (Princess Margaret Hospital) and industry (Precision X-Ray, Inc. [PXI]) and is currently a commercial product. The overall geometry of the system resembles the SARRP system in part but overall is closer to a conventional human treatment machine. It consists of the same x-ray tube (Varian/Comet 225 kVp) and flat panel (different generations of the system use either the 512 × 512 PerkinElmer flat panel detector with a Gd_2O_2S scintillator or the 1024 × 1024 PerkinElmer flat panel detector with a CsI scintillator) as the SARRP.

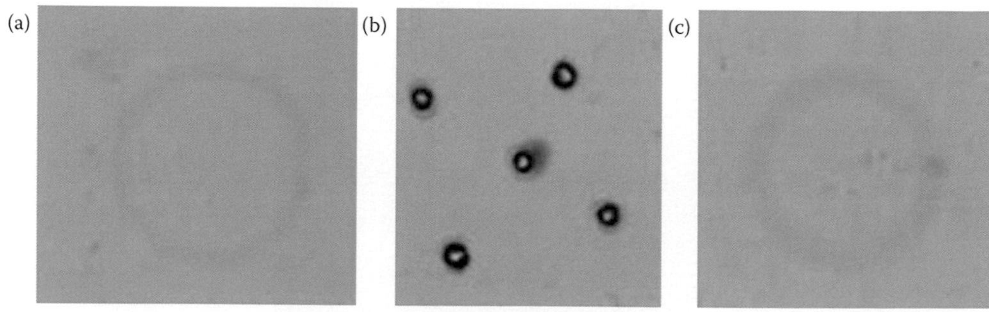

Figure 6.15 EBT2 film above, at, and below the isocenter. Two rings with identical diameters are expected on the films (a) above and (c) below the isocenter. The radius of the ring is expected to be the same as the thickness of the spacer. (b) There are five pinholes in the central film to assist visualization in the CBCT image.

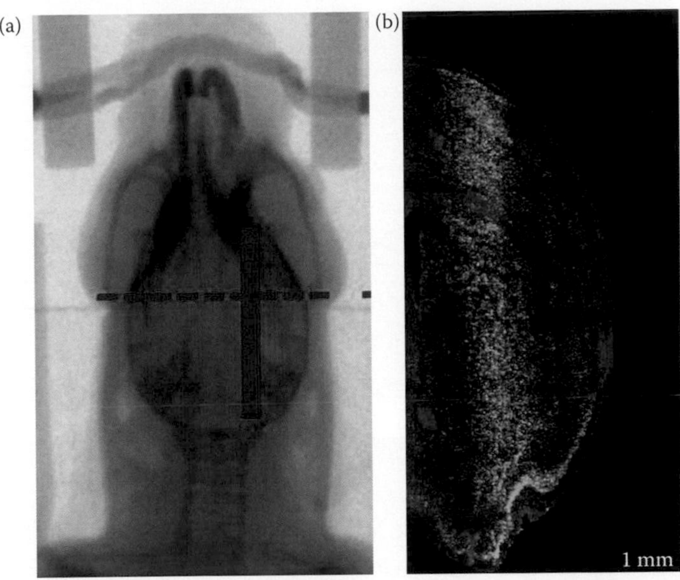

Figure 6.16 (a) A narrow strip (1 mm width) of brain tissue received 10 Gy. (b) DNA strand break is shown by γH2Ax staining.

The primary difference in terms of basic geometry is that the x-ray tube and flat panel are mounted orthogonally to each other on a c-arm gantry and rotate for both imaging and treatment. In this case, images are acquired under a "conventional" clinical radiation therapy treatment geometry in which the object is stationary and the source and detector rotate around the object. The animal stage is a 3° of freedom (translation-only) stage with encoded Parker-Daedal motors with an accuracy of 82 μm, allowing for only co-planar placement of radiation beams. Figure 6.18 shows photographs of the first PXI system installed at Princess Margaret Hospital.

System calibration, image acquisition, image reconstruction, treatment, and database management are provided with accompanying software (called PILOT). Radiation treatment fields are shaped with manually placed interchangeable collimators (shown in Figure 6.1c), which range in size from 0.1 cm diameter to 4 × 4 cm square. The uncollimated field is 10 × 10 cm at the isocenter. The source-to-axis distance of 30.7 cm and source-to-detector distance of 64.5 cm provide a factor of 2× magnification in imaging. The source-to-collimator distance is 23 cm, leaving 7 cm clearance between the end of the collimator and the isocenter. The system is capable of delivering radiation from 5–225 kVp with a maximum power output of 3 kW for the large focal spot and 1.5 kW for the small focal spot. Currently, two calibrated beam qualities are being used for irradiation: 100 kVp with 2 mm Al added filtration, yielding a HVL of 2.8 mm Al, and 225 kVp with 0.3 mm Cu added filtration, yielding a HVL of 0.93 mm Cu.

Dose rates have been quantified at isocenter at a 0.5-cm depth (a typical treatment depth for an isocentric treatment prescribed to mid-plane in a murine model). Dose rates have been calibrated following the TG-61 protocol. Relative output factors and depth dose curves have been measured in water and solid water with both small volume ion chambers and EBT2 Gafchromic film. For 225 kVp, 13 mA, dose rates range from 2.95 Gy/min (dose to water) for a 0.5-cm field to 4.1 Gy/min for an open field. Figure 6.19 shows a table of output for different collimator sizes as well as profiles (acquired with EBT2 film) for a 1 × 1 cm square collimator. There is a 7 cm separation between the collimator end and the surface of the solid water, given a penumbra (80%–20%) of 0.5–0.9 mm (depending on field size and depth).

Flex in the c-arm as a function of rotation is present during both image acquisition and radiation delivery. For image acquisition, the flex of the system is compensated for during the process of image reconstruction. A flexmap is acquired by imaging a ball bearing as a function of gantry rotation, and

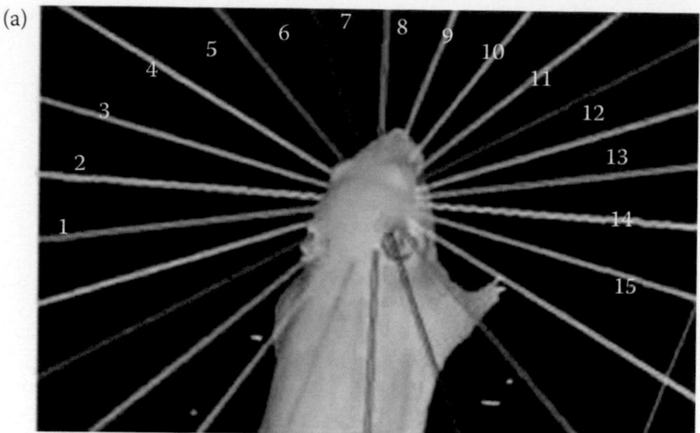

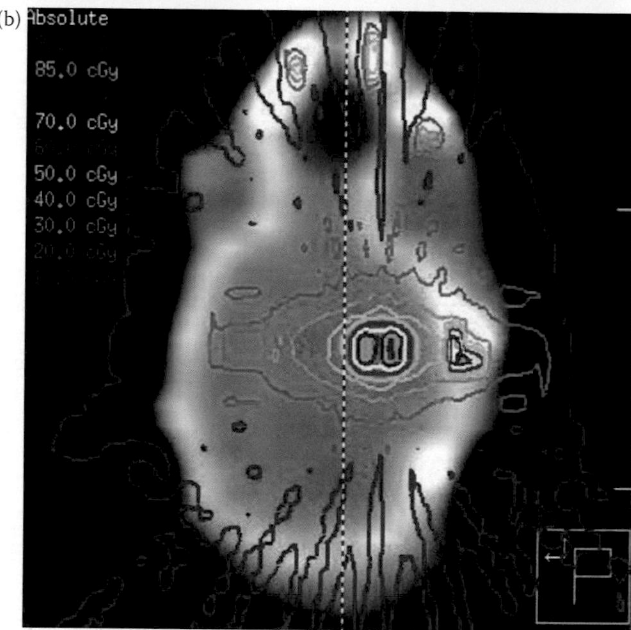

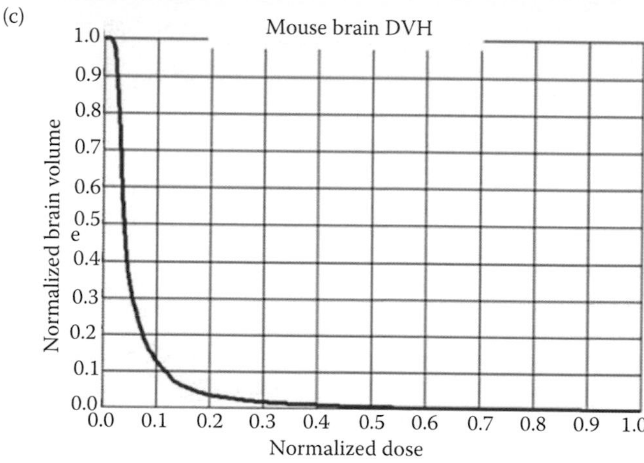

Figure 6.17 (a) Paths of arc beams by rotating the robotic stage; (b) isodose distribution; (c) DVH of the brain.

(a)

(b)

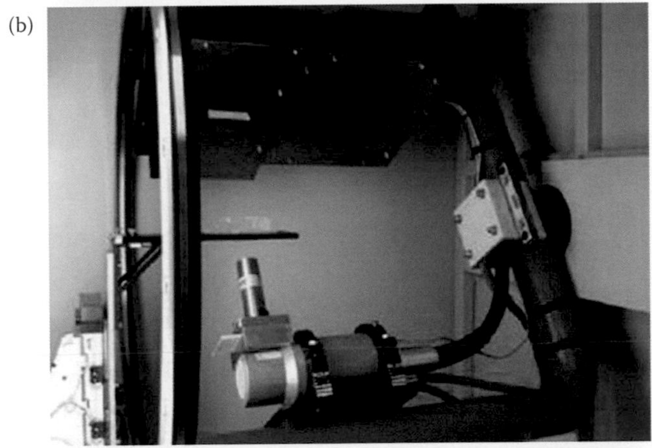

(c)

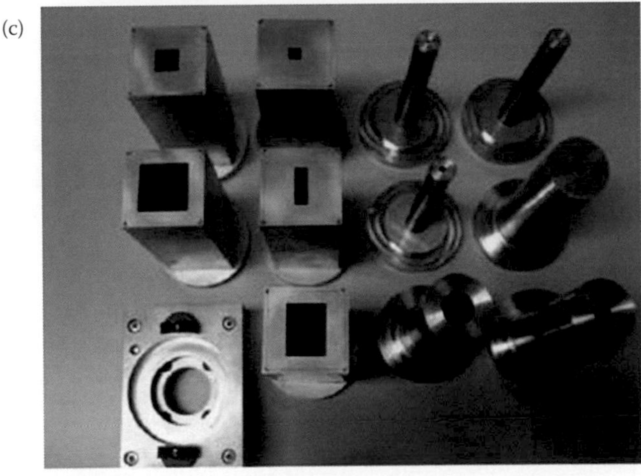

Figure 6.18 PXI system showing the (a) cabinet, (b) interior, and (c) interchangeable collimators.

this information is used to offset the projection images. Flex in the system during treatment delivery is compensated for by applying corrections to the stage position as a function of gantry angle. The magnitude of this flex is characterized by acquisition of a Winston-Lutz test as a function of gantry angle, in which the location of a ball bearing relative to the center of a collimator is determined. This offset is then uploaded to the hardware controller and the stage position track with the gantry angle. The flex in the system is

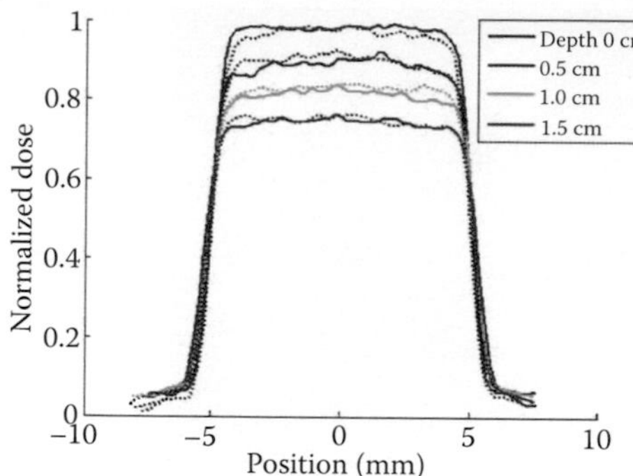

Equivalent diameter (cm)	Dose rate (Gy/min)
0.5	2.95
1.12	3.04
1.5	3.12
2.5	3.27
11.16 (uncollimated)	4.10

Figure 6.19 Profiles acquired with EBT2 film for a 1 × 1 cm field as well as dose rates for various field sizes.

<1 mm at isocenter with the magnitude being the largest in the direction along the axis of rotation. After compensation for system flex, a targeting accuracy of 0.2 mm can be achieved (Clarkson et al. 2011). The first PXI system at the Princess Margaret Hospital has been in use for more than 2 years. In that time, more than 3000 fractions of animal irradiation have been performed, and the system showed robust stability in the targeting accuracy. The effect of the dynamic stage tracking can be seen in Figure 6.20 in the star-shot images acquired without and with stage tracking.

Images are typically acquired at beam qualities from 40–100 kVp and 3–30 mAs. Typical image acquisition times are 40–60 s although both shorter and longer acquisition times are possible. Figure 6.21 shows images of a male C3H/HeJ mouse, imaged under inhaled (isofluorane) anesthetic. The image acquisition time for each image was 60 s—the images on the left are approximately 1 cGy dose, and the images on the right are approximately 4 cGy. Due to the conventional geometry of imaging acquisition, the PXI system is able to avoid the dramatic change of path length and the resulting streak artifacts and bone saturation present in the SARRP system.

Extensive assessment of image quality and image-guidance capabilities of the system have been performed (Clarkson et al. 2011). It has been shown that the imaging noise of the PXI system is 30 HU in a 2.5-cm-diameter solid water phantom with a 1 cGy imaging dose.

6.6 OTHER INVESTIGATIONAL SMALL ANIMAL IRRADIATORS

Besides the two relatively mature small animal systems, there are several other platforms under investigation. They may not be as mature compared with the first two systems, but they are unique in one or more engineering or physical aspects that can be inspiring for the future development of the small animal irradiator.

A ring gantry–mounted small animal CT/RT system was developed at Stanford University (Motomura et al. 2010; Rodriguez et al. 2009; Zhou et al. 2010) based on a GE eXplore RS 120 micro CT scanner. The original imaging x-ray tube was replaced by an x-ray tube with higher capacity for dual CT/RT purposes, but the highest energy remained at 120 kVp. A unique feature of the system is the variable aperture from 0.1 to 6 cm that allows greater flexibility to modulate the beams for varying target sizes. The collimator consists of two-stage hexagonal irises (Figure 6.22). The two irises are offset by 30° to create dodecagonally shaped beams. Commissioning of the system is similar to that of the SARRP system in that an ion chamber was used to determine the absolute air dose first, followed by solid water and EBT film measurement to determine the depth dose and beam profile. Maximum dose rate for the 6-cm field is more than 200 cGy/min on the system, but overheating of the tube can happen for experiments requiring 8 Gy or higher.

The small animal irradiator developed in the Washington University is drastically different from most x-ray tube–based systems (Kiehl et al. 2008; Stojadinovic et al. 2007) in that it is a teletherapy system. It utilizes a high dose rate Ir-192 brachytherapy source with a half life of 74 days and average gamma-ray

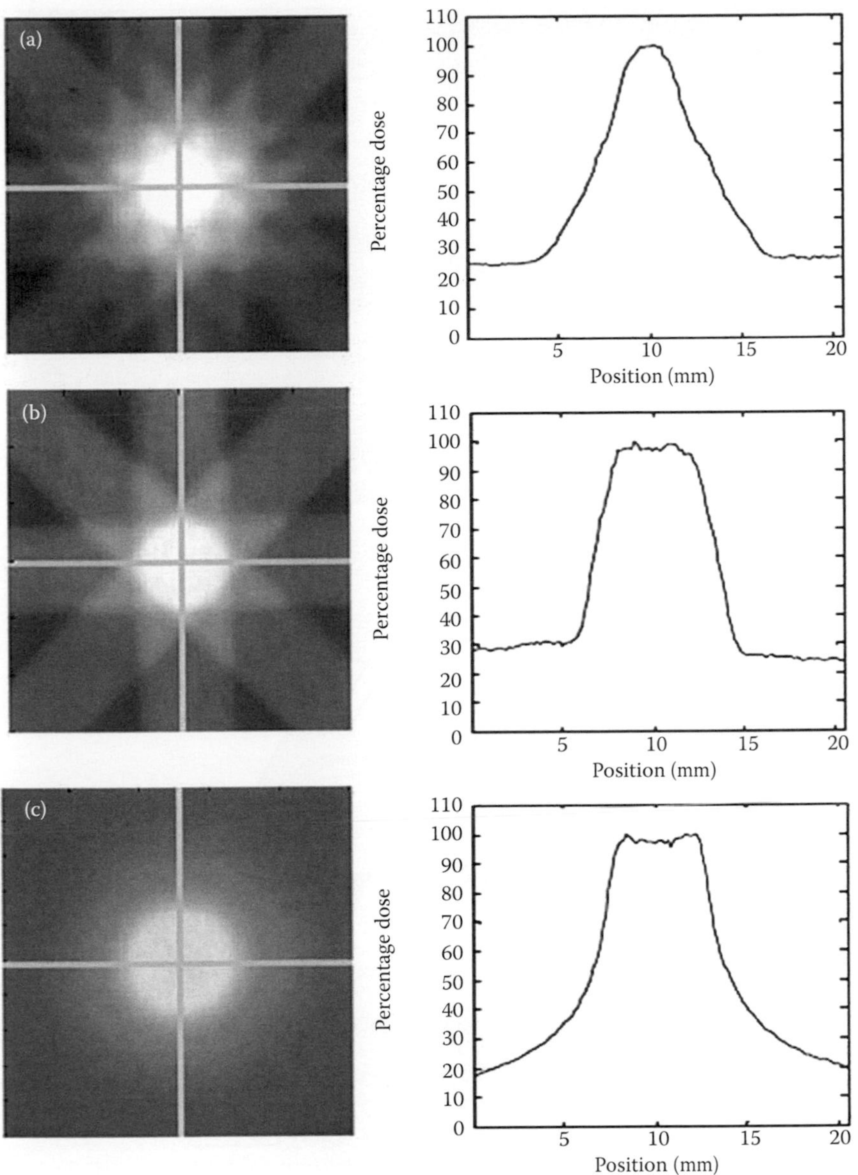

Figure 6.20 (a) Star-shot without stage tracking. (b) Star-shot with stage tracking and 8 equally spaced beams. (c) Conformal arc shot with dynamic stage tracking showing excellent iso-centricity.

energy of 309 kV (Nucletron Selectron) as the source. As shown in Figure 6.23, the animal can be placed on a motor stage that is shielded by an aluminum alloy collimator assembly. Four catheters are placed outside of the collimator assembly for the Ir-193 source to travel and dwell. An articulated arm (MicroScribe3DX, Immersion Corp, San Jose, CA) that can index the 3-D position is used to calibrate the alignment between the motor and the collimation system. There are four openings equally spaced on the aluminum collimator assembly that can fit customer tungsten collimators. Positioning of the animal is based on the external fiducial markers. A treatment-planning system was developed using MATLAB® based on CT images acquired separately. Using the planning software, the dwell time of the I-192 seed at each of the four openings can be determined.

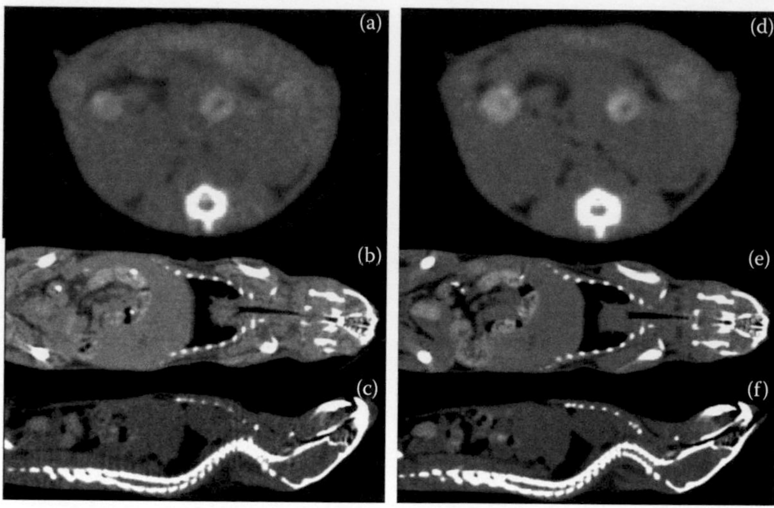

Figure 6.21 CBCT images of a sedated mouse with (a–c) 2cGy and (d–f) 4 cGy imaging doses.

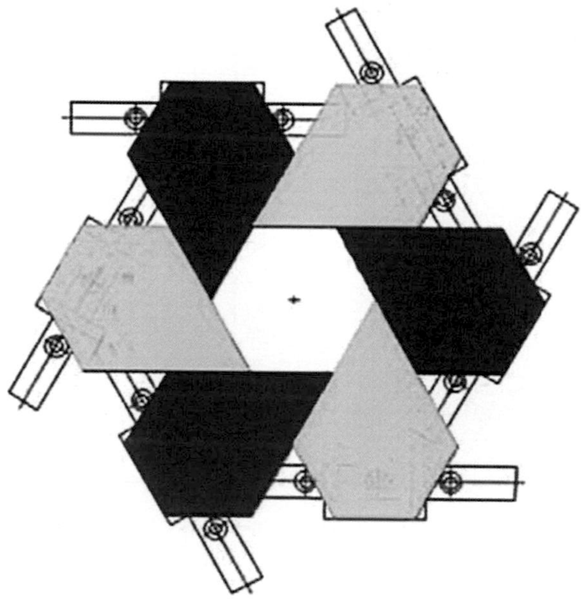

Figure 6.22 An innovative iris for continuously variable aperture from 1 to 60 mm.

A prototype small animal irradiator was developed at the University of Texas, Southwest (Cho et al. 2010). The system utilizes a higher energy x-ray tube (X-RAD 320 x-ray source, Precision x-ray, North Branford, CT) operating at 320 kVp. Other major components of the system include an x-ray fluorescent screen, a collimator, and a 2-D manual positioning stage. The collimator is made of 5-cm-thick Cerrobend with interchangeable disks for variable collimator sizes. Due to the higher tube voltage, a higher output is achieved for up to 980 cGy/min using 5 mm collimation. Positioning and localization of the animal is guided by the x-ray fluorescent images and radio opaque ball bearings (Figure 6.24).

A highly innovative small animal imaging and irradiation platform was developed using carbon nanotube (CNT) field emission technology (Wang et al. 2007). A prototype micro-RT CNT field emission

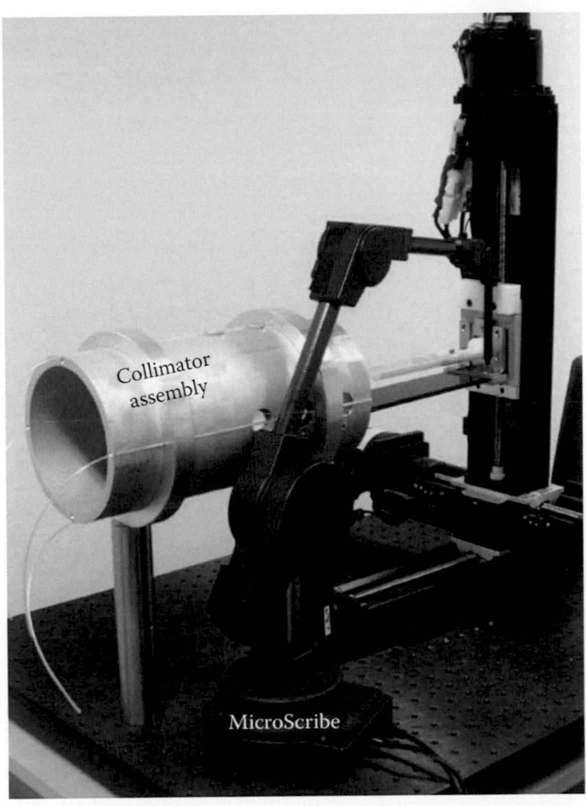

Figure 6.23 Design of a small animal irradiator using 192Ir HDR source and remote after loader.

Figure 6.24 A prototype small animal irradiator using higher energy x-ray tube and 2-D fluorescent images for animal positioning.

cathode array chip with 25 individually addressable cathodes was fabricated. The novel radiation source is potentially advantageous in high temporal resolution (~1 ms) and creation of radiation fields with intensity modulation without multileaf collimators. A maximum dose rate of 100 cGy/min was estimated from Monte Carlo simulation. Improvements in the device lifespan, output, and cost are still needed before the technology can be used for production (Figure 6.25).

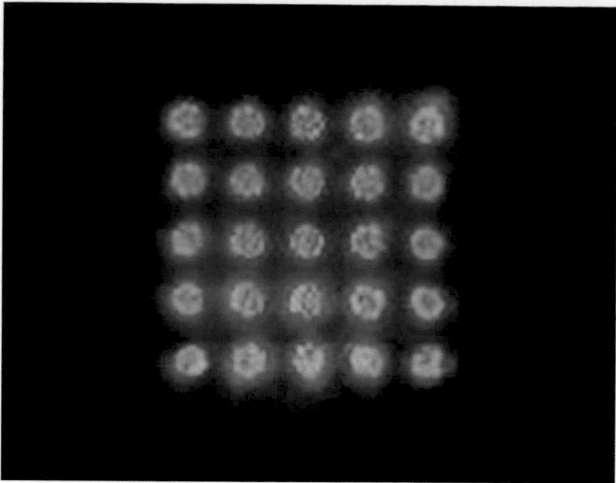

Figure 6.25 5 × 5 cm CNT x-ray emitter for potential intensity modulated small animal irradiation.

6.7 CONCLUSIONS

Small animal irradiation was historically performed on equipment not designed for the task. Rigorous radiation dosimetry was not performed, resulting in crude estimates of the actual dose delivered to the animals despite the fact that a precise (but uncertain) value of the radiation dose was reported in most radiobiological studies. In the meantime, the accuracy of human treatment was improved tremendously in both dosimetry and localization. This disparity was clearly realized by many investigators, and the relevance of the animal experiments to human radiobiology was questioned. To bridge the gap, significant efforts have been made as summarized in the chapter. Larger mammals were treated on human treatment machines for conformal or image-guided radiotherapy. Custom modifications of existing small animal irradiators were implemented for partial volume irradiation of larger rodents. However, the most remarkable progress has been the technology translation from the lab to commercialization of dedicated image-guided small animal irradiators. With these recent inventions in the small animal irradiator and commercialization, image-guided small animal irradiators have begun to play a more prominent role in radiobiological research. This development is likely to have an important impact in understanding SBRT radiobiology. The following discoveries are likely to be made in the near future on these machines:

1. Quantification of the normal tissue injury versus radiation dose and volume for hypofractionated (2 Gy to 8 Gy per fraction) and extreme hypofractionated (>8 Gy/fraction) radiotherapy.
2. Optimization of treatment strategies, including temporally and spatially modulated radiotherapy that may be otherwise difficult or unethical to perform on patients.
3. Development of novel radiosensitizers and radioprotectors that are effective in the SBRT treatment.
4. Understanding the immunological response of animals to the SBRT treatment and the mechanisms to utilize them for higher tumor control probability and lower normal tissue toxicity.
5. Isolation of the intrinsic tumor response to radiation and the environmental factors that arise from collateral radiation dose.

In the meantime, small animal irradiators in their current form are not perfect. Many technological and engineering challenges are to be overcome for them to match the proportional accuracy of human-scaled treatment machines.

A common challenge to these irradiators is the dose calculation and treatment-planning platform. Dose calculation for kV x-rays (gamma rays) are uniquely challenging compared with the MV beams used for human treatment because of the higher percentage of photoelectric interactions that not only depend on the density, but also on the atomic composition of the tissue. Even with such information available, dose calculation is not trivial. Monte Carlo can correctly take the heterogeneity in tissue density and atomic composition into consideration (Chow et al. 2010). However, its application is limited by the lack of a commercial software program. To properly model the x-ray source and collimators for a customer treatment-planning program, substantial expertise in Monte Carlo simulation is needed. The expertise may not be available for a typical radiobiology lab. The time to calculate dose for each individual animal can also be prohibitively long. Convolution/superposition based on Monte Carlo calculated dose kernels is a more practical way to calculate the dose. Accuracy of this method under kilovoltage energy and small field conditions has not been fully tested although some improvement in the dose calculation can be expected from using finer energy steps in the dose kernels (Verhaegen, Granton, and Tryggestad 2011). For these reasons, a realistic goal in the uncertainty of dose calculation for small animal studies should be set lower than the 3%–5% for human patients. A dedicated guideline or task group would help to minimize the discrepancies between different labs and manufacturers and eventually lead to more reliable results from animal experiments.

Compared with the human system, another important component missing from the majority of small animal irradiators, with the exception of the CNT cathodes, is the ability to modulate dose within a field and easily create complex field shapes. A multileaf collimator (MLC) for intensity modulation and field shaping are generally not available on these systems. Therefore, only targets with simple shapes can be treated conformally by available small animal irradiators. Although it is technologically feasible to add an MLC for more flexible treatment planning and dose shaping, it would significantly increase the complexity of the machine and subsequently the cost for both fabrication and maintenance. Given the limitation in the dose accuracy, the benefit from adding an MLC is not clear. The final challenge arising from these novel small animal irradiators is treatment throughput. A direct side effect from an increase in conformality and accuracy in treatment delivery is a much longer time to prepare and treat an animal. As a typical experiment involves hundreds of small animals, a significant increase in manpower is needed for the additional activities involved in the experiment. Improvement in the automation of the process, such as devices that can perform batch treatment, is highly desired.

ACKNOWLEDGMENT

The authors would like to thank Dr. David Schlesinger for proofreading and editing.

REFERENCES

Abdollahi, A., M. Li, G. Ping et al. 2005. Inhibition of platelet-derived growth factor signaling attenuates pulmonary fibrosis. *J. Exp. Med.* 201(6): 925–35.

Adler, J. R., Jr., M. J. Murphy, S. D. Chang et al. 1999. Image-guided robotic radiosurgery. *Neurosurgery* 44(6): 1299–306; discussion 1306–7.

Antonadou, D., A. Petridis, M. Synodinou et al. 2003. Amifostine reduces radiochemotherapy-induced toxicities in patients with locally advanced non-small cell lung cancer. *Semin. Oncol.* 30(6 Suppl 18): 2–9.

Armour, M., E. Ford, I. Iordachita et al. 2010. CT guidance is needed to achieve reproducible positioning of the mouse head for repeat precision cranial irradiation. *Radiat. Res.* 173(1): 119–23.

Barcellos-Hoff, M. H., C. Park, and E. G. Wright. 2005. Radiation and the microenvironment – Tumorigenesis and therapy. *Nat. Rev. Cancer* 5(11): 867–75.

Beavis, A. W. 2004. Is tomotherapy the future of IMRT? *Br. J. Radiol* 77(916): 285–95.

Bijl, H. P., P. van Luijk, R. P. Coppes et al. 2002. Dose-volume effects in the rat cervical spinal cord after proton irradiation. *Int. J. Radiat. Oncol. Biol. Phys.* 52(1): 205–11.

Cai, J., J. F. Mata, M. D. Orton et al. 2009. A rabbit irradiation platform for outcome assessment of lung stereotactic radiosurgery. *Int. J. Radiat. Oncol. Biol. Phys.* 73(5): 1588–95.

Castiglione, F., G. Porcile, and O. C. Gridelli. 2000. The potential role of amifostine in the treatment of non small cell lung cancer. *Lung Cancer* 29(1): 57–66.

Ch'ang, H. J., J. G. Maj, F. Paris et al. 2005. ATM regulates target switching to escalating doses of radiation in the intestines. *Nat. Med.* 11(5): 484–90.

Cho, J., R. Kodym, S. Seliounine, J. A. Richardson, T. D. Solberg, and M. D. Story. 2010. High dose-per-fraction irradiation of limited lung volumes using an image-guided, highly focused irradiator: Simulating stereotactic body radiotherapy regimens in a small-animal model. *Int. J. Radiat. Oncol. Biol. Phys.* 77(3): 895–902.

Choi, N. C. 2003. Radioprotective effect of amifostine in radiation pneumonitis. *Semin. Oncol.* 30(6 Suppl 18): 10–7.

Chow, J. C. L., M. K. K. Leung, P. E. Lindsay, and D. A. Jaffray. 2010. Dosimetric variation due to the photon beam energy in the small-animal irradiation: A Monte Carlo study. *Med. Phys.* 37(10): 5322–9.

Clarkson, R., P. E. Lindsay, S. Ansell et al. 2011. Characterization of image quality and image-guidance performance of a preclinical microirradiator. *Med. Phys.* 38(2): 845–56.

DesRosiers, C., M. S. Mendonca, C. Tyree et al. 2003. Use of the Leksell Gamma Knife for localized small field lens irradiation in rodents. *Technol. Cancer Res. Treat.* 2(5): 449–54.

Garcia-Barros, M., F. Paris, C. Cordon-Cardo et al. 2003. Tumor response to radiotherapy regulated by endothelial cell apoptosis. *Science* 300(5622): 1155–9.

Guerrero, M., and X. A. Li. 2004. Extending the linear-quadratic model for large fraction doses pertinent to stereotactic radiotherapy. *Phys. Med. Biol.* 49(20): 4825–35.

Hopewell, J. W., M. Rezvani, and H. F. Moustafa. 2000. The pig as a model for the study of radiation effects on the lung. *Int. J. Radiat. Biol.* 76(4): 447–52.

Hranitzky, E. B., P. R. Almond, H. D. Suit, and E. B. Moore. 1973. A cesium-137 irradiator for small laboratory animals. *Radiology* 107(3): 641–4.

Khan, M. A., J. Van Dyk, I. W. Yeung, and R. P. Hill. 2003. Partial volume rat lung irradiation: Assessment of early DNA damage in different lung regions and effect of radical scavengers. *Radiother. Oncol.* 66(1): 95–102.

Kiehl, E. L., S. Stojadinovic, K. T. Malinowski et al. 2008. Feasibility of small animal cranial irradiation with the microRT system. *Med. Phys.* 35(10): 4735–43.

Komaki, R. 2005. What should the optimal timing be for Amifostine administration relative to radiation and chemotherapy? *J. Clin. Oncol.* 23(28): 7232–3; author reply 7233–5.

Kutcher, G. J., L. Coia et al. 1994. Comprehensive QA for radiation oncology: Report of AAPM Radiation Therapy Committee Task Group 40. *Med. Phys.* 21(4): 581–618.

Lawrence, J. A., and L. J. Forrest. 2007. Intensity-modulated radiation therapy and helical tomotherapy: Its origin, benefits, and potential applications in veterinary medicine. *Vet. Clin. North Am. Small Anim. Pract.* 37(6): 1151–65; vii–iii.

Liu, L., D. A. Bassano, C. Prasad, S. Hahn, and C. Chung. 2003. The linear-quadratic model and fractionated stereotactic radiotherapy. *Int. J. Radiat. Oncol. Biol. Phys.* 57(3): 827–32.

Mackie, T. R. 2006. History of tomotherapy. *Phys. Med. Biol.* 51(13): R427–53.

Matinfar, M., E. Ford, I. Iordachita, J. Wong, and P. Kazanzides. 2009. Image-guided small animal radiation research platform: Calibration of treatment beam alignment. *Phys. Med. Biol.* 54(4): 891–905.

Matinfar, M., I. Iordachita, E. Ford, J. Wong, and P. Kazanzides. 2008. Precision radiotherapy for small animal research. *Med. Image Comput. Comput.-Assist. Interv.* 11(Pt. 2): 619–26.

Medin, P. M., R. D. Foster, A. J. van der Kogel et al. 2011. Spinal cord tolerance to single-fraction partial-volume irradiation: A swine model. *Int. J. Radiat. Oncol. Biol. Phys.* 79(1): 226–32.

Mori, Y., D. Kondziolka, J. Balzer et al. 2000. Effects of stereotactic radiosurgery on an animal model of hippocampal epilepsy. *Neurosurgery* 46(1): 157–65; discussion 165–8.

Motomura, A. R., M. Bazalova, H. Zhou, P. J. Keall, and E. E. Graves. 2010. Investigation of the effects of treatment planning variables in small animal radiotherapy dose distributions. *Med. Phys.* 37(2): 590–9.

Novakova-Jiresova, A., P. van Luijk, H. van Goor, H. H. Kampinga, and R. P. Coppes. 2005. Pulmonary radiation injury: Identification of risk factors associated with regional hypersensitivity. *Cancer Res.* 65(9): 3568–76.

Park, C., L. Papiez, S. Zhang et al. 2008. Universal survival curve and single fraction equivalent dose: Useful tools in understanding potency of ablative radiotherapy. *Int. J. Radiat. Oncol. Biol. Phys.* 70(3): 847–52.

Poulson, J. M., Z. Vujaskovic, S. M. Gillette et al. 2000. Volume and dose-response effects for severe symptomatic pneumonitis after fractionated irradiation of canine lung. *Int. J. Radiat. Biol.* 76(4): 463–8.

Powers, B. E., H. D. Thames, S. M. Gillette, C. Smith, E. R. Beck, and E. L. Gillette. 1998. Volume effects in the irradiated canine spinal cord: Do they exist when the probability of injury is low? *Radiother. Oncol.* 46(3): 297–306.

Psarros, T. G., B. Mickey, K. Gall et al. 2004. Image-guided robotic radiosurgery in a rat glioma model. *Minim. Invasive Neurosurg.* 47(5): 266–72.

Quinn, A. M. 2002. CyberKnife: A robotic radiosurgery system. *Clin. J. Oncol. Nurs.* 6(3): 149–56.

Rodriguez, M., H. Zhou, P. Keall, and E. Graves. 2009. Commissioning of a novel microCT/RT system for small animal conformal radiotherapy. *Phys. Med. Biol.* 54(12): 3727–40.

Semenenko, V. A., R. C. Molthen, C. Li et al. 2008. Irradiation of varying volumes of rat lung to same mean lung dose: A little to a lot or a lot to a little? *Int. J. Radiat. Oncol. Biol. Phys.* 71(3): 838–47.

Stojadinovic, S., D. A. Low, A. J. Hope et al. 2007. MicroRT-small animal conformal irradiator. *Med. Phys.* 34(12): 4706–16.

Tailor, R. C., W. F. Hanson, and G. S. Ibbott. 2003. TG-51: Experience from 150 institutions, common errors, and helpful hints. *J. Appl. Clin. Med. Phys.* 4(2): 102–11.

Timmerman, R., L. Papiez, R. McGarry et al. 2003. Extracranial stereotactic radioablation: Results of a phase I study in medically inoperable stage I non-small cell lung cancer. *Chest* 124(5): 1946–55.

Timmerman, R. D., C. Park, and B. D. Kavanagh. 2007. The North American experience with stereotactic body radiation therapy in non-small cell lung cancer. *J. Thorac. Oncol.* 2(7 Suppl 3): S101–12.

Tryggestad, E., M. Armour, I. Iordachita, F. Verhaegen, and J. W. Wong. 2009. A comprehensive system for dosimetric commissioning and Monte Carlo validation for the small animal radiation research platform. *Phys. Med. Biol.* 54(17): 5341–57.

Vardy, J., E. Wong, M. Izard et al. 2002. Life-threatening anaphylactoid reaction to Amifostine used with concurrent chemoradiotherapy for nasopharyngeal cancer in a patient with dermatomyositis: A case report with literature review. *Anticancer Drugs* 13(3): 327–30.

van der Kogel, A. J. 1993. Dose-volume effects in the spinal cord. *Radiother. Oncol.* 29(2): 105–9.

Verhaegen, F., P. Granton, and E. Tryggestad. 2011. Small animal radiotherapy research platforms. *Phys. Med. Biol.* 56(12): R55–83.

Vigneulle, R. M., J. Herrera, T. Gage et al. 1990. Nonuniform irradiation of the canine intestine. I. Effects. *Radiat. Res.* 121(1): 46–53.

Wang, S., Z. Liu, S. Sultana, E. Schreiber, O. Zhou, and S. Chang. 2007. A novel high resolution micro-radiotherapy system for small animal irradiation for cancer research. *Biofactors* 30(4): 265–70.

Werner-Wasik, M., R. S. Axelrod, D. P. Friedland et al. 2002. Phase II: Trial of twice weekly Amifostine in patients with non-small cell lung cancer treated with chemoradiotherapy. *Semin. Radiat. Oncol.* 12(1 Suppl 1): 34–9.

Wong, J., E. Armour, P. Kazanzides et al. 2008. A high resolution small animal radiation research platform with x-ray tomographic guidance capabilities. *Int. J. Radiat. Oncol. Biol. Phys.* 71(5): 1591–9.

Zeman, G. H., T. H. Mohaupt, P. L. Taylor, T. J. MacVittie, A. Dubois, and R. M. Vigneulle. 1990. Nonuniform irradiation of the canine intestine. II. Dosimetry. *Radiat. Res.* 121(1): 54–62.

Zhou, H., M. Rodriguez, F. Van Den Haak et al. 2010. Development of a micro-computed tomography-based image-guided conformal radiotherapy system for small animals. *Int. J. Radiat. Oncol. Biol. Phys.* 78(1): 297–305.

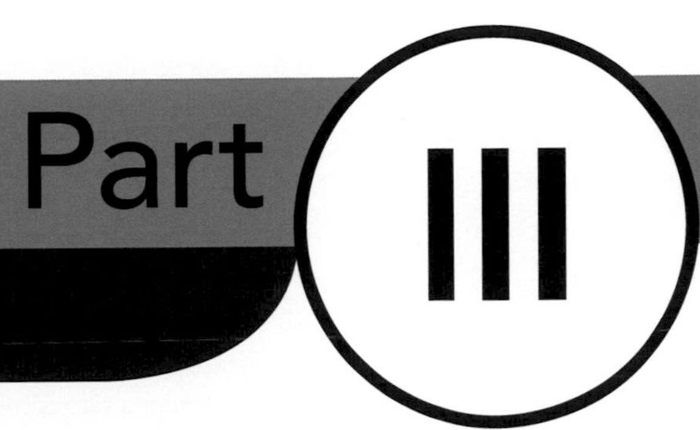

Part III

Stereotactic radiation therapy, precision patient positioning, and immobilization

Part III of this volume is devoted to the general topics of immobilization, through frames and body fixation, and what to do when you have a "moving target." The final chapter deals with the rapidly expanding field of frameless stereotactic radiosurgery and SBRT.

As the first chapter of this book describes, stereotactic radiosurgery quickly evolved from stereotactic neurosurgery. In the first two decades after Leksell's seminal 1951 paper, additional stereotactic frames were developed, and improvements were made. This path led to the incorporation of advanced radiological techniques, including angiography, computed tomography, and magnetic resonance imaging. In 1985, Laitinen in Stockholm developed a removable (relocatable) frame for adapting CT imaging to stereotaxis. Shortly after that, Gill and others adapted the BRW frame for fractionated treatments. Neither of these frames required anesthesia or surgical assistance, bringing stereotaxis into the realm of radiation oncology.

It was clear that the next advancement in stereotaxis would be extracranial, and Allan Hamilton and others at the University of Arizona created a hard docking system with temporary spinal clamps to do spinal radiosurgery in 1995. This stereotaxic methodology was almost identical to cranial radiosurgery: A hard docking device immobilized the spinal target, and a fiducial system was visible on a CT scanner. At nearly the same time, Blomgren and Lax created a full body immobilization and localization device for stereotactic body radiotherapy. Again, this device mimicked the functionality of earlier cranial SRS frames by attempting to immobilize the torso and providing a set of fiducial markers visible on CT scans. Later stereotactic body radiotherapy relied on body landmarks (such as bony anatomy), implanted gold markers, or external reflective markers. The issue of target motion for these early SBRT devices was somewhat finessed by choosing targets (e.g., spine, kidney, liver) that were thought not to move with respect to external markers.

True progress in the treatment of extracranial targets did not come until imaging capabilities were fully integrated into the treatment devices in systems such as the first CyberKnife and the first version of the Novalis. These and later systems allowed the observation of targets, such as lung tumors, that were not rigidly fixed to bony structures. Strategies to minimize motion during breathing were developed, such as plates to compress the lower abdomen and guided breathing. Introduction of cone-beam CT (using either the megavoltage treatment beam or an add-on kV x-ray source) provided excellent comparison with CT simulation images taken before treatment. Early users of the TomoTherapy device, which required daily megavoltage computed tomography (MVCT) images to align the patient with the simulation CT, noted, at the treatment console, the anatomical changes that occurred during treatment, especially the shrinkage of lung tumors and sometimes profound changes in head and neck tumors. The most recent advances utilize 4-D CT images of lung patients to characterize the breathing cycle. Predictive algorithms are used at the time of treatment to assure that the treatment beam is *on* only when the target is correctly aligned with the field. Outstanding clinical results for lung SBRT demonstrate that these techniques can be very effective.

Developing stereotactic frames for cranial treatment

Habeeb Saleh and Bassel Kassas

Contents

7.1 HISTORICAL PERSPECTIVE

Stereotactic frames used in radiosurgery evolved from apparatus used in stereotactic neurosurgery. Neurosurgeons and scientists were interested in developing stereotactic surgery apparatus as early as the 19th century (Al-Rodhan and Kelly 1992; Gildenberg 1990; Hassler and Riechert 1957). British neurosurgeon Victor Horsley and his surgeon colleague Robert Clarke described the first stereotactic apparatus for targeting lesions in a monkey brain in 1908 (Horsley and Clarke 1908). The Horsley-Clarke system used Cartesian coordinates to locate internal brain structures. Following Horsley and Clarke's work, many other neurosurgeons and scientists have described, experimented, or developed stereotactic apparatus for animals and humans. These apparatus had one thing in common: They were all lacking accurate localization of internal brain structure with external markers. This accuracy problem was the research focus of the American neurosurgeons Ernest Spiegel and Henry Wycis, who developed the first human stereotactic apparatus in 1947 (Spiegel et al. 1947). Their instrument was custom fitted to each patient and head in place with a plaster cast. They used positive contract ventriculography and the pineal body to localize intracranial targets. The Spiegel-Wycis apparatus also was based on Cartesian coordinates. Swedish neurosurgeon Lars Leksell, who was studying under the supervision of Spiegel and Wycis in Philadelphia, Pennsylvania, developed his first stereotactic frame in 1949. Lars Leksell's frame was different from that of Spiegel and Wycis in that it used the concept of the arc-centered quadrant. Leksell first used his frame in 1951 at the Karolinska Hospital in Stockholm where he worked as the head of the Neurosurgery Department. He used the frame to treat patients with a 250 kVp x-ray unit. Unlike the

Spiegel-Wycis apparatus, the Leksell frame uses a polar coordinate system, which is easier to use and calibrate in the operating room than the Cartesian-based apparatus. Neurosurgeon Edwin Todd and engineer Trent Wells adapted the Leksell arc-radius frame and developed their Todd-Wells stereotactic system in 1965 (Heilbrun et al. 1983). Their first frame was attached to the operating table. Others also adapted the arc-radius system, which eventually led to the invention of the widely used Brown-Roberts-Wells frame (Benedict et al. 2008). There were several other frames that were developed over the years. Some of those frames were used for radiosurgery and neurosurgery, and some were used for neurosurgery only. Some of these frames are listed here: the Narabayashi frame (Japan, 1949); the Cooper Stereotactic Device; Traugott Riechert and Fritz Mundinger (Germany, 1955); McKinney Stereotactic frame (United States, 1958); the University of Tokyo type stereotactic device developed by Professor Sano; Hitchcock frame (UK) (Bullard and Nashold 1995; Jensen, Stone, and Hayne 1996).

Cranial stereotactic radiosurgery frames continued to evolve with the advances of imaging modalities, such as CT and MRI. The major milestone in the development of the cranial frame has been the migration from invasive to noninvasive and relocatable frames with the perseverance of the same submillimeter accuracy. This migration was achieved by the application of online x-ray imaging devices.

7.2 INVASIVE FRAMES

Invasive frames are mostly non-relocatable and used for delivering a single fraction stereotactic treatment. However, due to interest in delivering stereotactic treatments in multiple fractions, relocatable invasive frames have also been developed. Both frame types that are commercially available are described below.

7.2.1 NONRELOCATABLE INVASIVE FRAMES

7.2.1.1 Leksell stereotactic system

The original Leksell stereotactic frame was developed in 1949 (Bullard and Nashold 1995; Jensen, Stone, and Hayne 1996). The Leksell Model G stereotactic frame, shown in Figure 7.1, uses the arc-quadrant

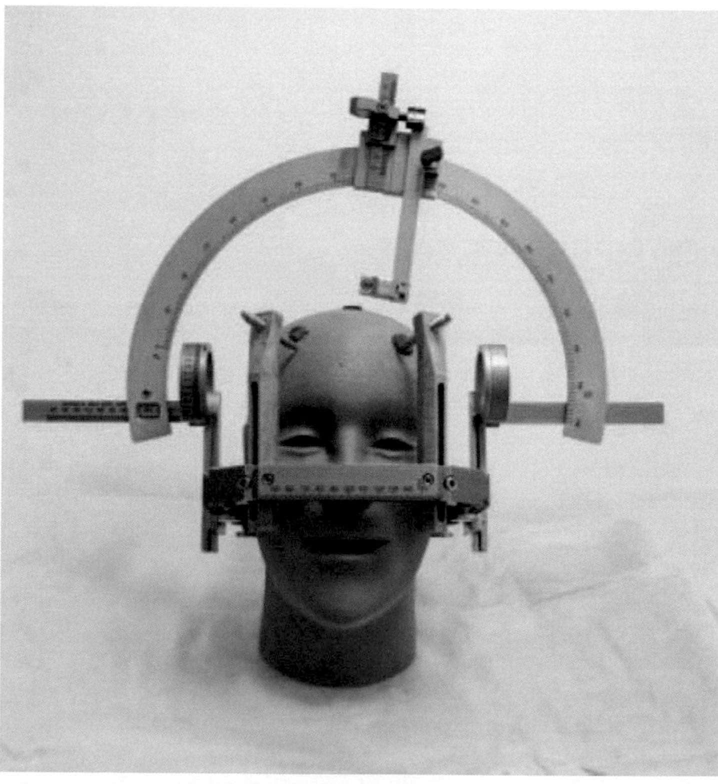

Figure 7.1 Arc-quadrant frame (original Leksell stereotactic instrument).

concept. It consists of a semicircular arc, which is fixed to the patient's skull using fixation pins and a movable instrument that slides on the semicircular arc. The arc quadrant can be swung along the left-right arc angle and in an anterior-posterior ring angle. With these degrees of freedom, the arc center can be positioned at the desired target point inside the brain.

The Leksell frame underwent many refinements over the years. The most commonly used Leksell frame is the G model, which is the frame used with the Gamma Knife radiosurgery treatment machine. The current commercially available Leksell stereotactic frame system (Elekta AB, Stockholm, Sweden) consists of the Leksell Coordinate Model G frame, Leksell CT indicator, Leksell MR indicator, x-ray indicator with markers, skull-scaling instruments, and other optional adapters. The Model G, shown in Figure 7.2, is the basic component of the system. In addition to its use in radiosurgery, the Model G frame can also be used for stereotactic microsurgery when used with the Leksell Multipurpose Stereotactic Arc or the Steiner-Lindquist Microsurgical Laser Guide (Elekta 1998a). The frame is both CT and MR compatible.

The Leksell frame consists of a rectangular base ring with four posts, which is attached to the patient's head using four pins. The pins are either aluminum screws with hard metal tips, titanium screws, or disposable aluminum screws (Lunsford, Kondziolka, and Leksell 2009). The base frame now has Teflon insulators to prevent eddy currents when the patient is scanned in an MRI scanner. Several incidents of mild to moderate skin burns were reported before this change was implemented. The four posts used with the Model G frame are available in five different designs to provide flexibility in treating tumors at different locations and for different size heads. The front piece of the frame is removable and comes in two different designs, either straight or curved. The curved front piece provides access for intubation. The origin of the Leksell coordinate system is located at a point outside the frame. This point (at which X, Y, and Z are numerically zero) is superior, lateral, and posterior to the frame on the patient's right side as shown in Figure 7.3. It is notable that this Cartesian coordinate system uses positive numbers only, making it impossible to interchange positive and negative coordinates, which has been blamed for wrong-side surgery with other stereotactic frames. The center of the frame is located at Leksell stereotactic coordinates 100, 100, 100.

The Leksell CT indicator, shown in Figure 7.4, is used in conjunction with the Model G frame for target localization (Elekta 1998b). The CT indicator contains fiducial markers in the anterior and the two lateral plates. The marker lines form an N-shape, which is made of dense material (copper) embedded in a Perspex plate. These markers are imposed on the CT image. Each CT image contains nine markers (three in the anterior side of the image and three on both the left and right sides of the image). The Leksell MR indicator, shown in Figure 7.5, is also used with the Model G frame as an imaging accessory (Elekta 1998c). The fiducials in the MR indicator are the same as those used in the CT indicator with three N-shaped line markers (filled with copper sulfate solution). These fiducials are used for determining target coordinates as well as for image slice alignment by the treatment-planning software. The original algorithm relied on determination of the stereotactic coordinates on an image-by-image basis, leading to poor results near the base frame and requiring the patient to be scanned while exactly level. By the late 1990s, the algorithm had improved to derive stereotactic coordinates from the entire canonical ensemble of axial (or coronal but *not* sagittal) images. This enabled error checking using least squares fitting.

In addition to CT and MR image indicators, an angiographic x-ray indicator with markers, shown in Figure 7.6, is also available for target localization (Elekta 1998d). The device is used when two x-ray images

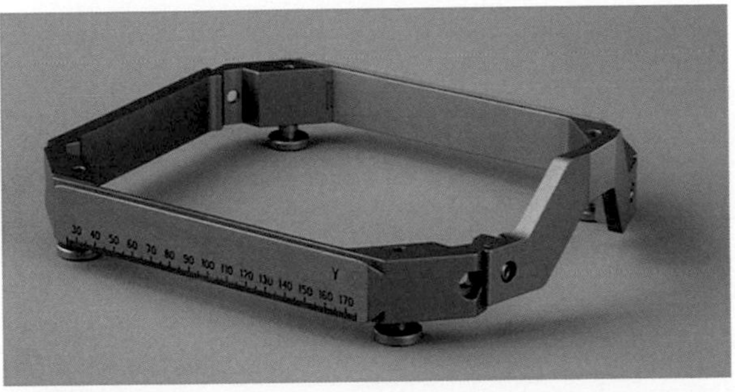

Figure 7.2 The Leksell Model G frame.

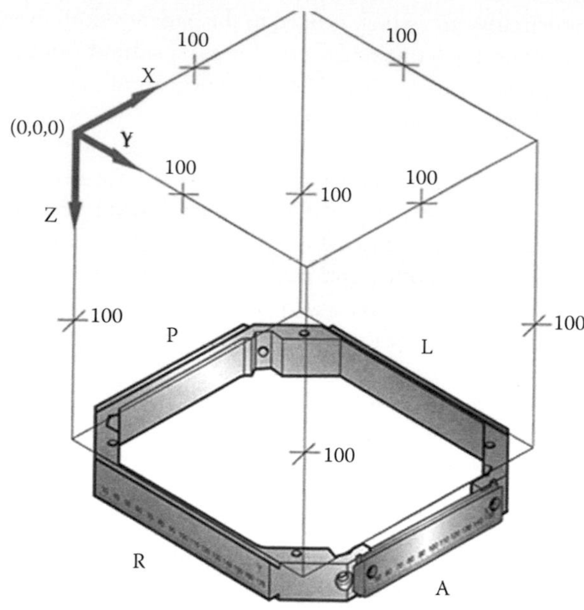

Figure 7.3 The Leksell coordinate system.

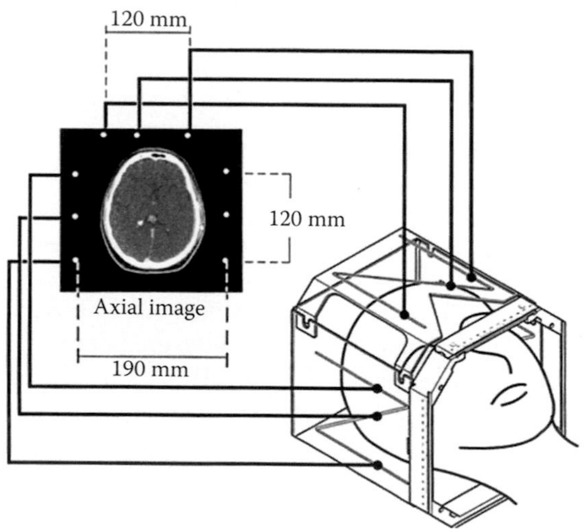

Figure 7.4 The Leksell CT indicator.

(left and right laterals) are applied for treatment planning in neurosurgery or radiosurgery. The left side panel of the device consists of five indicators, and the right side consists of four indicators. These indicators are arranged in a way to distinguish the plane at the x-ray source side from the plane near the x-ray film.

A skull-scaling instrument is another device used with the Leksell stereotactic system. The instrument is a partial sphere manufactured from transparent material that can be attached to the Model G frame. It is used to measure distances between the center of the Leksell stereotactic space and the surface of the patient's skull. These measurements are needed for the GammaPlan treatment-planning software used with the Leksell Gamma Knife.

7.2.1.2 Brown-Roberts-Wells frame

The development of the Brown-Roberts-Wells (BRW) frame started in 1977 at the University of Utah when Dr. Ted Roberts, chairman of neurosurgery, and a third-year medical student, Dr. Russell Brown,

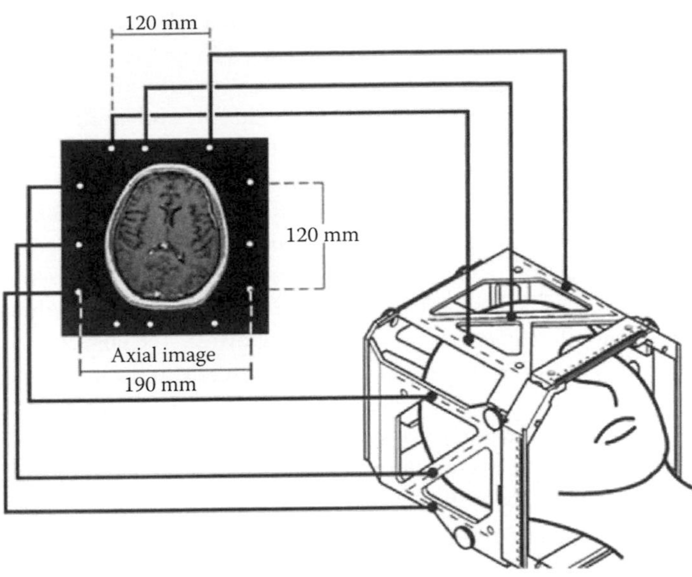

Figure 7.5 The Leksell MR indicator.

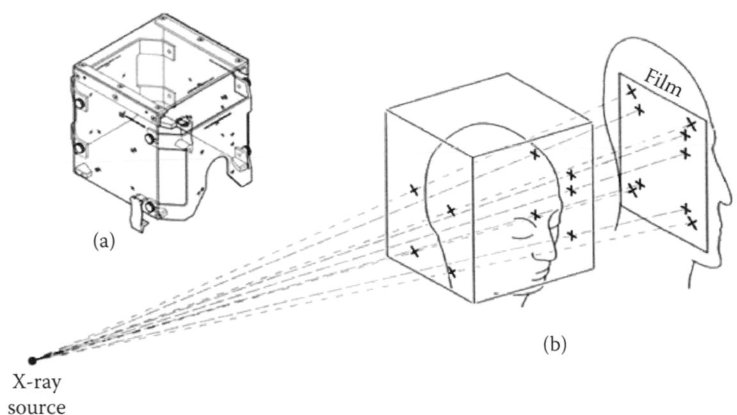

Figure 7.6 The Leksell x-ray indicator with markers: (a) parts of x-ray indicator with markers, (b) principles of operation.

worked together to improve the accuracy of the CT-based biopsy technique (Brown 1979). At that time, CT was a new imaging modality, but it suffered from inaccurate axial movements. The two investigators collaborated to design a system that would obtain relative spatial coordinates from one axial image to another, independent of the CT coordinate system and human anatomy. The design, hence, would provide a method of localizing the target based on an external coordinate system relative to the immobilization device, the rigidly fixed head ring. Brown designed the N-shaped vertical and diagonal semi-opaque rods that would be visible on each CT slice and serve as the image localizer system (Roberts 1998). They introduced their design to Trent Wells in 1979. Wells was an inventor who had his own engineering firm and was involved in developing stereotactic instruments for animals at UCLA. Wells and Dr. Edwin Todd, a neurosurgeon at UCLA, had successfully developed the classic Todd-Wells stereotactic apparatus in the 1960s, used with two x-ray tubes orthogonally arranged (Wells 1998). Thus, from the innovative design of Brown and Roberts to the ingenuity and engineering experience of Wells, the BRW frame was born and became an essential part of the armamentarium of stereotactic surgery.

The neurosurgery component of the BRW frame was a semicircular arc with a probe holder situated on a base ring attached to the head ring. This component was not a target-centered device and had a semi-polar coordinate system that required complicated calculations of four angles to reach the desired target. In 1988, the Cosman-Roberts-Wells (CRW) frame was introduced commercially and replaced the BRW's neurosurgery component with one that is target-centered: a semicircular arc-radius coordinate system neurosurgery device (Arle 2009). The CRW uses the same head ring and localizer as the BRW. Therefore, for the purpose of SRS and SRT procedures, which do not use the neurosurgery component, the two frames are similar.

The commercially available SRS BRW frame system (Integra Radionics, Burlington, MA) shown in Figure 7.7a consists of a head ring, four posts and pins, an imaging localizer, a rectilinear phantom base for quality assurance and verification, and a laser target localizer frame for localization on a treatment machine using lasers. The ring takes four post supports around its diameter, and they can be adjusted for depth to accommodate various head sizes using the head ring drives. The posts are attached to the head ring drives, and the four pins fix the posts and thus the frame rigidly to the outer table of the patient's skull (calvarium). The ring has three ball extension sockets and locks that accommodate the ball clamps on the BRW localizer for its secure attachment to the ring. Two clamp screws are available to attach the ring to the patient support system whether it is a couch or a pedestal stand. Both the rectilinear phantom base and laser target localizer frame are equipped with a Vernier scale to accurately set the coordinates of the target. In addition, a couch mount equipped with a Vernier scale for fine adjustments is provided.

The origin of the localizer is in the center of the head ring 80 mm from its top surface. The CT localizer shown in Figure 7.7b has nine fiducial rods that can be seen on each axial image. The location of these rods is known within the frame space. The inter-distances of these rods determine the relative frame coordinates of any point in the image independent of the imaging coordinate system. MRI and angiography localizers are also available and utilize the same fiducial rod concept.

Localization on the treatment machine is established using the laser target localizer frame shown in Figure 7.7c. First, the isocenter of the machine is verified using a mechanical isocenter stand system that is precalibrated to point to the isocenter. A tungsten ball is attached to the mechanical isocenter stand,

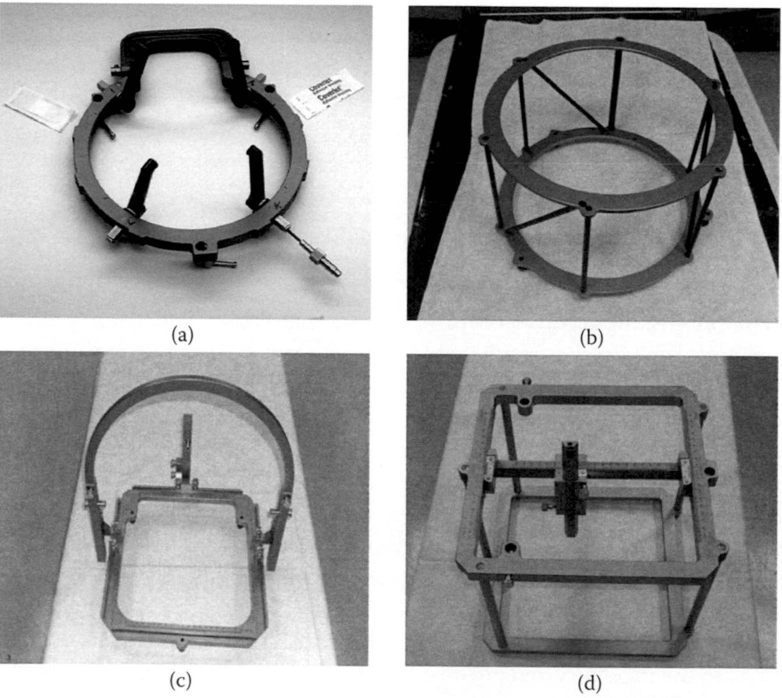

Figure 7.7 The BRW frame system: (a) BRW frame, (b) BRW CT imaging localizer, (c) laser target localizer, (d) phantom base.

pointing to the exact location of the isocenter, and a series of films are taken at various gantry angles (Lutz, Winston, and Maleki 1988). Lasers are then adjusted to accurately point to the verified isocenter. The phantom base shown in Figure 7.7d is set with the target coordinates using Vernier scales and attached to the treatment table. A tungsten ball is attached to the phantom base and points to the location of the target. The target isocenter is then verified through a series of films taken at various angles of both the gantry and table. The laser target localizer frame is set to the coordinates of the target and then attached to the base phantom. Fine adjustments can be made to the target frame so that the coordinate indicators are all aligned with the lasers. The reference frame is attached to the patient's head ring at treatment time between couch rotations. The couch mount Vernier scales can be adjusted to align the target frame coordinate indicators with the lasers.

The accuracy of the BRW frame was investigated in several studies and generally was within 1 to 2 mm. (Galloway and Maciunas 1990; Lutz, Winston, and Maleki 1988; Maciunas, Galloway, and Latimer 1994). Lutz, Winston, and Maleki (1988) reported the average treatment distance error (defined as the displacement of the center of the radiation distribution from the center of the target) for a target localized using the BRW frame with CT imaging to be 1.33 ± 0.64 mm (mean ± one standard deviation). Maciunas, Galloway, and Latimer (1994) reported a mean mechanical error within 1 mm and overall accuracy of 1.9 ± 1.0 mm that includes clinically relevant steps of image reconstruction, coordinate system transformation, and mechanical errors. Yeung et al. (1994b) reported a cumulative mean error, including inaccuracies associated with treatment setup of 1.2 ± 0.5 mm for CT localization with a 2-mm-slice thickness.

7.2.1.3 Brainlab head ring

The Brainlab head ring (Brainlab AG, Feldkirchen, Germany) consists of a ring, fixation posts, and set of fixation pins as shown in Figure 7.8a. The ring is made of an aluminum alloy, which is lightweight and strong, thus eliminating the possibility of bending as well as error during patient setup. The ring is designed with an interchangeable curved frontal piece for intubation. It is CT compatible but not compatible with MR, PET, or SPECT. Four carbon fiber fixation posts are used with the Brainlab head ring. Two posts are short, and two posts are long. The short post must be installed on the posterior side of the head ring, and the long post must be installed on the anterior side of the head ring. With the posts inserted into the head ring, the ring is attached to the patient using fixation pins. The fixation pins are ceramic tipped and self-penetrating. The pins are reusable and are CT compatible. It is rigidly fixed to the patient's head during simulation and treatment.

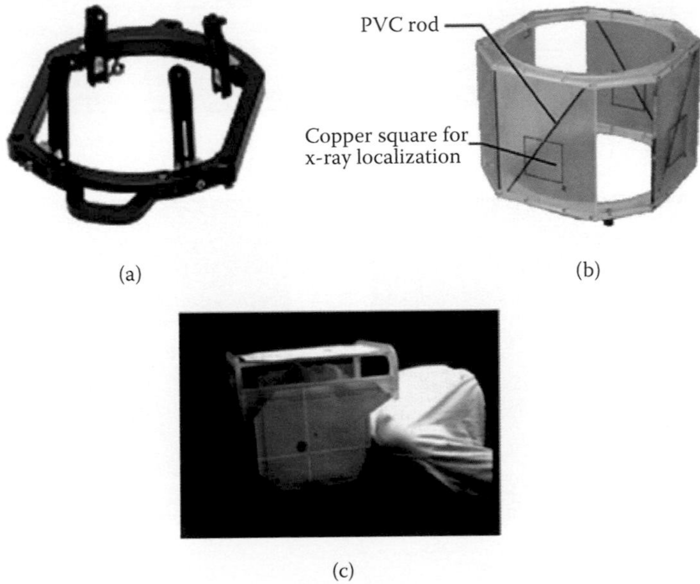

Figure 7.8 The Brainlab frame system: (a) head ring, (b) CT localizer, (c) treatment localizer.

For image localization, the Brainlab image localizer, shown in Figure 7.8b, is used with the Brainlab head ring. The localizer can be used for localizing CT images as well as x-ray radiography. CT images are localized using six embedded localizer rods on the left, right, and anterior sides. X-ray localization can be achieved using the four copper squares. The localizer is made of acrylic glass, and the rods are made of polyvinyl chloride. The six rods looks like six index marks in axial CT images. These marks correlate the location of any anatomical structure to the head ring. The copper boxes are used to localize the position of the anatomical structure in relation to the head ring in x-ray images. When localizing x-ray images, one sagittal and one coronal image of the head must be used. In each x-ray image, the two copper boxes must be visible.

The localizer is fixed to the head ring using the three mounting spheres attached to three locks on the surface of the upper side of the localizer. It can be used with the Radionics BRW frame and the Brainlab mask system. It can also be used with the Leksell Model G frame attached with four fixation rods.

Before treatment, a target positioner shown in Figure 7.8c is attached to the head ring. It consists of a skeletal aluminum frame and four target positioner plates. The plates are attached to four sides (left, right, superior, and anterior). During the treatment-planning phase, target positioned overlays are generated. These overlays show the position of the treatment isocenter. The patient is set up such that the isocenter of the target positioner overlays coincides with the treatment machine isocenter. The overlays are attached to the corresponding target positioner plates. Both plates and overlays are labeled with letters A to D. Overlay A is attached to plate A and so on for the rest of the plates and overlays. In addition to the Brainlab head ring, the target positioner can also be used with the Leksell head ring using the same procedure for attaching overlays to plates.

7.2.1.4 Varian medical system optically guided frame

The Varian medical system (VMS) frame (Varian, Palo Alto, CA) is a commercial variation of the BRW frame, shown in Figure 7.9. The head ring is similar to that of the BRW system. It uses four posts and pins for rigid attachment to the patient skull. The CT imaging localizer uses the same fiducial rod principle as

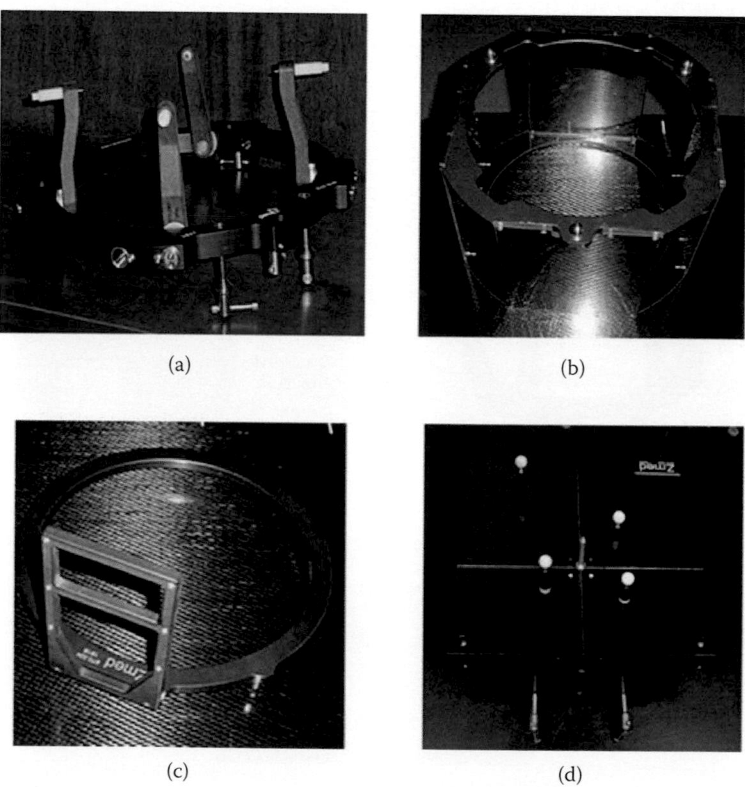

Figure 7.9 The VMS frame system: (a) head ring, (b) CT imaging localizer, (c) optical target localizer frame, (d) optical calibration jig.

the BRW system with the exception of using six fiducial rods instead of nine. The VMS CT localizer is longer in the superior-inferior direction than the BRW, allowing for more of the head anatomy to be within the stereotactic coordinate space.

Localization on the treatment machine is established using an optically guided target localizer frame. The optical positioning and tracking system was developed at the University of Florida, Gainesville, and was later available commercially through ZMED (Holliston, MA), which was acquired by VMS in 2003.

The optically guided positioning system of the University of Florida was developed to improve the accuracy of couch-mounted stereotactic frames to be comparable with floor-mounted systems and to be used for frameless applications (Bova et al. 1997; Meeks et al. 1998; Meeks et al. 2000; Tome et al. 2000; Yang 1994). The system uses an array of six infrared light emitting diodes mounted on a ring that can be attached to the patient's head ring in the same manner as the imaging localizer does, employing the ball-socket locking mechanism. An optical camera, mounted in the ceiling, is calibrated to locate the treatment machine isocenter in the room. Therefore, by accurately locating the diodes on the array in the room using the ceiling-mounted optical camera, the target isocenter can be localized and tracked in real time during the entire treatment delivery through the use of a computer system that processes the camera's output. Fine-tuning of the patient's position can be done using the couch-mount adjustment mechanism, which has knobs for 3° of freedom adjustment as well as using the couch movement mechanism with feedback from the optical camera computer system.

The calibration of the optical camera is performed using an optical calibration jig that can be mounted on the treatment couch in the same manner as the patient head ring is attached. The optical calibration jig has four reflective markers that can be seen by the optical camera. The jig also has an isocenter pointer, the location of which relative to the reflective markers is known. By accurately aligning the isocenter pointer to the isocenter location of the treatment machine using a calibrated radiation isocenter pointer attached to the machine collimator, the camera can be calibrated to determine the isocenter location in the room.

Meeks et al. (1998) showed that coordinate localization using the infrared light emitting diode system can provide translational accuracy to within 0.2 ± 0.1 mm (mean ± one standard deviation) of the absolute isocenter location. They also compared this system to target localization using wall-mounted lasers and found the average vector error between the two systems to be 1.7 ± 0.7 mm.

The VMS frame, therefore, replaces the laser target localizer frame of the BRW system with an optically guided target localizer frame allowing for improvement in positional accuracy in couch-mounted systems. The optical frame stays on during treatment, giving the system the advantage of real-time tracking, ease of use, and convenience to the patient because it does not require the placement and removal of the localizer frame between couch rotations in non-coplanar arc delivery.

7.2.2 RELOCATABLE INVASIVE FRAMES

7.2.2.1 TALON® removable head frame system

The TALON frame (Best NOMOS, Pittsburgh, PA, now part of Best Medical) combines the accuracy of an invasive frame with the flexibility of a relocatable system for multiple stereotactic treatments. The frame, shown in Figure 7.10, is invasive because it requires a one-time placement of two self-tapping titanium screws in the patient's skull by a neurosurgeon. This procedure takes approximately 20 to 30 min and can be done in the operating room or, more commonly, in an outpatient surgical suite with local anesthesia. The TALON device is then attached to the titanium screws and a table-mount adapter, called Nomogrip, for CT imaging. The TALON can be finely adjusted through the use of locking ball joints and extension rods, allowing for relatively comfortable patient positioning on the table. The device is then tightened and locked into the desired patient position. This locked TALON is now patient-specific and will be used for accurate repositioning of the patient on the treatment table. The origin of the frame's stereotactic coordinate system is established using a localization pointer that gets attached to the assembly at the time of CT imaging. A target box can be mounted on the frame for patient setup guidance by providing laser alignment, light-field alignment, and fiducial markers for film verification. The TALON/Nomogrip assembly is removed after each treatment delivery, and the patient is left with only the titanium screws in the skull.

The repositioning accuracy of the TALON frame has been investigated by Salter et al. (2001), using measurements of anatomical landmark coordinates on repeated CT scans. They reported a mean isocenter

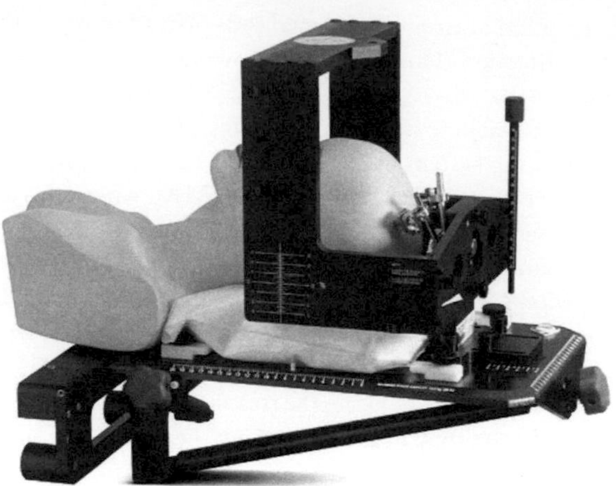

Figure 7.10 TALON removable head frame system. (Courtesy of Best NOMOS.)

translation of 0.99 ± 0.28 mm (mean ± one standard deviation) for a single fraction SRS and 1.38 ± 0.48 mm for fractionated SRT over a 6-week treatment course. Their 95% confidence interval (mean error + 2 standard deviations) is 1.55 mm for SRS and 2.34 mm for SRT.

The TALON provides a unique solution for SRS procedures by allowing ample time for treatment planning to be done prior to scheduling patients for treatments. The pressure on staff-associated invasive non-relocatable frame-based procedures in which the frame placement, imaging, planning, QA testing, and treatment delivery are all to be done in a few hours can potentially be alleviated with the use of the TALON frame without compromising treatment accuracy. Perhaps because it is an invasive frame, the TALON frame has not received wide acceptance by the radiation oncology community (Leybovich et al. 2002).

7.3 NONINVASIVE RELOCATABLE FRAMES

7.3.1 GILL-THOMAS-COSMAN FRAME

The Gill-Thomas-Cosman (GTC) frame is a noninvasive relocatable frame that is compatible with the BRW system. It accommodates the same imaging and treatment localizers. It can be used on a table mount or a floor stand. The original design, known as the Gill-Thomas (GT) frame, was developed in 1990 (Gill et al. 1991). It consists of an aluminum base ring compatible with the BRW system, a dental plate that is firmly attached to the ring, an occipital headrest plate, and three nylon straps that hold the dental plate and the occipital plate tightly together with a quick-release mechanism (Gill et al. 1991). The dental plate was individually constructed for each patient from the impression of the upper and lower dentitions using a polymer material that hardens in ultraviolet light. Three sizes of dental trays are available to accommodate various adult jaw sizes. The occipital headrest plate uses a moldable material applied during patient fitting and simulation. The position of the plate can be locked. By fixing the dental plate to the head ring and firmly holding the occipital plate to the dental tray with straps, the cranium is fixed to the head ring, and the system then can be used for stereotactic procedures using the BRW imaging localizer.

Kooy et al. (1994) reported on the improvement and modification of the frame that became the commercially available GTC frame (Integra Radionics, Burlington, MA), shown in Figure 7.11a. Their modifications included the replacement of the nylon straps with Velcro that allow for quick tightening and release. The strap lengths can be marked at each side for daily reproducibility. The occipital plate is modified to mount on two upright bars to allow the central support to be placed at the occipital protuberance for better and firmer positioning. The mounting upright bars also allow for movement of the plate in the anterior-posterior direction, making it more adaptable to various patient head sizes. The locking mechanism of the dental plate to the head ring also was improved. In addition, to overcome the need to rely on portal orthogonal images to verify the daily repositioning of the frame that the original

(a)

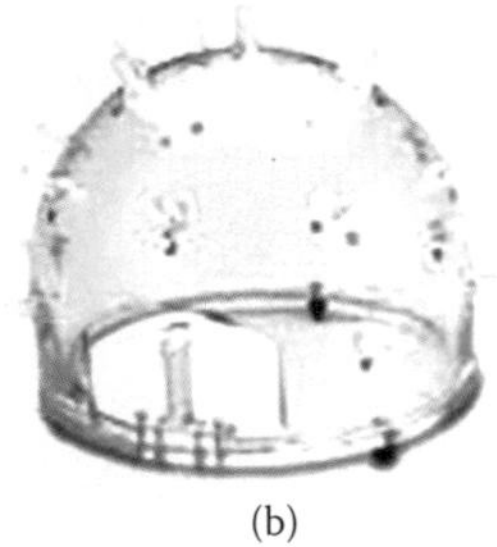

(b)

Figure 7.11 GTC head frame and depth confirmation helmet: (a) GTC frame, (b) depth helmet. (Courtesy of Integra Radionics.)

design used, a depth confirmation helmet was developed to measure distance from the BRW coordinate origin to the cranial surface at 25 fixed positions. The depth confirmation helmet, shown in Figure 7.11b, is a hemispherical dome that is attached to the head ring in the same manner as the BRW imaging localizer. The distance reading is obtained using a probe that has a millimeter scale, which is inserted into the 25 holes in the helmet to reach the cranial surface. The distances are recoded at fitting simulation and compared to measurements obtained for each daily treatment. A good fit is established if all point measurements fall within an acceptable tolerance of 80% of points within ±0.5 mm.

Reproducibility of fixation of the original GT frame was reported to be within 1 mm with a mean displacement of 0.5 mm using repositioning measurements on 10 patients (Gill et al. 1991). Kooy et al. (1994) reported a mean displacement of 0.71 ± 0.06 mm (mean ± one standard deviation) with respect to patient anatomy derived for 20 patients. However, the observed range in their worst-case situation was reported to be 3.4 ± 0.6 mm, and hence, they concluded that the frame is not adequate for SRS but is generally acceptable for fractionated stereotactic radiotherapy. Graham et al. (1991) reported reproducibility of the GTC frame to range from 0.7 to 1.2 mm with a mean of 1 mm in the anterior-posterior dimension and from 0.4 to 1.6 mm with a mean of 1 mm in the lateral dimension. It is important to note that the above-reported reproducibility numbers do not reflect the entire simulation and treatment process and, therefore, are not to be considered as the overall accuracy of a stereotactic treatment using the GTC frame.

Some reported practical problems with the GTC frame are related to poor dentition. Ashamalla, Ross, and Ikoro (1999) studied the effects of oral pain and discomfort associated with the use of the GTC frame. They reported that poor dentition, associated with full or partial dentures, can cause various degrees of discomfort, ranging from mild to complete intolerance of the frame. Furthermore, the comfort level was highly correlated with the accuracy of reproducibility. They found that inaccuracy can rise to 2 to 3 mm in patients with a high level of discomfort. Therefore, patient comfort and cooperation is essential to the successful use of the GTC frame in stereotactic applications.

Another practical issue related to the use of the GTC frame is that the daily variation of the pull forces applied each day with the Velcro straps may affect localization in relationship to the head ring (Bova et al. 1997). The adequacy of using the localization helmet as a measure of accuracy has also been challenged (Miyabe et al. 2004; Yeung et al. 1994a) when systematic errors may occur and rotational displacement might be missed. Additional imaging localization may be warranted with the now widely available portal imaging, on-board imaging, and cone-beam CT systems.

7.3.2 TARBELL-LOEFFLER-COSMAN PEDIATRIC FRAME

The Tarbell-Loeffler-Cosman (TLC) pediatric frame (Integra Radionics, Burlington, MA) is a noninvasive relocatable frame that is compatible with the BRW system (Dunbar et al. 1994). It accommodates the same imaging and treatment localizers. It can be used on a table mount or a floor stand. It is a modification of the GTC frame specifically designed for pediatric patients who require anesthesia and, therefore, access to patient airways. The frame makes use of a glabellar mask instead of the dentition plate for patient fixation along with an occipital plate. Kooy et al. (1994) reported on the frame development as the Boston Children's Hospital frame prior to its commercial adaptation as the TLC pediatric frame. The frame is shown in Figure 7.12.

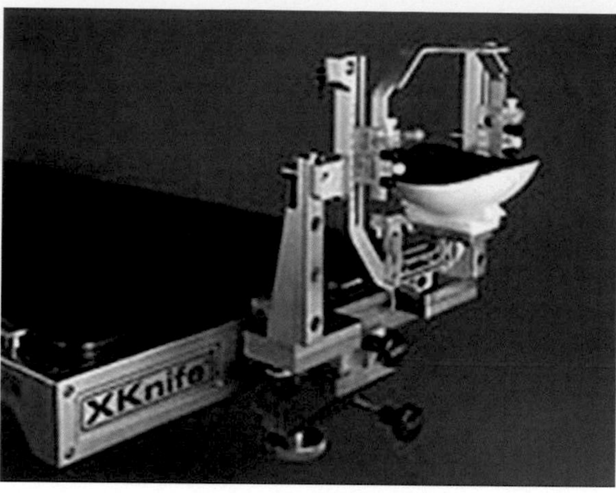

Figure 7.12 The TLC pediatric frame. (Courtesy of Integra Radionics.)

The modification includes ear canal supports for the purpose of frame localization and repositioning in the inferior-superior and left-right positions. They are not used for frame fixation and, therefore, they do not have to be pressed in firmly as in the case of the Laitinen frame. The occipital support consists of a head cup and a moldable material fitted at time of simulation. The glabellar mask is repositioned using a spectacle frame. The spectacle frame can move in the anterior-posterior and allows the glabellar mask to firmly fixate the cranium against the occipital support. By noting the spectacle frame position, the anterior-posterior direction, and the position of ear canal supports in the inferior-superior and the left-right, the patient fixation can be reproduced for daily treatment. Significant material around the cranium in the occipital plate is typically applied for better patient positioning, which may restrict access for posterior treatment arcs (Kooy et al. 1994). The frame makes use of the depth confirmation helmet, as in the case of the GTC frame, for verification of daily repositioning.

Kooy et al. (1994) reported on the repositioning measurements of 20 patients, 17 of them treated with the GTC frame and three treated with the TLC frame, using the depth confirmation helmet. The mean reported repositioning error was 0.35, 0.52, and 0.34 in the lateral, superior-inferior, and anterior-posterior directions, respectively. The overall mean displacement was 0.71 ± 0.06 mm (mean ± one standard deviation) with respect to patient anatomy. It is important to note that the above-reported repositioning errors are not specific to the TLC frame as only three patients among the 20 reported on were treated with this frame. Furthermore, these reproducibility numbers do not reflect the entire simulation and treatment process and, therefore, are not to be considered as the overall accuracy of a stereotactic treatment using the TLC frame.

The TLC frame, therefore, applies the technology of the GTC frame and offers a unique solution for the special needs of pediatric patients by ensuring unrestricted access to the patient's airways for anesthesia requirements.

7.3.3 BRAINLAB MASK SYSTEM

The Brainlab mask system (Brainlab AG, Feldkirchen, Germany) is based on the GTC frame. The Brainlab mask system consists of five components. These components are the mask ring, headrest, vertical post, screws, and cam locks as shown in Figure 7.13. The mask is made of thermo-transformable material. The hardware is reusable, but the mask is disposable, and it is patient-specific.

The mask system is used along with Brainlab CT/MR/x-ray imaging localizers, and treatment target positioner shown in Figure 7.8b and 7.8c, respectively. These localizers were previously described in Section 7.2.1.3.

During simulation, a patient-specific mask is constructed. Then a CT or MR imaging localizer is attached to the ring to acquire CT or MR data. During treatment, the mask system is used with the target positioner for patient setup.

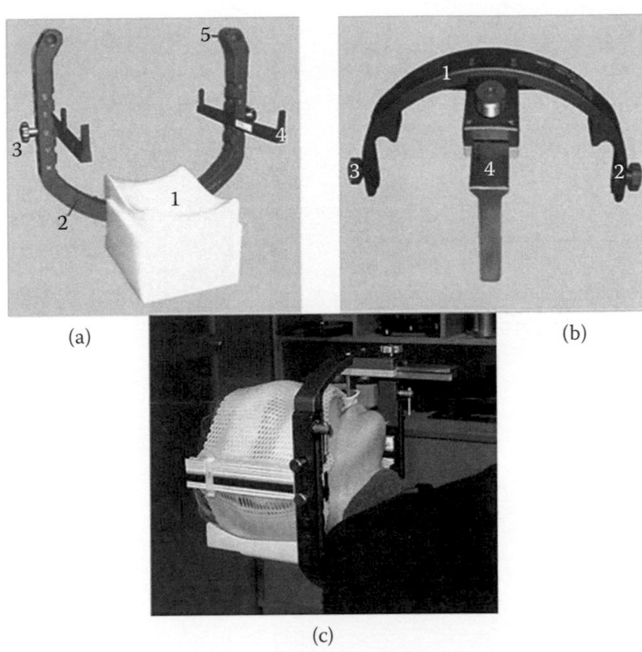

Figure 7.13 Brainlab mask system: (a) U frame, (b) upper jaw support, and (c) frame fitted on patient.

As an optional accessory, an upper jaw support, shown in Figure 7.13b, can be used with the Brainlab mask system to reduce longitudinal movement of the patient. Bite plates are used along with the upper jaw support. There are three different sizes of bite plates: normal, short, and lower mask fixation.

The mask system is used for both single-fraction stereotactic treatments as well as multifraction treatments. With fractionated treatments, the frame must be repositioned with every treatment. This could lead to interfraction setup uncertainty due to head position changes inside the mask and daily setup error. Alheit et al. (2001) reported that the positional uncertainty of the patient's head inside the thermo-transformable mask can be as much as 2 mm if no image guidance is used. Even with image guidance, other investigators reported setup uncertainty of 2 mm using the mask system (Ali et al. 2010).

7.3.4 LAITINEN FRAME

The Laitinen Stereoadapter® 5000 frame (Sandstrom Trade & Technology Inc., Ontario, Canada) was designed by Lauri V. Laitinen in Sweden in the 1980s (Laitinen et al. 1985; Laitinen 1987). It is a noninvasive relocatable frame that provides a high degree of reproducibility through the use of three reference points in the skull: the left and right external auditory meatus and the bridge of the nose. The frame, shown in Figure 7.14, is made of aluminum alloy and reinforced plastic and, hence, is CT and MRI compatible. It consists of two lateral triangular plastic components, two earplugs, a connector plate, a nasion support assembly, a frontal laterality scale, a frontal midline pin, two plastic target plates, and a couch adapter plate (Hariz and Laitinen 2009).

The lateral triangular plate has four 2-mm-thick transverse bars that are 25 mm apart from each other. The lateral triangular plates are attached to the earplugs and held together using a strap that wraps posteriorly around the head. The nasion support assembly has two arms with millimeter scales. Each arm attaches to a triangular plate by a cogwheel. By tightening the cogwheel, the nasion support assembly is pressed against the bridge of the nose. The assembly is equipped with an adjustment thumbscrew at the nasion that serves to push in the earplugs and tighten the triangular plates against them. The connector plate has a millimeter scale and serves to join the triangular lateral plates at the vertex as well as lock them tightly against the skull. The frontal midline pin joins the center of the nasion support assembly to the connector plate and serves as a lateral reference structure. The frontal laterality scale attaches to the midline marker and can be equipped with a cone that serves as an indicator of the isocentricity of the table rotation. The plastic target plates attach to the lateral triangular plates and are marked with lines indicating the location of the transverse bars and the posterior side of the triangular plates.

(a)

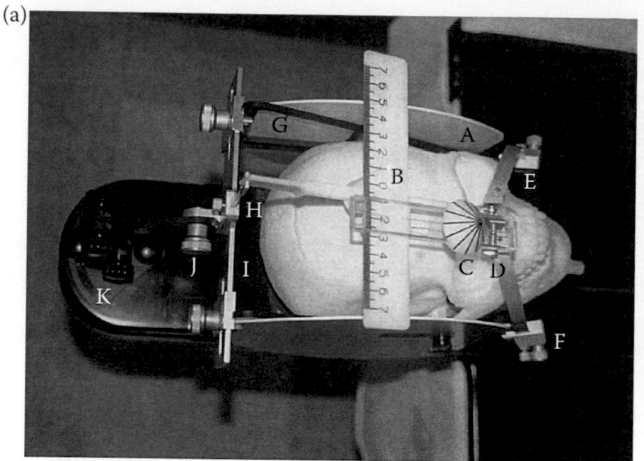

(b)

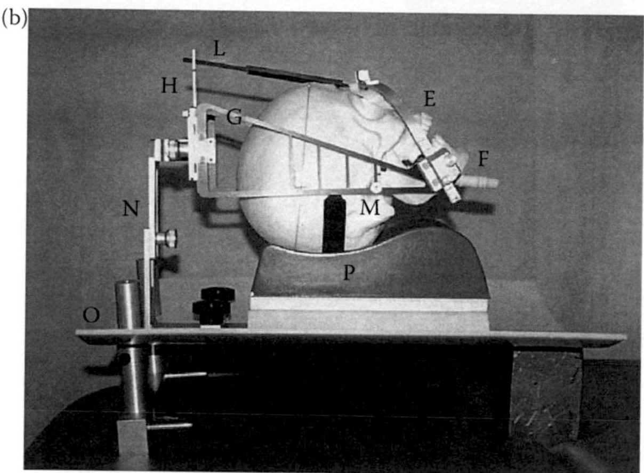

(c)

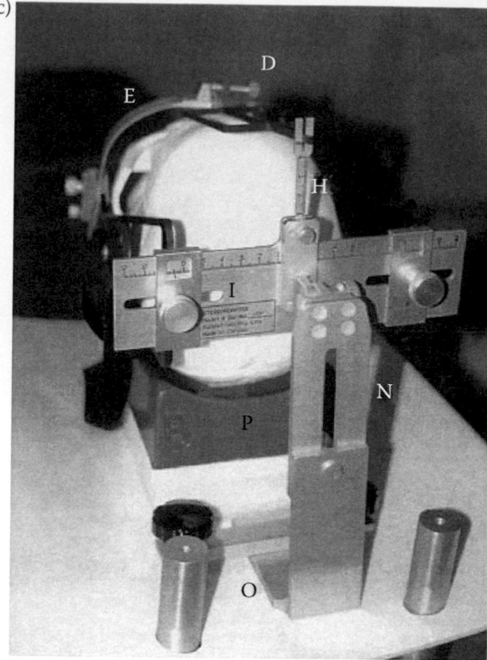

Figure 7.14 Laitinen frame: (a) anteroposterior view of the stereotactic frame on the treatment couch of the linear accelerator, (b) lateral view of the stereotactic frame in the CT adapter apparatus for a quick CT scanner, (c) superoinferior view of stereotactic frame in the CT adapter apparatus, (A) target plate, (B) scale for x coordinate, (C) pyramid to confirm isocentricity of table rotation, (D) thumbscrew for earplug, (E) nasion support assembly, (F) cogwheel with scale for support assembly, (G) triangular sidebars, (H) adjustable fork, (I) connector plate with scale, (J) couch adapter fixation screw, (K) couch adapter device, (L) midline marker, (M) earplugs, (N) adapter plate attachment, (O) CT adapter plate, (P) headrest. (Reproduced from Kalapurakal, J. A. et al., *Radiology* 218(1), 157–61, 2001.)

These plates are used to set up the patient to the treatment isocenter with the use of the room lasers. The frame is attached to the imaging or treatment couch through the use a multijoint couch adapter plate that connects the connector plate at the vertex to the couch. An adjustable head extender couch mount equipped with an adapter plate is also available, and it can be attached to a treatment couch.

The reference coordinates for this frame are established using the mid-sagittal plane passing through the midline frontal pin for lateral direction, the coronal plane passing through the posterior side of the triangular plates for anterior-posterior direction, and the transverse plane between a pair of transverse bars in the triangular plates. Using distances from these planes, a treatment isocenter can be uniquely identified and marked on the target plates and the frontal laterality scale.

Reproducibility of patient setup for each fractional delivery is achieved by recording the readings of the lateral positions of the triangular plates at the millimeter-scaled connector plate at the vertex, the positions of the two millimeter-scaled nasal support arms at the cogwheels, and the number of turns on the adjustment thumbscrew at the nasion set during simulation. The frame settings need to be firm against the patient nose bridge, earplugs, and patient skull and maintain comfortable immobilizations.

Repositioning accuracy of the Laitinen frame has been investigated by several authors (Delannes et al. 1991; Golden et al. 1998; Hariz 1990; Hariz and Erikss0 1986; Kalapurakal et al. 2001; Weidlich et al. 1998) and generally ranged from 1 to 2 mm. Delannes et al. (1991) reported a mean distance error of less than 1 mm in all three directions. Golden et al. (1998) reported a general accuracy of 2 mm using orthogonal images. Kalapurakal et al. (2001) also used portal images and reported a mean isocenter shift ± one standard deviation of 1 ± 0.7 mm in the lateral direction, 0.8 ± 0.8 mm in the anterior/posterior direction, and 1.7 ± 1 mm in the superior-inferior direction.

The Laitinen frame can be used for adult and pediatric patients with various earplug sizes to accommodate both large and small external auditory meatus. The frame offers a unique solution in that it does not require any custom-made devices for each patient and provides a simple localization reference structure. It is important to note that a high degree of repositioning accuracy requires good cooperation by patients unless sedation is used (Hariz and Laitinen 2009). Therefore, patient comfort with the earplugs plays an important role in the successful use of this frame in stereotactic applications. Topical anesthesia can be applied to ease the discomfort in the ears.

A modification of the Laitinen frame was later developed as the Beverly frame. The Beverly frame relies on the same immobilization techniques as the Laitinen frame but replaces the L-shaped couch adapter plate with two lateral ones. This design modification allows for more stable attachment to the couch and provides clearance for the use of sagittal beams through the vertex (Delannes et al. 1994).

Another modification of the Laitinen frame is the Gildenberg-Laitinen adapter device (GLAD) (Ohio Medical Corporation, Cincinnati, OH) (Gildenberg 1998). The GLAD system combines the Laitinen frame with the traditional BRW or Leksell head ring and localizer. The GLAD-S uses the invasive BRW head ring, and the GLAD-X uses the noninvasive head ring with noninvasive probes. The head ring is attached to the patient but does not interfere or come into contact with any part of the Laitinen frame. The ring is attached to the table using a BRW couch mount instead of using the L-shaped couch adapter plate of the Laitinen frame. The GLAD has its own lateral triangular plates that are aligned with those of the Laitinen frame during setup. This alignment of the two systems allows superimposing the Laitinen coordinates on the BRW or Leksell coordinates. Consequently, commercially available treatment-planning systems for BRW and Leksell systems can be used, and the patient can be set up for treatment using either coordinate systems. A study by Ashamalla et al. (2003) reported on the use of GLAD-X. Their reported mean relocation accuracy was 1.5 ± 0.8 mm (mean ± one standard deviation). They found the device to generally be well tolerated by patients.

7.4 FRAMELESS SYSTEMS

Traditionally, stereotactic radiosurgery has been performed using invasive frames, such as the frames described earlier in this chapter. These invasive frames were the only frames used when stereotactic treatment was delivered in a single fraction. However, with the shift from a single-fraction stereotactic treatment to multiple fractionated treatments, a need for alternative noninvasive frames became critical. Several noninvasive relocatable frames were developed, some of which were described in this chapter. In recent years, because of the rapid development of image-guided radiotherapy and the interest in extracranial stereotactic radiotherapy, several frameless intracranial systems were introduced for stereotactic radiosurgery and radiotherapy. Two commercially available frameless systems are described below.

7.4.1 OPTICALLY GUIDED BITE-PLATE FRAMELESS SYSTEM

The optically guided bite-plate frameless system, shown in Figure 7.15, was developed at the University of Florida and was described by Bova et al. (1997). It is available commercially through VMS (Palo Alto, CA). It consists of a thermoplastic mask and head cushion for immobilization and an upper-jaw bite-plate tray with an attached optical fiducial array for positioning and tracking using an optical camera system (Bova et al. 1997; Meeks et al. 2000; Phillips et al. 2000; Tome et al. 2000; Yang 1994). This system separates the immobilization from localization as the bite block is not attached to the mask.

The bite-plate tray comes in various sizes. Typically, the patient is fitted for the proper size after removing nonpermanent dental devices as these can affect the accuracy of patient setup. The teeth

(a)
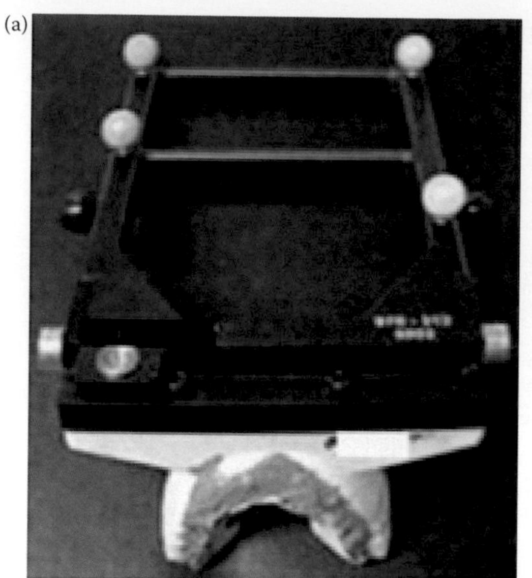
(b)
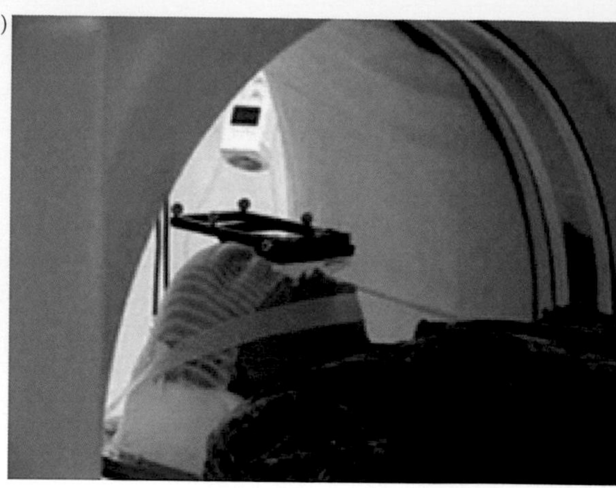
(c)
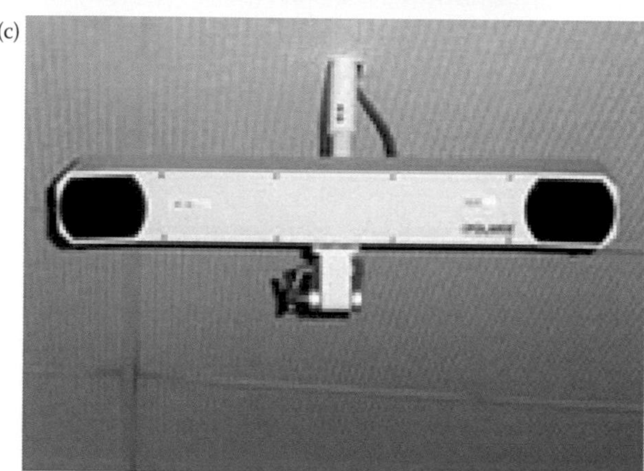

Figure 7.15 The optically guided bite-plate frameless system: (a) optical array, (b) frame fitted on patient, and (c) IR camera.

should fit within the plate without touching its edges. Dental impression material is then used to fill the plate completely and then inserted into the patient's mouth. The patient is instructed to bite down for a few minutes to obtain the dental impression. The optical fiducial array attaches to the bite plate using hex screws. The array can be angled for optimum viewing by the camera and be locked in any desired tilt position. It is essential that reseat verification is done prior to acceptance of the bite-plate tray for treatment. This is accomplished by comparing the position of the bite plate fiducial array relative to a reseat verification array attached to the head with a headband over several mouth insertions.

The system uses an array of six fiducials, four of which are reflective and can be used for optical tracking. This array replaces the infrared light emitting diode array of the original University of Florida design. The patient is imaged and treated with the bite plate and optical fiducial array in place. The six fiducials are localized in the CT space in the treatment-planning system. By determining the fiducial positions relative to the treatment isocenter, a stereotactic coordinate system is defined (Kamath et al. 2005). The planning system compares the CT coordinates of the fiducials of the optical array to the known geometry of the reference array to insure internal consistency in the CT scan. A lack of consistency with the known geometry of the reference array indicates a movement during image acquisition (Ryken et al. 2001). An optical camera mounted in the ceiling is calibrated to locate the treatment machine isocenter in the room.

Therefore, by accurately locating diodes on the array in the room using the ceiling-mounted optical camera, the target isocenter can be localized and tracked in real time during the entire treatment delivery through the use of a computer system that processes the camera's output. Fine adjustments of the patient's position can be done using the couch mount adjustment mechanisms, which has 3° of freedom adjustment knobs as well as using the couch movement mechanism with feedback from the optical camera computer system.

The calibration of the optical camera is performed using an optical calibration jig that can be mounted on the treatment couch in the same manner as the patient's head ring attached to the couch. The optical calibration jig has four reflective markers that can be seen by the optical camera. The jig also has an isocenter pointer. The precise location of the isocenter pointer relative to the reflective markers of the jig is known. By accurately aligning the isocenter pointer to the isocenter location of the treatment machine using a calibrated radiation isocenter pointer attached to the machine collimator, the camera can be calibrated to determine the isocenter location in the room.

The reproducibility of placement of the bite plate in patients was investigated by Bova et al. (1997) and found to be 0.5 ± 0.3 mm (mean ± one standard deviation). The accuracy of positioning a target point in the radiation field using this system was reported by Phillips et al. (2000) as 1.0 ± 0.2 mm.

The optically guided bite-plate frameless system offers the advantage of real-time tracking. Couch rotation errors can be reduced using the system. Reproducibility of the bite plate, however, is essential for the successful and accurate implementation of the system.

7.4.2 BRAINLAB FRAMELESS MASK IMMOBILIZATION SYSTEM

The Brainlab frameless mask immobilization system (Brainlab AG, Feldkirchen, Germany), shown in Figure 7.16, has been in clinical use since 2006. The mask system consists of six optical marking spheres used on top of a thermoplastic mask. Patients are fitted with the traditional cranial or head-and-neck thermoplastic mask during simulation. The frameless system is then used on top of the thermoplastic mask during treatment for localization and tracking using an infrared camera system. The mask is used in combination with the 6° of freedom x-ray ExacTrac system and robotic couch for real-time tracking. With the utilization of an infrared camera, the system allows real-time tracking of externally marked points to accurately determine any departures from the treatment isocenter and the correct patient positioning during treatment. The tracking information can be used to determine the frequency of x-ray imaging during treatment for verification.

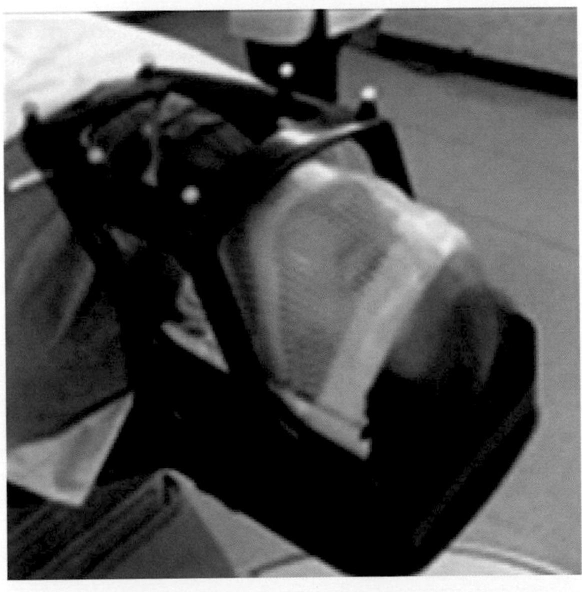

Figure 7.16 The Brainlab frameless mask system.

The positional accuracy of the Brainlab frameless system was investigated by Verbakel et al. (2010), who conducted a head phantom and daily setup of 46 patients. They found that the ExacTrac system positioning accuracy is approximately 0.3 mm in each direction (one standard deviation) and reported intrafractional motion of 0.35 ± 0.21 mm. Blokzijl et al. (2008) compared this frameless system with the Brainlab invasive head ring frame. Their study was conducted on a head phantom using the two frame systems. The frameless system was positioned using the ExacTrac imaging system, and the invasive frame was positioned using the target positioner and room lasers. The imaging system was then used to assess the positional accuracy of both frames. They reported that the frameless system is significantly better than the invasive system. However, shifts calculated for the invasive head frame were defined as unadjusted deviations because these deviations were not adjusted in their daily practice.

REFERENCES

Alheit, H., S. Dornfeld, M. Dawel et al. 2001. Patient position reproducibility in fractionated stereotactically guided conformal radiotherapy using BrainLab mask system. *Strahlenther. Onkol.* 177(5): 264–8.

Ali, I., J. Tubbs, K. Hibbitts et al. 2010. Evaluation of the setup accuracy of a stereotactic radiotherapy head immobilization mask system using kV on-board imaging. *J. Appl. Clin. Med. Phys.* 11(3): 26–37.

Al-Rodhan, N. R., and P. J. Kelly. 1992. Pioneers of stereotactic neurosurgery. *Stereotact. Funct. Neurosurg.* 58: 60–6.

Arle, J. 2009. Development of a classic: The Todd-Wells apparatus, the BRW, and the CRW stereotactic frames. In *Textbook of Stereotactic and Functional Neurosurgery*, 2nd ed., eds. A. M. Lozano, P. L. Gildenberg, and R. R. Tasker, 454–67. Berlin/Heidelberg: Springer-Verlag.

Ashamalla, H., D. Addeo, N. C. Ikoro, P. Ross, M. Cosma, and N. Nasr. 2003. Commissioning and clinical results utilizing the Gildenberg-Laitinen adapter device for x-ray in fractionated stereotactic radiotherapy. *Int. J. Radiat. Oncol. Biol. Phys.* 56(2): 592–8.

Ashamalla, H., P. Ross, and N. C. Ikoro. 1999. Considerations of fractionated stereotactic radiotherapy in patients with dentures. *J. Radiosurg.* 2: 89–93.

Benedict, S. H., F. J. Bova, B. Clark et al. 2008. Anniversary paper: The role of medical physicists in developing stereotactic radiosurgery. *Med. Phys.* 35: 4262–77.

Bullard D. E., and B.S. Nashold, Jr. 1995. Evolution of principles of stereotactic neurosurgery. *Neurosurg. Clin. N. Am.* 6(1): 27–41.

Blokzijl, E., M. Hollander, R. Bolt, A. Borden, H. Langendijk, and H. Bijl. 2008. Frameless fixation system for linac based intracranial radiosurgery has superior positioning accuracy compared with the established invasive frame system. *Proceedings of the European Society for Therapeutic Radiology and Oncology (ESTRO)*, Göteborg, Sweden.

Bova, F. J., J. M. Buatti, W. A. Friedman, W. M. Mendenhall, C. Yang, and C. Liu. 1997. The University of Florida frameless high-precision stereotactic radiotherapy system. *Int. J. Radiat. Oncol. Biol. Phys.* 38: 875–82.

Brainlab. 2004. RT/RS stereotactic hardware, Germany, Brainlab AG.

Brown, R. A. 1979. A computerized tomography-computer graphics approach to stereotaxic localization. *J. Neurosurg.* 50: 715–20.

Delannes, M., N. J. Daly, J. Bonnet, J. Sabatier, and M. Tremoulet. 1991. Fractionated radiotherapy of small inoperable lesions of the brain using a non-invasive stereoadapter. *Int. J. Radiat. Oncol. Biol. Phys.* 21(3): 749–55.

Delannes, M., N. Daly-Schweitzer, J. Sabatier et al. 1994. Fractionated brain stereotactic irradiation using a non-invasive frame: Technique and preliminary results. *Radiat. Oncol. Investig.* 2: 92–8.

Dunbar, S. F., N. J. Tarbell, H. M. Kooy et al. 1994. Stereotactic radiotherapy for pediatric and adult brain tumors: A preliminary report. *Int. J. Radiat. Oncol. Biol. Phys.* 30: 531–9.

Elekta. 1998a. Leksell coordinate frame model G technical manual. Stockholm: Elekta Instrument.

Elekta. 1998b. Leksell CT indicator technical manual. Stockholm: Elekta Instrument.

Elekta. 1998c. Leksell MR indicator technical manual. Stockholm: Elekta Instrument.

Elekta. 1998d. Leksell x-ray indicator with markers technical manual. Stockholm: Elekta Instrument.

Galloway, R. L., and R. J. Maciunas. 1990. Stereotactic neurosurgery. *Crit. Rev. Biomed. Eng.* 18(3): 181–205.

Gildenberg, P. 1990. The history of stereotactic neurosurgery. *Neurosurg. Clin. N. Am.* 1(4): 765–80.

Gildenberg, P. L. 1998. In *Textbook of Stereotactic and Functional Neurosurgery*, eds. P. L. Gildenberg and R. R. Tasker, 169–90. New York: McGraw-Hill.

Gill, S. S., D. G. Thomas, A. P. Warrington, and M. Brada. 1991. Relocatable frame for stereotactic external beam radiotherapy. *Int. J. Radiat. Oncol. Biol. Phys.* 20: 599–603.

Golden, N. M., T. Tomita, A. G. Kepka, T. Bista, and M. H. Marymont. 1998. The use of the Laitinen stereoadapter for three-dimensional conformal stereotactic radiotherapy. *J. Radiosurg.* 1(3): 191–200.

Graham, J. D., A. P. Warrington, S. S. Gill, and M. Brada. 1991. A non-invasive, relocatable stereotactic frame for fractionated radiotherapy and multiple imaging. *Radiother. Oncol.* 21(1): 60–2.

Hariz, M. I. 1990. A non-invasive adaptation system for computed tomography-guided stereotactic neurosurgery. Thesis, University of Umea, Umea, Sweden.

Hariz, M. I., and A. T. Eriksso. 1986. Reproducibility of repeated mountings of noninvasive CT/MRI stereoadapter. *Appl. Neurophysiol.* 49(6): 336–47.

Hariz, M. I., and L. V. Laitinen. 2009. Laitinen stereotactic appartus. In *Textbook of Stereotactic and Functional Neurosurgery*, 2nd ed., eds. A. M. Lozano, P. L. Gildenberg, and R. R. Tasker, 511–20. Berlin/Heidleberg: Springer-Verlag.

Hassler, R., and T. Riechert. 1957. A case of bilateral fornicotomy in so-called temporal epilepsy. *Acta Neurochir.* 5: 330–40.

Heilbrun, M. P., T. S. Roberts, M. L. J. Apuzzo, T. H. Wells, and J. K. Sabshin. 1983. Preliminary experience with Brown-Robert-Wells (BRW) computerized tomography stereotactic guidance system. *J. Neurosurg.* 59: 217–22.

Horsley, V., and R. H. Clarke. 1908. The structure and functions of the cerebellum examined by a new method. *Brain* 31: 45–124.

Jensen, R. L., J. L. Stone, and R. A. Hayne. 1996. Introduction of human Horsley-Clarke stereotactic frame. *Neurosurgery* 38(3): 563–7.

Kalapurakal, J. A., I. Zainab, A. G. Kepka et al. 2001. Repositioning accuracy with the Laitinen frame for fractionated stereotactic radiation therapy in adult and pediatric brain tumors: Preliminary report. *Radiology* 218(1): 157–61.

Kamath, R., T. C. Ryken, S. L. Meeks, E. C. Pennington, J. Ritchie, and J. M. Buatti. 2005. Initial clinical experience with frameless radiosurgery for patients with intracranial metastases. *Int. J. Radiat. Oncol. Biol. Phys.* 61(5): 1467–72.

Kooy, H. M., S. F. Dunbar, N. J. Tarbell et al. 1994. Adaptation and verification of the relocatable Gill-Thomas-Cosman frame in stereotactic radiotherapy. *Int. J. Radiat. Oncol. Biol. Phys.* 30(3): 685–91.

Laitinen, L. V. 1987. Noninvasive multipurpose stereoadapter. *Neurol. Res.* 9(2): 137–41.

Laitinen, L. V., B. Liliequist, M. Fagerlund, and A. T. Eriksson. 1985. An adapter for computed tomography-guided stereotaxis. *Surg. Neurol.* 23: 559–66.

Leybovich, L. B., A. Sethi, N. Dogan, E. Melian, M. Krasin, and B. Emami. 2002. An immobilization and localization technique for SRT and IMRT of intracranial tumors. *J. Appl. Clinical. Med. Phys.* 3(4): 317–22.

Lunsford, L. D., D. Kondziolka, and D. Leksell. 2009. Leksell stereotactic apparatus. In *Textbook of Stereotactic and Functional Neurosurgery*, eds. A. M. Lozano, P. L. Gildenberg, and R. R. Tasker, 496–512, Berlin Heidelberg: Springer-Verlag.

Lutz, W., K. R. Winston, and N. Maleki. 1988. A system for stereotactic radiosurgery with a linear accelerator. *Int. J. Radiat. Oncol. Biol. Phys.* 14: 373–81.

Maciunas, R. J., R. L. Galloway, and J. W. Latimer. 1994. The application accuracy of stereotactic frames. *Neurosurgery* 35(4): 682–94.

Meeks, S. L., F. J. Bova, W. A. Friedman, J. M. Buatti, R. D. Moore, and W. M. Mendenhall. 1998. IRLED-based patient localization for linac radiosurgery. *Int. J. Radiat. Oncol. Biol. Phys.* 41(2): 433–9.

Meeks, S. L., F. J. Bova, T. H. Wagner, J. M. Buatti, W. A. Friedman, and K. D. Foote. 2000. Image localization for frameless stereotactic radiotherapy. *Int. J. Radiat. Oncol. Biol. Phys.* 46(5): 1291–9.

Miyabe, Y., S. Yano, T. Okada et al. 2004. Evaluation of positioning reproducibility using a relocatable head frame for stereotactic radiotherapy. *Nippon Hoshasen Gijutsu Gakkai Zasshi* 60(10): 1444–51.

Phillips, M. H., K. Singer, E. Miller, and K. Stelzer. 2000. Commissioning an image-guided localization system for radiotherapy. *Int. J. Radiat. Oncol. Biol. Phys.* 48(1): 267–76.

Roberts, T. S. 1998. The BRW/CRW stereotactic apparatus. In *Textbook of Stereotactic and Functional Neurosurgery*, eds. P. L. Gildenberg and R. R. Tasker, 65–71. New York: McGraw-Hill.

Ryken, T. C., S. L. Meeks, E. C. Pennington et al. 2001. Initial clinical experience with frameless stereotactic radiosurgery: Analysis of accuracy and feasibility. *Int. J. Radiat. Oncol. Biol. Phys.* 51(4): 1152–8.

Salter, B. J., M. Fuss, D. G. Vollmer et al. 2001. The TALON removable head frame system for stereotactic radiosurgery/radiotherapy: Measurement of the repositioning accuracy. *Int. J. Radiat. Oncol. Biol. Phys.* 51: 555–62.

Spiegel, E. A., H. T. Wycis, E. G. Szekely, M. Marks, and A. S. Lee. 1947. Stereotaxic apparatus for operations on the human brain. *Science* 106: 349–50.

Tome, W. A., S. L. Meeks, J. M. Buatti, F. J. Bova, W. A. Friedman, and Z. Li. 2000. A high-precision system for conformal intracranial radiotherapy. *Int. J. Radiat. Oncol. Biol. Phys.* 47(4): 1137–43.

Verbakel, W., F. Lagerwaard, L. Verdui, S. Heukelom, and B. Slotman. 2010. The accuracy of frameless stereotactic intracranial radiosurgery. *Radiother. Oncol. Radiother. Oncol.* 97(3): 390–4.

Weidlich, G. A., J. A. Gebert, and J. J. Fuery. 1998. Clinical commissioning of Laitinen stereoadapter for fractionated stereotactic radiotherapy. *Med. Dosim.* 23(4): 302–6.

Wells, T. H. 1998. The Todd-Wells apparatus. In *Textbook of Stereotactic and Functional Neurosurgery*, eds. P. L. Gildenberg and R. R. Tasker, 65–71. New York: McGraw-Hill.

Yang, C. C. 1994. High precision treatment for fractionated stereotactic radiotherapy. Ph.D. dissertation, University of Florida, Gainesville, FL.

Yeung, D., V. A. Frouhar, J. Fontanesi, and J. Palta. 1994a. Repositional accuracy of a commercial relocatable head frame for fractionated stereotactic radiotherapy (abstr.). *Med. Phys.* 2(1): 920.

Yeung, D., J. Plata, J. Fontanesi, and L. Kun. 1994b. Systematic analysis of errors in target localization and treatment delivery in stereotactic radiosurgery (SRS). *Int. J. Radiat. Oncol. Biol. Phys.* 28(2): 493–8.

8 Frames for extracranial treatment

D. Michael Lovelock

Contents

8.1 OBJECTIVES

Central to the implementation of stereotactic body radiation therapy (SBRT) is the hypothesis that the improved spatial accuracy of dose delivery to the target structures permits the use of smaller margins than those used in traditional radiotherapy. This improves the therapeutic ratio, making possible the large increases in the biologically effective doses characteristic of SBRT.

To achieve a safe reduction in the margins surrounding both target and critical structures, however, careful attention must be given to patient positioning and setup. Ideally,

- The relative positions of the target and critical structures at the time of the planning CT scan are reproduced in the treatment room just prior to dose delivery.
- The anatomical reference point is positioned at the radiation isocenter or treatment machine reference point.
- The positions of the target and critical structures remain fixed throughout the dose delivery or, in the case of targets influenced by respiratory motion, the pattern of respiratory motion remains constant.

In practice, the dose distribution delivered to the patient may differ from that planned. A change in the pose of the patient between the planning scan and treatment may make registration difficult, soft issues may move and deform, and patients are unlikely to keep perfectly still during the long treatment times associated with SBRT. These problems can be addressed or at least ameliorated with the use of a patient immobilization system specifically designed for SBRT treatment.

These systems, variously called whole body immobilization systems, stereotactic body frames, or patient immobilization cradles, perform several functions:

- Constrain the patient's body, thereby minimizing the chance of inadvertent patient motion during treatment
- Maximize patient comfort

- Provide a means to ensure that the patient position and pose on the expanded foam or vacuum cushions used to support the patient are the same as those set at the time of the simulation planning CT scan
- Through the use of external scales or a stereotactic localizer frame, provide a means of positioning the body frame and patient in the treatment room in preparation for setup using image guidance
- Optionally, provide a means of applying abdominal compression to reduce respiratory motion

8.2 STEREOTAXY AND IMAGE GUIDANCE

The use of a body frame and its use in the setup of patients undergoing hypofractioned RT was first described by Ingmar Lax and Henric Blomgren at Karolinska Hospital, Stockholm (Lax et al. 1994). Their approach was strictly stereotactic; the body frame had both an internal and an external coordinate system (Figure 8.1). The internal system consisted of a set of wires, some of which were set obliquely into the base and sides of the body frame that were visible in the CT. To use the stereotactic approach to target positioning, a reference point, possibly a soft tissue feature seen in the CT, is chosen. Its position with respect to the positions of the wires seen in the CT slice containing the feature is measured, thus the three-dimensional (3-D) coordinates of the feature in the internal system are determined. The coordinates of this reference point are then expressed in the external coordinate system. This consists of a set of scales affixed to the body frame or an accessory positioning arch. Setup of this soft tissue reference point to the machine isocenter is achieved in part by positioning the body frame on the couch of the treatment machine such that

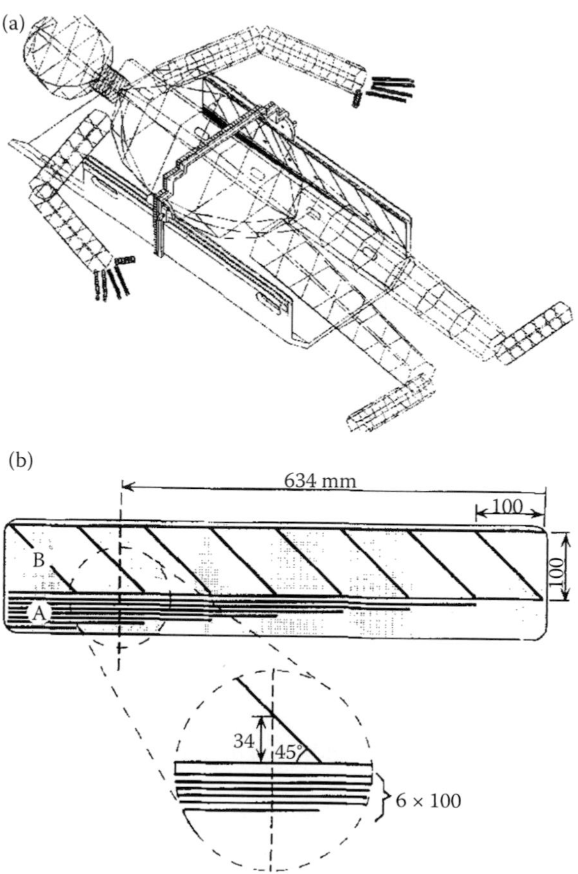

Figure 8.1 (a) A schematic view of the Karolinska stereotactic body frame and position of the patient. (b) The pattern of counting wires and oblique wires embedded in the side of the frame allowed the longitudinal position of the target to be determined. (From Lax, I. et al., *Acta Oncol.*, 33, 6, 677–83, 1994.)

the isocenter room lasers strike the external scales at the correct coordinates. What remains, however, is to ensure that the target is positioned correctly with respect to the radiation beams. This was approached by attempting to reproduce the position of the patient in the body frame. To do this, isocenter and alignment tattoos were placed on the trunk (preferably the sternum) and both tibial tuberosities. A laser mounted on the body frame was used to ensure optimum patient alignment. In the initial implementations, the patient was set up in the body frame and the body frame positioned in the treatment room using the room lasers. This approach of setting up to coordinates and treating was the best that could be done in the early 1990s as in-room image guidance was not available. Several investigators have investigated the setup accuracy of bony targets using this approach (Fuss et al. 2004; Lohr et al. 1999; Negoro et al. 2001; Wulf et al. 2000). The average 3-D displacements of the bony target from the simulation position ranged from 3.6 to 4.9 mm. Note that soft tissue targets will have poorer reproducibility. The setup errors seen in individual treatments, of prime concern to hypofractionated and single-fraction SBRT, may be much greater.

The advent of treatment machines with image guidance has greatly improved the setup accuracy of both bony and soft tissue targets. Targets that can be imaged or implanted with radio-opaque markers can be localized with respect to the radiation beams directly. The use of image guidance for all SBRT setups is mandated in the American Association of Medical Physicists Task Group Report TG101 (Benedict et al. 2010, p. 4085):

> For SBRT, image-guided localization techniques shall be used to guarantee the spatial accuracy of the delivered dose distribution with a high confidence level. Body frames and associated fiducial systems may be used for immobilization and coarse localization; however, they shall not be used as a sole localization technique.

Thus, the stereotactic body frame still has an important role in image-guided treatment procedures.

8.3 PATIENT POSITIONING

The goal of the patient positioning procedure prior to each treatment delivery is to reproduce the setup and pose of the patient within the body frame that was established at the time of simulation. By minimizing patient translations and rotations within the immobilization system being used, the risk of target rotation or deformation and any change in the spatial relationship of nearby critical structures with the target is reduced. Patient positioning within the immobilization system is generally accomplished by placing skin marks or tattoos on the patient at the time of simulation. Their positions in the immobilization frame coordinate system can be recorded using either room lasers or a laser attached to the frame itself. Typically, tattoos are placed to triangulate the isocenter with additional superior, inferior, and lateral tattoos to help with patient alignment in the immobilization system. Using the lasers, therapists can readily verify there is no twist or rotation of the patient during the setup procedure. At the time of treatment, in preparation for the localization step, the position of the patient in the immobilization frame is adjusted to bring the tattoos to their original coordinates. Although this is done to make localization less susceptible to error, it must be recognized that significant changes in the positions of the target and critical structures may still exist even if all tattoos are correctly positioned. Shifts in skin position, the variable filling of bladder and rectum, the dynamic nature of bowel gas, difficulties in repositioning the scapulae, and changes in diaphragm position during a gated treatment (von Siebenthal et al. 2007), for example, cannot be addressed by patient positioning.

8.4 PATIENT COMFORT

The time the patient spends in the treatment position can be very long, 1 to 1.5 hr is not unusual for high-dose single-fraction treatments. Even the most highly motivated patient is likely to move if he or she becomes uncomfortable or is in pain. During both simulation and treatment, attention must be given to ensuring the patient is comfortable. The measures to be taken are often simple; the use of a knee cushion helps to reduce strain on the lower back, for example. When face masks are used, particular attention needs to be given to the technique used to make them. If the time the mask is left in the water bath is incorrect or if it is removed from the patient too quickly, it may shrink. In effect, this compresses the patient's head, which may result in severe discomfort, nausea, numbness, and other problems. Not all head shapes will be accommodated by the standard set of head cushions included with most immobilization systems. Several manufacturers offer customizable head cushions that can be molded to the patient's head contour to improve comfort and support for the lower cervical spine (Figure 8.2) (Bentel et al. 1995; Houweling et al. 2010). The position of the arms must also be considered. An arms-up position may be advantageous when planning lung and liver treatments, for example, but it may also be very difficult for the patient to maintain over long treatment delivery times (Figure 8.3). Furthermore, the ability to maintain such a position varies with individual patients, and problems may not become apparent until after treatment has begun. The commercial immobilization systems may offer various accessories that support a patient's raised arms. A careful assessment of the clinical requirements and how well the arm-support accessories meet them must be made.

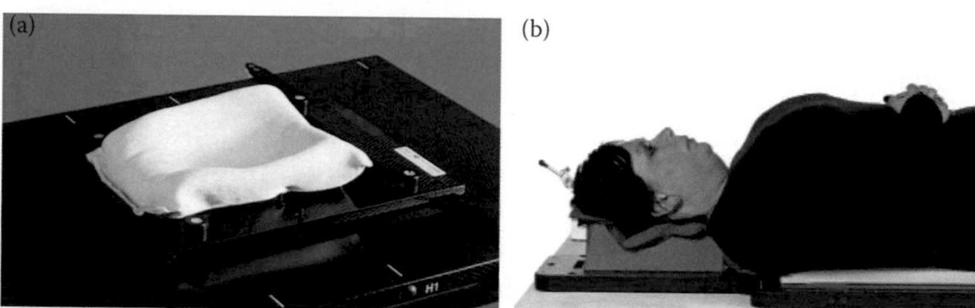

Figure 8.2 Custom head cushions. Each system allows the cushion to be molded to the shape of the patient's head during simulation, reducing the likelihood that the patient will experience pain or numbness. (a) The Accuform cushion (Civco Medical Solutions) is a soft fabric bag filled with expanded polystyrene beads shown placed over a standard Silverman head cushion. Moisture is used to set the shape. (Image courtesy of Civco Medical Solutions.) (b) A vacuum-bag system (Orfit Industries) designed to fit over a standard head cushion, an option with Orfit immobilization systems. (Image courtesy of Orfit Industries.)

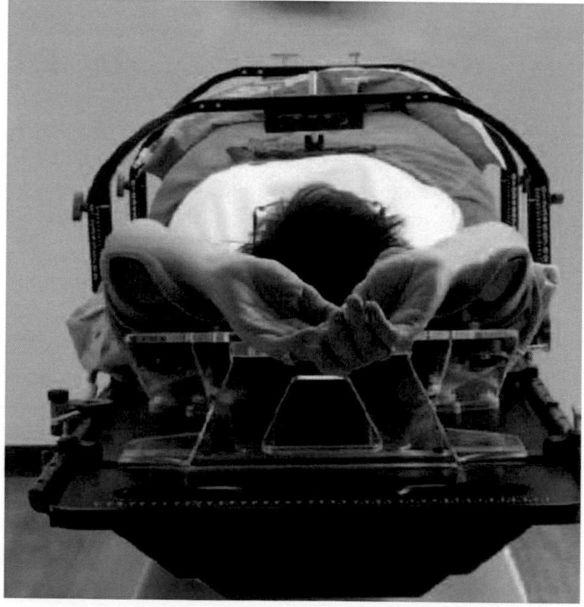

Figure 8.3 Patient comfort. Support devices integral to the immobilization system are used to make the procedure less tiring for the patient. Arm support for the Body Pro-Lok from Civco. (Image courtesy of Civco Medical Solutions.)

8.5 PATIENT IMMOBILIZATION

Because of the high doses delivered during each fraction, the mistargeting resulting from patient motion during treatment delivery will be impossible to correct for by modification of subsequent fractions; thus, it is generally advisable to immobilize the patient for SBRT treatment. The immobilization systems cannot positively immobilize either the bony anatomy or soft tissue of the patient. The approach that is taken by most manufacturers is to support the torso of the patient with a cushion that has been molded to his or her body contour during the simulation procedure. Two systems in use are the vacuum cushion and the expanded foam cushion. The vacuum cushion (or "beanbag") is a large bag filled with small polystyrene spheres. The therapists first distribute the polystyrene spheres within the bag appropriately and then place the bag under the patient. A slight vacuum is applied, causing the cushion to shrink slightly. The bag is then sealed, and the partial vacuum within causes the bag to retain its shape for the duration of the simulation and treatments. The expanded foam cushions are constructed by mixing together the two components of the foaming agent and pouring the mix into a thin plastic bag. The bag is taped shut; the therapists spread the mix uniformly around the bag and then place the bag under the patient. The foam expands, causing the bag to mold itself around the patient. After about 10–15 min, the foam sets, resulting in the final support cushion. Each system has its advantages. The vacuum bags are reusable after the end of a patient's treatment, and the expanding foam system never suffers from a loss of the partial vacuum at any point during the treatment process. Many of the commercial systems permit the use of either type of cushion.

Additional measures are provided to further stabilize the patient position. See, for example, the cushioned arch in the Qfix system (Figure 8.4) placed over the thighs to stabilize the pelvis and the lateral paddles in the Aktina system (Figure 8.5) to help prevent inadvertent patient motion.

8.6 ABDOMINAL COMPRESSION

Abdominal compression to reduce the respiratory motion of targets, such as lung and liver, has been used since the beginning of SBRT (Lax et al. 1994). Although there are a number of newer motion management techniques, such as gating and tracking (Keall et al. 2006), it retains some advantages: It is easy to implement, it does not require a 4DCT planning scan, it is compatible with a conventional C-arm gantry linac, and initial setup can be done with a standard (nongated or non-4D) cone-beam scan. Daly, Perks, and Chen (2013) report that abdominal compression is the most commonly used motion-management technique in thoracic SBRT in the United States, being used by 51% of those surveyed in their patterns-of-care study. Two methods are used to apply pressure to the abdomen. In one approach, a small plate is placed on the patient's abdomen, usually just below the ribs (see Figures 8.4, 8.6, 8.7, and 8.8). This is attached to an arch-and-screw system that permits the force on the abdomen to be adjusted. The other technique uses a tight belt fitted to the patient. Here the pressure is applied pneumatically. An air bladder integral to the belt is positioned over the patient's abdomen. The air pressure within the bladder

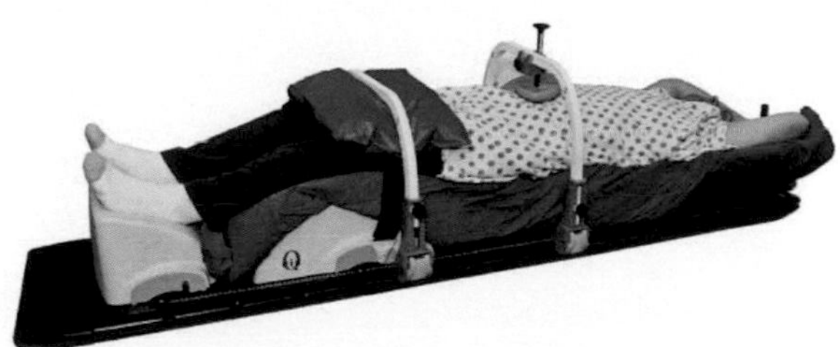

Figure 8.4 KVue™ Stradivarius SBRT—QFix. The Stradivarius insert system is designed to be used with the KVue couch top. It features a central cutout for unobstructed treatment paths. It is optimized to significantly reduce secondary electron production from beams that would otherwise pass through the couch. The Stradivarius Overlay system is designed to be used with existing CT and Linac couches. A 230-cm-long whole body vacuum bag supports the patient. Abdominal compression is applied using an arch with an adjustable screw. The arch with custom cushions over the knees or thighs provides additional immobilization for the patient. (Image courtesy of QFix, Avondale, PA.)

Figure 8.5 Memorial Body Cradle—Aktina. The system uses no baseplate, thus eliminating baseplate bolusing. Patient-support cushions index to and are supported by the couch top. A unique feature of this system is the use of four radiolucent paddles positioned on either side of the hips and chest, which are used to comfortably immobilize the patient. The system can be used with either vacuum or expanding foam cushions. (Image courtesy of Aktina Medical Corporation.)

is controlled with a pump and gauge, similar to that used in a sphygmomanometer. An advantage of the pneumatic systems used in both pressure plate and belt approaches is that the pressure can be applied reproducibly at simulation and at each treatment. The approach is effective; several groups report motion reduced to an average peak-to-peak excursion of 5–8 mm (Heinzerling et al. 2008; Hof et al. 2003; Negoro et al. 2001; Wunderink et al. 2008).

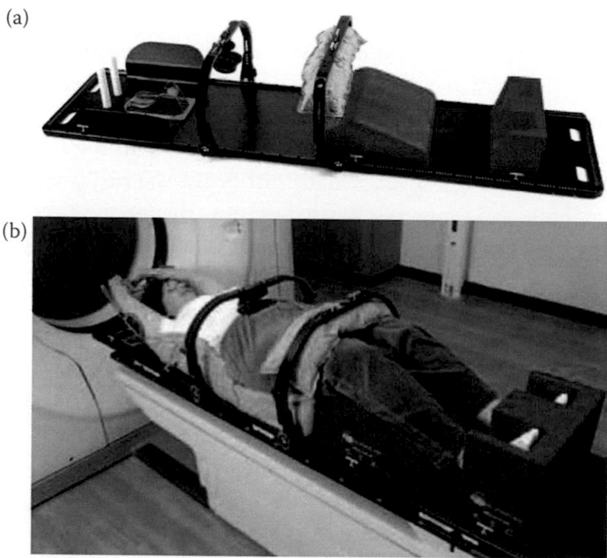

Figure 8.6 Body-Pro-Lok—Civco Medical Solutions. (a) A variety of attachments including respiratory belts and plates, forehead restraint, hand grips, and a shoulder-restraint bridge can be attached to the company's universal couch-top. Shown are a wing-board with hand-grips, a bridge with respiratory plate, a second bridge with a cushion used to assist with immobilization, a knee cushion and a foot localizer. (b) The system with a patient, showing the use of the arm support and vacuum cushion. (Images courtesy of Civco Medical Solutions.)

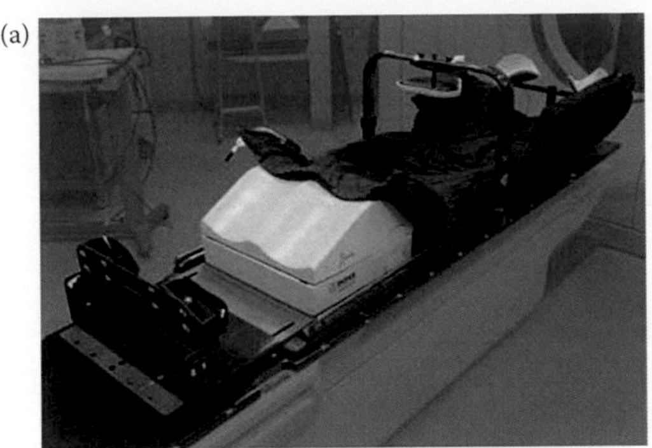

Figure 8.7 Omni V Positioning System—Bionix Radiation Therapy. (a) The system consists of a lightweight radiolucent baseplate and a set of accessories such as an arch used for abdominal compression, thigh/knee cushions and a foot positioner that are indexed to the baseplate. A set of external scales provide the coordinate system used for setup in the room. (b) For stereotactic location of structures including soft issues, an external fiducial arch can be used. Vacuum cushions used for patient support are indexed to the baseplate. (Images courtesy of Bionix Radiation Therapy.)

Figure 8.8 SBRT Solution—Orfit. (a) A variety of custom cushions that provide hand grips, support for the head, arms, knees, and a foot localizer can be attached and indexed to a baseplate. Abdominal compression is done either with an arch with adjustable screw (illustrated), or with a compression belt with inflatable air bladder. The system can also be used with vacuum bags under the patient for additional patient comfort. (b) A unique feature of this system is the ability to use a range of body masks to help immobilize the patient. These attach to slots that run the length of the base. (Images courtesy of Orfit Industries, Belgium.)

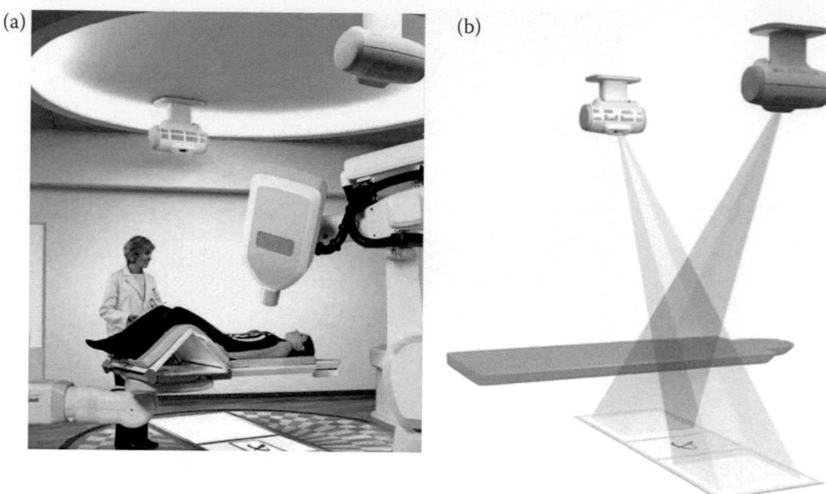

Figure 8.9 The Cyberknife system. (a) The Linac and patient support are mounted on robotic arms. (b) Dual X-ray imaging systems, with the sources mounted on the ceiling and the imaging panels below the floor permit rapid 3D localization of the target or implanted fiducial markers in real time throughout treatment delivery.

8.7 INTRAFRACTIONAL IMAGING

The need for immobilization is also influenced by the ability of the dose-delivery system to both detect and correct for the changes in target or critical structure position that may occur during treatment. Systems such as the CyberKnife that use dual in-room kV x-ray units are capable of acquiring 50 or more pairs of orthogonal radiographs during a single fraction delivery (Figure 8.9). Using registration of the radiographs, small changes in target position and rotation are determined. The robotic arm compensates automatically for the observed translations and rotations by adjusting the delivery position of the subsequent beams. This system is designed to track targets influenced by respiratory motion (Hoogeman et al. 2009) as well as targets, such as spinal tumors, that are subject to small random shifts in patient position (Furweger et al. 2010). Although there is a small time delay, the patient is effectively tracked throughout the treatment procedure, thereby reducing the need for immobilization. Such frequent 3-D target localization using a conventional linac without in-room imaging would be prohibitively time consuming; thus, the usual approach is to immobilize the patient.

8.8 HEAD-AND-NECK SYSTEMS

Patients with targets in the head and neck or upper cervical spine can be set up using the commercial face-mask systems used to immobilize head-and-neck patients. Although initial setup errors can be up to 8 mm, the intrafractional motion is much smaller, typically 3 mm or less (Kang et al. 2011; Linthout et al. 2006). Thus, the face-mask immobilization systems, when coupled with image guidance, can provide a means to set up a patient for SBRT to the head, neck, and upper cervical spine. The use of systems in which the mask extends down to cover the shoulders is favored by some clinics (Houweling et al. 2010; Linthout et al. 2006) although Lena Sharp and coworkers report little benefit (Sharp et al. 2005). Some systems permit the use of a head cushion customized to each patient (Figure 8.2). This has been reported to reduce both the inter- and intrafractional motion and deformation (Bentel et al. 1995; Houweling et al. 2010).

8.9 COMMERCIAL SYSTEMS

There are several additional considerations in selecting a system to best fit the needs of a clinic. A clinic with a high throughput may need several systems for placement in the simulation suite and treatment rooms. Some systems may be better suited for specific disease sites. The bolusing effect of any baseplate and support cushions, along with the couch itself, may be important when treating posterior lesions, such as on a rib, for example (Hoppe et al. 2008). This can result in skin doses being much higher than indicated by the treatment-planning system. Measurements using thermoluminescent dosimetry or optically stimulated dosimetry may be necessary. The upcoming AAPM report of Task Group 176, "Dosimetric Effects Caused by Couch Tops and Immobilization Devices," will review the bolusing effects of immobilization systems. The importance of immobilizing the head and the use of a face mask, vacuum cushion, or expanded foam all need to be considered. The systems may also have a large number of accessories, so storage space will be needed in the treatment vault and simulation suite.

Several commercial patient immobilization systems are currently available, seven of which are illustrated in Figures 8.4 through 8.8, 8.10, and 8.11.

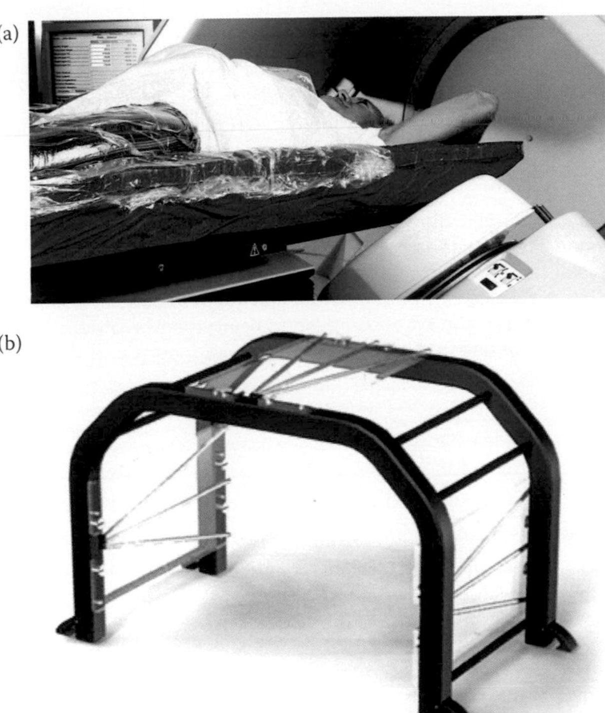

Figure 8.10 BodyFIX – Elekta. (a) This system is unique in that two separate vacuum systems are used, one for the patient support vacuum bag, and another to maintain a partial vacuum between a cover sheet that covers the patient from the upper chest to the feet, and the vacuum support cushion. The pressure from the cover sheet helps to gently immobilize the patient. (b) Stereotactic localizers equipped with either a CT fiducial system at simulation or patient specific marking sheets at treatment can be used for stereotactic target positioning.

(a)

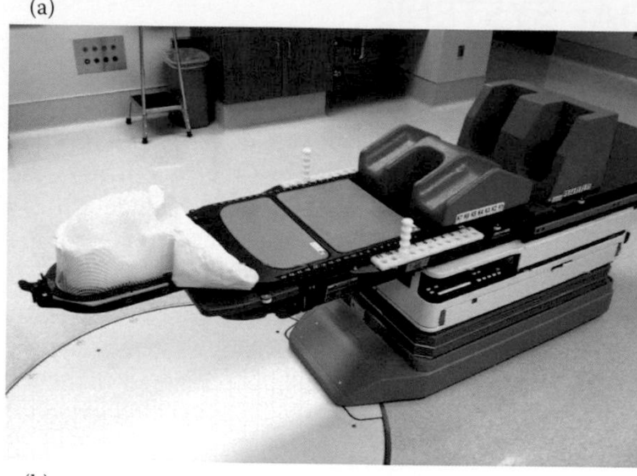

(b)

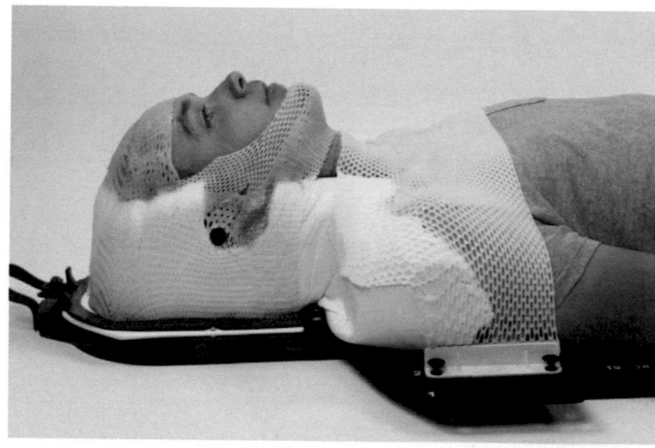

Figure 8.11 Freedom Patient Immobilization System—CDR Systems. (a) This system is designed to accommodate both SRS and SBRT treatment sites. This system is lightweight (< 10 lbs) and compatible with all common treatment couches. An arch and adjustable screw system, or pneumatic bladder system, are available for abdominal compression. The system is compatible with both vacuum bags and expanding foam cushioning. (b) The head and neck immobilization can be accomplished using the MayoMold™, an expanding foam system that is formed around the posterior and lateral aspects of the head, neck and shoulders.

8.10 SUMMARY

An immobilization system and its associated clinical procedures form an important component of an SBRT program. It provides a means by which therapists can reproducibly set up and comfortably immobilize patients for treatment and can perform the initial positioning of the frame and patient on the treatment machine couch prior to image guidance.

REFERENCES

Benedict, S. H., K. M. Yenice, D. Followill et al. 2010. Stereotactic body radiation therapy: Report of AAPM Task Group 101. *Med. Phys.* 37(8): 4078–101.

Bentel, G. C., L. B. Marks, G. W. Sherouse, and D. P. Spencer. 1995. A customized head and neck support system. *Int. J. Radiat. Oncol. Biol. Phys.* 32(1): 245–8.

Daly, M. E., J. R. Perks, and A. M. Chen. 2013. Patterns-of-care for thoracic stereotactic body radiotherapy among practicing radiation oncologists in the United States. *J. Thorac. Oncol.* 8: 202–7.

Furweger, C., C. Drexler, M. Kufeld, A. Muacevic, B. Wowra, and A. Schlaefer. 2010. Patient motion and targeting accuracy in robotic spinal radiosurgery: 260 single-fraction fiducial-free cases. *Int. J. Radiat. Oncol. Biol. Phys.* 78(3): 937–45.

Fuss, M., B. J. Salter, P. Rassiah, D. Cheek, S. X. Cavanaugh, and T. S. Herman. 2004. Repositioning accuracy of a commercially available double-vacuum whole body immobilization system for stereotactic body radiation therapy. *Technol. Cancer Res. Treat.* 3(1): 59–67.

Heinzerling, J. H., J. F. Anderson, L. Papiez et al. 2008. Four-dimensional computed tomography scan analysis of tumor and organ motion at varying levels of abdominal compression during stereotactic treatment of lung and liver. *Int. J. Radiat. Oncol. Biol. Phys.* 70(5): 1571–8.

Hof, H., K. K. Herfarth, M. Münter, M. Essig, M. Wannenmacher, and J. Debus. 2003. The use of the multislice CT for the determination of respiratory lung tumor movement in stereotactic single-dose irradiation. *Strahlenther. Onkol.* 179(8): 542–7.

Hoogeman, M., J. B. Prévost, J. Nuyttens, J. Poll, P. Levendag, and B. Heijmen. 2009. Clinical accuracy of the respiratory tumor tracking system of the CyberKnife: Assessment by analysis of log files. *Int. J. Radiat. Oncol. Biol. Phys.* 74(1): 297–303.

Hoppe, B. S., B. Laser, A. V. Kowalski et al. 2008. Acute skin toxicity following stereotactic body radiation therapy for stage I non-small-cell lung cancer: Who's at risk? *Int. J. Radiat. Oncol. Biol. Phys.* 72(5): 1283–6.

Houweling, A. C., S. van der Meer, E. van der Wal, C. H. Terhaard, and C. P. Raaijmakers. 2010. Improved immobilization using an individual head support in head and neck cancer patients. *Radiother. Oncol.* 96(1): 100–3.

Kang, H., D. M. Lovelock, E. D. Yorke, S. Kriminski, N. Lee, and H. I. Amols. 2011. Accurate positioning for head and neck cancer patients using 2D and 3D image guidance. *J. Appl. Clin. Med. Phys.* 12(1): 3270.

Keall, P. J., G. S. Mageras, J. M. Balter et al. 2006. The management of respiratory motion in radiation oncology: Report of AAPM Task Group 76. *Med. Phys.* 33(10): 3874–900.

Lax, I., H. Blomgren, I. Naslund, and R. Svanstrom. 1994. Stereotactic radiotherapy of malignancies in the abdomen. Methodological aspects. *Acta Oncol.* 33(6): 677–83.

Linthout, N., D. Verellen, K. Tournel, and G. Storme. 2006. Six dimensional analysis with daily stereoscopic x-ray imaging of intrafraction patient motion in head and neck treatments using five points fixation masks. *Med. Phys.* 33(2): 504–13.

Lohr, F., J. Debus, C. Frank et al. 1999. Noninvasive patient fixation for extracranial stereotactic radiotherapy. *Int. J. Radiat. Oncol. Biol. Phys.* 45(2): 521–7.

Negoro, Y., Y. Nagata, T. Aoki et al. 2001. The effectiveness of an immobilization device in conformal radiotherapy for lung tumor: Reduction of respiratory tumor movement and evaluation of the daily setup accuracy. *Int. J. Radiat. Oncol. Biol. Phys.* 50(4): 889–98.

Sharp, L., F. Lewin, H. Johansson, D. Payne, A. Gerhardsson, and L. E. Rutqvist. 2005. Randomized trial on two types of thermoplastic masks for patient immobilization during radiation therapy for head-and-neck cancer. *Int. J. Radiat. Oncol. Biol. Phys.* 61(1): 250–6.

von Siebenthal, M., G. Székely, A. J. Lomax, and P. C. Cattin. 2007. Systematic errors in respiratory gating due to intrafraction deformations of the liver. *Med. Phys.* 34(9): 3620–9.

Wulf, J., U. Hadinger, U. Oppitz, B. Olshausen, and M. Flentje. 2000. Stereotactic radiotherapy of extracranial targets: CT-simulation and accuracy of treatment in the stereotactic body frame. *Radiother. Oncol.* 57(2): 225–36.

Wunderink, W., A. Mendez Romero, W. de Kruijf, H. de Boer, P. Levendag, and B. Heijmen. 2008. Reduction of respiratory liver tumor motion by abdominal compression in stereotactic body frame, analyzed by tracking fiducial markers implanted in liver. *Int. J. Radiat. Oncol. Biol. Phys.* 71(3): 907–15.

9 Motion management

Krishni Wijesooriya, Rohini George, and Mihaela Rosu

Contents

9.1 MOTION MANAGEMENT AND TUMOR TRACKING

9.1.1 INTRODUCTION

Intrafraction organ motion, predominantly caused by patient breathing, can compromise the treatment outcome of radiation therapy for tumors in the thorax and abdomen, either by reducing the tumor-control probability with insufficient safety margins or by increasing the normal tissue toxicity by setting excessively large margins. Respiratory motion affects all tumor sites in the thorax and in the abdominal region, resulting in temporal anatomic changes. Such motion may distort the tumor volume and position (Chen, Kung, and Beaudette 2004; Shimizu et al. 2000). It has been shown in the literature that tumor motion due to respiration can be as large as 3 cm, especially in the superior/inferior direction, and heavily depends on the location of the tumor and the individual patient (Table 9.1).

Table 9.1 Lung tumor–motion data

OBSERVER	DIRECTION		
	SI	AP	LR
Barnes et al. (2001)			
Lower lobe	18.5 (9–32)	—	—
Middle, upper lobe	7.5 (2–11)	—	—
Chen et al. (2001)	(0–50)	—	—
Ekberg et al. (1998)	3.9 (0–12)	2.4 (0–5)	2.4 (0–5)
Engelsman et al. (2001)			
Middle, upper lobe	(2–6)	—	—
Lower lobe	(2–9)	—	—
Erridge et al. (2003)	12.5 (6–34)	9.4 (5–22)	7.3 (3–12)
Ross et al. (1990)			
Upper lobe	—	1 (0–5)	1 (0–3)
Middle lobe	—	0	9 (0–16)
Lower lobe	—	1 (0–4)	10.5 (0–13)
Plathow et al. (2004)			
Lower lobe	9.5 (4.5–16.4)	6.1 (2.5–9.8)	6.0 (2.9–9.8)
Middle lobe	7.2 (4.3–10.2)	4.3 (1.9–7.5)	4.3 (1.5–7.1)
Upper lobe	4.3 (2.6–7.1)	2.8 (1.2–5.1)	3.4 (1.3–5.3)
Seppenwoolde et al. (2002)	5.8 (0–25)	2.5 (0–8)	1.5 (0–3)
Shimizu et al. (2000)	—	6.4 (2–24)	—
Sixel et al. (2001)	(0–13)	(0–5)	(0–4)
Stevens et al. (2001)	4.5 (0–22)	—	—

Source: Reproduced from Table 2, AAPM Task Group 76 (Keall, P. J., Mageras, G. S., Balter, J. M. et al., *Med. Phys.*, 33, 10, 3874–3900, 2006). References for each measurement are given in AAPM Task Group 76.

Note: The mean range of motion and the (minimum-maximum) ranges in millimeters for each cohort of subjects.

An example of extreme tumor motion is shown in Figure 9.1. The hysteresis of lung tumor motion is illustrated in Figure 9.2, in which the tumor motion path is changed from inhalation to exhalation and back to inhalation by as much as 5 mm.

Intrafraction motion can introduce significant errors in delivery, such as exposing more healthy tissue to high doses while under-dosing the tumor (Vedam et al. 2001). Tumor motion is most relevant to lung cancer more than other sites, and the prognosis of inoperable lung cancer remains the worst. Even when using a stereotactic body frame to suppress breathing-induced motion, Wulf et al. (2000) showed that with a margin for target variability of 5 mm (A/P and L/R) and 10 mm (cranio-caudal) about 12%–16% of the targets might be partially missed. However, dose escalation to lung tumors is currently ongoing with promising results (Guckenberger et al. 2009b; Nagata et al. 2005; Xia et al. 2006). To escalate target doses to very high levels utilizing stereotactic body radiation therapy (SBRT) while keeping the normal tissue dose (such as mean lung dose and V_{20}) at a minimum, one needs motion management.

There are various techniques currently available in clinics for motion management during simulation, planning, and treatment. It is possible to minimize motion by abdominal compression (Lax et al. 1994); breath hold, which includes voluntary deep inspiratory breath hold (DIBH) (Mah et al. 2000) or active

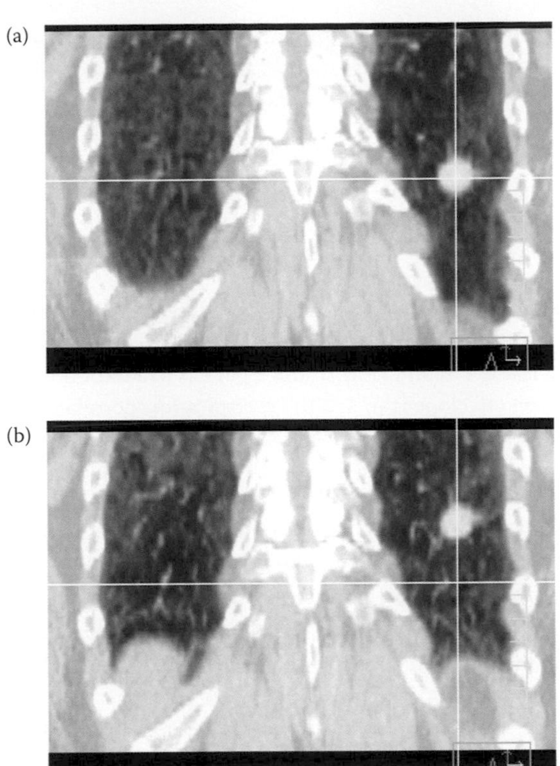

Figure 9.1 Tumor motion of 2.5 cm in cranio-caudal dimension. (a) Inhalation phase image and (b) exhalation phase image.

breathing control (ABC) breath hold using an air volume control device, such as a spirometer (Garcia et al. 2002; Lu et al. 2005; Wong et al. 1999); or respiratory synchronized techniques, such as respiratory gating or real-time tumor tracking with MLC tracking, couch tracking, and using robotics (Guckenberger et al. 2009a; Murphy 2004; Shirato et al. 2000a; Shirato et al. 2000b). For all these motion-management techniques to be successful, one needs a reliable form of image-guided radiation therapy (IGRT). The

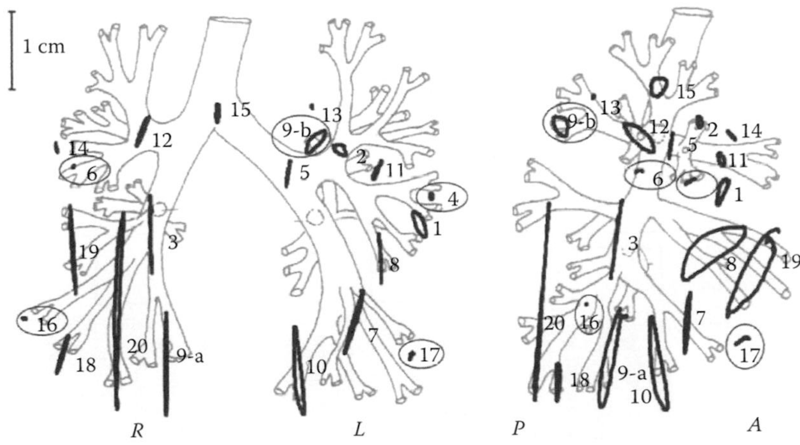

Figure 9.2 1–5 mm hysteresis of breathing trajectories measured. (From Seppenwoolde, Y. et al., *Med. Phys.*, 34, 2774–84, 2007.)

motion can be monitored using IGRT by multiple surrogates: the tumor itself, an artificial marker implanted in or near the tumor, or a surrogate such as the diaphragm (de Mey et al. 2005; Mageras et al. 2001; Mah et al. 2000; Schweikard et al. 2000; Shirato et al. 2000a).

9.1.2 RESPIRATORY MOTION MANAGEMENT

Motion management in the clinic has to be implemented at three different points: simulation, planning, and treatment.

9.1.2.1 4-D simulation

Fast, multislice computed tomography (CT) scanners can "freeze" the tumor in one location at any given moment in the breathing cycle. Because this does not represent the full motion envelope of the tumor, delineating the tumor on a fast CT scan can introduce partial irradiation of the tumor. One technique used to obtain the full motion envelope is to acquire a respiration-correlated scan or 4-D CT (Bosmans et al. 2006; Ford et al. 2003; Keall et al. 2006; Mageras and Yorke 2004; Nehmeh et al. 2004; Vedam et al. 2003). In 4-D CT, the images are taken over several breathing cycles. This type of scan has the patient's individual respiratory motion signal recorded simultaneously so that the CT acquisition and the respiratory signal are synchronized (Figure 9.3). This signal can be from multiple surrogates: The Varian real-time position management (RPM) system (Varian Medical Systems, Palo Alto, CA) uses an external marker block on the patient's chest with an infrared (IR) camera mounted onto the CT couch to capture the RPM signal. The Siemens Anzai pressure belt uses abdominal straps with a pressure sensor, measuring airflow with thermocouples or spirometers in the patient's mouth (Bosmans et al. 2006; Ford et al. 2003; Lu et al. 2005). There are two types of 4-D CT acquisitions: Cine CT, with which the couch stays stationary while repeat CT images are acquired corresponding to different phases of the respiratory cycle and then the couch is incremented (Low et al. 2003; Pan et al. 2004), or helical CT, with which the pitch of the CT scan is varied between 0.5–0.1 and the CT parameters are adjusted such that the CT beam is on for at least one respiratory cycle at each couch position (Keall et al. 2004; Pan 2005). Helical CT is more common and the more efficient of the two techniques.

Retrospective sorting is used by most CT scanners to obtain the individual 3-D CT images that correspond to the respective motion phases or motion amplitudes (Abdelnour et al. 2007; Lu et al. 2006) of a 4-D CT scan. There are two image-binning approaches used to create 4-D CT images: Phase binning (PB) assigns each image according to the phase of the breathing cycle at the moment the image was generated, and amplitude binning (AB) assigns each image according to the breathing signal's amplitude at the moment

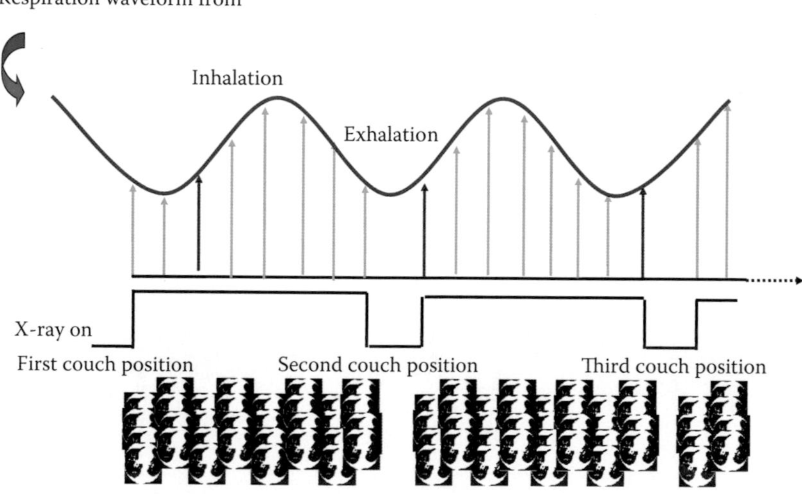

Figure 9.3 Cine mode 4-D CT acquisition and retrospective sorting.

Stereotactic radiation therapy, precision patient positioning, and immobilization

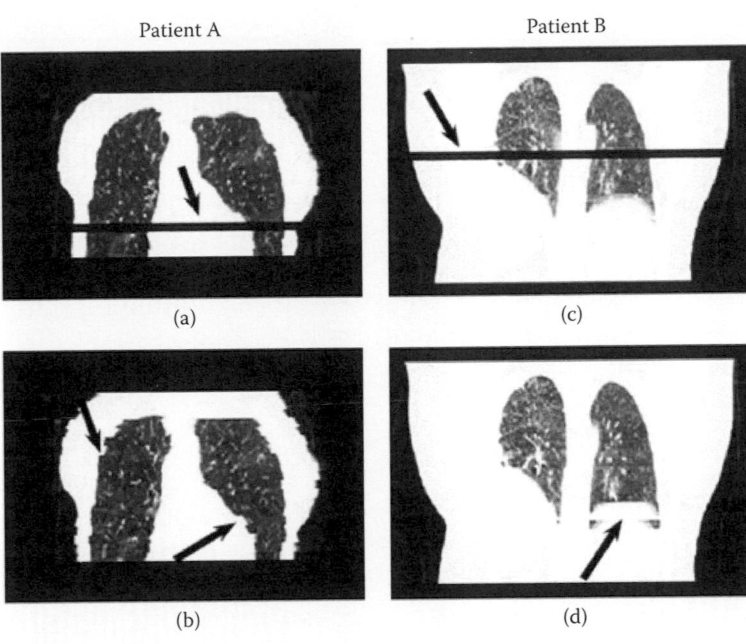

Figure 9.4 Coronal 4-D CT lung images for patient A (a and b) and patient B (c and d). Top images show a case of amplitude binning, and the bottom images show phase binning at the same respective bins. Artifacts are shown in arrows. (From Abdelnour, A. F. et al., *Phys. Med. Biol.*, 52, 3515–29, 2007.)

the image was generated. It has been shown by authors Abdelnour et al. (2007) that using AB to sort the 4-D CT results in missing slices in certain images if the patient's breathing is irregular so that the amplitude changes from one cycle to another as shown in Figure 9.4a and c. PB images tend to populate the bins evenly regardless of the variation in the patient's actual breathing amplitude but suffers visible artifacts, such as "mushroom effects" on the diaphragm due to mis-binning as shown in Figure 9.4b and d.

Audiovisual signals will help the patient breathe reproducibly to minimize 4-D CT artifacts and also to match the breathing at the time of treatment delivery. Audiovisual coaching can improve a patient's breathing reproducibility and baseline shift as shown in Figure 9.5. The patient's respiratory signal period and CT scan rotation time should be similar to avoid differing amounts of motion included in the CT images compared with actual motion.

9.1.2.2 Internal target volume (ITV) definition

Incorporating the 4-D CT image information into treatment planning can be performed with multiple methods, depending on the treatment technique. Maximum intensity projections (MIPs) can be used to obtain the full motion envelope for tumors that are bright in intensity relative to their background, such as lung tumors, and minimum intensity projections (MinIPs) can be used to view the full motion envelope of the tumors that are darker compared to their background, such as liver tumors. MIPs reflect the highest pixel value encountered from all CTs along the viewing ray for each pixel, giving rise to an artificial intensity display of the brightest object along each ray on the projection image (Figure 9.6).

Another treatment planning method is to contour the gross tumor volume (GTV) in each of the phases in the respiratory cycle and use the overlap volume. Currently one can automate this process by auto-segmentation tools, which would use a deformable image registration algorithm to propagate a GTV contour drawn on an end exhalation image to all the other phases as shown in Figure 9.7.

Particular care should be taken when contouring tumors that are attached to the diaphragm or the chest wall. Utilizing just the MIP image for all phases will not guarantee proper contouring of the full ITV.

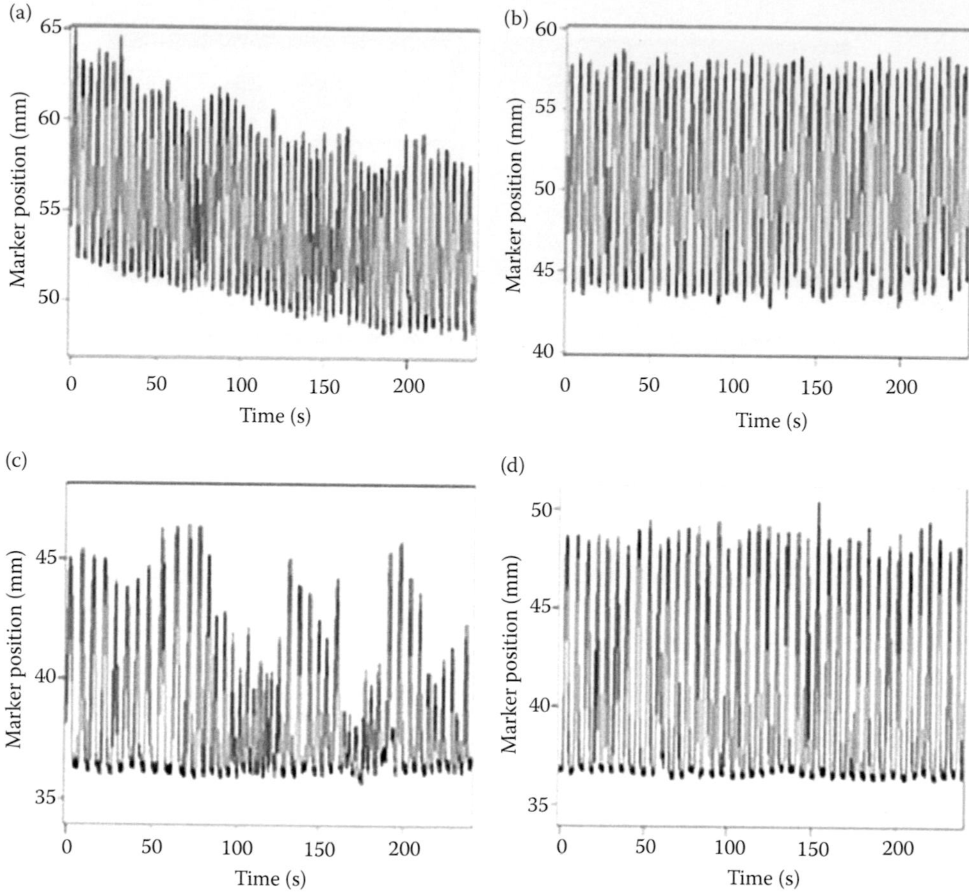

Figure 9.5 Example of how audio-visual coaching could help patients breath reproducibly. Breathing trace acquired during free breathing (a and c) and audio-visual coaching (b and d). Data sets (a) and (b) belong to one volunteer and (c) and (d) belong to another. The free breathing data exhibits baseline shifts (a) and irregular breathing (c). Those problems were eliminated with breath coaching (b and d). (From Neicu, T. et al., *Phys. Med. Biol.*, 51, 617–36, 2006.)

In such cases, the physician should compare all phases individually to obtain the full MIP as shown in Figure 9.8.

Another uncertainty for ITV definition is contouring smaller tumor volumes (less than 5 cc in volume) with larger motion amplitudes (larger than 1.5 cm). Due to the sampling resolution, discrete volumes are visible even in the full MIP image. Note the locations with the lowest velocity, that is, the highest probability density function has the highest intensity even in the MIP image as shown by Figure 9.9.

The above-mentioned techniques will give an ITV where PTV margins still need to be added. This technique is only useful if the patient will be treated on a machine that does not have gating or tumor-tracking capability. If the treatment delivery system has the capability of gating the beam, one should use an ITV that corresponds to the treatment window. Respiratory gating allows the beam to be *on* during a particular interval of the breathing cycle, typically around the exhalation phase, with motion amplitude less than a particular threshold. This is where the most amount of time is spent in respiration and, hence, has the most reproducible and minimum motion. The choice of the gating window is a trade-off between the amount of tumor motion and treatment time. American Association of Physicists in Medicine (AAPM) Task Group 76 (Keall et al. 2006) recommends using a value of 5 mm as the motion amplitude within the gating window. In delineating the ITV for a gated treatment, one needs to use the image set that corresponds to the gating window. For example, if the patient's motion in 3-D is less than 5 mm for the 30%–70% phases, the MIP should be created from the image sets 30, 40, 50, 60, and 70 phases. AP and

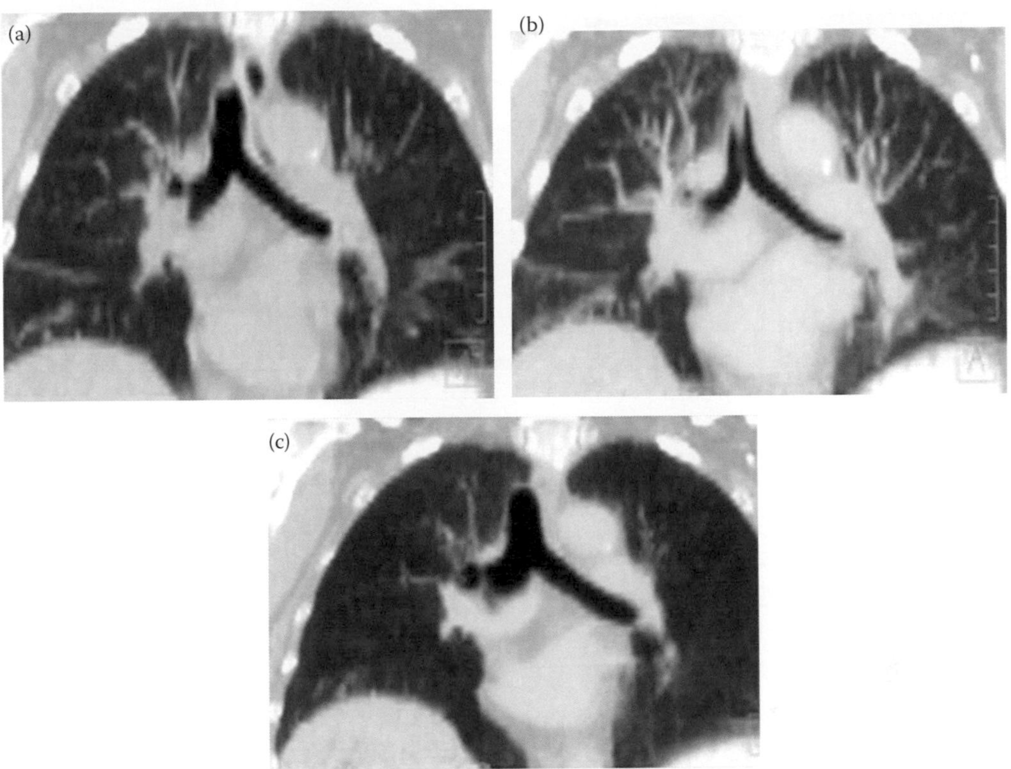

Figure 9.6 Variations of 4-D CT images useful in planning moving tumors. (a) Time-averaged image over all 4-D CT images. (b) MIP image for 4-D CT showing the enhanced high-contrast areas. (c) MinIP image in which the low-contrast areas are enhanced with respect to the time-averaged image.

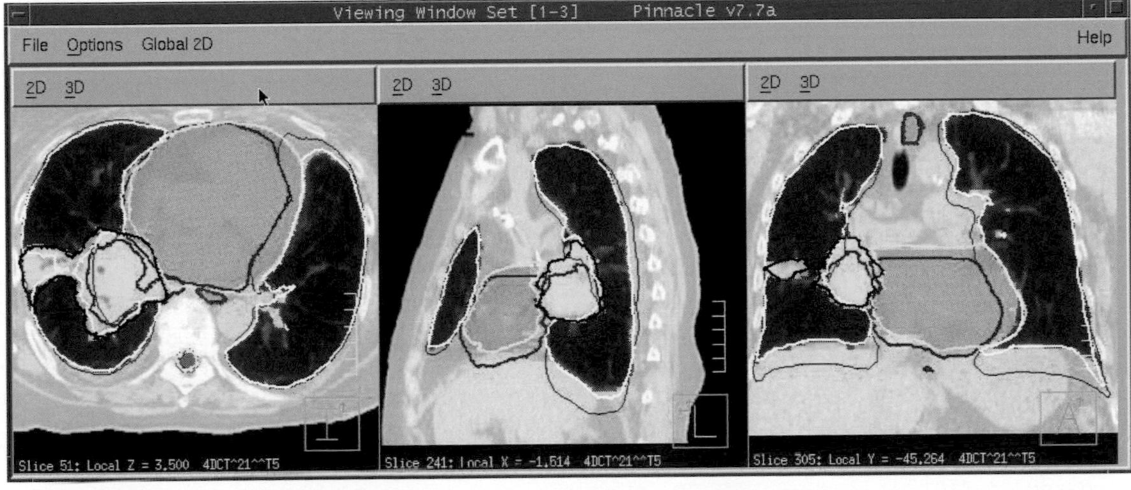

Figure 9.7 Manual versus automated contouring results for a single patient axial, sagittal, and coronal views from Pinnacle 7.7. Red contours are for the inhale phase. Color wash contours are for the manually drawn exhale phase. Auto contours from inhale to exhale are black (GTV), yellow (cord, heart), pink (esophagus), and white (lungs). (From Wijesooriya, K. et al., *Med. Phys.*, 35, 1251–60, 2008.)

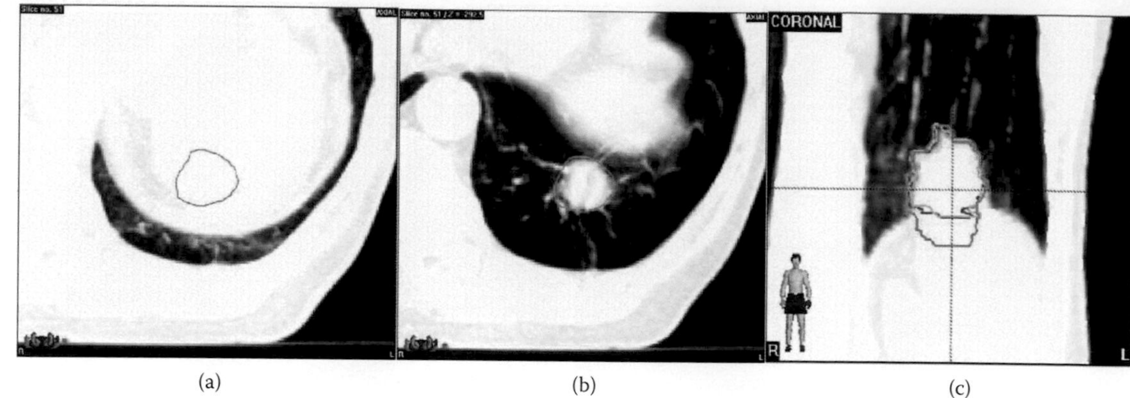

Figure 9.8 (a) Example of tumor located adjacent to the diaphragm, which was not fully visible on maximum intensity projection maps as a result of the overlap with the diaphragm. (b) The extreme end-inspiratory tumor position was visible on the corresponding single-phase scan. (c) MIP image would underestimate the caudal extent of the internal target volume (orange and pink contours) compared with an ITV reconstructed from individual 10 phases (green contour). (From Underberg, R. W. et al., *Int. J. Radiat. Oncol. Biol. Phys.*, 63, 253–60, 2005.)

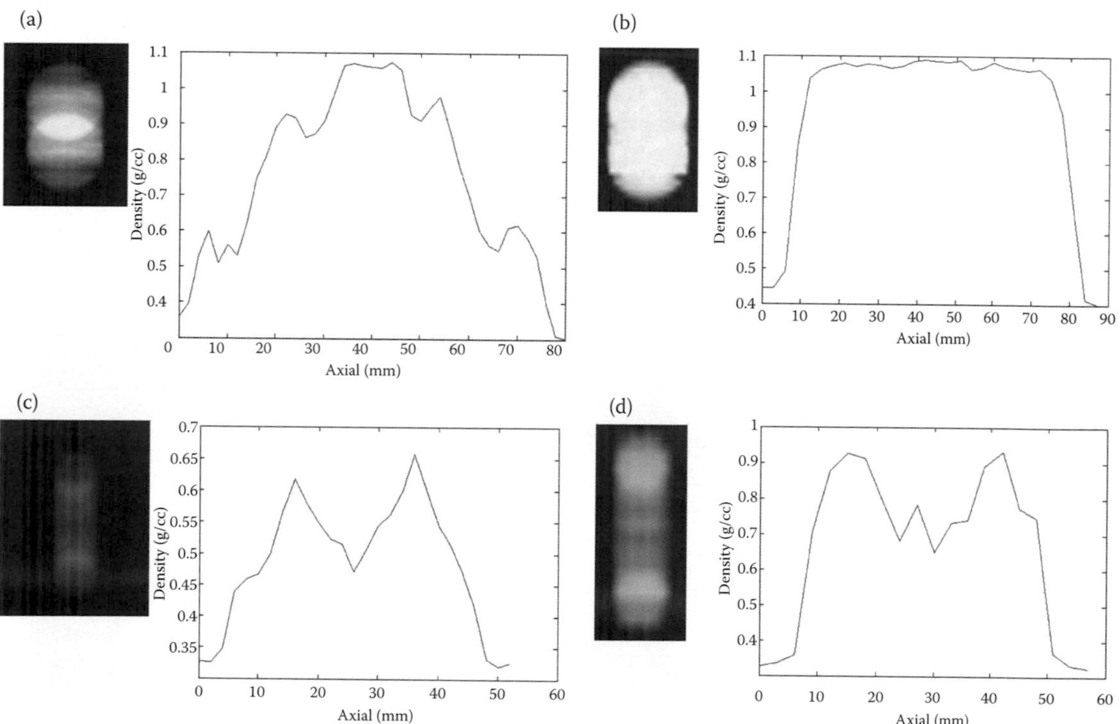

Figure 9.9 Time-average image (LHS) and the corresponding MIP image (RHS) for a larger tumor with smaller motion amplitude (a and b) and for a smaller tumor (5 cc) with larger motion amplitude (1.5 cm inferior/superior) (c and d).

lateral digitally reconstructed radiographs (DRRs) should be created with the diaphragm and possibly the tumor in view with the isocenter using the image set that corresponds to the treatment gate for internal verification of the treatment gate. This is in addition to Cone beam kVCT (CBCT) images, and AP/lateral images that will be used for setting up prior to verifying the gate. It is important to realize facilities that currently do not have access to techniques to account for respiratory motion can evaluate the maximum motion envelope by fluoroscopy or 4-D CT and use a PTV margin for the treatment.

9.1.2.3 Treatment

For treatment delivery units with no motion-management capabilities, it is still important to verify the location of the tumor in three dimensions prior to each day of SBRT treatment delivery. CBCT or megavoltage CT (MVCT), due to the long acquisition time (about 1 min), will encompass the full motion envelope, and hence, these images are very appropriate to compare to the planning CT, which is typically the time-averaged image. Having the ITV and PTV contours and the organs at risk (OAR) contours superimposed on the CBCT images allows the physician to align the treatment beam to the ITV while maintaining a healthy distance to the OARs.

For treatment delivery units with the capability to perform motion management, such as gating, more time should be allowed at the treatment machine prior to treatment to capture extra images. Due to the possibility of patient breathing changes from the planning CT, daily CBCT acquisition, and the fact that most of these gated treatments are routinely performed with external signals, it is important to verify the location of the tumor and also that the other anatomical landmarks during the treatment phases are in phase with the planning CT. This can be achieved by some form of instant or real time IGRT, such as kV digital imaging, portal imaging, or fluoroscopy with the treatment gate on, but not a CBCT or a MVCT due to the long acquisition time (about 1 min) that will encompass the full motion envelope, including the phases outside the treatment gate. During patient setup, the tumor home position at this fraction should be matched to the reference home position. Pretreatment imaging with the treatment gate on is the only way to verify that the tumor is within the treatment volume and is therefore crucial to accurate motion management. There are two types of scenarios: when fiducial markers are implanted (internal tracking) or when external respiratory surrogates, such as markers, are placed on the surface of the patient (external tracking). In the case of fiducial markers, one needs to verify that there is no marker migration. It has been shown that liver and prostate tumor marker migration is less than 1.5 mm, and fiducials implanted in lung tumors (especially central tumors) could have large migration issues. Kilovoltage or megavoltage planar imaging could be used to catch large fiducial migrations, and CBCT or MVCT could be used to catch small fiducial migrations. One could also utilize the fluoroscopic system to verify the location of the markers (Harada et al. 2002).

In the case of external surrogates, it is important to recalibrate the correlation between the external surrogate signal and the internal target position to eliminate the interfraction variation of correlation. This can be achieved using an on-board x-ray imaging system, an ultrasound system, or implanted electromagnetic transponders. Implanted fiducial markers for liver tumor patients can be used to indicate the tumor position. Tumor mass can often be used directly for lung cancer patients for matching if it's visible in the projection x-ray images or the relevant anatomic features, such as the diaphragm. Constancy of diaphragm position of the side of the lung to be treated with respect to a stationary landmark or isocenter in S/I and lat with the treatment gate on could be verified at a minimum with an AP radiograph (Figure 9.10c and d). Often, the lung tumor itself is visible from the radiograph as shown by Figure 9.10a and b. In such cases, it is advantageous to display the MLC pattern corresponding to the treatment on the DRR for comparison. Direct tumor imaging and verification can be difficult due to poor image quality, low target contrast, and no clear tumor shape. To maintain a constant tumor home position when the beam is turned on throughout the treatment, patient breath coaching should be performed (Figure 9.11).

9.1.2.4 Breath-hold treatment

Breath-hold treatment is also employed with lung, liver, and breast cancer patients to manage motion (Dawson et al. 2001; Hanley et al. 1999; Sixel, Aznar, and Ung 2001; Wong et al. 1999). An ABC device or voluntary DIBH with the RPM system controls the beam with a visual signal from goggles for the patient to observe and thereby control the breath-hold amplitude such that it agrees with the breath-hold

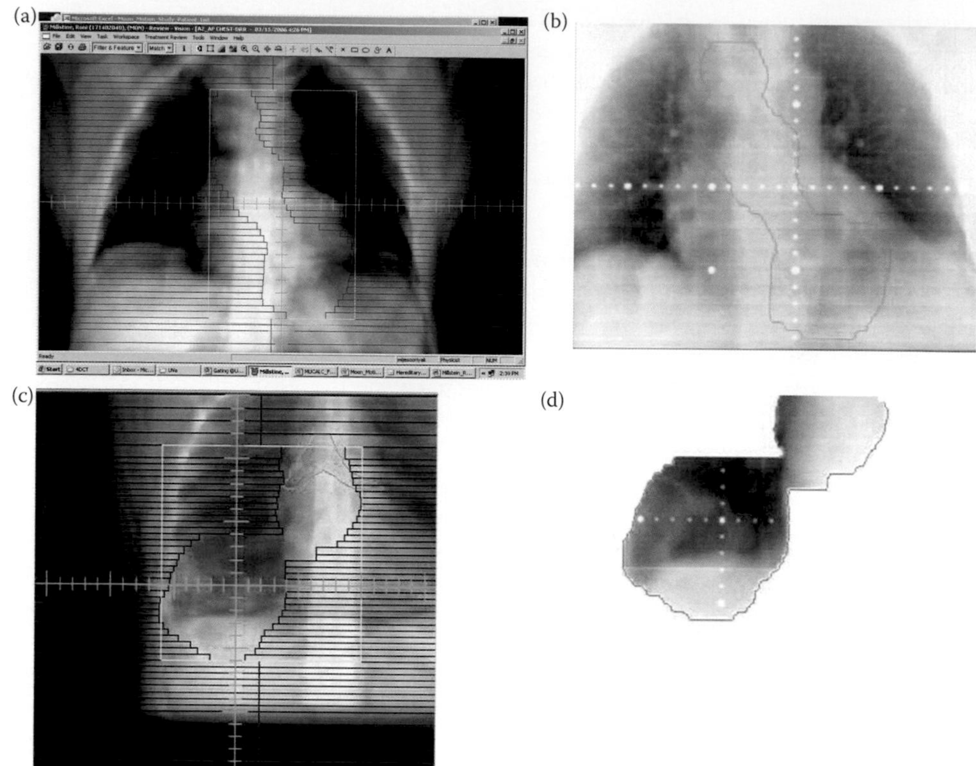

Figure 9.10 MV portal images to verify the pretreatment lung tumor location. Top images (a and b) are for a case of actual tumor registration and bottom images (c and d) are with both the tumor and the diaphragm on the same side for tumor home position verification. Left-hand-side images (a and c) are from the planning DRRs in the gating window and the right-hand-side images (b and d) are the pretreatment MV portal images taken during the gating window.

amplitude at the simulation. The main goal of these methods is to exploit anatomical immobilization to minimize the effects of breathing motion. The additional benefit of this technique for lung and breast patients is to move the dose-limiting organs away from the target by utilizing deep to moderate breath holding. This minimizes the dose to the heart and lung. In all these breath-hold techniques, a reproducible baseline is an essential step to achieving reproducible breath holds. Similar to respiratory gating, a second key issue is the accuracy of externally placed breath-hold monitors in predicting internal positions of the tumor and nearby organs. McIntosh et al. (2011) describes a technique to verify the internal anatomy based on breath-hold amplitude reproducibility using pre-treatment kV orthogonal imaging (Figure 9.12).

9.1.3 INTERPLAY EFFECTS FOR LUNG SBRT

For static beam nongated treatments, intrafraction motion only causes an averaging or blurring of the dose distribution around the beam edges along the direction of the motion that is perpendicular to beam direction. This is due to the homogeneous dose distributions of static beams. However, for nongated intensity modulated radiation theraphy (IMRT) treatments, a motion artifact follows from a possible interplay between motion of the leaves of the collimator and the component of target motion perpendicular to the beam (Bortfeld et al. 2002; Chui, Yorke, and Hong 2003; Duan et al. 2006; Engelsman et al. 2001; Seco et al. 2007; Yu, Jaffray, and Wong 1998). This effect is particularly magnified for few-fraction treatments, such as SBRT.

9.1.4 QA FOR MOTION MANAGEMENT

There are QA measures that need to be established in a clinic to ensure a safe motion-management program (Jiang, Wolfgang, and Mageras 2008). QA measures should be adapted in multiple stages of

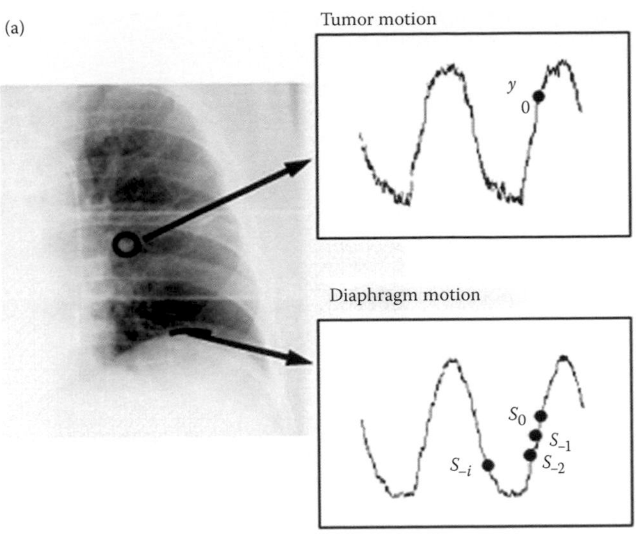

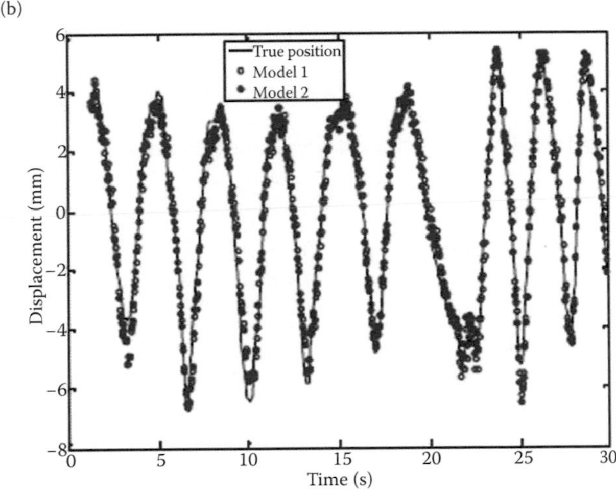

Figure 9.11 (a) Tumor motion and diaphragm motion correlation. (b) Real tumor position (solid line) and tumor position predicted by two models using the diaphragm (hollow and solid circles) for a given patient. Mean and maximum errors are 1 and 2 mm. (From Cervino, L. I. et al., *Phys. Med. Biol.*, 54, 3529–41, 2009.)

a motion-management program: initial testing of equipment and clinical procedures and frequent QA examination during early stages of implementation. These procedures should verify the accuracy of each component: simulation, planning, patient positioning, and treatment delivery and verification. CT simulation errors could occur due to three reasons: irregular patient breathing, CT image reconstruction algorithm, and resorting of reconstructed CT images with respiratory signal. Irregular patient breathing can be addressed by using audio-visual coaching, and estimation of errors due to CT image reconstruction and sorting can be accomplished by using a motion phantom capable of moving in amplitude and frequency in the ranges of patient breathing signals. Hurkmans et al. (2011) showed that there is no correlation between specific scan protocol parameters and accuracy of tumor volume estimate for SBRT lung patients from a nine-center multiscanner study. Because most of the gated treatments are performed in the form of external surrogates, the relationship between the tumor motion and the surrogate signal may change over time, both inter- and intrafractionally. Five QA steps have been proposed to ensure a clinically acceptable accuracy:

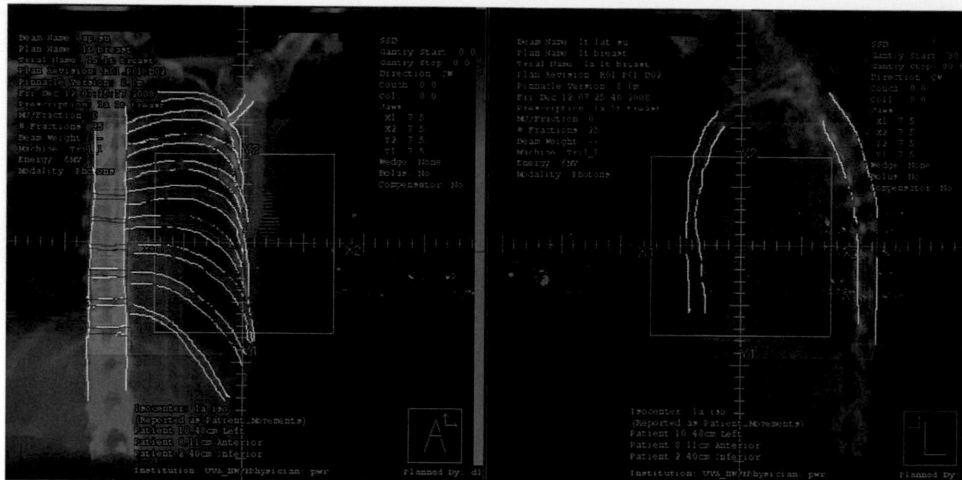

Figure 9.12 Bony anatomy contours used for the registration of the AP/lateral kV images. The left-hand side shows the AP projections, and the right-hand side shows the lateral projections. The left-hand side gives the planning DRR, and the right-hand side shows the KV onboard image prior to treatment. This figure shows the registration results of planning DRR (white contours) to kV projection images (purple contours) for the bony anatomy. (From McIntosh, A. et al., *Int. J. Radiat. Oncol. Biol. Phys.*, 81, e569–76, 2011.)

- During simulation, the reference home position should be accurately measured (4-D CT).
- During treatment planning, patient and tumor geometry corresponding to the gating window should be used.
- During patient setup, tumor home position at this fraction should be matched to the reference home position: image guidance (x-ray, ultrasound, implanted EM transponders); lung tumor or diaphragm; liver, implanted fiducial markers.
- During treatment delivery, measures should be taken to ensure constant tumor home position (the tumor should be at the same position when the beam is on) breath coaching, visual aids, stable end of expiration position by two straight lines for amplitude gating.
- During treatment delivery, the tumor positions corresponding to the gating window should be measured and compared with the reference home position, either on or offline. Electronic portal imaging device (EPID) in cine mode for small patients and non-IMRTs (Berbeco et al. 2005) and on-board imaging (OBI) KV images for others.

9.1.5 TRACKING WITH MLC AND ROBOTICS

Respiratory synchronized or real-time treatments techniques offer the optimal solution for motion corrections, especially for SBRT. Organ motion causes an averaging or blurring of the static dose distribution along the path of motion. To avoid interplay effects and to make the treatments more efficient and accurate, clinics need to move toward real-time tumor tracking systems. Furthermore, to individualize the radiation therapy by deliberately delivering inhomogeneous dose distributions to cope with tumor heterogeneity (dose painting by numbers and biologically conformal radiation therapy) also require real-time tumor tracking.

One method of correcting for tumor motion is repositioning the radiation beam and the robotic couch in real time so that it follows the tumor motion (Keall et al. 2001; Tacke et al. 2010). For linac-based radiotherapy, tumor motion can be compensated for by using a dynamic multileaf collimator (MLC) (Keall et al. 2001; Keall et al. 2005; Neicu et al. 2003; Neicu et al. 2006; Papiez and Rangaraj 2005; Rangaraj and Papiez 2005; Suh et al. 2004; Webb 2005a; Webb 2005b; Wijesooriya et al. 2005). This is still only 2-D compensation. To achieve full 3-D compensation, one needs a linac mounted on a robotic arm, such as CyberKnife (Depuydt et al. 2011; Kamino et al. 2006; Murphy 1997; Murphy et al. 2002; Schweikard et al. 2000). All these techniques require some kind of IGRT to reach full potential. It should be noted that these tracking systems rely upon an accurate predictive model of the breathing motion to predict the future position of the tumor, which has proven to be very challenging (Murphy 2004). Compared to the gating technique, the tracking technique potentially offers higher delivery efficiency and less residual target motion provided there is no system latency in the full predictive system. The beam tracking technique of following the target dynamically with the radiation beam was first implemented in a robotic radiosurgery system (Adler et al. 1999; Murphy 2002; Murphy et al. 2003; Schweikard et al. 2000). There are three techniques of IGRT for real-time tumor tracking: deriving the tumor position based on external surrogates: fluoroscopic tracking of radio-opaque fiducial markers implanted inside the tumor and fluoroscopic tracking of the lung tumor without implanted fiducial markers. When the tumor is distinct and visible in fluoroscopy, then conventional motion-tracking methods, such as optical flow, template matching, and the active shape model (Cui et al. 2007; Xu et al. 2007; Xu et al. 2008) can be used. Cho, Poulsen, and Keall (2010) compared estimation accuracy for three different imaging systems. As shown in Figure 9.13, the estimated traces are all very similar, even in the places where they differ from the tumor motion trace. They claim that the similarity suggests that the limiting factor in motion estimation is not due to the IGRT system geometry, but to limitations in the model itself or the underlying unaccounted-for variations in the intrinsic internal-external correlation. However, both sequential and synchronous estimation methods

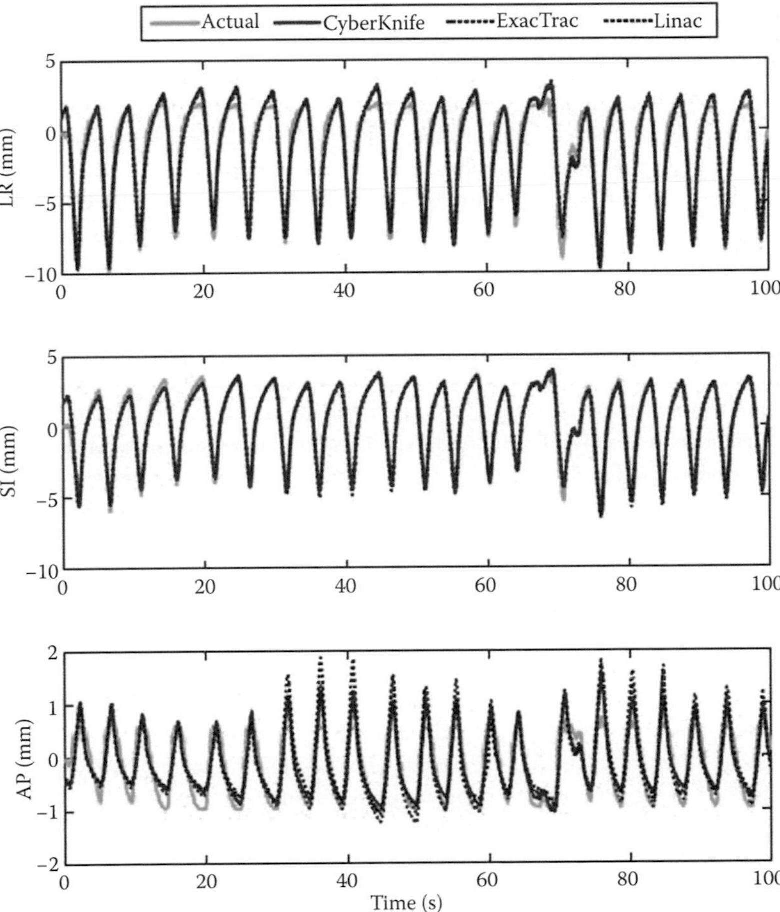

Figure 9.13 Typical tumor motion estimation results for the CyberKnife geometry (synchronous image-based method), ExacTrac, and linac geometries (sequential image-based method). Estimation results are similar for all methods. The kV imaging frequency was one image per second. (From Cho, B. et al., *Phys. Med. Biol.*, 55, 12, 3299–316, 2010.)

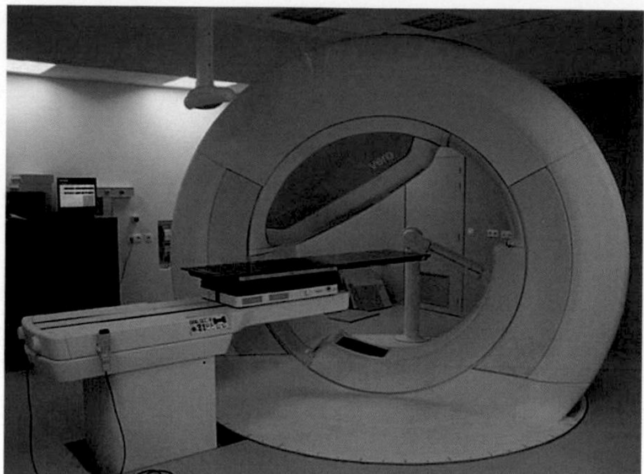

Figure 9.14 The Vero system consisting of an O-ring gantry allowing noncoplanar irradiation, gimbaled linac, EPID, dual source x-ray imaging system, 5-D robotic treatment couch, and infrared tracking system.

have been shown to achieve mean 3-D errors less than 1 mm for all three imaging systems at all imaging intervals.

The Vero system from Brainlab (Brainlab AG, Feldkirchen, Germany) and MHI (Mitsubishi Heavy Industries, Tokyo, Japan) is another system that allows performing real-time tracking of moving tumors to perform IGSBRT (Verellen et al. 2010) (Figure 9.14). For on-board imaging, the Vero system is equipped with both an EPID for MV portal imaging as well as two orthogonal kV imaging systems attached to the O-ring at 45° from the MV beam axis. The latter allows for cone-beam CT acquisition, simultaneous acquisition of orthogonal kV images, and fluoroscopy for tumor tracking. The system latency for the IR marker tracking was 47.7 ms, and the root mean square error (RMSE) and E90% was below 0.95 mm and 1.37 mm without forward prediction and 0.2 mm and 0.37 mm for 50 ms with forward prediction.

9.2 IMPLANTED MARKERS

The position of the tumor in lung cancer patients can be determined prior to treatment planning using imaging techniques, such as computed tomography, ultrasound, magnetic resonance, and nuclear imaging. Motion information for the tumor can be obtained using 4-D simulation. In the process of 4-D simulations, the scans are obtained throughout the entire breathing trace and correlated to the various phases of the patients breathing (inhalation or exhalation) to provide localization of the tumor with respect to respiration motion.

However, during treatment delivery, it is important to know where the tumor is in real time so that the information gathered during simulation and used for the treatment plan can be used to accurately treat the patient. In order to track the tumor while the treatment is being delivered, it is necessary to know the current position of the tumor. In addition, it is necessary also to transfer the information on the position of the tumor to the radiation system so that the x-ray beam may be turned on at the correct time point.

Several studies have been performed to determine the real-time position of the tumor as it moves with respiratory motion, including using fluoroscopic imaging and predictions from external breathing patterns. Among newly developed techniques to track tumor motion, one of the most promising is the use of fiducial markers placed in the tumor site or the surrounding tissue.

This section will concentrate on the use of radiographic fiducials, electromagnetic fiducials, and other detectors that are implanted in the tumor for the purpose of tracking the tumor and thereby providing feedback for the radiation treatment beam to accurately follow the tumor. This section on implanted markers is divided into (1) passive radio-opaque markers and (2) emitting markers such as electromagnetic markers.

Both passive and emitting markers have been used in radiation therapy not only for monitoring and tracking intrafraction motion caused by respiratory motion, bowel movement, bladder filling, and rectal gas, but also for the purpose of initial setup and localization of the patients.

9.2.1 PASSIVE RADIO-OPAQUE SEEDS

One example of passive radio-opaque seeds are gold fiducials. Sterile gold fiducials have been inserted in and around lung tumors bronchoscopically and percutaneously since 1995 in order to measure lung tumor motion. Several studies have been performed to investigate the feasibility of implanted gold markers in estimating tumor motion. Morice et al. (2005) concluded that gold markers were a safe and useful method to determine lung tumor motion that was more accurate than using external respiratory trace.

9.2.1.1 Use of passive radio-opaque seeds in the lung

In some cases, fiducial markers are implanted in and around the fiducials percutaneously. Kothary et al. (2009) discussed the optimal fiducial marker placement for the CyberKnife Synchrony system in order to track the tumor in real time. Typically, three to five (minimum of three) radio-opaque gold fiducial markers are implanted percutaneously using a 19-gauge introducer needle. Because the CyberKnife system uses orthogonal x-ray systems to identify and track the implanted markers, it is important that the markers are not placed superimposed on each other in 45° oblique views. Hence, it is best if the fiducials are placed in different octants, which can be accomplished by torquing the needle or two skin entries may be required. If two markers appear superimposed in an orthogonal view, it is necessary to place a third marker to gain accurate positional information of the tumor. It is also recommended that markers not be clustered together but spaced a minimum distance of 1.5 cm between markers. For small tumors, artifacts from the markers may pose issues, and hence, the markers may be placed in the tumor periphery. Migration of the markers due to a high incidence of pneumothorax after percutaneous interventions is often seen. One solution suggested for this issue is using platinum micro coils. Bronchoscopic implantation of fiducial markers for lung tumors also reduces the risk of pneumothorax. Imura et al. (2008) investigated the implantation of three fiducials for patients and found that bronchoscopic insertions were useful for setup for peripheral lung cancer and had an accuracy of 2 mm. Bronchoscopic insertion was also shown to be more desirable than percutaneous implantation due to the reduced risk of pneumothorax.

Bronchoscopic insertions of 1–2 mm gold markers were used to evaluate the feasibility of using a real-time tumor-tracking system by Harada et al. (2002). It was observed that markers were inserted and fixation was good (88%) for peripheral lung cases, but not viable for central-type lung cases.

Seppenwoolde et al. (2002), in their report, used a 2-mm marker and measured the tumor motion. Two fluoroscopy image processor units in the treatment room were used as part of the real-time tumor-tracking system, and tumor position was measured in all three directions.

9.2.1.2 Use of passive radio-opaque seeds in other sites (summary of various studies)

In the liver, Gierga et al. (2005) performed an internal-external tumor motion-correlation analysis using radio-opaque fiducial markers of diameter 1–2 mm. It was concluded from this study that treatment margins should be determined based on measurement of internal marker motion and not relying only on external marker information.

In another study, three gold seeds (1 mm diameter, 5 mm length) were implanted in the liver of 12 patients undergoing compression-supported SBRT to evaluate the feasibility of using liver implanted fiducials for guiding liver stereotactic body radiation therapy. Wunderink et al. (2010) concluded that high-precision tumor radiation setup could be accomplished using marker guidance. However if the markers were not implanted in the tissue surrounding the tumor, marker guidance may induce significant errors caused due to the rotation and deformation of the organ.

Kitamura et al. (2002) used internal gold markers to study the migration and registration accuracy for liver and prostate patients. Internal fiducial markers were found to be useful for real-time tumor tracking. However a margin was still needed to account for issues such as marker migration and measurement uncertainty.

In the pancreas, motion data was collected using a fluoroscopic video signal for 30 s tracking surgically implanted clips for pancreatic patients (Gierga et al. 2005). Quantitative fluoroscopic analyses of patients with radio-opaque tumor markers were useful in determining whether motion mitigation or interventional strategies were necessary and in characterizing the tumor motion.

In the prostate, radio-opaque markers, such as gold fiducials, have been used in the prostate mainly for daily setup and localization. However, recently, gold fiducial markers have been implanted in the prostate to track the prostate in real time. Litzenberg et al. (2002) used radio-opaque markers implanted in the periphery of the prostate to improve the accuracy of daily localization. The use of implanted markers in the prostate allowed reduction of treatment margins and also showed intratreatment prostate movement.

Chen et al. (2007) provided a single institution experience for daily localization of the prostate using three implanted fiducials. They concluded that daily portal imaging with implanted fiducials improved the localization of the prostate and was necessary for reduction in treatment margins.

Welsh et al. (2004) implanted gold seeds during brachytherapy. These seeds were then used for daily setup and localization and found to be an easy and practical means of prostate localization during the subsequent image-guided external beam radiotherapy. With the use of these markers, the conformality of the treatment was improved without the installation of expensive additional equipment.

Xie at al. (2008) used three gold fiducial markers implanted in the prostate of patients to study the intrafractional prostate motion. The prostate movement could be characterized as stable, continuous drift, transient excursion, persistent excursion, high-frequency excursion, and irregular movement. Based on the prostate motion data collected, it was concluded that real-time image guidance and motion-compensation techniques were important especially during hypofractionated prostate irradiation. Given the magnitude and random nature of prostate motion, real-time monitoring of motion and compensation is important to ensure target coverage and critical structure sparing.

Fiducial marker size, composition, location, and orientation are important to determining the usefulness of markers in patients receiving proton therapy. To this affect, Giebeler et al. (2009) studied the dosimetric impact of small, medium, and large helical wire markers for proton therapy. Small markers were found to be too fragile for transrectal implantation. Medium and large markers caused proton dose perturbations up to 10%.

9.2.2 EMITTING MARKERS: ELECTROMAGNETIC (EM), CALYPSO, DETECTORS

9.2.2.1 Use of EM markers in lung cancer patients for motion compensation

Mayse et al. (2008) studied the fixation rates in canine cases for both electromagnetic transponders as well as gold markers. It was found from this study that although bronchoscopic implantation of transponders was feasible, fixation rates were low. Stable fixation rates at 60 days were six out of 15 electromagnetic transponders and seven out of 12 gold markers.

9.2.2.2 Use of EM markers in other sites for motion compensation (summary of various studies)

In the prostate, even though radio-opaque markers have been implanted in the prostate, the main purpose of those markers was daily setup and localization. However, with the introduction of electromagnetic markers for prostate patients, there are several studies that report on the motion of prostate during treatment fractions.

Langen et al. (2009) used electromagnetic beacons (Calypso Technology, now Varian Medical) implanted in the prostate to obtain motion information. The electromagnetic markers are excited with an array of source coils that is placed in a known position above the patient for the entire duration of the treatment. After excitation, the markers emit an electromagnetic signal that is detected by a set of receiver coils that are located in the same array that houses the source coils. Three markers with unique resonances are implanted, and the expected position of the markers with respect to the treatment isocenter is determined in the treatment-planning system and then imported to the beacon-tracking system. It was concluded that although there was severe degradation of dose coverage for individual fractions, the effect over the cumulative dose distribution over the fractions was minimal.

Noel et al. (2009) investigated the necessity of continuous real-time tracking in prostate cancer radiation therapy. The electromagnetic beacons (Calypso Technology) were used to assess prostate motion data. Imaging models were created to simulate pre- and posttreatment imaging as well as intermittent imaging. It was concluded that snapshot pre- and posttreatment imaging is not a sensitive test of intrafraction prostate motion, and thus, continuous real-time tracking is valuable in reducing the uncertainty of surrounding prostate motion during radiation treatment delivery.

9.2.2.3 Phantom studies for real-time position monitoring of implanted markers

For electromagnetic markers, several studies have measured and quantified daily patient setup, localization, and tumor-position monitoring. However, there are still several issues that need to be addressed toward completing the feedback loop so that the linear accelerator radiation beam can follow the tumor trajectories during treatment delivery. Sawant et al. (2008) have integrated the position-monitoring system with a dynamic multileaf collimator–based 4-D intensity-modulated radiotherapy delivery system. A moving phantom with electromagnetic markers embedded was tracked with the MLC to quantify the latency of the system. Geometric accuracy was also evaluated. The system developed by these authors represents a significant step toward achieving the goal of accurately delivering the dose to moving targets using an implanted marker and a feedback loop to the linear accelerator.

Smith et al. (2009) integrated the linear accelerator with an electromagnetic tracking system that used implanted wireless transponders to study the feasibility of gating. It was found that the latencies with this real-time electromagnetic system with the implanted wireless markers were short enough to be suitable for beam gating. This phantom study used realistic lung trajectories and film experiments to show the dosimetric improvement in using gating.

Smith et al. (2009) successfully integrated the 3-D position-monitoring system (Calypso 4-D localization system) with the dynamic multileaf collimator system to deliver dose with accuracy comparable to gating and with improved efficiency.

9.2.3 CONCLUSIONS

Implanted fiducial markers can be an effective technique for directing real-time image guidance for SBRT and allowing for a reduction in treatment margins. The fiducials act as a radio-visible surrogate for tumors that are otherwise difficult to track on orthogonal radiographs, fluoroscopy, CT, or CBCT. However, markers do have some potential limitations, including complications from the marker implantation, marker migration, and delivered dose perturbation. The risks, benefits, and uncertainties surrounding the use of implanted markers should be carefully considered when developing a program for SBRT.

9.3 REVIEW OF THE CURRENT COMMERCIAL SYSTEMS

Breathing-induced motion management is an integral part of the emerging hypofractionated radiation therapy regimens that require and rely upon very accurate tumor definition. The first step toward addressing respiratory motion is the detection of movements induced by respiration. The tumor location can be inferred from external breathing surrogates or can be detected directly by tracking markers implanted in the tumor.

9.3.1 RESPIRATORY MOTION DETECTION SYSTEMS

Several respiratory-sensing devices are available for use in radiation therapy.

The most widely spread respiratory motion-detection commercial system is the RPM from Varian Oncology Systems (Palo Alto, CA) (Mageras and Yorke 2004). The breathing motion is detected with an infrared camera that monitors the movement of a marker block with infrared reflecting dots that is placed on the patient's chest or abdomen.

An alternative to the RPM system is the air bellows belt available from Philips Medical Systems (Milpitas, CA). The system is a pneumatic device placed around the patient's chest or abdomen, and it detects changes in pressure caused by respiratory motion. A pressure transducer converts the pressure waveform into a voltage signal that is then digitized and displayed as breathing trace.

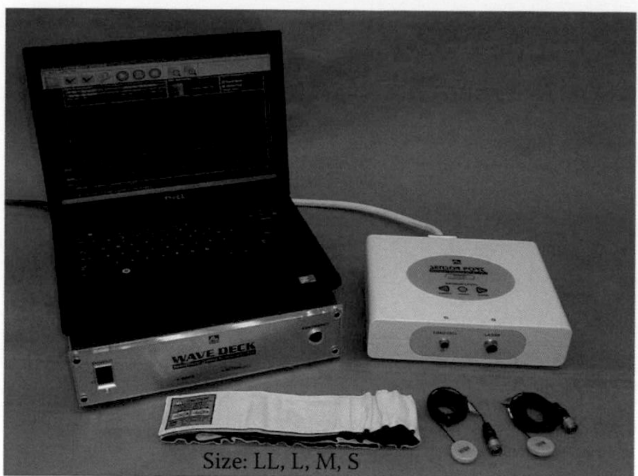

Figure 9.15 The Anzai Respiratory System (Anzai Medical Co., Ltd., Tokyo). The system detects pressure changes as a surrogate for respiratory motion using a pressure sensor fitted into a belt that wraps around the patient.

Similarly, the Anzai Respiratory System from Anzai Medical Co., Ltd. (Tokyo, Japan), uses a pressure sensor inserted into the pocket of a belt that wraps around the patient (Figure 9.15) (Biederer et al. 2009). The system detects pressure changes, which are the surrogates for the respiratory signals.

The systems described above detect the patient's surface motion (as surrogate for the tumor movements) through an "interpreter" (i.e., the RPM box and the belts). However, the patient's external surface motion can be detected directly using stereoscopic optical surface imaging. One such system is GateCT/GateRT, an extension of AlignRT, available from VisionRT, Ltd. (London, UK) (Bert et al. 2006). The surface-imaging device consists of two or three camera pods mounted to the ceiling; each pod houses two stereoscopically positioned cameras that enable the acquisition of images with slightly different perspectives as required for the computation of the 3-D surface. A speckle flash and a speckle projector are used to project a pseudo-random speckle pattern onto the surface of the patient during image acquisition. The speckle pattern permits the digital reconstruction of a 3-D surface model of the patient. The corresponding regions in both images acquired by the cameras of one pod are identified by the pseudo-random speckle pattern, and a 3-D model is reconstructed for this region using triangulation. The 3-D models obtained from each pod are merged to create the patient's external surface model. When the system is operated in the respiration-gated capture mode, a sequence of images is acquired; then a tracking point is defined on the surface of the first frame, and a breathing curve is generated for this point from the frames recorded (Figure 9.16).

Figure 9.16 Photograph of a patient on the linac treatment couch where the speckle flash and a speckle projector are used to project a pseudo-random speckle pattern to the surface of the patient during image acquisition. This system is GateCT/GateRT, an extension of AlignRT, available from VisionRT, Ltd. (London, UK).

The respiratory pattern can also be derived from monitoring lung volume changes as used in the active breathing coordinator device offered by Elekta (Stockholm, Sweden) (Sarrut et al. 2005). The tidal flow is measured with a turbine spirometer via a mouthpiece; a nose clip is also used to increase the accuracy of the spirometry measurement. The relative airflow is displayed as a breathing pattern. The device can be employed for assisted, repeatable breath holds at a certain tidal volume by blocking the airflow using a small balloon valve. The system provides visual feedback for the patients, allowing them to see their breathing process and thus helping them choose when to initiate a breath hold.

Unlike the above device, for which the breath hold is triggered by the operator, the Abches device, manufactured by APEX Medical, Inc. (Tokyo, Japan), in collaboration with Yamanashi University (Onishi et al. 2006), is a respiratory monitoring device that allows the patient to self-control the respiratory motion of chest and abdomen (Figure 9.17). The breathing-induced motion is detected via three contacts placed on the abdomen and thorax and is converted to a rotational movement of a needle on an indicator unit. The patient can observe the level indicator panel on which three markers are available that define the full exhalation, the full inhalation, and the breath-holding range. The device has a switch as an auxiliary component, used by the patient to inform the therapist of the breath-holding state.

None of the systems described above visualize the tumor directly. Instead, they use the patient's surface as a surrogate. The tumor motion can be determined directly if metallic markers that can be detected through x-ray imaging are implanted in or near the tumor. The imaging systems can be room-mounted or linac-mounted.

Shirato et al. (2004) have developed, at the University of Hokkaido, in collaboration with Mitsubishi Electronics Co., Ltd. (Tokyo, Japan), a real-time tracking system based on pulsed fluoroscopy. Two x-ray tubes located on a circular track on the floor rotate synchronously with two detectors mounted on a circular track on the ceiling. The fluoroscopic images are used to detect the 3-D location of radio-opaque markers. The system is synchronized with the linear accelerator, and the beam can be automatically suspended if the fiducials fall outside of a predefined motion window.

ExacTrac is another room-mounted system, developed by Brainlab AG (Feldkirchen, Germany), which operates based on the synergy between two imaging systems: a stereoscopic kV x-ray system and a real-time infrared tracking system (Soete et al. 2002). The system has the ability to correlate internal tumor motion with external markers using x-ray imaging: the infrared tracking system monitors the motion of several reflective markers placed on the surface of the patient, and the kV x-ray system performs further tumor alignment (Figure 9.18).

The Synchrony Respiratory Tracking System, a subsystem of the CyberKnife robotic treatment device (Accuray, Inc., Sunnyvale, CA), operates based on a room-mounted stereoscopic x-ray system (Seppenwoolde et al. 2007). The patient wears a vest-like garment designed for use with LEDs as tracking

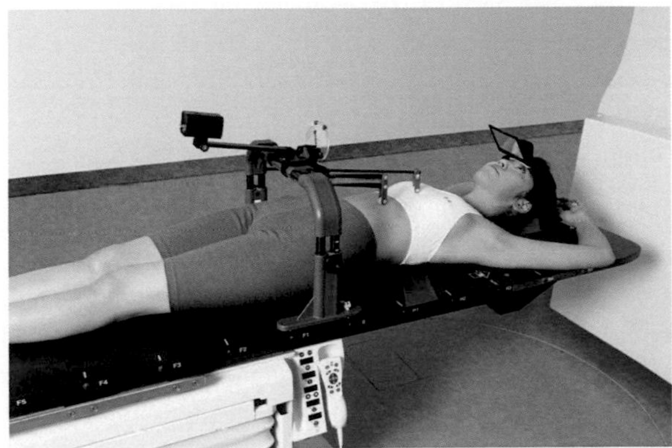

Figure 9.17 Photograph of the Abches device, manufactured by APEX Medical, Inc. (Tokyo, Japan), in collaboration with Yamanashi University, for respiratory monitoring that allows the patient to self-control the respiratory motion of the chest and abdomen. The breathing-induced motion is detected via three contacts placed on the abdomen and thorax and is converted to a rotational movement of a needle on an indicator unit.

(a)

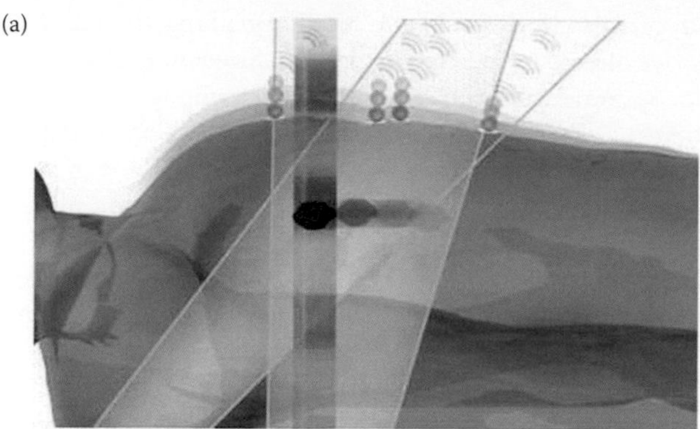

(b)

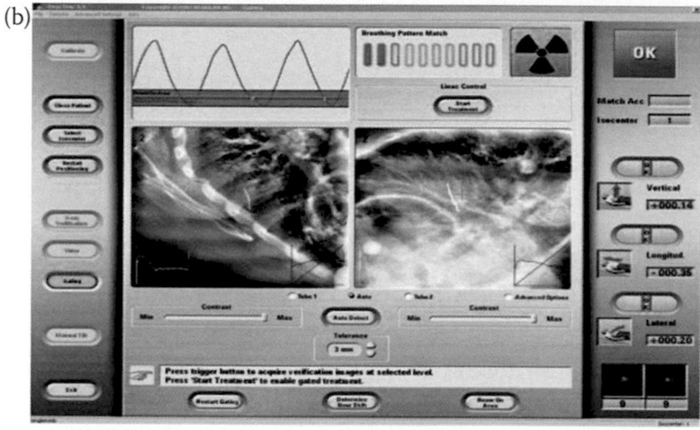

Figure 9.18 (a) Animation and (b) screen capture from the ExacTrac room-mounted system, developed by Brainlab AG (Feldkirchen, Germany), which operates based on the synergy between two imaging systems: a stereoscopic kV x-ray system and a real-time infrared tracking system. The system has the ability to correlate internal tumor motion with external markers using x-ray imaging: the infrared tracking system monitors the motion of several reflective markers placed on the surface of the patient.

markers. Three charge-coupled device cameras mounted to the ceiling are continuously recording the location of the LED markers. The x-ray system consists of two diagnostic x-ray sources and two corresponding orthogonal flat-panel amorphous silicon digital x-ray detectors. Prior to treatment, several gold fiducial markers are implanted in the tumor. The Synchrony system takes periodic images of these internal fiducials and predicts their location at a future time using the motion of the markers that are located on the patient's skin; thus, the body's surface movement correlates with the movement of the tumor.

Linac-mounted imaging systems with one kV x-ray unit orthogonal to the MV treatment beam are available from Varian (On-Board Imager) and Elekta (Synergy). Artiste from Siemens is a dual kV-MV imaging system with the kV x-ray system placed on a retractable shelf and the kV detection system mounted below the MV collimator.

An alternative to the radiographic detection of metallic markers is the monitoring of "wireless" transponders implanted inside tumors. The technology is available from Calypso Medical Technologies, Inc. (Seattle, WA) (Balter et al. 2005). An AC magnetic array contains source coils, which generate electromagnetic fields that excite the Beacon transponders as well as sensor coils that detect the response signals returned by transponders. The system has been used successfully for the real-time targeting of the prostate. Recently, the Beacon transponders have been redesigned to include an anchoring feature that provides stability in the airways of the lung to accommodate the use of the system for tracking lung tumors.

Another novel technology for target tracking without ionizing radiation is available from ViewRay, Inc. (Cleveland, OH). The Renaissance system consists of cobalt radiative sources integrated with a MRI real-time volumetric imaging system that provides superior soft-tissue visualization (the drawback, however, is limited treatment plan quality due to the increased penumbra and the low energy of the cobalt source).

9.3.1.1 Radiation therapy for moving targets

The acquisition of high-quality CT data for locations that undergo movements due to respiration is of paramount importance before the onset of the radiation treatment. The most widespread solution to this challenge is 4-D CT or respiration-correlated CT. The RPM system, the Anzai, and the bellows belt described above are the most common choices to determine the patient's breathing pattern during CT, which is then used to bin by phase or by amplitude the sinograms acquired while the patient is breathing freely. Nonetheless, the image quality is adversely affected by variations in the breathing pattern, and artifacts are still present even when breathing-training techniques are employed.

Managing breathing during radiation delivery can be accomplished by gating the beam (that is, the radiation is delivered only at certain points during the breathing cycle) or by tracking the tumor over its respiratory-induced trajectory. The practical realization of gating usually involves using the RPM, ABS, or Abches systems. The disadvantages of gated treatments include prolonged treatment times, patient discomfort, and lack of proper patient cooperation, deeming the implementation and the widespread use of such techniques still very challenging.

Following the tumor in real-time can be accomplished in four major ways as summarized by Murphy (Murphy 2004): moving the couch, moving the beam, moving the linear accelerator, and moving the dynamic multileaf collimator. The common demand for all approaches is the ability to localize the target in real time, using tumor motion surrogates or active or passive motion detection markers as described above. One major challenge on the path of tracking tumors in real-time is the latency in the system feedback. The issue could, in theory, be overcome by proper tumor motion prediction—an area still under investigation.

REFERENCES

Abdelnour, A. F., S. A. Nehmeh, T. Pan et al. 2007. Phase and amplitude binning for 4D-CT imaging. *Phys. Med. Biol.* 52: 3515–29.

Adler, J. R., Jr., M. J. Murphy, S. D. Chang, and S. L. Hancock. 1999. Image-guided robotic radiosurgery. *Neurosurgery* 44: 1299–306; discussion 1306–7.

Balter, J. M., J. N. Wright, L. J. Newell et al. 2005. Accuracy of a wireless localization system for radiotherapy. *Int. J. Radiat. Oncol. Biol. Phys.* 61: 933–7.

Barnes, E. A., B. R. Murray, D. M. Robinson, L. J. Underwood, J. Hanson, and W. H. Roa. 2001. Dosimetric evaluation of lung tumor immobilization using breath hold at deep inspiration. *Int. J. Radiat. Oncol. Biol. Phys.* 50(4): 1091–8.

Berbeco, R. I., T. Neicu, E. Rietzel, G. T. Chen, and S. B. Jiang. 2005. A technique for respiratory-gated radiotherapy treatment verification with an EPID in cine mode. *Phys. Med. Biol.* 50: 3669–79.

Bert, C., K. G. Metheany, K. P. Doppke, A. G. Taghian, S. N. Powell, and G. T. Chen. 2006. Clinical experience with a 3D surface patient setup system for alignment of partial-breast irradiation patients. *Int. J. Radiat. Oncol. Biol. Phys.* 64: 1265–74.

Biederer, J., J. Dinkel, G. Remmert et al. 2009. 4D-imaging of the lung: Reproducibility of lesion size and displacement on helical CT, MRI, and cone beam CT in a ventilated ex vivo system. *Int. J. Radiat. Oncol. Biol. Phys.* 73: 919–26.

Bortfeld, T., K. Jokivarsi, M. Goitein, J. Kung, and S. B. Jiang. 2002. Effects of intra-fraction motion on IMRT dose delivery: Statistical analysis and simulation. *Phys. Med. Biol.* 47: 2203–20.

Bosmans, G., J. Buijsen, A. Dekker et al. 2006. An "in silico" clinical trial comparing free breathing, slow and respiration correlated computed tomography in lung cancer patients. *Radiother. Oncol.* 81: 73–80.

Cervino, L. I., A. K. Chao, A. Sandhu, and S. B. Jiang. 2009. The diaphragm as an anatomic surrogate for lung tumor motion. *Phys. Med. Biol.* 54: 3529–41.

Chen, G. T., J. H. Kung, and K. P. Beaudette. 2004. Artifacts in computed tomography scanning of moving objects. *Semin. Radiat. Oncol.* 14: 19–26.

Chen, J., R. J. Lee, D. Handrahan, and W. T. Sause. 2007. Intensity-modulated radiotherapy using implanted fiducial markers with daily portal imaging: Assessment of prostate organ motion. *Int. J. Radiat. Oncol. Biol. Phys.* 68: 912–9.

Chen, Q. S., M. S. Weinhous, F. C. Deibel, J. P. Ciezki, and R. M. Macklis. 2001. Fluoroscopic study of tumor motion due to breathing: Facilitating precise radiation therapy for lung cancer patients. *Med. Phys.* 28(9): 1850–6.

Cho, B., P. R. Poulsen, and P. J. Keall. 2010. Real-time tumor tracking using sequential kV imaging combined with respiratory monitoring: A general framework applicable to commonly used IGRT systems. *Phys. Med. Biol.* 55(12): 3299–316.

Chui, C. S., E. Yorke, and L. Hong. 2003. The effects of intra-fraction organ motion on the delivery of intensity-modulated field with a multileaf collimator. *Med. Phys.* 30: 1736-46.

Cui, Y., J. G. Dy, G. C. Sharp, B. Alexander, and S. B. Jiang. 2007. Multiple template-based fluoroscopic tracking of lung tumor mass without implanted fiducial markers. *Phys. Med. Biol.* 52: 6229–42.

Dawson, L. A., K. K. Brock, S. Kazanjian et al. 2001. The reproducibility of organ position using active breathing control (ABC) during liver radiotherapy. *Int. J. Radiat. Oncol. Biol. Phys.* 51: 1410–21.

de Mey, J., J. Van de Steene, F. Vandenbroucke et al. 2005. Percutaneous placement of marking coils before stereotactic radiation therapy of malignant lung lesions. *J. Vasc. Interv. Radiol.* 16: 51–6.

Depuydt, T., D. Verellen, O. Haas et al. 2011. Geometric accuracy of a novel gimbals based radiation therapy tumor tracking system. *Radiother. Oncol.* 98(3): 365–72.

Duan, J., S. Shen, J. B. Fiveash, R. A. Popple, and I. A. Brezovich. 2006. Dosimetric and radiobiological impact of dose fractionation on respiratory motion induced IMRT delivery errors: A volumetric dose measurement study. *Med. Phys.* 33: 1380–7.

Ekberg, L., O. Holmberg, L. Wittgren, G. Bjelkengren, and T. Landberg. 1998. What margins should be added to the clinical target volume in radiotherapy treatment planning for lung cancer? *Radiother. Oncol.* 48: 71–7.

Engelsman, M., E. M. Damen, K. De Jaeger, K. M. van Ingen, and B. J. Mijnheer. 2001. The effect of breathing and set-up errors on the cumulative dose to a lung tumor. *Radiother. Oncol.* 60: 95–105.

Erridge, S. C., Y. Seppenwoolde, S. H. Muller, M. van Herk, K. De Jaeger, J. S. Belderbos, L. J. Boersma, and J. V. Lebesque. 2003. Portal imaging to assess set-up errors, tumor motion and tumor shrinkage during conformal radiotherapy of non-small cell lung cancer. *Radiother. Oncol.* 66(1): 75–85.

Ford, E. C., G. S. Mageras, E. Yorke, and C. C. Ling. 2003. Respiration-correlated spiral CT: A method of measuring respiratory-induced anatomic motion for radiation treatment planning. *Med. Phys.* 30: 88–97.

Garcia, R., R. Oozeer, H. Le Thanh et al. 2002. Radiotherapy of lung cancer: The inspiration breath hold with spirometric monitoring. *Cancer Radiother.* 6(1): 30–8.

Giebeler, A., J. Fontenot, P. Balter, G. Ciangaru, R. Zhu, and W. Newhauser. 2009. Dose perturbations from implanted helical gold markers in proton therapy of prostate cancer. *J. Appl. Clin. Med. Phys.* 10(1): 2875.

Gierga, D. P., J. Brewer, G. C. Sharp, M. Betke, C. G. Willett, and G. T. Chen. 2005. The correlation between internal and external markers for abdominal tumors: Implications for respiratory gating. *Int. J. Radiat. Oncol. Biol. Phys.* 61: 1551–8.

Guckenberger, M., J. Wilbert, T. Krieger, A. Richter, K. Baier, and M. Flentje. 2009a. Mid-ventilation concept for mobile pulmonary tumors: Internal tumor trajectory versus selective reconstruction of four-dimensional computed tomography frames based on external breathing motion. *Int. J. Radiat. Oncol. Biol. Phys.* 74: 602–9.

Guckenberger, M., J. Wulf, G. Mueller et al. 2009b. Dose-response relationship for image-guided stereotactic body radiotherapy of pulmonary tumors: Relevance of 4D dose calculation. *Int. J. Radiat. Oncol. Biol. Phys.* 74: 47–54.

Hanley, J., M. M. Debois, D. Mah et al. 1999. Deep inspiration breath-hold technique for lung tumors: The potential value of target immobilization and reduced lung density in dose escalation. *Int. J. Radiat. Oncol. Biol. Phys.* 45: 603–11.

Harada, T., H. Shirato, S. Ogura et al. 2002. Real-time tumor-tracking radiation therapy for lung carcinoma by the aid of insertion of a gold marker using bronchofiberscopy. *Cancer* 95: 1720–7.

Hurkmans, C. W., M. van Lieshout, D. Schuring et al. 2011. Quality assurance of 4D-CT scan techniques in multicenter phase III trial of surgery versus stereotactic radiotherapy (radiosurgery or surgery for operable early stage (stage 1A) non-small-cell lung cancer [ROSEL] study). *Int. J. Radiat. Oncol. Biol. Phys.* 80(3): 918–27.

Imura, M., K. Yamazaki, K. C. Kubota et al. 2008. Histopathologic consideration of fiducial gold markers inserted for real-time tumor-tracking radiotherapy against lung cancer. *Int. J. Radiat. Oncol. Biol. Phys.* 70: 382–4.

Jiang, S. B., J. Wolfgang, and G. S. Mageras. 2008. Quality assurance challenges for motion-adaptive radiation therapy: Gating, breath holding, and four-dimensional computed tomography. *Int. J. Radiat. Oncol. Biol. Phys.* 71(Suppl): S103–7.

Kamino, Y., K. Takayama, M. Kokubo et al. 2006. Development of a four-dimensional image-guided radiotherapy system with a gimbaled x-ray head. *Int. J. Radiat. Oncol. Biol. Phys.* 66: 271–8.

Keall, P. J., S. Joshi, S. S. Vedam, J. V. Siebers, V. R. Kini, and R. Mohan. 2005. Four-dimensional radiotherapy planning for DMLC-based respiratory motion tracking. *Med. Phys.* 32: 942–51.

Keall, P. J., V. R. Kini, S. S. Vedam, and R. Mohan. 2001. Motion adaptive x-ray therapy: A feasibility study. *Phys. Med. Biol.* 46: 1–10.

Keall, P. J., G. S. Mageras, J. M. Balter et al. 2006. The management of respiratory motion in radiation oncology report of AAPM Task Group 76. *Med. Phys.* 33(10): 3874–900.

Keall, P. J., G. Starkschall, H. Shukla et al. 2004. Acquiring 4D thoracic CT scans using a multislice helical method. *Phys. Med. Biol.* 49: 2053–67.

Kitamura, K., H. Shirato, S. Shimizu et al. 2002. Registration accuracy and possible migration of internal fiducial gold marker implanted in prostate and liver treated with real-time tumor-tracking radiation therapy (RTRT). *Radiother. Oncol.* 62(3): 275–81.

Kothary, N., S. Dieterich, J. D. Louie, D. T. Chang, L. V. Hofmann, and D. Y. Sze. 2009. Percutaneous implantation of fiducial markers for imaging-guided radiation therapy. *AJR Am. J. Roentgenol.* 192: 1090–6.

Langen, K. M., W. Lu, T. R. Willoughby et al. 2009. Dosimetric effect of prostate motion during helical tomotherapy. *Int. J. Radiat. Oncol. Biol. Phys.* 74: 1134–42.

Lax, I., H. Blomgren, I. Naslund, and R. Svanstrom. 1994. Stereotactic radiotherapy of malignancies in the abdomen. Methodological aspects. *Acta Oncol.* 33: 677–83.

Litzenberg, D., L. A. Dawson, H. Sandler et al. 2002. Daily prostate targeting using implanted radiopaque markers. *Int. J. Radiat. Oncol. Biol. Phys.* 52: 699–703.

Low, D. A., M. Nystrom, E. Kalinin et al. 2003. A method for the reconstruction of four-dimensional synchronized CT scans acquired during free breathing. *Med. Phys.* 30: 1254–63.

Lu, W., P. J. Parikh, I. M. El Naqa et al. 2005. Quantitation of the reconstruction quality of a four-dimensional computed tomography process for lung cancer patients. *Med. Phys.* 32: 890–901.

Lu, W., P. J. Parikh, J. P. Hubenschmidt, J. D. Bradley, and D. A. Low. 2006. A comparison between amplitude sorting and phase-angle sorting using external respiratory measurement for 4D CT. *Med. Phys.* 33: 2964–74.

Mageras, G. S., and E. Yorke. 2004. Deep inspiration breath hold and respiratory gating strategies for reducing organ motion in radiation treatment. *Semin. Radiat. Oncol.* 14: 65–75.

Mageras, G. S., E. Yorke, K. Rosenzweig et al. 2001. Fluoroscopic evaluation of diaphragmatic motion reduction with a respiratory gated radiotherapy system. *J. Appl. Clin. Med. Phys.* 2: 191–200.

Mah, D., J. Hanley, K. E. Rosenzweig et al. 2000. Technical aspects of the deep inspiration breath-hold technique in the treatment of thoracic cancer. *Int. J. Radiat. Oncol. Biol. Phys.* 48: 1175–85.

Mayse, M. L., P. J. Parikh, K. M. Lechleiter et al. 2008. Bronchoscopic implantation of a novel wireless electromagnetic transponder in the canine lung: A feasibility study. *Int. J. Radiat. Oncol. Biol. Phys.* 72: 93–8.

McIntosh, A., A. N. Shoushtari, S. H. Benedict, P. W. Read, and K. Wijesooriya. 2011. Quantifying the reproducibility of heart position during treatment and corresponding delivered heart dose in voluntary deep inhalation breath hold for left breast cancer patients treated with external beam radiotherapy. *Int. J. Radiat. Oncol. Biol. Phys.* 81: e569–76.

Morice, R. C., L. Keus, C. A. Jimenez et al. 2005. Bronchoscopic implantation of gold fiducials for estimating lung tumor motion during gated radiation therapy. *Chest* 128: 163S.

Murphy, M. J. 1997. An automatic six-degree-of-freedom image registration algorithm for image-guided frameless stereotaxic radiosurgery. *Med. Phys.* 24: 857–66.

Murphy, M. J. 2002. Fiducial-based targeting accuracy for external-beam radiotherapy. *Med. Phys.* 29: 334–44.

Murphy, M. J. 2004. Tracking moving organs in real time. *Semin. Radiat. Oncol.* 14: 91–100.

Murphy, M. J., S. D. Chang, I. C. Gibbs et al. 2003. Patterns of patient movement during frameless image-guided radiosurgery. *Int. J. Radiat. Oncol. Biol. Phys.* 55: 1400–8.

Murphy, M. J., D. Martin, R. Whyte, J. Hai, C. Ozhasoglu, and Q. T. Le. 2002. The effectiveness of breath-holding to stabilize lung and pancreas tumors during radiosurgery. *Int. J. Radiat. Oncol. Biol. Phys.* 53: 475–82.

Nagata, Y., K. Takayama, Y. Matsuo et al. 2005. Clinical outcomes of a phase I/II study of 48 Gy of stereotactic body radiotherapy in 4 fractions for primary lung cancer using a stereotactic body frame. *Int. J. Radiat. Oncol. Biol. Phys.* 63: 1427–31.

Nehmeh, S. A., Y. E. Erdi, T. Pan et al. 2004. Quantitation of respiratory motion during 4D-PET/CT acquisition. *Med. Phys.* 31: 1333–8.

Neicu, T., R. Berbeco, J. Wolfgang, and S. B. Jiang. 2006. Synchronized moving aperture radiation therapy (SMART): Improvement of breathing pattern reproducibility using respiratory coaching. *Phys. Med. Biol.* 51: 617–36.

Neicu, T., H. Shirato, Y. Seppenwoolde, and S. B. Jiang. 2003. Synchronized moving aperture radiation therapy (SMART): Average tumour trajectory for lung patients. *Phys. Med. Biol.* 48: 587–98.

Noel, C., P. J. Parikh, M. Roy et al. 2009. Prediction of intrafraction prostate motion: Accuracy of pre- and post-treatment imaging and intermittent imaging. *Int. J. Radiat. Oncol. Biol. Phys.* 73: 692–8.

Onishi, H., K. Marino, N. Sano et al. 2006. 2456: A simple and efficient irradiation system for a lung tumor with small internal margin: Patient's self-breath-hold using a newly developed respiratory indicator (Abches) and self-turning radiation-beam on and off. *Int. J. Radiat. Oncol. Biol. Phys.* 66(Suppl): S462–3.

Pan, T. 2005. Comparison of helical and cine acquisitions for 4D-CT imaging with multislice CT. *Med. Phys.* 32: 627–34.

Pan, T., T. Y. Lee, E. Rietzel, and G. T. Chen. 2004. 4D-CT imaging of a volume influenced by respiratory motion on multi-slice CT. *Med. Phys.* 31: 333–40.

Papiez, L., and D. Rangaraj. 2005. DMLC leaf-pair optimal control for mobile, deforming target. *Med. Phys.* 32: 275–85.

Plathow, C., S. Ley, C. Fink, M. Puderbach, W. Hosch, A. Schmahl, J. Debus, and H. U. Kauczor. 2004. Analysis of intrathoracic tumor mobility during whole breathing cycle by dynamic MRI. *Int. J. Radiat. Oncol. Biol. Phys.* 59(4): 952–9.

Rangaraj, D., and L. Papiez. 2005. Synchronized delivery of DMLC intensity modulated radiation therapy for stationary and moving targets. *Med. Phys.* 32: 1802–17.

Ross, C. S., D. H. Hussey, E. C. Pennington, W. Stanford, and J. F. Doornbos. 1990. Analysis of movement of intrathoracic neoplasms using ultrafast computerized tomography. *Int. J. Radiat. Oncol. Biol. Phys.* 18(3): 671–7.

Sarrut, D., V. Boldea, M. Ayadi et al. 2005. Nonrigid registration method to assess reproducibility of breath-holding with ABC in lung cancer. *Int. J. Radiat. Oncol. Biol. Phys.* 61: 594–607.

Sawant, A., R. Venkat, V. Srivastava et al. 2008. Management of three-dimensional intrafraction motion through real-time DMLC tracking. *Med. Phys.* 35: 2050–61.

Schweikard, A., G. Glosser, M. Bodduluri, M. J. Murphy, and J. R. Adler. 2000. Robotic motion compensation for respiratory movement during radiosurgery. *Comput. Aided Surg.* 5: 263–77.

Seco, J., G. C. Sharp, J. Turcotte, D. Gierga, T. Bortfeld, and H. Paganetti. 2007. Effects of organ motion on IMRT treatments with segments of few monitor units. *Med. Phys.* 34: 923–34.

Seppenwoolde, Y., R. I. Berbeco, S. Nishioka, H. Shirato, and B. Heijmen. 2007. Accuracy of tumor motion compensation algorithm from a robotic respiratory tracking system: A simulation study. *Med. Phys.* 34: 2774–84.

Seppenwoolde, Y., H. Shirato, K. Kitamura et al. 2002. Precise and real-time measurement of 3D tumor motion in lung due to breathing and heartbeat, measured during radiotherapy. *Int. J. Radiat. Oncol. Biol. Phys.* 53: 822–34.

Shimizu, S., H. Shirato, K. Kagei et al. 2000. Impact of respiratory movement on the computed tomographic images of small lung tumors in three-dimensional (3D) radiotherapy. *Int. J. Radiat. Oncol. Biol. Phys.* 46: 1127–33.

Shirato, H., M. Oita, K. Fujita, Y. Watanabe, and K. Miyasaka. 2004. Feasibility of synchronization of real-time tumor-tracking radiotherapy and intensity-modulated radiotherapy from viewpoint of excessive dose from fluoroscopy. *Int. J. Radiat. Oncol. Biol. Phys.* 60: 335–41.

Shirato, H., S. Shimizu, K. Kitamura et al. 2000a. Four-dimensional treatment planning and fluoroscopic real-time tumor tracking radiotherapy for moving tumor. *Int. J. Radiat. Oncol. Biol. Phys.* 48: 435–42.

Shirato, H., S. Shimizu, T. Kunieda et al. 2000b. Physical aspects of a real-time tumor-tracking system for gated radiotherapy. *Int. J. Radiat. Oncol. Biol. Phys.* 48: 1187–95.

Sixel, K. E., M. C. Aznar, and Y. C. Ung. 2001. Deep inspiration breath hold to reduce irradiated heart volume in breast cancer patients. *Int. J. Radiat. Oncol. Biol. Phys.* 49: 199–204.

Smith, R. L., K. Lechleiter, K. Malinowski et al. 2009. Evaluation of linear accelerator gating with real-time electromagnetic tracking. *Int. J. Radiat. Oncol. Biol. Phys.* 74: 920–7.

Soete, G., J. Van de Steene, D. Verellen et al. 2002. Initial clinical experience with infrared-reflecting skin markers in the positioning of patients treated by conformal radiotherapy for prostate cancer. *Int. J. Radiat. Oncol. Biol. Phys.* 52: 694–8.

Stevens, C. W., R. F. Munden, K. M. Forster, J. F. Kelly, Z. Liao, G. Starkschall, S. Tucker, and R. Komaki. 2001. Respiratory-driven lung tumor motion is independent of tumor size, tumor location, and pulmonary function. *Int. J. Radiat. Oncol. Biol. Phys.* 51(1): 62–8.

Suh, Y., B. Yi, S. Ahn et al. 2004. Aperture maneuver with compelled breath (AMC) for moving tumors: A feasibility study with a moving phantom. *Med. Phys.* 31: 760–6.

Tacke, M. B., S. Nill, A. Krauss, and U. Oelfke. 2010. Real-time tumor tracking: Automatic compensation of target motion using the Siemens 160 MLC. *Med. Phys.* 37(2): 753–61.

Underberg, R. W., F. J. Lagerwaard, B. J. Slotman, J. P. Cuijpers, and S. Senan. 2005. Use of maximum intensity projections (MIP) for target volume generation in 4D CT scans for lung cancer. *Int. J. Radiat. Oncol. Biol. Phys.* 63: 253–60.

Vedam, S. S., P. J. Keall, V. R. Kini, and R. Mohan. 2001. Determining parameters for respiration-gated radiotherapy. *Med. Phys.* 28: 2139–46.

Vedam, S. S., P. J. Keall, V. R. Kini, H. Mostafavi, H. P. Shukla, and R. Mohan. 2003. Acquiring a four-dimensional computed tomography dataset using an external respiratory signal. *Phys. Med. Biol.* 48: 45–62.

Verellen, D., T. Depuydt, T. Gevaert et al. 2010. Gating and tracking, 4D in thoracic tumours. *Cancer Radiother.* 14(6–7): 446–54.

Webb, S. 2005a. The effect on IMRT conformality of elastic tissue movement and a practical suggestion for movement compensation via the modified dynamic multileaf collimator (dMLC) technique. *Phys. Med. Biol.* 50: 1163–90.

Webb, S. 2005b. Limitations of a simple technique for movement compensation via movement-modified fluence profiles. *Phys. Med. Biol.* 50: N155–61.

Welsh, J. S., C. Berta, S. Borzillary et al. 2004. Fiducial markers implanted during prostate brachytherapy for guiding conformal external beam radiation therapy. *Technol. Cancer Res. Treat.* 3: 359–64.

Wijesooriya, K., C. Bartee, J. V. Siebers, S. S. Vedam, and P. J. Keall. 2005. Determination of maximum leaf velocity and acceleration of a dynamic multileaf collimator: Implications for 4D radiotherapy. *Med. Phys.* 32: 932–41.

Wijesooriya, K., E. Weiss, and V. Dill. 2008. Quantifying the accuracy of automated structure segmentation in 4D CT images using a deformable image registration algorithm. *Med. Phys.* 35: 1251–60.

Wong, J. W., M. B. Sharpe, D. A. Jaffray et al. 1999. The use of active breathing control (ABC) to reduce margin for breathing motion. *Int. J. Radiat. Oncol. Biol. Phys.* 44(4): 911–9.

Wulf, J., U. Hadinger, U. Oppitz, B. Olshausen, and M. Flentje. 2000. Stereotactic radiotherapy of extracranial targets: CT-simulation and accuracy of treatment in the stereotactic body frame. *Radiother. Oncol.* 57: 225–36.

Wunderink, W., A. Méndez Romero, Y. Seppenwoolde, H. de Boer, P. Levendag, and B. Heijmen. 2010. Potentials and limitations of guiding liver stereotactic body radiation therapy set-up on liver-implanted fiducial markers. *Int. J. Radiat. Oncol. Biol. Phys.* 77(5): 1573–83.

Xia, T., H. Li, Q. Sun et al. 2006. Promising clinical outcome of stereotactic body radiation therapy for patients with inoperable Stage I/II non-small-cell lung cancer. *Int. J. Radiat. Oncol. Biol. Phys.* 66: 117–25.

Xie, Y., D. Djajaputra, C. R. King, S. Hossain, L. Ma, and L. Xing. 2008. Intrafractional motion of the prostate during hypofractionated radiotherapy. *Int. J. Radiat. Oncol. Biol. Phys.* 72: 236–46.

Xu, Q., R. J. Hamilton, R. A. Schowengerdt, B. Alexander, and S. B. Jiang. 2008. Lung tumor tracking in fluoroscopic video based on optical flow. *Med. Phys.* 35: 5351–9.

Xu, Q., R. J. Hamilton, R. A. Schowengerdt, and S. B. Jiang. 2007. A deformable lung tumor tracking method in fluoroscopic video using active shape models: A feasibility study. *Phys. Med. Biol.* 52: 5277–93.

Yu, C. X., D. A. Jaffray, and J. W. Wong. 1998. The effects of intra-fraction organ motion on the delivery of dynamic intensity modulation. *Phys. Med. Biol.* 43: 91–104.

10 Image-guided radiation therapy and frameless stereotactic radiation therapy

Ryan McMahon and Fang-Fang Yin

Contents

10.1 ADAPTING DIAGNOSTIC IMAGING ONTO THERAPY EQUIPMENT

This chapter gives an overview of how to adapt diagnostic imaging technologies on to linear accelerators for image-guided frameless stereotactic radiation therapy (SRT) when the primary focus is on target localization and verification inside the treatment room. Basic applications of in-room imaging techniques are discussed along with a description of several commercially available image-guided radiation therapy (IGRT) solutions. The remainder of the chapter is devoted to current trends in frameless SRT treatment techniques with an emphasis on the role of image-guidance.

10.1.1 IMAGING TECHNIQUES

10.1.1.1 Imaging techniques for in-room image guidance

Image guidance has been an integral part of radiation therapy for decades. Imaging technology that has played a role in the simulation, planning, treatment, and verification of radiation therapy includes 2-D radiography, fan-beam CT and cone-beam CT (CBCT), positron emission tomography/single photon emission computed tomography (PET/SPECT), magnetic resonance imaging (MRI), ultrasound, and optical imaging. Comprehensive reviews of these technologies and their applications to radiation therapy can be found elsewhere (Curran, Balter, and Chetty 2006). The information provided here is intended to give the reader a perspective on how technological developments in IGRT have made frameless SRT a clinical reality.

10.1.1.1.1 2-D radiography

The use of 2-D projection imaging for treatment verification has a long history, starting with the use of radiographic film. Film is able to accurately represent fine anatomical detail with simple imaging geometry. The geometry represented in the image is primarily dependent on the relationship between the source, object, and film. Limitations, such as exposure level, chemical sensitivity, and the need for an offline developer, have limited its application as an efficient tool for online image guidance. The introduction of computed radiography eliminated some of film's limitations, but it still requires offline readout with a dedicated device.

While most modern departments still have the ability to perform film-based verification, it has fallen out of favor due to the widespread introduction of digital imaging technology. Compared to film, the image quality of digital radiography is somewhat degraded, but advances in detector technology have closed this gap (Antonuk 2002).

In terms of their application to patient setup and treatment target verification, film and digital imaging systems have been shown to be comparable (Herman et al. 2001; Yin et al. 1996). The primary advantage of digital imaging is not related to image quality or localization accuracy, but instead to its ability to be integrated with the treatment device. Digital imaging can perform within an online closed-loop system, and image acquisition, image analysis, and patient repositioning are all integrated. This online system integration is perhaps the key component of the modern definition of IGRT.

In-room image guidance can be performed with either megavoltage (MV) or kilovoltage (kV) energies. MV imaging uses the treatment beam, and kV imaging necessitates the addition of a separate kV x-ray source. MV and kV imaging are, in large part, parallel systems that offer many of the same capabilities for treatment verification and patient setup. MV imaging is limited by its poor soft-tissue contrast and relatively high dosage, but it can be useful when bony anatomy or implanted fiducials can be used as reliable surrogates. kV imaging provides higher soft-tissue contrast with less dosage but still often relies on bony anatomy or fiducials for patient alignment. In either case, the most common correction strategy is to compare 2-D images acquired online to 2-D digitally reconstructed radiographs generated from the planning CT. Fully integrated systems use online analysis software to automatically calculate and apply appropriate corrections to the treatment couch.

10.1.1.1.2 CT

The introduction of digital imaging made it possible to integrate imaging and treatment systems, but the addition of in-room volumetric imaging made it possible to develop turnkey IGRT solutions. Compared to 2-D planar imaging, in-room CT offers substantial improvements in soft-tissue contrast. This makes it possible to directly visualize soft-tissue targets rather than relying on surrogates, such as bony anatomy or implanted fiducials.

One of the first clinical applications of integrated in-room CT was in Japan in 1996 when a treatment room was developed that contained a linac and CT scanner with a shared treatment couch (Uematsu et al. 1996). The initial motivation for this configuration was to facilitate frameless SRT. Since then, this concept has matured into what is generically referred to as the "CT on rails" system. While these systems may be too cumbersome for routine clinical use, they have allowed for pioneering work in the application of IGRT to adaptive therapy (Barker et al. 2004; de Crevoisier et al. 2004). Because these systems use diagnostic-quality scanners, daily imaging can be directly compared to pretreatment planning CTs, and dose distributions can be computed on these images with little uncertainty in the calculation (relative to the calculation performed on the initial planning CT).

CBCT has also recently become a standard technique for in-room volumetric imaging. This approach began with the introduction of a mobile C-arm system into the treatment room (Kriminski et al. 2005; Siewerdsen et al. 2005). Modern systems employ kV or MV sources and detectors that are directly mounted on the linac gantry. This configuration has low space requirements, which makes it easy to retrofit onto an existing linac. CBCT image quality is poorer than fan-beam CT due to increased scatter. However, relative to 2-D radiographic images, CBCT provides a substantial improvement in target visualization. Two configurations are available for on-board CBCT. The first uses the MV treatment beam with a flat panel imager (MV-CBCT). The second uses a separate kV source-detector pair mounted on the gantry (kV-CBCT). Both configurations have been successfully implemented into the clinic. From an image-quality standpoint, the most important difference is that kV-CBCT tends to deliver better image quality with lower patient dosage (Groh et al. 2002).

10.1.2 NEW DEVELOPMENTS AND FUTURE POTENTIAL

10.1.2.1 MRI

The use of MRI in radiation oncology as an offline tool has been increasing rapidly, and several researchers have been exploring the idea of in-room MR imaging (Lagendijk et al. 2008; Raaijmakers et al. 2007; Raaijmakers, Raaymakers, and Lagendijk 2007, 2008). Technologically, this is a challenge due to the need to separate the magnetic environments of the Linac and the MR device. One group has addressed this issue by using Co-60 for the treatment beam (Dempsey et al. 2006), which eliminates the magnetic field produced by a linac and reduces the number of components that have to be built out of non-ferromagnetic materials (see Chapter 15). Other groups are coupling MR units to a linac using passive (Fallone et al. 2009) or active (Raaymakers et al. 2009) magnetic shielding.

The primary benefits of in-room MR imaging are three-fold. First, soft-tissue contrast of MR is substantially better than CT, which is a clear advantage for many tumor sites. Secondly, MR can provide fast, dynamic, volumetric imaging, which could make it possible to explore advanced motion-management strategies that wouldn't be possible with x-ray imaging alone. Last, MR imaging does not contribute to patient dosage, which makes it an ideal tool for adaptive therapy.

While MR may provide superior target visualization, it is not currently suitable as a stand-alone imaging modality for radiation therapy. For example, more work needs to be done to understand the impact of MR artifacts on patient setup and MR-based dosage calculations (Kirkby et al. 2008; Kirkby, Stanescu, and Fallone 2009).

10.1.2.2 PET and SPECT

The use of PET and SPECT in an offline setting for diagnosis, staging, and treatment planning has been rapidly increasing in radiation oncology. In-room online functional imaging for patient setup and treatment verification is also being investigated (Roper, Bowsher, and Yin 2009). Due to the limited spatial resolution and anatomical information of these images, it is unlikely that PET/SPECT could provide adequate setup information alone. However, in conjunction with CT or MR, in-room functional imaging would offer the ability to incorporate a biological target volume into daily setup. This ability to visualize physiological processes in the treatment room could become important as biological target volumes begin to play a more important role in treatment planning. Furthermore, these systems could be used to collect data for ongoing assessment of treatment response.

10.1.3 COMMERCIALLY AVAILABLE SYSTEMS

While all of the aforementioned imaging modalities offer unique advantages for patient setup, the remainder of this chapter will focus primarily on in-room kV/MV x-ray imaging simply because it is currently the most widely available in-room imaging modality. The commercially available x-ray IGRT solutions are available in three variations: Gantry-mounted systems, ceiling- or floor-mounted systems, and rail track-mounted systems. A comprehensive review of this technology has been compiled by the American Association of Physicists in Medicine (AAPM) Task Group 104 (Yin et al. 2009). A brief overview is presented in the following sections. Varian on-board imaging technology is shown in Figures 10.1 and 10.2, Siemens in Figures 10.3 and 10.4, and Elekta in Figures 10.5 and 10.6. See Chapter 15 for a discussion of new systems.

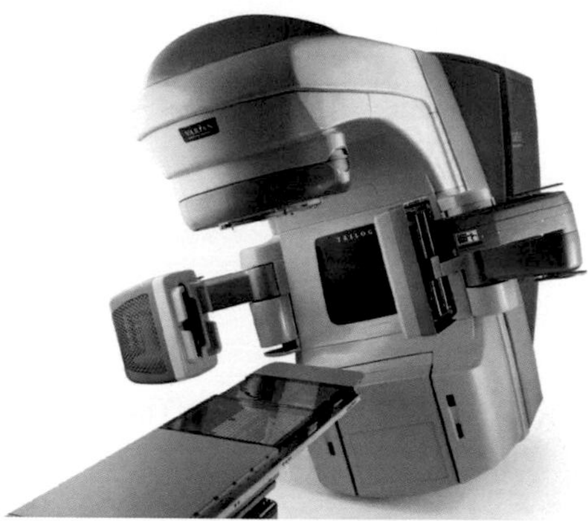

Figure 10.1 Varian Trilogy, kV source is mounted to a gantry via robotic arms. MV detector is shown retracted at bottoms of gantry. (Image provided courtesy of Varian Medical Systems.)

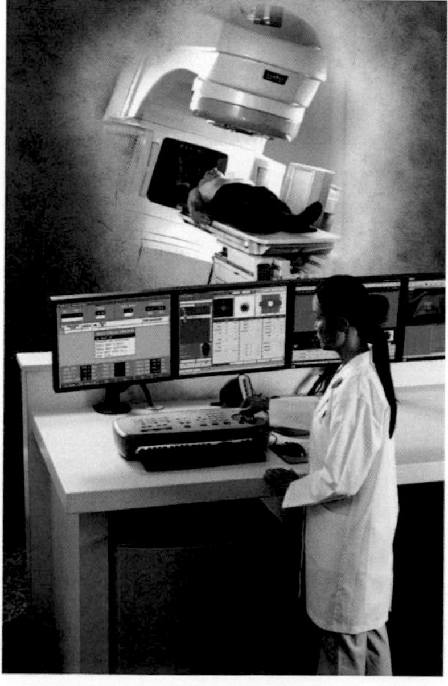

Figure 10.2 Varian Trilogy, integrated workstations for treatment delivery and on-board image analysis. (Image provided courtesy of Varian Medical Systems.)

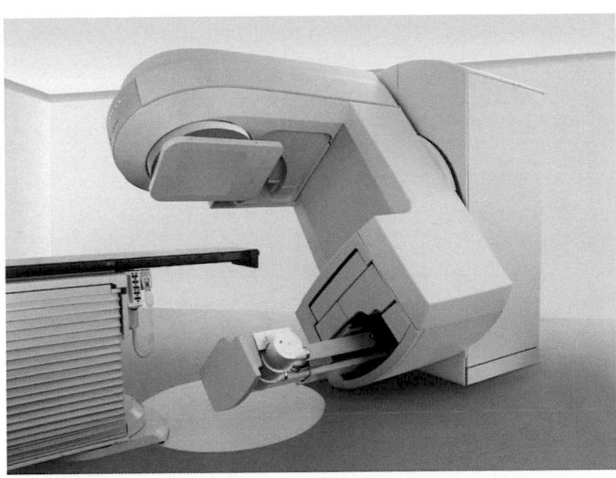

Figure 10.3 Siemens Artiste. kV imaging system also shown in image although not commercially released at present. kV source is extended at the bottom of the gantry, and retractable kV flat-panel detector is mounted on gantry head. MV detector shown retracted at bottom of gantry above kV source. (Image provided courtesy of Siemens Healthcare.)

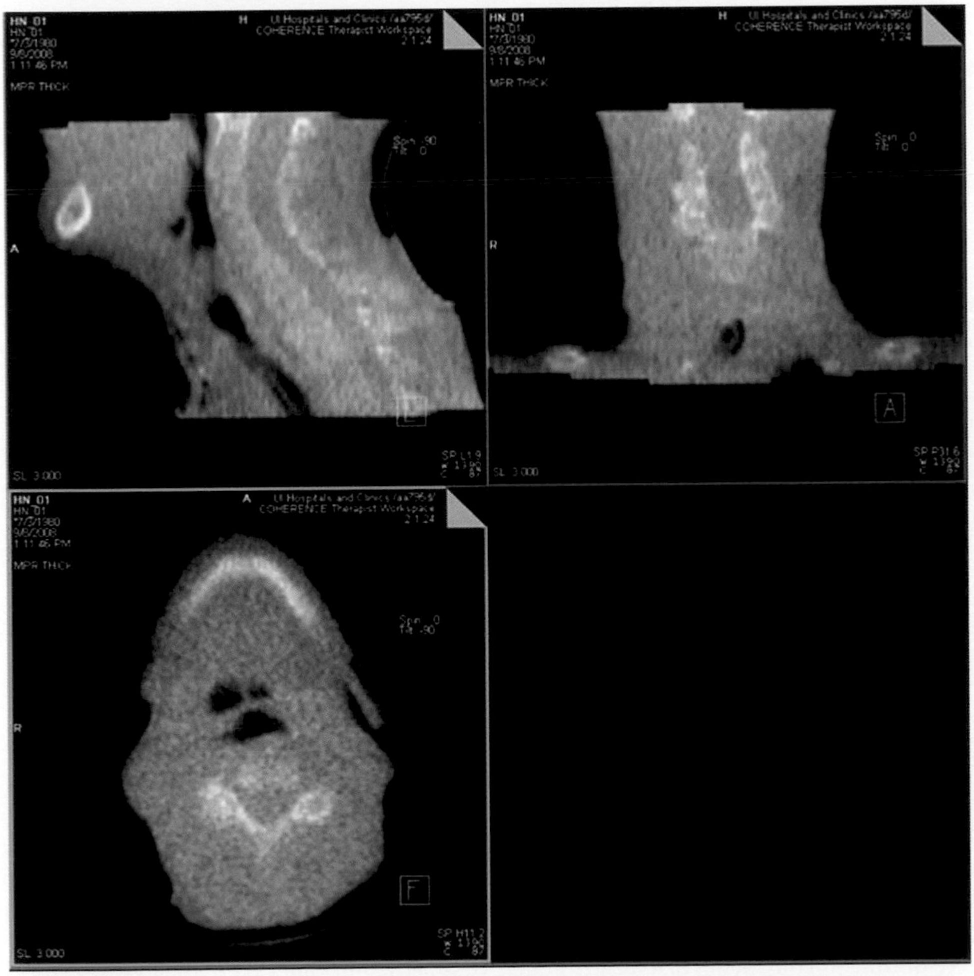

Figure 10.4 Siemens MVision™. Online software environment for evaluation on-board MV-CBCT images. (Image provided courtesy of Siemens Healthcare.)

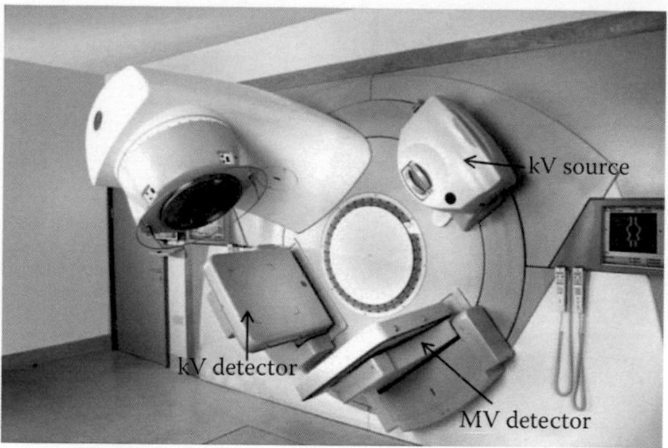

Figure 10.5 Elekta Synergy™. kV source (right) and detector (left) and MV detector (bottom) are all mounted to gantry. (Image provided courtesy of Elekta.)

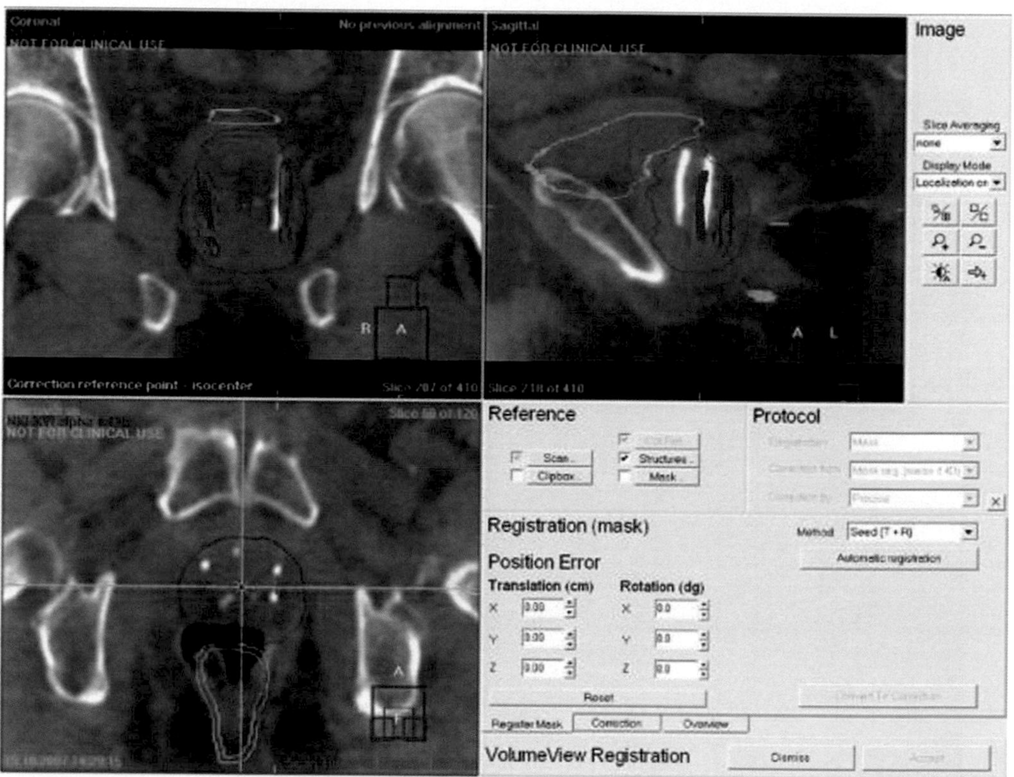

Figure 10.6 Elekta Intuity™. Online software environment for evaluation on-board 2-D/3-D MV images. (Image provided courtesy of Elekta.)

10.1.3.1 Gantry-mounted MV imaging systems

MV imaging based on amorphous silicon electronic portal imaging devices can play several roles in radiation treatment, including patient setup, treatment verification, patient-specific QA, and machine-specific QA. Elekta, Varian, and Siemens all offer on-board MV imaging systems with the primary difference in functionality being that Siemens offers volumetric imaging (MV-CBCT) while the Elekta and Varian systems only have 2-D planar imaging capabilities at present.

10.1.3.2 Gantry-mounted kV imaging systems

The most common configuration for accelerators designed for IGRT is a gantry-mounted kV source and detector. Elekta's Synergy X-Ray Volume Imaging (XVI) and Varian's On-Board Imager® (OBI) are two commercially available gantry-mounted kV systems. While these systems have some key design differences, they are very similar in terms of performance and clinical workflow. In the Elekta Synergy® system, the kV x-ray source and detector are mounted on retractable arms on the machine's drum gantry. The Varian OBI consists of a kV source and flat panel detector that are mounted to the gantry on robotic arms. kV-CBCT capabilities are available in each system, and both offer integrated software tools for online image registration and remote patient repositioning. On-board kV imaging is currently being developed by Siemens, but the system is not commercially available at present (Figure 10.3).

10.1.3.3 Rail track-mounted tomographic systems

Two commercially available "CT on rails" systems are available. The first is a collaboration between Varian and GE, and the second is marketed by Siemens (Figure 10.7). In both systems, the CT scanner is translated toward the treatment couch on a rail system, but the couch must be moved to different positions for imaging and treatment. Typically the couch is rotated by 180° to transition between imaging and treatment positions. Both systems have software tools for online image reconstruction, registration, and patient repositioning.

10.1.3.4 Ceiling- or floor-mounted systems

Ceiling- or floor-mounted systems have kV source-detector assemblies mounted onto the walls, ceiling, or floor of the treatment room. Because the imaging components are fixed, two pairs of source-detectors are required for orthogonal imaging, and tomographic imaging is not available. Two commercially available floor- or ceiling-mounted systems have been developed by Brainlab (Novalis ExacTrac®) and Accuray (CyberKnife). Both of these systems offer 6-D patient repositioning capabilities using a robotic couch. See Chapter 15 for the new Varian integrated system.

The Novalis ExacTrac system uses a combination of kV and infrared imaging. The kV system consists of two x-ray tubes recessed into the floor and two corresponding flat panel detectors mounted on the ceiling. The imaging system is oriented at an oblique angle relative to the gantry's plane of rotation. The infrared system consists of two infrared cameras mounted on the ceiling and several reflective markers attached to the patient or treatment couch. This kV system is used to image the patient and determine corrections to patient setup, and the infrared system is used to guide the couch motion when repositioning the patient.

The CyberKnife® Robotic Radiosurgery system consists of a small x-band linac mounted on a robotic arm. The imaging system consists of two kV x-ray tubes mounted on the ceiling and two corresponding flat

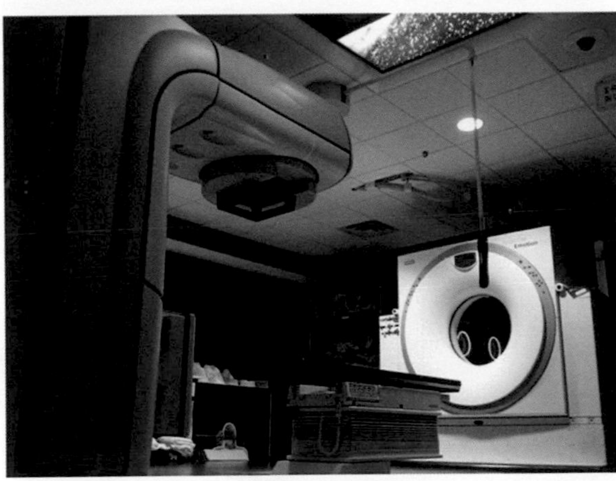

Figure 10.7 Siemens Primatom™. Diagnostic CT scanner installed inside treatment room. (Image provided courtesy of Elekta.)

Stereotactic radiation therapy, precision patient positioning, and immobilization panel detectors that are mounted next to the couch or recessed into the floor. The two source-detector pairs are orthogonal to each other with each being symmetrically offset 45° from the mid-sagittal plane.

10.1.4 OTHER COMMERCIAL SYSTEMS

10.1.4.1 Tomotherapy

Early efforts to develop turnkey CT-based IGRT solutions are perhaps best represented by the evolution of helical tomotherapy (Langen et al. 2006). From its initial conception, in-room tomographic imaging was a central component of the device. The first generation of helical tomotherapy units used a 6-MV beam for both CT localization imaging and intensity-modulated radiation treatment. The new device (Vera system) added a kV source mounted on a fixed ring-gantry to provide on-board kV CT imaging. The tomotherapy IGRT solution is shown in Figures 10.2 and 10.8.

10.1.4.2 Vision RT

The AlignRT system (Vision RT, Ltd., London, UK) is a 3-D surface mapping system that can be installed as an add-on in a standard linear accelerator vault. The system uses two ceiling-mounted cameras to reconstruct the 3-D surface of a patient prior to treatment and then maps it to a reference surface to

(a)

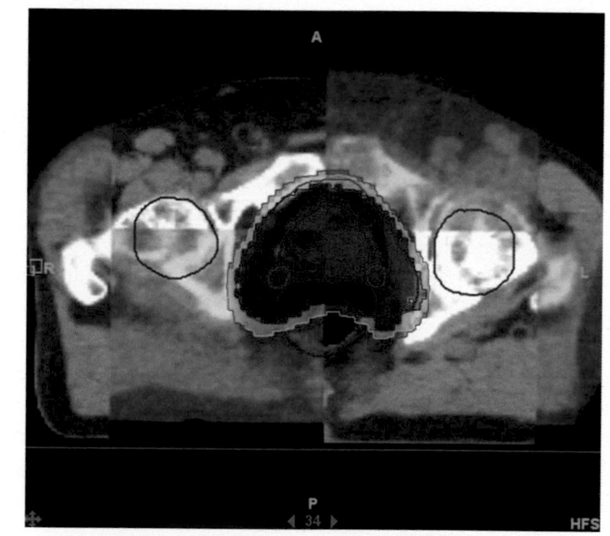

(b)

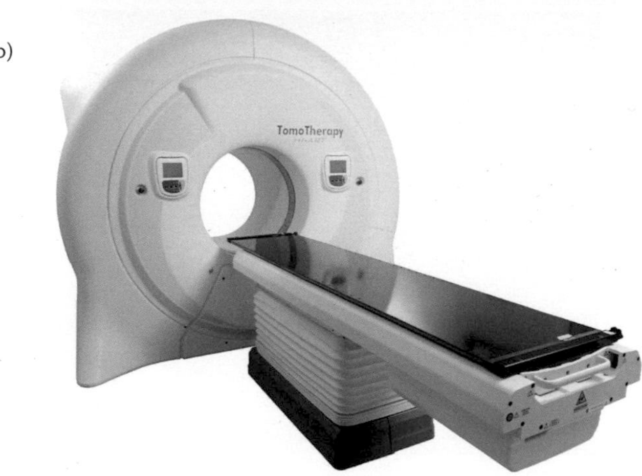

Figure 10.8 (a) Online software environment for evaluation on-board MV-CT with tomotherapy. (b) Tomotherapy HiArt. Rotational treatment geometry provides built-in MV-CT imaging capabilities. (Image provided courtesy of Tomotherapy, Inc.)

determine the couch shifts required to align the patient. The system also monitors the patient in real time throughout the treatment and can be configured for use with respiratory gating. This system has been utilized in several treatment sites, including cranial FSRT and SRS, breast, and lung (Bert et al. 2005, 2006; Cervino et al. 2010; Hughes et al. 2009).

Figure 10.9 Novalis Tx™ system marketed jointly by Varian Medical Systems and Brainlab. All Varian on-board imaging capabilities are included. Brainlab's ExacTrac™ system is also included, which consists of two kV x-ray tubes recessed into the floor and two ceiling-mounted flat-panel detectors. (Image provided courtesy of Brainlab.)

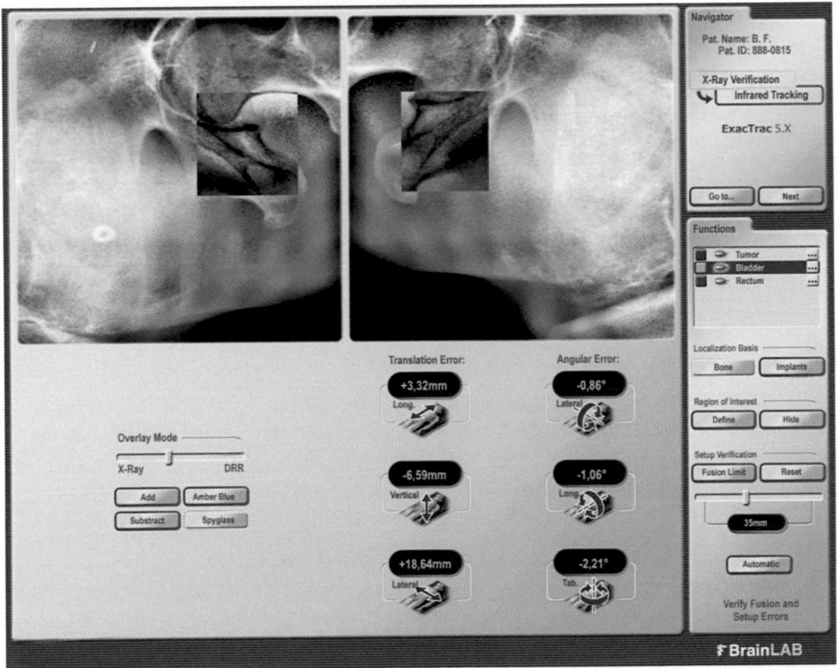

Figure 10.10 Online image-guided analysis software for Brainlab ExacTrac. 6-D setup corrections are calculated from a pair of oblique kV x-ray images. (Image provided courtesy of Brainlab.)

10.1.4.3 Hybrid kV systems

Each of the above categories of in-room imaging has inherent advantages and limitations. If it is technologically feasible, combining multiple imaging systems into one integrated unit can provide more flexibility for a variety of image-guided clinical scenarios. One commercially available hybrid delivery unit is the NovalisTx™, now marketed at Novalis powered by TrueBeam, which is jointly marketed by Varian Medical Systems and Brainlab (shown in Figures 10.9 and 10.10). The hybrid imaging system combines the Varian OBI system with the Brainlab ExacTrac x-ray system to provide both 3-D tomographic imaging and dual planar imaging for patient positioning (Chang et al. 2008).

10.2 SELECTION OF TECHNOLOGIES FOR IGRT

10.2.1 CLINICAL NEED AND JUSTIFICATION

When selecting and implementing an image-guided system, it is important to evaluate the system in terms of clinical goals and desired outcomes. The primary advantage of IGRT is the potential to reduce treatment margins through improvements in patient setup accuracy. A second, related advantage of IGRT is treatment verification, that is, verification that treatment margins are adequate. Generally speaking, determining which patients will benefit from IGRT is a decision that has to be made on a case-by-case basis, so it is important to have a clear understanding of the applications and limitations of each IGRT system being considered.

While patient-related objectives should be the motivation for implementing IGRT, there are many other factors to consider, including product availability, existing infrastructure, manpower, and future development. Because integration is a key step in image-guided processes, the entire clinical workflow must be examined when discussing the feasibility of IGRT implementation. Implementation requires an understanding of all steps involved in the IGRT workflow, including data collection, data analysis, intervention, and patient safety.

10.2.2 SYSTEM INTEGRATION

The integration of the imaging system with the treatment-planning system and treatment device is the key to safe and efficient implementation of IGRT in the clinic. This importance can be understood by restating the three primary tasks involved in image-guided processes: data collection, data analysis, and intervention. Lack of integration between any of these tasks will decrease the efficiency of image-guided treatments.

In this context, data collection is simply the image acquisition. The integrity of this process must be continually monitored with a rigorous quality-assurance (QA) program (Yin et al. 2009). In an integrated workflow, this data must be automatically stored in a location that is accessible to online and offline analysis systems. This is not a trivial task, considering most patients will have multiple image sets to store. A single CBCT data set, for example, can require nearly 1 GB of storage for the raw projection data and reconstructed image. A procedure must also be in place for rapid and efficient backup of critical image data and periodic removal of unnecessary data.

Data analysis is typically performed using online and offline software tools. Online analysis tools serve as a direct link between image acquisition and intervention. For full system integration, this software needs to be able to load the acquired image, compare it to a reference image, and translate the observed differences into adjustable machine parameters (e.g., couch shifts). Offline analysis tools, on the other hand, typically function better as a link between image acquisition and the treatment-planning system (e.g., for replanning). Integrated offline systems need to be able to efficiently handle multiple data sets for single patients and allow for inter-comparison of all acquired and reference data.

Intervention can be based on offline or online image analysis. For online interventions, the analysis software should be able to communicate directly with the treatment console to automatically apply setup corrections. The automatic shifts should be verified by careful observation during their application and, potentially, repeat verification imaging. Periodic QA tests should also be designed to test the end-to-end performance of the image-guidance system. In the absence of such integration, the setup corrections can be reported by the software and applied manually. In this case, particular care must not only be taken to ensure that corrections are applied correctly, but also that they are applied consistently throughout the entire treatment.

Interventions based on offline image analysis typically take the form of replanning. In this situation, the largest potential for error lies in the mismanagement of large amounts of imaging data. Offline tools often have access to image sets from different treatment sessions as well as multiple sets within each session. Thus, great care must be taken to ensure that offline interventions are based on a clear understanding of how the stored images reflect the actual treatment.

10.2.3 QUALITY-ASSURANCE CONSIDERATIONS

Recommendations for QA of linear accelerators equipped with imaging subsystems have been published by AAPM Task Group 142 (Klein et al. 2009). QA of in-room kV imaging has also been discussed by AAPM Task Group 104 (Yin et al. 2009). A brief conceptual overview of IGRT QA is given here, and specific guidelines for tolerances and frequency can be found in these reports.

When developing a QA program, it is important to consider how image guidance fits in with other technology. QA of in-room imaging, for example, will necessarily tie in with the QA program of the treatment device itself. However, as more imaging technology becomes available in the clinic, it is often necessary to develop a dedicated imaging QA protocol. This allows for uncertainties in imaging to be addressed consistently across all modalities, which is particularly important when multimodality imaging is used to guide treatment decisions.

IGRT QA can be broadly divided into two categories: individual component tests and end-to-end tests. Individual component tests will be dependent on the imaging system under consideration. In practice, these tests are best used to monitor the stability of each of the components without giving a direct indication of how patient treatment is affected. End-to-end tests are intended to quantify how all the uncertainties in each component add up to affect patient treatment. They typically test the ability of the IGRT system to acquire an image, calculate setup corrections, and correctly apply those corrections.

10.3 IGRT IS THE GOLD STANDARD FOR PATIENT SETUP

10.3.1 RATIONALE FOR FRAMELESS SRT

Given the variety of SRT immobilization devices in use, it is difficult to clearly state a defining difference between frame-based and frameless SRT. Here, we make the distinction based on the concepts of *immobilization* and *localization*. Immobilization refers to the ability to keep the patient anatomy stationary without regard to targeting accuracy of treatment. Localization refers to the identification of the target within a defined coordinate frame.

In a frame-based system, immobilization and localization are combined into one device (i.e., the coordinate system used to localize the target is defined with respect to the immobilization device itself). This clearly implies that a frame-based system, without any supplementary information, will only be accurate when the relationship between the target and frame is rigidly fixed. This is the case, for example, when a stereotactic head ring is rigidly fixed to a patient's skull. When such a rigid relationship between target and immobilization holds, localization based on the frame alone provides a high degree of targeting accuracy and precision. Frame-based SRS systems have been used for intracranial treatments for more than 50 years (Benedict et al. 2008).

In a frameless system, the immobilization and localization steps are not directly linked. The first motivation for this separation was fractionated SRT (Bova et al. 1997), with which it is impractical to attach an invasive stereotactic head frame for each treatment session. A second motivation came from extracranial applications because most sites outside of the skull cannot be rigidly fixed to the immobilization device. In these cases, the coordinate frame used for target localization must be defined in another manner, and in-room image guidance has emerged as the gold standard for this task.

Once the feasibility of frameless SRT treatments was established for intracranial lesions, two developments led to the widespread application of frameless technology outside of the skull. First, sufficient clinical interest arose after several pioneering SBRT clinical trials reported favorable results. Second, the rapid introduction of in-room imaging technology allowed these procedures to be performed more efficiently and confidently in a wider variety of tumor sites.

10.3.2 RELATIVE ACCURACY OF FRAMELESS SRT

When discussing the accuracy of frameless SRT, the appropriate questions to ask are the following:

1. How well can the target be immobilized?
2. How does target localization accuracy compare to frame-based systems?
3. How accurate is patient repositioning?

10.3.2.1 Frameless immobilization

Immobilization addresses both inter- and intrafractional uncertainties in anatomy and patient position. The discussion here primarily pertains to the interfraction reproducibility of frameless immobilization. Intrafraction motion management is a critical component of frameless SRT, covered in detail in Chapter 9.

The gold standard in cranial immobilization is a rigid head frame that is affixed to the patient's skull. Frame-based systems are available for Gamma Knife and linac-based radiosurgery. With these systems, submillimeter immobilization is readily achievable if the frame is not removed between imaging and treatment.

Rigid fixation has also been applied to extracranial sites, most notably the spine (Hamilton et al. 1995, 1996). Due in part to the invasive nature of these techniques, they never achieved widespread clinical use. Rigid fixation of the target is simply not possible for most tumor sites in the thorax, abdomen, and pelvis.

Frameless immobilization was first applied to intracranial sites. Unlike frame-based immobilization, frameless masks are typically removed between simulation and treatment. This introduces some inherent uncertainty in patient setup. A large body of literature has accumulated examining the reproducibility of frameless cranial immobilization, independent of any other treatment uncertainties. The most direct measurements have come from studies that use repeat CT scans of a single patient. Table 10.1 summarizes the results of several representative studies. While it is difficult to separate the uncertainties of the measurement technique from the inherent uncertainty of the immobilization, these studies seem to agree that several commercially available frameless immobilization systems have inherent reproducibility limits that are on the order of 1–2 mm (Alheit et al. 2001; Baumert et al. 2005; Kumar et al. 2005; Minniti et al. 2010; Willner Flentje, and Bratengeier 1997). When interpreting these studies, it is important to note the following:

1. Translational errors will depend on whether or not rotational alignment is also considered.
2. Depending on how errors were quantified, the mean error may be a reflection of systematic errors that are not related to the reproducibility of immobilization.
3. Immobilization uncertainty is only one component of the overall treatment uncertainty.

Table 10.1 Summary of several studies examining the reproducibility of frameless cranial immobilization

INVESTIGATOR	FRAMELESS IMMOBILIZATION	TRANSLATIONAL REPRODUCIBILITY
Minniti et al. 2010	Thermoplastic mask	Mean = 0.5 mm StDev = 0.4 mm Max = 1.4 mm
Alheit et al. 2001	Thermoplastic mask	Mean = 0.7 mm Max = 2.5 mm
Willner et al. 1997	Thermoplastic mask	Mean = 2.4 mm StDev = 1.3 mm
Kumar et al. 2005	Relocatable frame	Median = 0.7 mm Max = 1.4 mm
Baumert et al. 2005	Thermoplastic mask ± bite blocks	Mean = 2.8 mm StDev = 1.7 mm

Compared to frameless immobilization of the skull, frameless extracranial immobilization has been shown to be far less accurate. For example, an early study using a prototype body frame showed a targeting accuracy of 3.6 mm for spine SBRT (Lohr et al. 1999). For tumors in the lung and liver, a larger setup uncertainty of 7–10 mm has been measured for a similar body frame (Lax et al. 1994). Studies like these demonstrated very early in SBRT practices that frameless extracranial immobilization was not reliable as a stand-alone method for treatment setup. Prior to the introduction of advanced in-room image guidance, a CT scan had to be performed prior to each fraction to compensate for this large uncertainty. Patients were not moved between CT and treatment so that the coordinate systems built into the body frames could be used for in-room target localization.

In summary, clinical experience has demonstrated that there can be large interfraction variation in patient positioning with frameless immobilization. Extracranial sites show the most variation. Reducing this uncertainty is where in-room image guidance has played a critical role in the advancement of frameless SRT. In the image-guided paradigm, the essential role of immobilization is to minimize intrafraction patient motion because interfraction variations in setup can be adjusted with pretreatment imaging. However, this relies on the assumption that both image-guided localization and patient repositioning are sufficiently accurate. The accuracy of several commercial IGRT systems has been well documented as will be discussed in the following sections.

10.3.2.2 Image-based localization

In regards to the accuracy of target localization (independent of immobilization), there is a large body of evidence that suggests several commercially available image-guidance systems are capable of localizing targets with accuracy and precision that is comparable to conventional frame-based systems. The discussion in this section is relevant to the accuracy of target localization independent of immobilization and patient repositioning, which is primarily determined by the alignment of imaging and treatment isocenters. In this respect, the important errors to consider are systematic (offset between isocenters) and random (stability of imaging isocenter). These uncertainties should be monitored at each facility with a daily QA program (Klein et al. 2009).

The stability of in-room kV imaging systems has been well documented. For gantry-mounted kV imaging, reproducibility of the isocenter has been shown to be <0.6 mm over long time scales (Meyer et al. 2007; Wiehle et al. 2009). Reproducibility of ceiling- or floor-mounted systems has not been widely reported but should be expected to be as good or better than gantry-mounted systems due to their lack of moving parts.

The systematic component has been evaluated by several investigators. While alignment of imaging and treatment isocenters can be tested with simple phantom studies, the most relevant way to assess image-guided localization accuracy in clinical situations is to compare it to a frame-based localization for a group of patients. The following is a discussion of representative results from phantom and patient studies for a variety of commercially available IGRT systems.

Of the available in-room image guidance, CBCT provides the most information for patient setup. The localization accuracy of the Elekta XVI system has been evaluated (Wiehle et al. 2009). It was shown that, with proper system calibration, the uncertainty in isocenter localization with CBCT is <1 mm. CBCT-based localization with the Varian OBI system has been shown to be comparable to a frame-based setup (Chang et al. 2007).

Imaged-guided localization via 2-D orthogonal imaging has been extensively studied for the Novalis and CyberKnife systems. Several studies have demonstrated that the average discrepancy between frame-based and 2-D image-guided localization with Novalis ExacTrac is ≤1 mm (Feygelman et al. 2008; Lamba, Breneman, and Warnick 2009; Ramakrishna et al. 2010) (Table 10.2). Direct comparisons between frame-based and frameless localization are not possible with the CyberKnife system, but studies have shown that its 2-D imaging system is also capable of submillimeter localization accuracy for frameless SRT (Antypas and Pantelis 2008).

All of these studies were based on phantoms and intracranial patient treatments and thus represent a best-case scenario for image-based localization. Additional clinical uncertainty in localization will always be present for patients, particularly for extracranial treatments in places where bony anatomy is not representative of soft-tissue position. For example, Wang et al. (2009) measured setup errors, relative to CBCT, on the order of 5–10 mm in SBRT patient setups using lasers or 2-D orthogonal imaging. Similar

Table 10.2 Summary of studies examining localization accuracy for image-guided SRT

INVESTIGATION	IMAGING TECHNOLOGY	EVALUATION METRIC	LOCALIZATION ERROR
Ramakrishna et al. 2010	2-D (ExacTrac)	Isocenter discrepancy between frame-based and image-based patient setups (102 patients)	Mean = 1.0 mm StDev = 0.5 mm
Lamba, Breneman, and Warnick 2009	2-D (ExacTrac)	Isocenter discrepancy between frame-based and image-based patient setup (25 patients)	RMS error = 1.2 mm StDev = 0.4 mm
Feygelman et al. 2008	2-D (ExacTrac)	Deviation between linac and imaging isocenter (phantom)	Mean = 0.95 mm StDev = 0.18 mm
Chang et al. 2007	CBCT (Varian OBI)	Isocenter discrepancy between frame-based and image-based phantom setup (single phantom setup with four targets)	Mean = 1.34 mm StDev = 0.33 mm
Wiehle et al. 2009	CBCT (Elekta XVI)	Deviation between linac and imaging isocenter (phantom)	Mean <1.1 mm Variation <0.6 mm
Wiehle et al. 2009	2-D (Elekta XVI)	Isocenter discrepancy between frame-based and image-based phantom setup (single phantom setup)	0.56 mm

Table 10.3 Summary of several studies examining the accuracy and precision of automated patient repositioning for intra- and extracranial sites

INVESTIGATION	IMAGING AND REPOSITIONING TECHNOLOGY	TRANSLATIONAL REPOSITIONING ACCURACY	ROTATIONAL REPOSITIONING ACCURACY	COMMENTS
Yan, Yin, and Kim 2003	ExacTrac with 4-D couch	For any one translational axis Mean <0.7 mm StDev <0.3 mm	Not reported	Phantom study head, chest, pelvis
Takakura et al. 2010	ExacTrac with 6-D couch	Mean = 0.07 mm StDev = 0.22 mm	For any one axis of rotation Mean <0.05° StDev <0.14°	Phantom study intracranial
Meyer et al. 2007	CBCT with Hexapod 6-D Couch	Mean <0.1 mm StDev <0.5 mm	Mean <0.06° StDev <0.4°	Phantom study intracranial
Guckenberger et al. 2007	CBCT with Hexapod 6-D Couch	Mean = 0.9 mm StDev = 0.5 mm	Not reported	Patient study head and neck
Guckenberger et al. 2007	CBCT with Hexapod 6-D Couch	Mean = 1.6 mm StDev = 0.8 mm	Not reported	Patient study pelvic

discrepancies between ExacTrac and CBCT have also been reported (Ma et al. 2009). This additional uncertainty has to be assessed on a patient-specific basis.

10.3.2.3 Automated repositioning

Once a target is localized or positioned, the next step is to calculate and apply treatment couch shifts or rotations to reposition the patient. The latest generation of dedicated SRT systems typically uses a 6-D couch for increased ability to apply rotational corrections. Mechanical positioning accuracy of a treatment couch should be verified with routine quality assurance (Klein et al. 2009), but additional phantom studies can be used to measure repositioning accuracy in clinical situations. Analysis of repositioning accuracy has been performed for several commercially available systems. Table 10.3 summarizes several relevant investigations. The results of these studies are a reflection of a system's ability to measure a known displacement, calculate appropriate couch shifts, and mechanically reposition the couch. Depending on the experimental design, these results may include the uncertainty due to a systematic offset in imaging isocenter.

The repositioning accuracy of the Novalis system has been shown to be better than 1 mm in several of its configurations. Yan et al. studied the repositioning accuracy of the Novalis Body system, which used IR and kV imaging to reposition patients on a 4° of freedom couch (Yan, Yin, and Kim 2003). The repositioning accuracy was <1 mm for a variety of anatomical sites when considering both translational and rotational errors. The accuracy of the Novalis system has also been studied with the addition of a 6-D robotic couch (Takakura et al. 2010). In this configuration (IR + kV + 6-D), the repositioning accuracy was also shown to be <1 mm for a range of translational and rotational setup errors.

The repositioning accuracy of the Elekta XVI with Hexapod 6-D table has also been evaluated. Meyer et al. (2007) demonstrated (with phantom studies) that the accuracy and reproducibility of repositioning with CBCT was <1 mm for a large range of translational and rotational errors. Guckenberger et al. (2007) used repeat CBCT scans to study the XVI system for actual patient setups (cranial and extracranial) and also showed that the mean repositioning error after 6-D CBCT correction was <1 mm for head and neck sites and 2 mm for pelvic sites (Guckenberger et al. 2007).

10.4 INTEGRATING IMAGE-GUIDED FRAMELESS SRT INTO THE CLINICAL WORKFLOW

Image guidance plays a critical role in patient setup for frameless SRT and can be categorized into four *potential* applications: initial setup, verification of repositioning, intrafraction monitoring, and posttreatment verification. The implementation details of these tasks will necessarily depend on the technology available, but some common features can be identified. Furthermore, different treatment sites will demand differing levels of image guidance, so procedures must be sufficiently flexible to allow for appropriate adaptation of the technology to different clinical goals (see Figure 10.11).

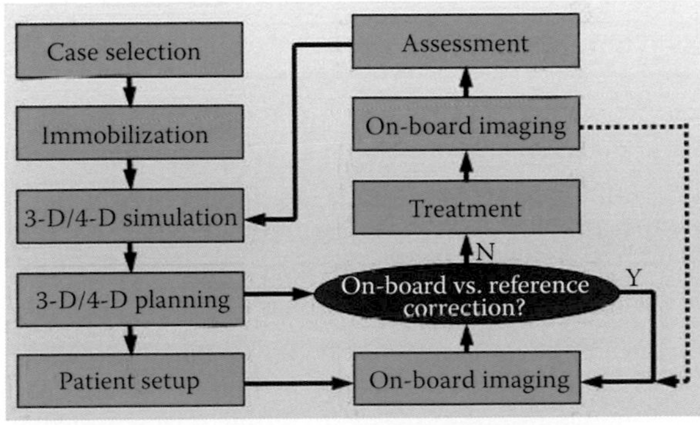

Figure 10.11 Integration of imaging-related tasks in the clinical workflow.

10.4.1 INITIAL PATIENT SETUP

As a general rule, the initial setup for frameless SRT should not be based on imaging alone. Even though the final patient position will be based on imaging information, the patient should still be carefully aligned to marks placed on the skin or immobilization during simulation. Once this is done, a quick low-dose image (e.g., orthogonal kV images) should be acquired to look for large setup variations. Even though 6-D treatment couches can correct for large setup variation, large discrepancies seen in these initial images should always be investigated before proceeding. Once the initial imaging indicates that the patient setup is adequate, a volumetric image may be acquired. This step is particularly important for nonrigid extracranial sites, where trade-offs may have to be made between the alignment of target and normal tissues. A volumetric image makes it easier to visually evaluate 6-D setup errors and changes in anatomy. Again, large discrepancies between any images should be immediately investigated.

10.4.2 PRETREATMENT VERIFICATION

Once the final patient position has been determined and the patient repositioned, a series of pretreatment verification images may be taken. First, depending on the magnitude of the correction needed for patient repositioning, an image may be taken to verify that the repositioning was accurate. Second, a quick, low-dose image may be taken to use as a reference for evaluating intrafraction variations in patient position.

10.4.3 INTRAFRACTION MONITORING

During treatment, the patient should be continually monitored for movement. This can be done with a combination of visual observation and imaging. Imaging may be integrated in real time with treatment. For example, portal imaging of the treatment beam may be used to continually monitor implanted fiducials. Intrafraction verification can also be retrospective, such as collecting and analyzing the fluence transmitted through the patient.

10.4.4 POSTTREATMENT VERIFICATION

Whether or not real-time imaging is used during treatment, a posttreatment image should be acquired and compared to pretreatment images. If this image reveals unexpected intrafraction motion, a more detailed investigation can be performed.

10.4.5 TREATMENT ASSESSMENT

Careful analysis of images taken prior to, during, and after each treatment session allows the clinical team to perform a detailed assessment of a treatment course. Based on this information, treatment plans or immobilization can be modified in response to anatomical changes, unexpected intrafraction motion, or other uncertainties.

10.5 LIMITATIONS AND SAFETY CONSIDERATIONS

10.5.1 INFORMATION OVERLOAD

Different clinical applications will require differing levels of image-guidance. Acquisition of image data should be a direct response to a clearly stated clinical question (i.e., imaging should not be done "just because you can"). The best prevention for redundant and/or excessive use of image-guidance is the presence of appropriate clinical policies, procedures, and protocols. For each new procedure being considered, an imaging protocol should be established and followed rigidly until adequate experience suggests it should be changed.

Because of the potential to acquire large amounts of image data over the course of a patient's treatment, image storage and database management become vitally important. Adequate IT support should be available to ensure that archived data is secure and readily accessible. To ensure an efficient clinical workflow, dedicated software should also be available for offline review and analysis of the images (such software is available from all major vendors).

10.5.2 LACK OF APPROPRIATE QUALITY ASSURANCE

Appropriate QA of an imaging system is essential to its clinical application. At present, much of this technology is at a relatively early stage of development and thus may be prone to variations and/or failures in mechanical or software components. An imaging system should not be used as the sole check of patient setup. An image-guided setup should always be checked against an independent system (i.e., lasers and skin marks) to safeguard against large treatment errors.

The latest recommendation for image-guided SRT QA is to implement an annual, monthly, and daily QA program to monitor mechanical stability and image quality. Of these, the daily checks of the imaging isocenter are perhaps the most important. By definition, imaging isocenter QA is as important to image-guided frameless SRT as laser QA is for frame-based SRS. It is important to test individual components as well as to perform end-to-end tests that combine errors in setup, localization, and repositioning.

10.5.3 PATIENT DOSE

Management of imaging dosage during IGRT procedures has been discussed in detail by AAPM task group 75 (Murphy et al. 2007). For 2-D kV imaging, entrance dosage levels are typically <1 mGy. For CBCT, single scan doses with most standard techniques are on the order of 10–30 mGy. The key to dosage management is to develop appropriate protocols and procedures to avoid excessive or unnecessary imaging.

10.5.4 STAFF TRAINING

A clinical workflow can be summarized in terms of the *devices* used and the *procedures* that are in place to ensure proper use of those devices. Over several decades, frame-based SRT has developed into a highly specialized treatment technique, for which the devices and procedures are clearly distinct from conventional therapy. This makes it easier for staff to recognize the differences.

Devices for frameless SRT, however, often have more in common with devices used for conventional therapy. It is important, therefore, that SRT procedures still be followed and that staff members are trained to recognize these differences and follow appropriate procedures.

10.6 CONCLUSION

Due to inherent limitations of frameless immobilization, in-room image guidance has emerged as the gold standard for frameless SRT patient setup. With proper training, quality assurance, and technical expertise, it is possible to develop a frameless SRT program with a targeting accuracy that is comparable to frame-based treatments. Rapid adoption of fractionated and extracranial SRT would not have been possible without the development of advanced image-guidance tools.

REFERENCES

Alheit, H., S. Dornfeld, M. Dawel et al. 2001. Patient position reproducibility in fractionated stereotactically guided conformal radiotherapy using the Brainlab mask system. *Strahlenther. Onkol.* 177: 264–8.

Antonuk, L. E. 2002. Electronic portal imaging devices: A review and historical perspective of contemporary technologies and research. *Phys. Med. Biol.* 47: R31–65.

Antypas, C., and E. Pantelis. 2008. Performance evaluation of a CyberKnife G4 image-guided robotic stereotactic radiosurgery system. *Phys. Med. Biol.* 53: 4697–718.

Barker, J. L., Jr., A. S. Garden, K. K. Ang et al. 2004. Quantification of volumetric and geometric changes occurring during fractionated radiotherapy for head-and-neck cancer using an integrated CT/linear accelerator system. *Int. J. Radiat. Oncol. Biol. Phys.*59: 960–70.

Baumert, B. G., P. Eglie, S. Studer, C. Dehing, and J. B. Davis. 2005. Repositioning accuracy of fractionated stereotactic irradiation: Assessment of isocentre alignment for different dental fixations by using sequential CT scanning. *Radiother. Oncol.* 74: 61–6.

Benedict, S. H., F. J. Bova, B. Clark et al. 2008. Anniversary paper: The role of medical physicists in developing stereotactic radiosurgery. *Med. Phys.* 35: 4262–77.

Bert, C., K. G. Metheany, K. Doppke, and G. T. Y. Chen. 2005. A phantom evaluation of a stereo-vision surface imaging system for radiotherapy patient setup. *Med. Phys.* 32: 2753–62.

Bert, C., K. G. Metheany, K. P. Doppke, A. G. Taghian, S. N. Powell, and G. T. Y. Chen. 2006. Clinical experience with a 3D surface patient setup system for alignment of partial-breast irradiation patients. *Int. J. Radiat. Oncol. Biol. Phys.* 64: 1265–74.

Bova, F. J., J. M. Buatti, W. A. Friedman, W. M. Mendenhall, C. C. Yang, and C. Liu. 1997. The University of Florida frameless high-precision stereotactic radiotherapy system. *Int. J. Radiat. Oncol. Biol. Phys.* 38: 875–82.

Chang, J., K. M. Yenice, A. Narayana, and P. H. Gutin. 2007. Accuracy and feasibility of cone-beam computed tomography for stereotactic radiosurgery setup. *Med. Phys.* 34: 2077–84.

Chang, Z., Z. Wang, Q. J. Wu et al. 2008. Dosimetric characteristics of Novalis Tx system with high definition multileaf collimator. *Med. Phys.* 35: 4460–3.

Cervino, L., T. Pawlicki, J. Lawson, and S. Jiang. 2010. Frame-less and mask-less cranial stereotactic radiosurgery: A feasibility study. *Phys. Med. Biol.* 55: 1863–73.

Curran, B. H., J. M. Balter, and I. J. Chetty, eds. 2006. *Integrating New Technologies into the Clinic: Monte Carlo and Image-Guided Radiation Therapy*. Madison: Medical Physics Publishing.

de Crevoisier, R., L. Dong, M. Bonnen et al. 2004. Quantification of volumetric changes and internal organ motion during radiotherapy for prostate carcinoma using an integrated CT/linear accelerator system. *Int. J. Radiat. Oncol. Biol. Phys.* 60: S227–8.

Dempsey, J., B. Dionne, J. Fitzsimmons et al. 2006. A real-time MRI guided external beam radiotherapy delivery system. *Med. Phys.* 33: 2254.

Fallone, B., B. Murray, S. Rathee et al. 2009. First MR images obtained during megavoltage photon irradiation from a prototype integrated linac-MR system. *Med. Phys.* 36: 2084–8.

Feygelman, V., L. Walker, P. Chinnaiyan, and K. Forster. 2008. Simulation of intrafraction motion and overall geometric accuracy of a frameless intracranial radiosurgery process. *J. Appl. Clin. Med. Phys.* 9: 68–86.

Groh, B. A., J. H. Siwerdsen, D. G. Drake, J. W. Wong, and D. A. Jaffray. 2002. A performance comparison of flat-panel imager-based MV and kV cone-beam CT. *Med. Phys.* 29: 967–75.

Guckenberger, M., J. Meyer, J. Wilbert, K. Baier, O. Sauer, and M. Flentje. 2007. Precision of image-guided radiotherapy (IGRT) in six degrees of freedom and limitations in clinical practice. *Strahlenther. Onkol.* 183: 307–13.

Hamilton, A. J., B. A. Lulu, H. Fosmire, and L. Gosset. 1996. Linac-based spinal stereotactic radiosurgery. *Stereotact. Funct. Neurosurg.* 66: 1–9.

Hamilton, A. J., B. A. Lulu, H. Fosmire, B. Stea, and J. R. Cassady. 1995. Preliminary clinical experience with linear accelerator-based spinal stereotactic radiosurgery. *Neurosurgery* 36: 311–9.

Herman, M. G., J. M. Balter, D. A. Jaffray et al. 2001. Clinical use of electronic portal imaging: Report of AAPM Radiation Therapy Committee Task Group 58. *Med. Phys.* 28: 712–37.

Hughes, S., J. McClelland, S. Tarte et al. 2009. Assessment of two novel ventilatory surrogates for use in the delivery of gated/tracked radiotherapy for non-small cell lung cancer. *Radiother. Oncol.* 91: 336–41.

Kirkby, C., T. Stanescu, and B. G. Fallone. 2009. Magnetic field effects on the energy deposition spectra of MV photon radiation. *Phys. Med. Biol.* 54: 243–57.

Kirkby, C., T. Stanescu, S. Rathee, M. Carlone, B. Murray, and B. G. Fallone. 2008. Patient dosimetry for hybrid MRI-radiotherapy systems. *Med. Phys.* 35: 1019–27.

Klein, E., J. Hanley, J. Bayouth et al. 2009. Task Group 142 report: Quality assurance of medical accelerators. *Med. Phys.* 36: 4197–212.

Kriminski, S., M. Mitschke, S. Sorensen et al. 2005. Respiratory correlated cone-beam computed tomography on an isocentric C-arm. *Phys. Med. Biol.* 50: 5263–80.

Kumar, S., K. Burke, C. Nalder et al. 2005. Treatment accuracy of fractionated stereotactic radiotherapy. *Radiother. Oncol.* 74: 53–9.

Lagendijk, J. J., B. W. Raaymakers, A. J. Raaijmakers et al. 2008. MRI/linac integration. *Radiother. Oncol.* 86: 25–9.

Lamba, M., J. C. Breneman, and R. E. Warnick. 2009. Evaluation of image-guided positioning for frameless intracranial radiosurgery. *Int. J. Radiat. Oncol. Biol. Phys.* 74: 913–9.

Langen, K., S. Meeks, P. Kupelian et al. 2006. Tomotherapy's implementation of image-guided and dose-guided adaptive radiation therapy. In *Integrating New Technologies into the Clinic: Monte Carlo and Image-Guided Radiation Therapy*, ed. B. H. Curran, J. M. Balter, and I. J. Chetty, 627–52. Madison: Medical Physics Publishing.

Lax, I., H. Blomgren, I. Naslund, and R. Svanstrom. 1994. Stereotactic radiotherapy of malignancies in the abdomen. Methodological aspects. *Acta Oncol.* 33: 677–83.

Lohr, F., J. Debus, C. Frank et al. 1999. Noninvasive patient fixation for extracranial stereotactic radiotherapy. *Int. J. Radiat. Oncol. Biol. Phys.* 45: 521–7.

Ma, J., Z. Chang, Z. Wang, Q. Jackie Wu, J. P. Kirkpatrick, and F. F. Yin. 2009. ExacTrac X-ray 6 degree-of-freedom image-guidance for intracranial non-invasive stereotactic radiotherapy: Comparison with kilo-voltage cone-beam CT. *Radiother. Oncol.* 93: 602–8.

Meyer, J., J. Wilbert, K. Baier et al. 2007. Positioning accuracy of cone-beam computed tomography in combination with a HexaPOD robot treatment table. *Int. J. Radiat. Oncol. Biol. Phys.* 67: 1220–8.

Minniti, G., M. Valeriani, E. Clarke et al. 2010. Fractionated stereotactic radiotherapy for skull base tumors: Analysis of treatment accuracy using a stereotactic mask fixation system. *Radiat. Oncol.* 5: 1–6.

Murphy, M. J., J. Balter, S. Balter et al. 2007. The management of imaging dose during image-guided radiotherapy: Report of the AAPM Task Group 75. *Med. Phys.* 34: 4041–63.

Raaijmakers, A. J., B. Hardemark, B. W. Raaymakers, C. P. Raaijmakers, and J. J. Lagendijk. 2007. Dose optimization for the MRI-accelerator: IMRT in the presence of a magnetic field. *Phys. Med. Biol.* 52: 7045–54.

Raaijmakers, A. J., B. W. Raaymakers, and J. J. Lagendijk. 2007. Experimental verification of magnetic field dose effects for the MRI-accelerator. *Phys. Med. Biol.* 52: 4283–91.

Raaijmakers, A. J., B. W. Raaymakers, and J. J. Lagendijk. 2008. Magnetic-field-induced dose effects in MR-guided radiotherapy systems: Dependence on the magnetic field strength. *Phys. Med. Biol.* 53: 909–23.

Raaymakers, B. W., J. J. Lagendijk, J. Overweg et al. 2009. Integrating a 1.5 T MRI scanner with a 6 MV accelerator: Proof of concept. *Phys. Med. Biol.* 54: N229–37.

Ramakrishna, N., F. Rosca, S. Friesen, E. Tezcanli, P. Zygmanszki, and F. Hacker. 2010. A clinical comparison of patient setup and intrafraction motion using frame-based radiosurgery versus a frameless image-guided radiosurgery system for intracranial lesions. *Radiother. Oncol.* 95: 109–15.

Roper, J., J. Bowsher, and F. F. Yin. 2009. On-board SPECT for localizing functional targets: A simulation study. *Med. Phys.* 36: 1727–35.

Siewerdsen, J. H., D. J. Moseley, S. Burch et al. 2005. Volume CT with a flat-panel detector on a mobile, isocentric C-arm: Pre-clinical investigation in guidance of minimally invasive surgery. *Med. Phys.* 32: 241–54.

Takakura, T., T. Mizowaki, M. Nakata et al. 2010. The geometric accuracy of frameless stereotactic radiosurgery using a 6D robotic couch system. *Phys. Med. Biol.* 55: 1–10.

Uematsu, M., T. Fukui, A. Shioda et al. 1996. A dual computed tomography linear accelerator unit for stereotactic radiation therapy: A new approach without cranially fixated stereotactic frames. *Int. J. Radiat. Oncol. Biol. Phys.* 35: 587–92.

Wang, Z., J. W. Nelson, S. Yoo et al. 2009. Refinement of treatment setup and target localization accuracy using three-dimensional cone-beam computed tomography for stereotactic body radiotherapy. *Int. J. Radiat. Oncol. Biol. Phys.* 73: 571–7.

Wiehle, R., H. J. Koth, N. Nanko, A. L. Grosu, and N. Hodapp. 2009. On the accuracy of isocenter verification with kV imaging in stereotactic radiosurgery. *Strahlenther. Onkol.* 185: 325–30.

Willner, J., M. Flentje, and K. Bratengeier. 1997. CT simulation in stereotactic brain radiotherapy—analysis of isocenter reproducibility with mask fixation. *Radiother. Oncol.* 45: 83–8.

Yan, H., F. F. Yin, and J. H. Kim. 2003. A phantom study on the positioning accuracy of the Novalis Body system. *Med. Phys.* 30: 3052–60.

Yin, F. F., P. Rubin, M. C. Schell et al. 1996. An observer study for direct comparison of clinical efficacy of electronic to film portal images. *Int. J. Radiat. Oncol. Biol. Phys.* 35: 985–91.

Yin, F. F., J. Wong, J. Balter et al. 2009. *The Role of In-Room kV X-ray Imaging for Patient Setup and Target Localization. Report of AAPM Task Group 104*. College Park: American Association of Physicists in Medicine.

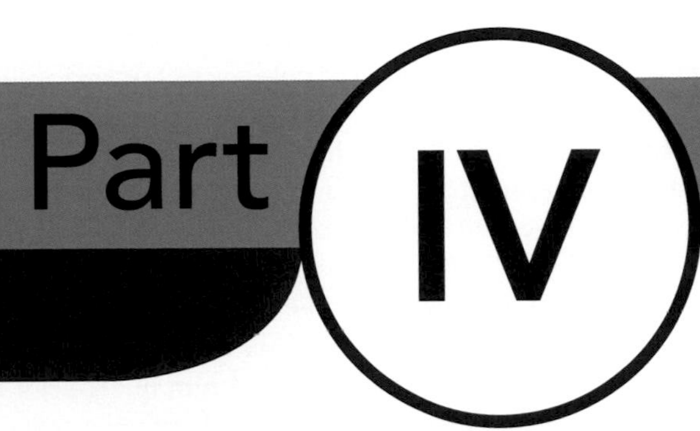

Stereotactic radiation therapy treatment planning and dosimetry

The development and advancement of technology used in stereotactic radiosurgery and stereotactic body radiotherapy was made possible by developments in three fields: the dose delivery equipment, radiological imaging, and the development of fast computer workstations capable of sophisticated treatment planning. Leksell's original treatment planning dated back to the era of slide rules and plotting charts, and the first dedicated treatment-planning computer systems were not introduced until around 1987. Some of the linear accelerator treatment plans evolved from conventional external beam planning systems, and others were created just for radiosurgery, such as the circular cone collimators used principally for cranial stereotactic radiosurgery (SRS). Gradually, some of the most popular general radiation therapy treatment-planning systems have evolved to include linac-based radiosurgery as an option. While some dedicated devices, such as the CyberKnife, Gamma Knife, and GammaPod, have their own proprietary treatment-planning systems, their use continues to this day. As the algorithms for dose calculations continue to improve and the optimization of various delivery strategies are investigated, modern optimization routines can now sample thousands of possible alternatives in a few minutes and present multiple plans to the clinicians in order to choose which one best satisfies the planner's criteria.

With the treatment-planning improvements possible in minimizing doses to surrounding normal structures coupled with escalating the dose to the targeted volume, the plan evaluation tools have evolved over the years as well. Dose volume histograms (DVH) were first introduced at highly sophisticated proton therapy centers in the early 1960s. However, entry of the DVH into the clinic had to await the arrival of powerful UNIX workstations in the mid-1990s. Many other indices comparing coverage, dose to organs at risk, and fall-off are now available and, in some cases, are updated in real time during the planning process.

Small field photon dosimetry is at the crux of the physics of the stereotactic radiosurgery and radiation therapy paradigm in ensuring that the doses in the planning system are in agreement with those that are actually delivered. Clearly, fields as small as 4 mm in diameter can be quite effective in the treatment of intracranial disease as 46 years of Gamma Knife experience has shown. However, high photon energies and extremely small fields, in which radiation equilibrium may not be fully achieved, are perplexing to clinical medical physicists. Partial volume effects and source occlusion challenge the wits of the clinical physicist and his or her available resources for measurement and confirmation from conventional radiation therapy. A plethora of small field devices, such as thermoluminescent dosimters (TLD), optically stimulated luminescent dosimeter (OSLD), metal-oxide semiconductor field-effect transistor (MOSFETS), diodes, and various kinds of film, have been employed to navigate these treacherous waters. Many papers have been written and continue to be published on this topic. Sadly, reported horrific miscalibrations and overdoses have made the entire radiosurgery community very cautious about small-field photon dosimetry, but it underscores the great importance of this area of medical physics within the field of SRT.

11 Stereotactic treatment planning

Bill J. Salter, Brian Wang, and Vikren Sarkar

Contents

11.1 INTRODUCTION

Treatment planning, as the name implies, entails the development of a *plan* of attack on a targeted tumor. The number and nature of treatment beams to be used as well as the shape, size, orientation, and direction of these beams all must be carefully determined in order to achieve the goal of doing maximum possible damage to the tumor while simultaneously minimizing damage to adjacent healthy organ systems and tissues. In most ways, treatment planning for stereotactic treatments has the same goals and challenges as any other form of treatment planning but with potentially higher stakes, and therein lies the most important difference.

Stereotactic treatments are, by definition, high-precision treatments of challenging tumor geometries with very high doses per fraction. The high-stakes environment, created by high levels of potential damage associated with stereotactic hypofractionated treatment approaches along with immediate proximity

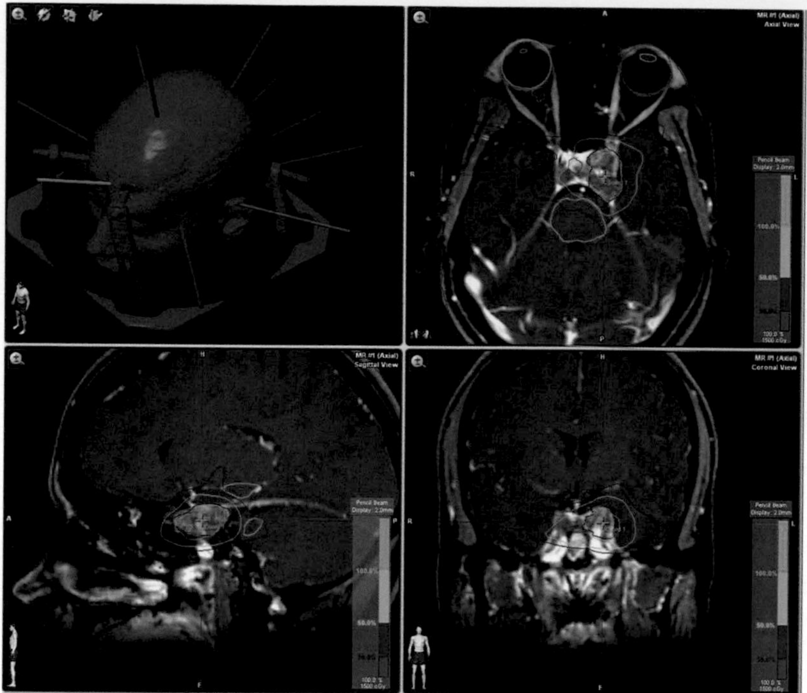

Figure 11.1 Ten non-coplanar beam IMRS, red contour = GTV, green contour = brainstem, orange isodose line = prescription 15 Gy, green isodose line = 50%, blue isodose = 30%. (Image courtesy of Brainlab.)

to sensitive healthy structures typically encountered in stereotactic treatments, make the "standard" treatment-planning problem even more challenging. For example, the percentage isodose line that can be tolerated to encroach on a nearby healthy structure is almost always less for a typical stereotactic treatment because the dose per fraction is so high, thus making a treatment plan design all the more difficult (Figure 11.1). The fact that the targeted lesion was designated for stereotactic treatment in the first place typically means that there are important and sensitive normal structures nearby. Simply put, treatment planning for stereotactic delivery environments faces all of the challenges of standard treatment planning with amplified levels of challenge and importance.

11.1.1 EVOLUTION OF STEREOTACTIC DELIVERY MODALITIES

Because of the challenges associated with delivering hypofractionated stereotactic treatments, numerous complex and very capable delivery approaches have evolved over the years. It was the Greek philosopher, Plato, who said that necessity is the mother of invention, and it can be reasonably argued that the challenges of the stereotactic environment have been the necessity for which many of the most important advances in radiation therapy have evolved. In order to plan safely and effectively for any of these modalities, one must first appreciate fully the unique and often subtle nuances of the treatment approach.

11.1.2 GAMMA KNIFE

The truly novel approach of using hundreds of very small "beamlets" to fire at a targeted tumor from many different angles of approach, as first developed by Lars Leksell in 1952 (Andrews et al. 2006), was necessitated by Leksell's desire to destroy challenging brain lesions with radiation while minimizing damage to eloquent areas of the brain (Figure 11.2). Interestingly, this beamlet-based approach can be argued to have been a key stepping stone on the way to one of the most important developments in radiation treatment delivery: intensity-modulated radiation therapy (IMRT).

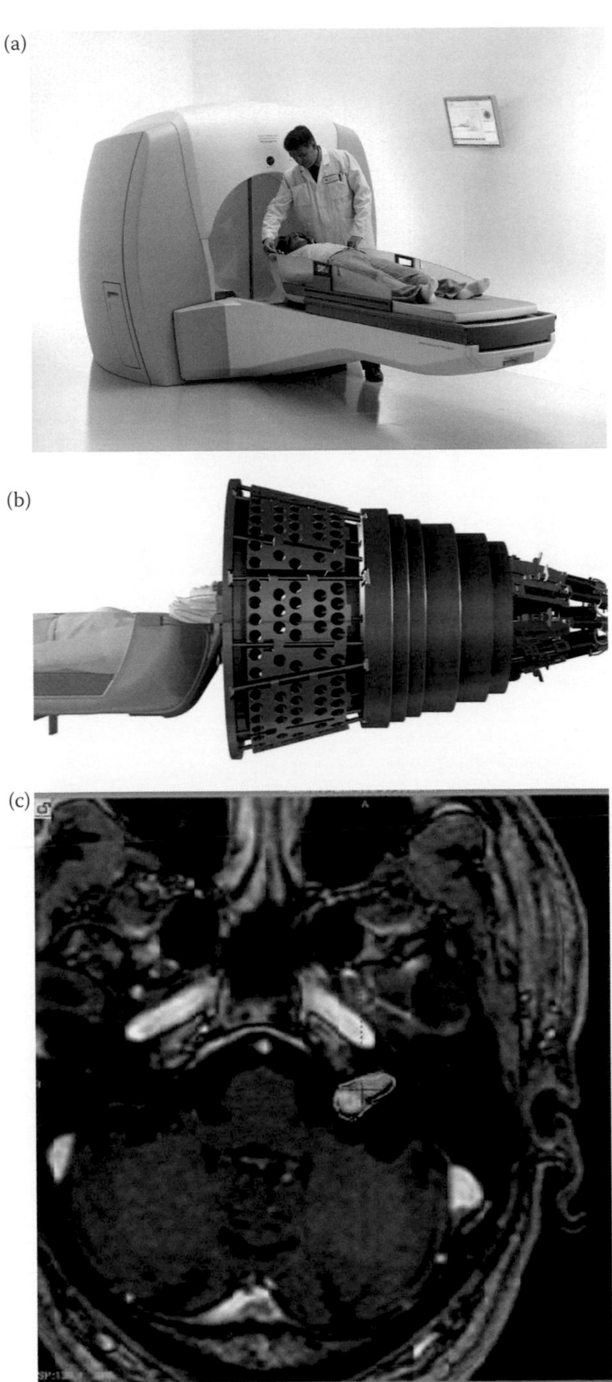

Figure 11.2 (a) Gamma Knife, (b) model of patient undergoing Gamma Knife radiosurgery showing beamlet geometry in the gantry, (c) resulting isodose distributions. (All images courtesy of Elekta website.)

The introduction of the Gamma Knife has been credited as the beginning of radiosurgery as we know it and also represented the introduction of so-called sphere-based (or sphere-packing) delivery approaches. Because the Gamma Knife uses numerous intersecting, circular-shaped beamlets delivered from many different directions to create its high-dose region, this high-dose region is essentially spherical in shape. Gamma Knife treatment approaches, therefore, use this spherically shaped dose distribution, called a

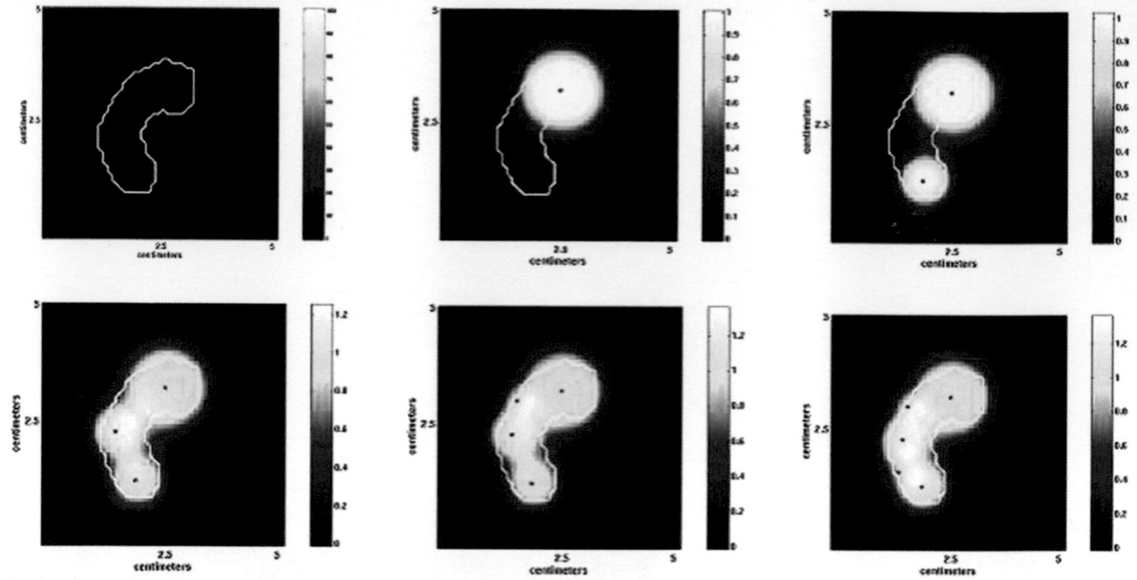

Figure 11.3 Example using sphere-shaped dose building block of Gamma Knife to treat nonspherically shaped lesion. (From Yu, C., and Shepard, D., *Technol. Cancer Res. Treat.*, 2, 2, 93–104, 2003.)

"shot," as the basis for treatment of all lesions. Lesions that are essentially round in shape are treated with a single shot, and more complex irregular lesions are treated with multiple intersecting shots or spheres of dose (Figure 11.3).

11.1.3 CONE-BASED DELIVERY

Cone-based linac radiosurgery techniques evolved to mimic the dose distributions of the Gamma Knife by using multiple intersecting arcs delivered through cylindrically shaped collimating "cones" (Figure 11.4). Cone-based radiosurgery also falls into the category of sphere-based delivery techniques because it produces a spherically shaped high-dose region similar to the Gamma Knife (Figure 11.5).

Figure 11.4 Circular collimator attached to linac. (Image courtesy of Brainlab.)

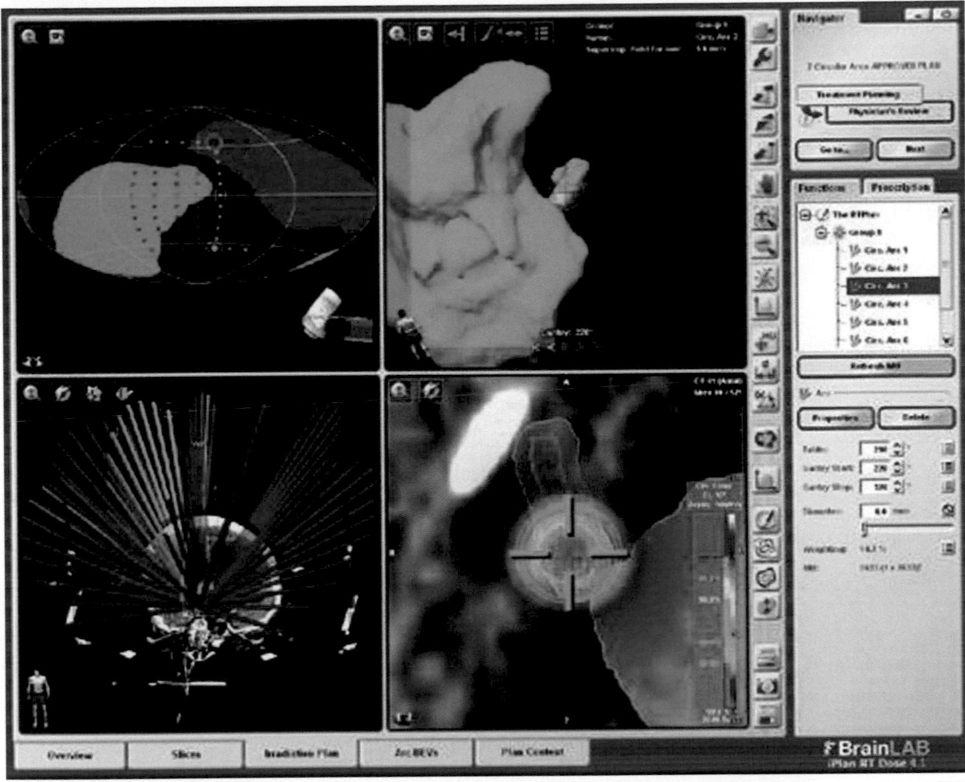

Figure 11.5 Spherical isodose distribution from cone-based delivery. (Image courtesy of Brainlab.)

11.1.4 INTENSITY-MODULATED RADIOSURGERY (IMRS)

IMRT's modulated beamlet-based delivery approach is typically referred to as IMRS when used to deliver a single fraction of a dose. By harnessing the power of intensity modulation to create isodose distributions that can be extremely conformal even to irregularly shaped targets (Figure 11.6), IMRS has become a highly utilized and valuable stereotactic delivery technique.

11.1.5 CYBERKNIFE

Another method of delivering "beamlet-based" IMRS is accomplished by the CyberKnife. The CyberKnife employs an x-band linear accelerator mounted on an articulated robotic arm to fire circular-shaped beamlets from numerous angles, or "nodes," about the targeted lesion. Because the intensity and the direction of each small circular beam can be varied from each node, the method is capable of producing very conformal, IMRT dose distributions. Recently, the original cylindrical collimators have been augmented by a variable aperture design that has been added (Iris) to the system and produces noncircular-shaped beams and improves conformity (Figure 11.7).

11.1.6 CONFORMAL ARC DELIVERY

In addition to the previously described stereotactic delivery techniques, methods for mimicking the Gamma Knife's concentration of high dose volumes and spreading of low-dose regions have evolved to center around the use of arcs. Much like the linac cone–based approach's evolution from the Gamma Knife, these approaches evolved further still to ultimately utilize dynamic MLCs to deliver conformal arc techniques. The Brainlab Novalis conformal arc delivery approach continuously reshapes the beam to

(a)

(b)

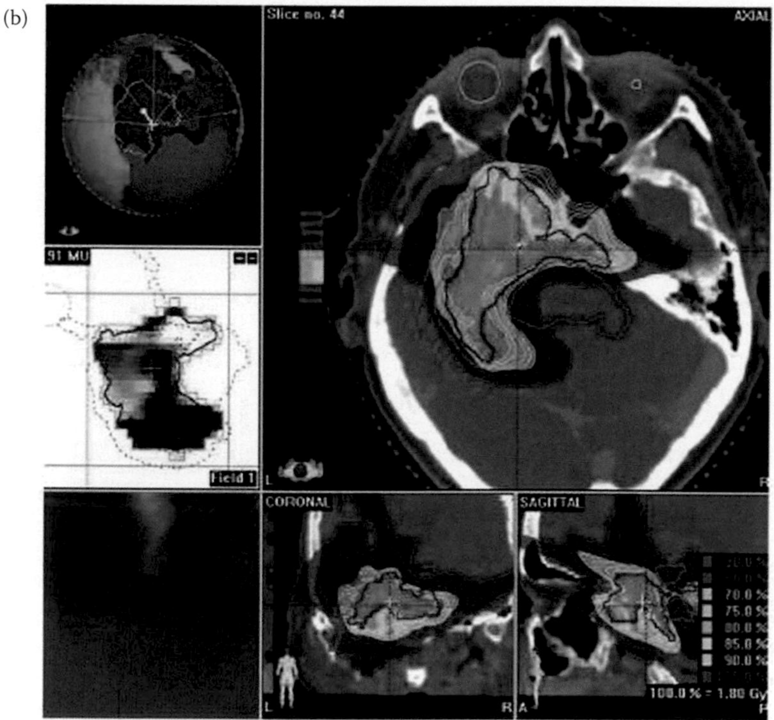

Figure 11.6 (a) Brainlab MM3 micro-multileaf collimator as used to create highly conformal beam geometries in IMRS. (Image courtesy of Brainlab website.) (b) An example of an IMRS treatment plan using six intensity-modulated beams for treating a concave target volume. (Image courtesy of Brainlab AG; From Yu, C., and Shepard, D., *Technol. Cancer Res. Treat.*, 2, 2, 93–104, 2003.)

match the beam's-eye-view shape of the target from each respective delivery angle (Figure 11.8). As such, this method represents a very efficient delivery approach, in that the dose is not "thrown away" as with intensity-modulation approaches, and also a very capable method for delivering highly conformal dose distributions that do not require a concave indention in the isodose distribution as can only be produced by intensity-modulation approaches.

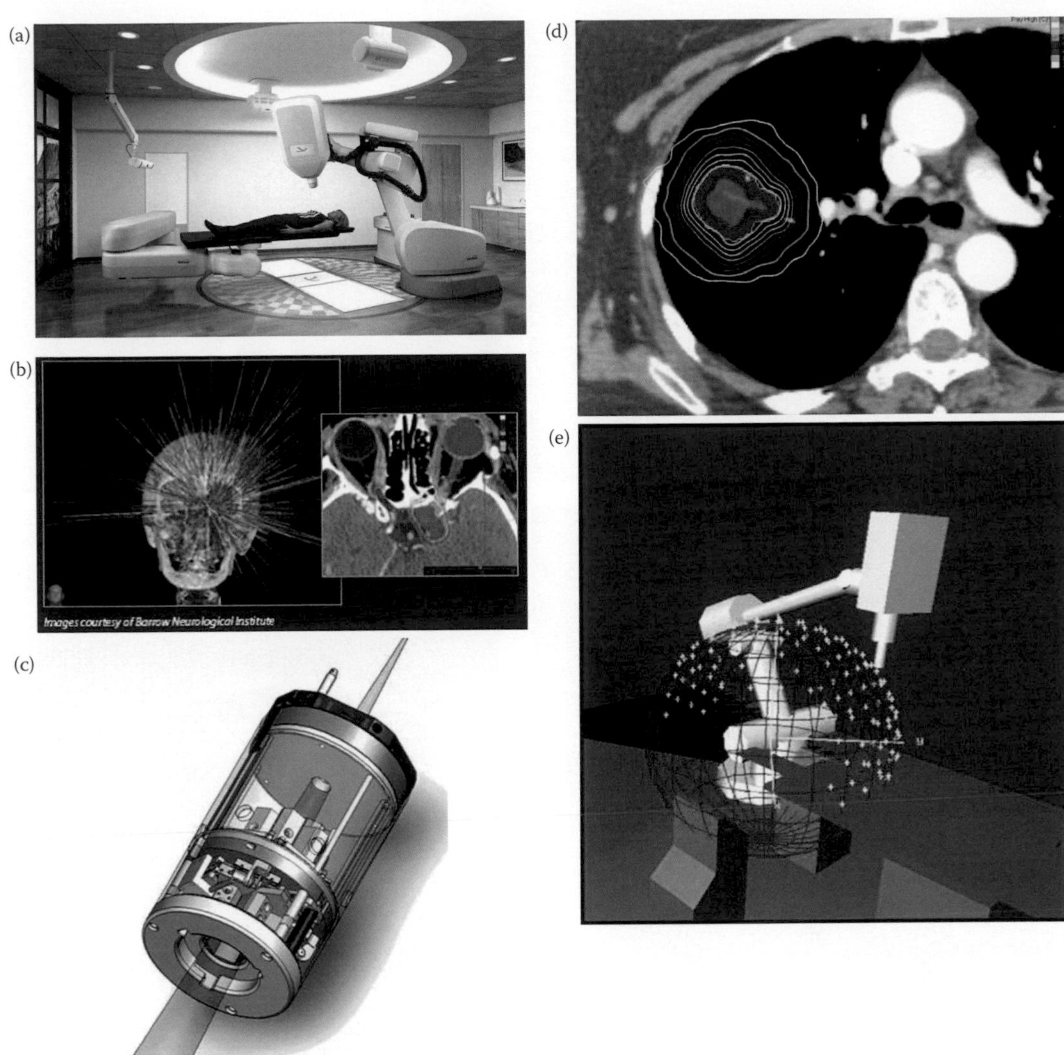

Figure 11.7 (a) CyberKnife treatment suite. (b) Isodose distribution and representation of beam centerlines from multiple nodes. (c) Iris variable aperture collimator and (d) improved isodose distribution that results. (e) representation of CyberKnife nodes. (All images courtesy of Accuray Calvin Maurer Jr.)

11.1.7 INTENSITY MODULATED ARC THERAPY (IMAT)

It seems only logical that the next step in the evolution of stereotactic delivery techniques would entail the marriage of the highly powered arc-based delivery approaches with the other powerful stereotactic delivery tool: intensity modulation. Commercially available solutions now exist for delivering IMAT (Figures 11.9 through 11.11). Such approaches continually reshape the beam aperture, similar to conformal arc techniques, but differ in that the aperture is not required to be shaped to treat the entire target volume from a particular gantry angle. Instead, an inverse planning optimization algorithm develops an optimal collection of beam apertures that are delivered from each respective gantry angle as the linac gantry arcs about the patient (Figure 11.12). Original solutions entailed the use of a binary multileaf collimator that could instantaneously shape individual "physical" pencil beams as the delivery head rotated either serially or helically about the patient. More recently, solutions utilizing conventional "sliding leaf" multileaf

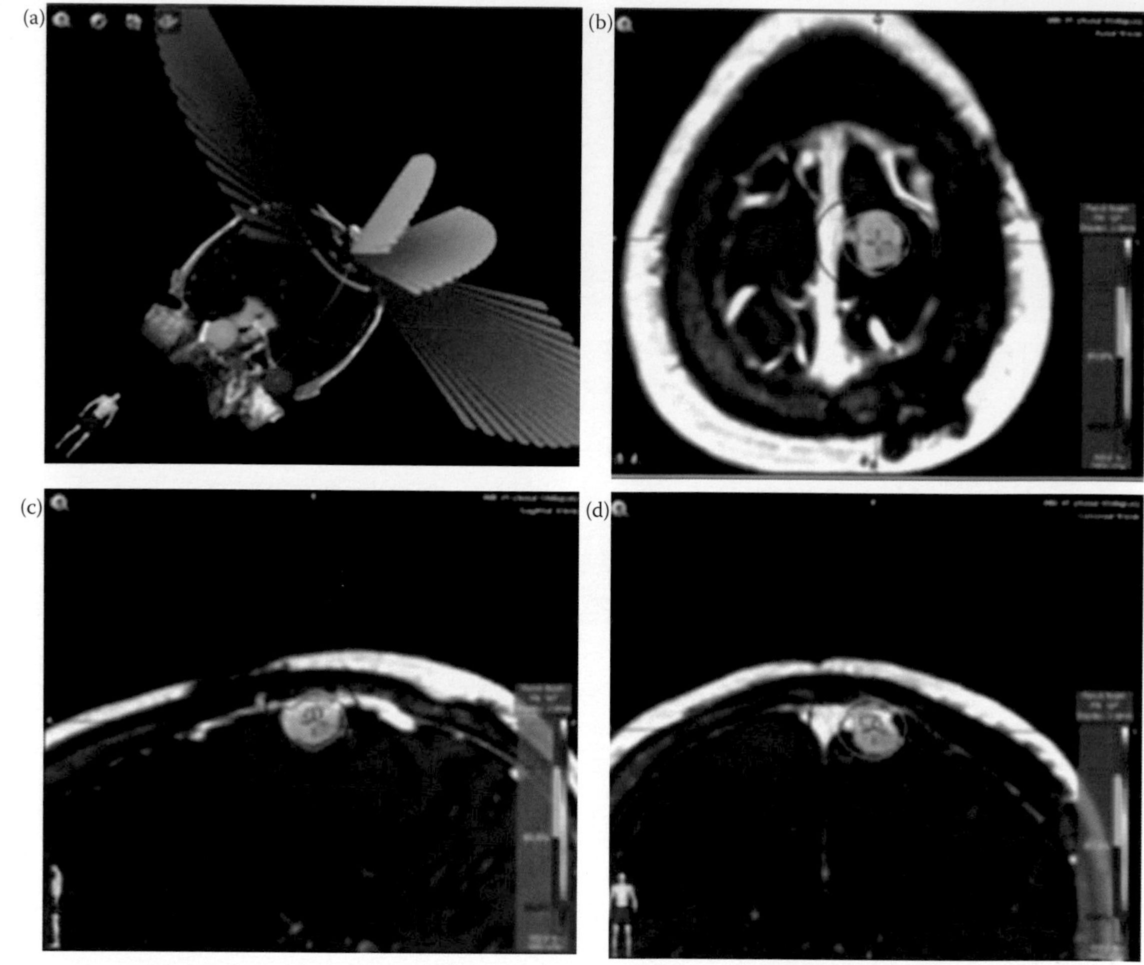

Figure 11.8 Novalis conformal arc plan and delivery suite; 5-arcs; red = GTV, green = prescription, blue = 30%. (a) Oblique cut image showing arc distribution, (b) axial cut, (c) sagittal cut, and (d) coronal cut. (Images courtesy of Brainlab.)

collimators have been introduced. As such, it is clear that the combination of arc and intensity-modulating approaches represent a potentially powerful and effective stereotactic delivery approach.

11.1.8 STEREOTACTIC BODY RADIATION THERAPY (SBRT)

Last, with regard to treatment approaches or modalities, SBRT has recently evolved to entail the migration of intracranial stereotactic successes to areas outside of the brain. It only stands to reason that the highly conformal delivery approaches developed for intracranial lesions could potentially benefit the targeting of challenging lesions outside the brain (e.g., lung and liver lesions), and, indeed, approaches utilizing high-dose, hypofractionated ablative delivery schemes have recently evolved to approach such lesions with encouraging success (Benedict et al. 2010) (Figure 11.13). Targeting of these moving lesions is, of course, more challenging than in the cranium where lesions are static, but that is a topic beyond the scope of this treatment-planning chapter.

Finally, while it stands to reason that treatment-planning specifics will vary greatly depending on the delivery "tool" being used, there are still important common themes that are consistent across the previously introduced delivery modalities, and these will be discussed in the next section relative to the previously introduced individual delivery modalities.

(a)

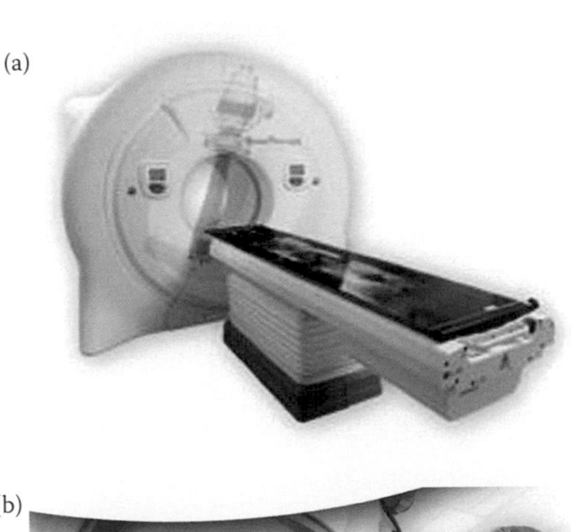

(b)

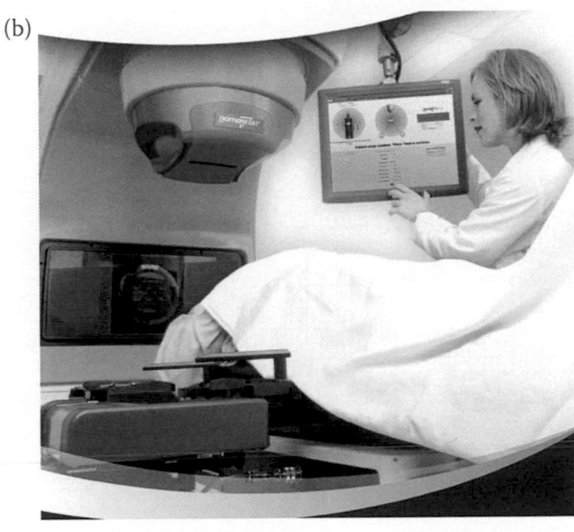

Figure 11.9 (a) Helical tomotherapy arc-based intensity-modulating system. (Image from tomotherapy.com.) (b) Nomos arc-based, serial tomotherapy intensity-modulating system. (Image courtesy of Nomos.)

(a)

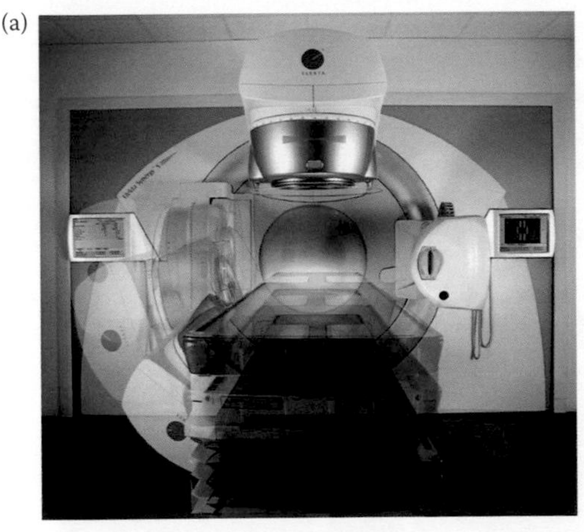

(b)

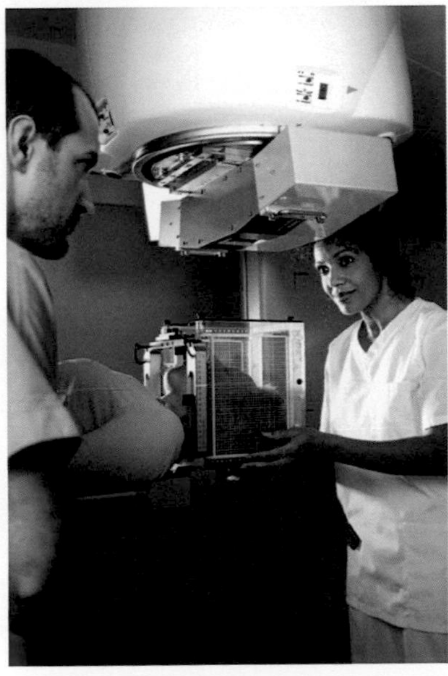

Figure 11.10 (a) Elekta Synergy S stereotactic linear accelerator with 80-leaf MLC (4 mm leaf thickness). (b) Elekta linac with an add-on micro dynamic MLC with leaves as small as 3 mm in thickness. (Images courtesy of Elekta website gallery.)

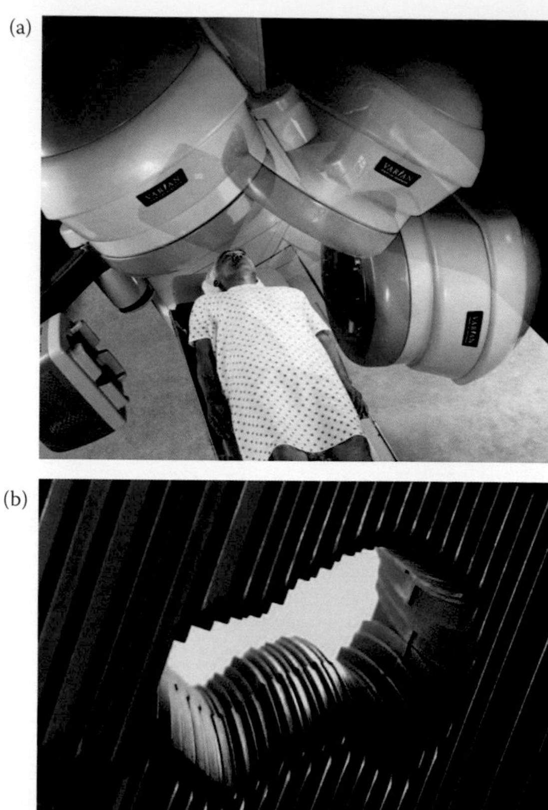

Figure 11.11 Freedom Patient Immobilization System - CDR Systems. (a) This system is designed to accommodate both SRS and SBRT treatment sites. This system is lightweight (< 10 lbs) and compatible with all common treatment couches. An arch and adjustable screw system, or pneumatic bladder system, are available for abdominal compression. The system is compatible with both vacuum bags and expanding foam cushioning. (b) The head and neck immobilization can be accomplished using the MayoMold™, an expanding foam system that is formed around the posterior and lateral aspects of the head, neck and shoulders.

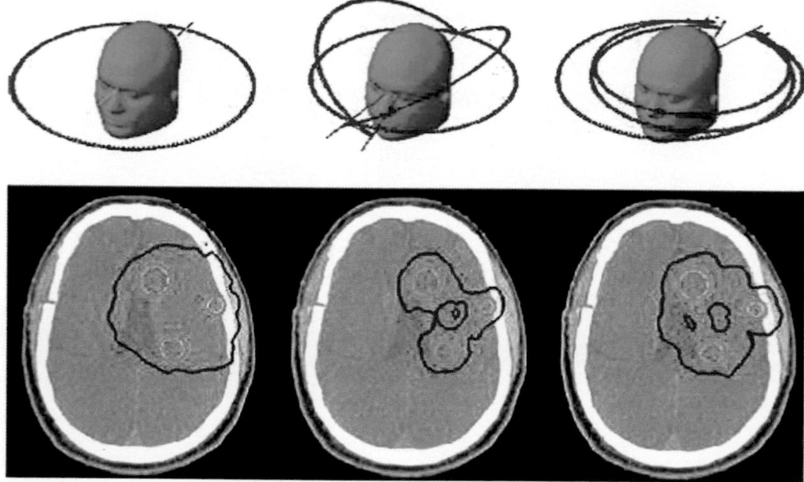

Figure 11.12 Single isocenter volume-modulated arc isodose distributions. Arc geometry and isodose lines of three treatment-planning scenarios with three tumors in same axial plane 3 cm apart. (Left) Single-arc/single-isocenter, (middle) triple-arc/single-isocenter, and (right) triple-arc/triple-isocenter. White indicates target volume; green, 100% isodose lines; and red, 50% isodose lines. (From Clark, G. M. et al., *Int. J. Radiat. Oncol. Biol. Phys.*, 76, 1, 296–302, 2010.)

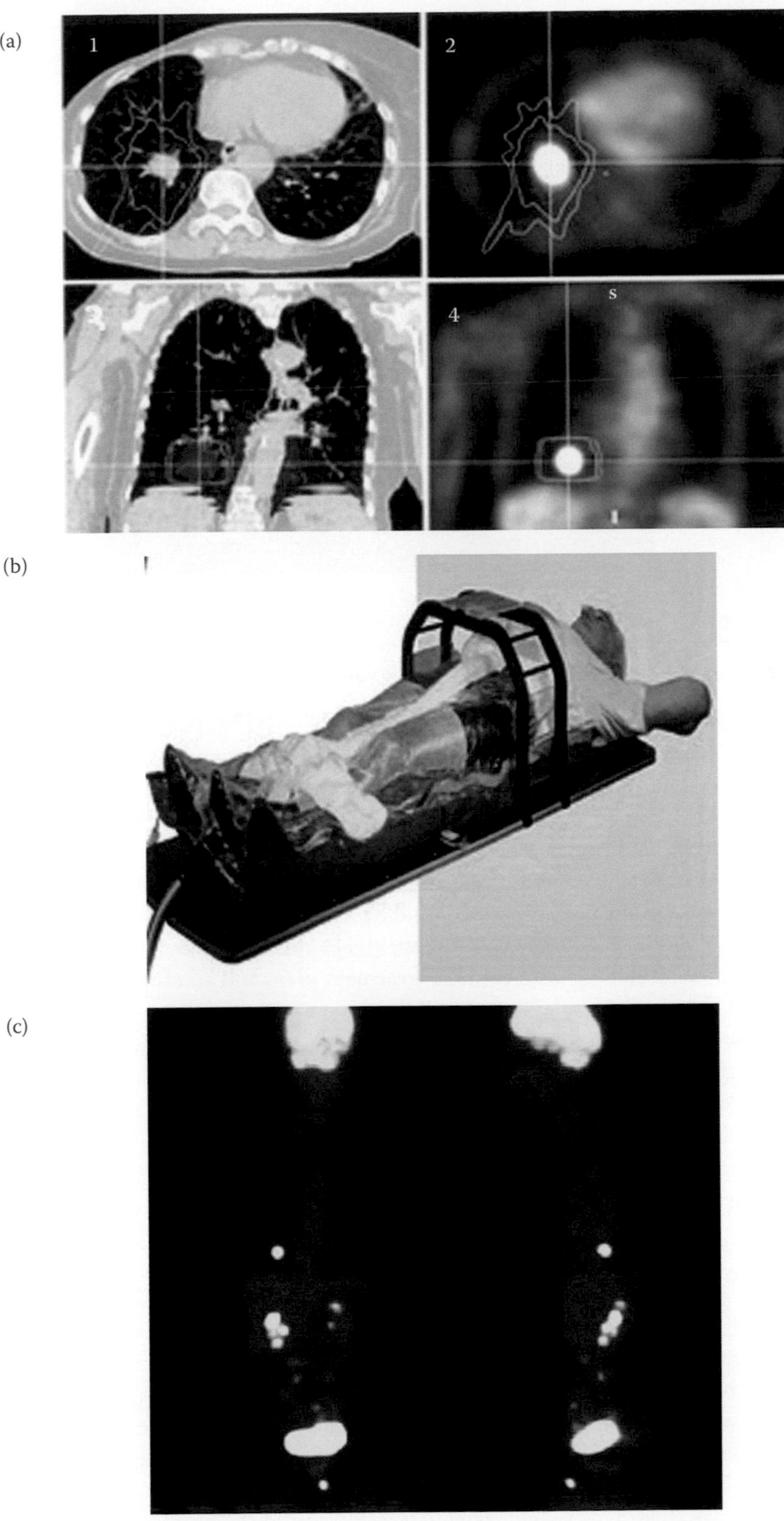

Figure 11.13 (a) Conformal isodose distribution on CT (left) and PET image (right) of lung lesion treated by SBRT. (b) Patient in Medical Intelligence full-body immobilization for SBRT treatment. (c) Pre-treatment PET image showing small right sided, enhancing lung lesion.

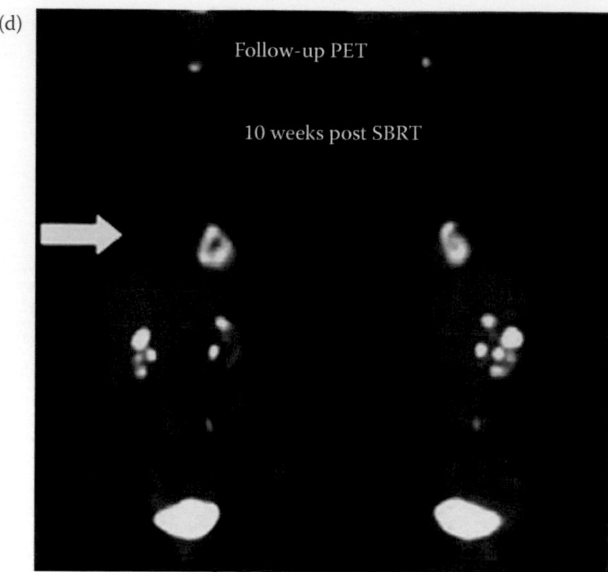

Figure 11.13 (Continued) (d) 10 week follow-up PET image of lesion showing complete resolution on PET (arrow indicates previous enhancing location of lesion). (All images courtesy of Bill J. Salter, University of Utah.)

11.2 STEREOTACTIC TREATMENT-PLANNING CONCEPTS AND SUBTLETIES

As mentioned in the Introduction section to this chapter, treatment planning for stereotactic treatments has similar goals and challenges to any other form of treatment planning but with potentially higher stakes. The high dose per fraction associated with radiosurgical treatments can make big problems out of subtle mistakes in planning. Percentage hot spots that could be easily tolerated in a fractionated delivery scheme can be catastrophic in a single-fraction delivery scenario. For this and many other similar reasons, great care must be taken when developing a stereotactic treatment plan, and careful attention must be paid to even subtle aspects of the treatment plan.

First and foremost in importance is the need to become familiar with the nuances of the particular delivery modality that is being employed. As described previously, there are numerous effective and capable delivery modalities that have evolved in recent years. Each approach can effectively deliver highly conformal, stereotactically targeted isodose distributions, but it is of paramount importance that the clinician performing and approving the treatment planning be familiar with the strengths and limitations of the particular delivery modality.

11.2.1 TREATMENT-PLANNING CONSIDERATIONS FOR SPHERE-BASED DELIVERY APPROACHES

The previously described Gamma Knife delivery approach has an extensive and well-tested track record in treating brain metastases. The roughly spherical shape of most metastatic brain lesions makes the sphere-based dose-building block of the Gamma Knife well suited to such treatments. Extremely steep dose gradients associated with this delivery approach allow for a rapid falloff of dose outside the boundaries of the targeted tumor. While more complex-shaped tumor geometries can also be effectively treated with the sphere-based approach of the Gamma Knife by utilizing multiple, strategically located, intersecting-dose spheres, the challenges of treating very large irregular lesions using the multiple-shot approach required for a sphere-based treatment of such a target are not trivial. Intensity-modulating approaches have been reported to excel over sphere-based approaches with regard to conformality of isodose distributions for such large and irregular lesions when clinically feasible numbers of spherical dose shots are compared (Nakamura et al. 2003).

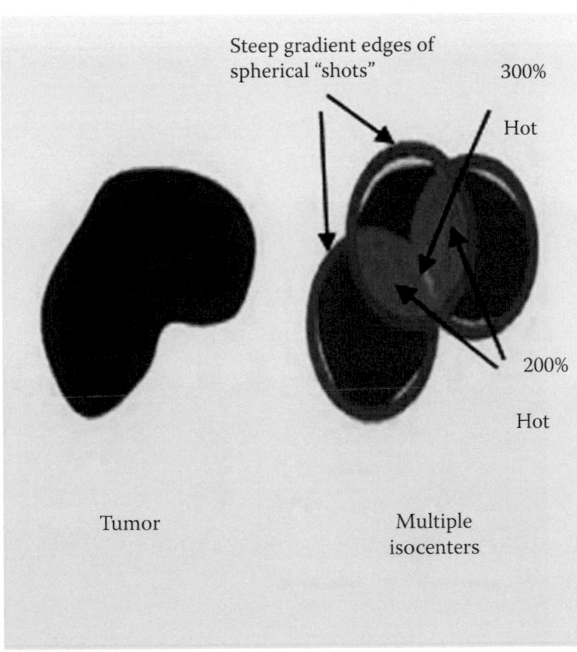

Figure 11.14 Diagram demonstrating inherent hot spots (200% isodose, 300% isodose) resulting from intersecting isodose spheres as used in cone-based or Gamma Knife delivery. (Image courtesy of Bill J. Salter, University of Utah.)

Additionally, when multiple intersecting isodose spheres are used, as for Gamma Knife or cone-based delivery approaches, the inherent hot spots that appear at the overlapping volumes where multiple-dose spheres intersect must be carefully managed (Figure 11.14). In particular, care must be taken to ensure that such hot spots, which can be as high as 300% of a prescribed dose, are ensured to reside within the boundaries of the targeted lesion.

Valuable peer-review work has also been published characterizing the value and importance of selecting the optimal collimator size for both cone-based and Gamma Knife delivery approaches (Li and Ma 2005; Ma et al. 2007). And so it becomes clear that, while sphere-based delivery approaches represent highly capable and time-tested delivery tools, it is imperative that the wielder of the tool be familiar with both the strengths and subtle challenges to these well-established treatment approaches.

11.2.2 TREATMENT-PLANNING CONSIDERATIONS FOR INTENSITY-MODULATING DELIVERY APPROACHES

Of course, the sphere-based approaches of Gamma Knife and cone-based delivery are not the only delivery approaches that, in addition to their inherent strengths, also possess limitations or challenges to treatment planning. Pencil beam–based intensity-modulating approaches, as mentioned previously, are highly capable of producing extremely conformal isodose distributions through optimized location and intensity of the intersecting pencil beams used to deliver the dose to the target. However, important and sometimes subtle challenges exist. For example, in the early days of implementation of IMRT techniques, it was soon realized, through sometimes undesirable clinical outcomes, that significant instances of skin erythema could appear from a combination of delineating a targeted lesion too close to the skin surface and the use of tangentially oriented delivery angles (Figure 11.15).

Prior to the advent of intensity-modulating techniques, dose buildup was limited by the underlying physics of the buildup region of a particular beam. With the advent of inverse optimized IMRT techniques, the optimizing software could now greatly increase the weighting of one of the tangentially oriented pencil beams intended to treat the target, such that the skin dose could actually be elevated to the requested prescription dose (Figure 11.16). It soon became clear that for intensity-modulating techniques, the clinician planner had to "be careful what they asked for" because the delivery and inverse planning

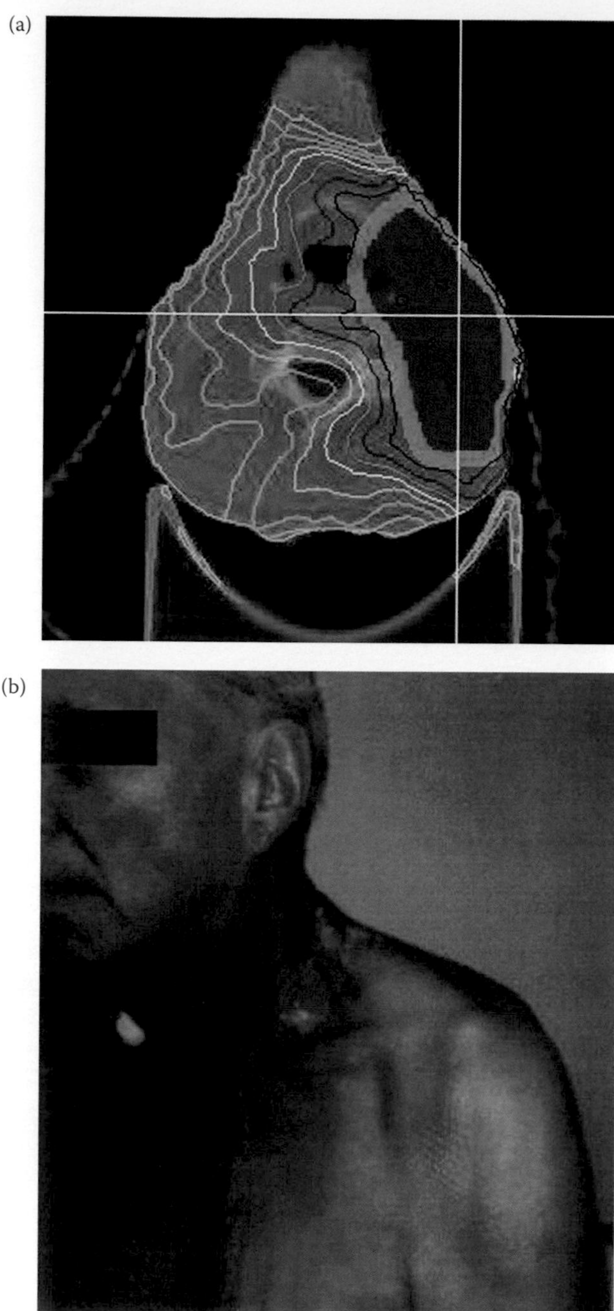

Figure 11.15 (a) Gold CTV expansion shown delineated from overdosage of skin to very near the surface. IMRT with tangentially oriented beams can achieve full prescribed dose at surface by overweighting tangential pencil beams as shown by blue 100% isodose extension to surface. This can result in significant skin erythema. (Image courtesy of Bill J. Salter, University of Utah.) (b) Example of skin desquamation. (From Lee, N. et al., *Int. J. Radiat. Oncol. Biol. Phys.*, 53, 3, 630–7, 2002.)

method were now capable of providing it. In similar fashion, it also soon became clear that, by requesting that the inverse planning optimizer deliver a high dose to a target bounded by nearby sensitive organs at risk (OARs), highly irregular dose distributions with high intensity "fingers" of dose could result (Figure 11.17). This could, in turn, result in doses being delivered to nearby sensitive structures that had not been considered "at risk" prior to the advent of intensity-modulating techniques (Figure 11.18). A

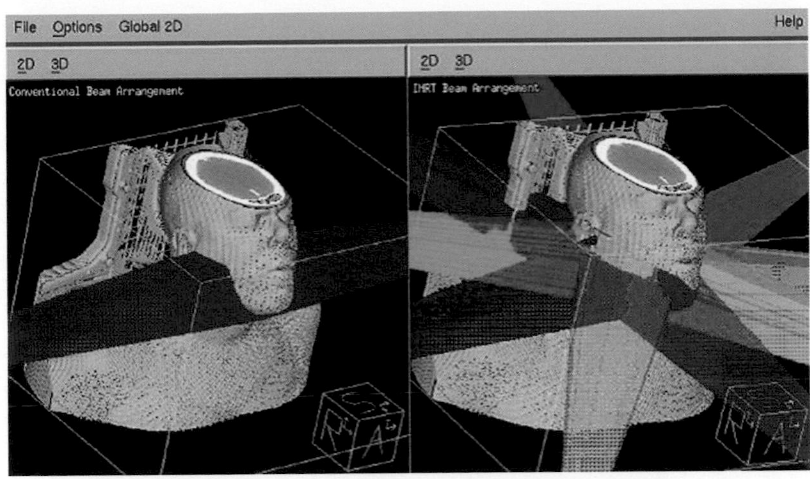

Figure 11.16 Comparison of conventional 3-D beam paths versus "nontarget" beam paths in IMRT treatment. (From Rosenthal, D. I. et al., *Int. J. Radiat. Oncol. Biol. Phys.*, 72, 3, 747–55, 2008.)

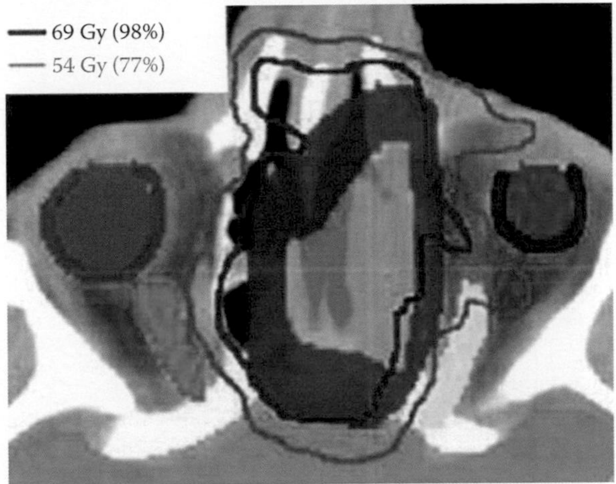

Figure 11.17 High-dose fingers extending outward from target along lines of highly valued pencil-beam angles. (Image courtesy of Bill J. Salter, University of Utah.)

valuable perspective on this topic was presented by the M.D. Anderson group (Rosenthal ct al. 2008). They observed the following:

> Dose reduction to specified structures during IMRT implies an increased beam path dose to alternate nontarget structures that may result in clinical toxicities that were uncommon with previous, less conformal approaches. These findings have implications for IMRT treatment planning and research, toxicity assessment, and multidisciplinary patient management.

For static gantry intensity-modulated radiosurgery approaches, it is also very important that enough static gantry portals be utilized so as to give the optimizer enough degrees of freedom to effectively spread the low-dose component of the treatment around to large volumes of tissue. This, in turn, prevents the system from being forced to weight certain high-value segments so much so as to create highly undesired hot spots in normal tissue or on the skin surface (Figure 11.19).

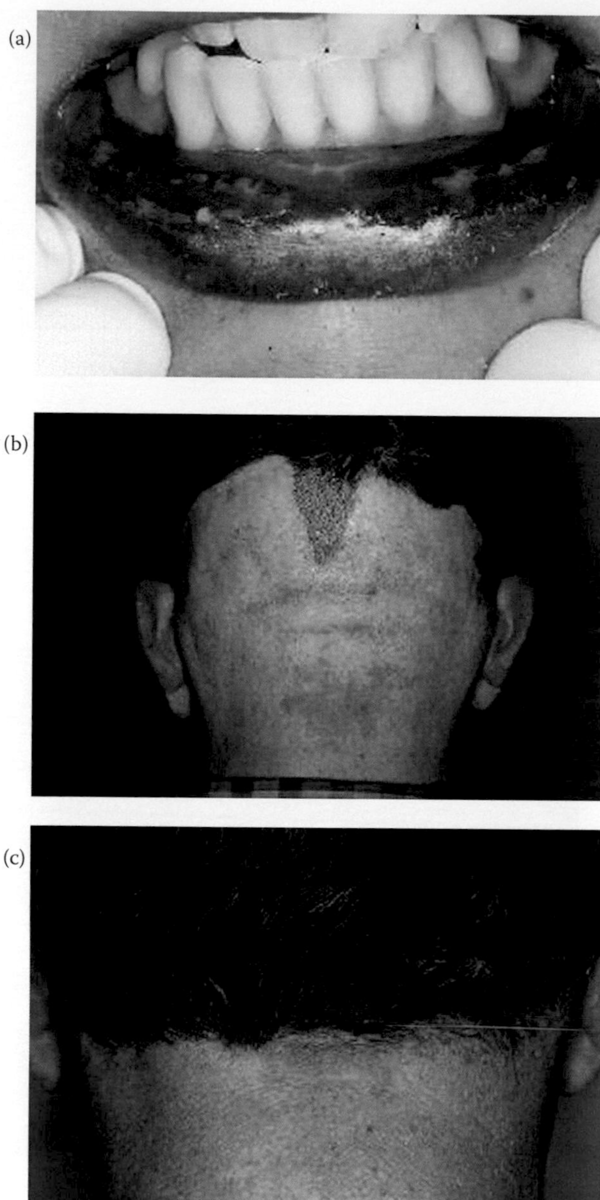

Figure 11.18 Consequences of "...increased beam path dose to alternate nontarget structures." (a) Anterior oral mucositis during IMRT. (b) Occipital scalp epilation after IMRT. (c) Scalp hair subsequent regrowth, same patient. (From Rosenthal, D. I. et al., *Int. J. Radiat. Oncol. Biol. Phys.*, 72, 3, 747–55, 2008.)

Other valuable peer-reviewed work has also been published demonstrating the importance of carefully optimizing beamlet size selection (Dvorak et al. 2005; Jin et al. 2005; Monk et al. 2003; Wu et al. 2009), collimator angle (Tobler, Leavitt, and Watson 2004), and isocenter location (Figure 11.20) (Salter et al. 2009).

Another subtle but important nuance of planning for IMRS delivery approaches includes the need for an appreciation of what is likely to be an "optimal" treatment plan. While the inverse planning optimizer will ultimately terminate its iterations and return with what is theoretically an "optimal" treatment plan, this optimality is defined as a function of the dose and volume criterion that were defined prior to optimization. Given that such criteria are typically conflicting, the inverse planning optimizer is forced to develop a compromise solution that may well lead to the best "score" for its objective function, but may also represent an entirely unacceptable clinical solution. For this reason, the clinician planner must pay

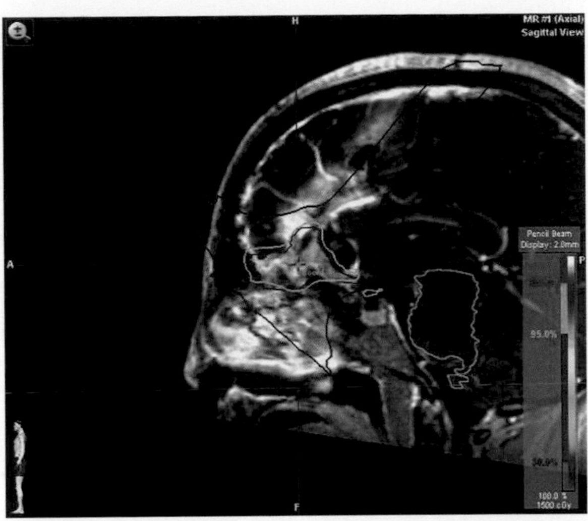

Figure 11.19 Example of high-valued IMRT segment being overweighted so as to cause high-dose "finger" extending to near skin surface. (Image courtesy of Bill J. Salter, University of Utah.)

careful attention to the "optimized" IMRS treatment plan, taking care to closely inspect all CT slices for acceptability of the isodose distribution. For this and numerous other similar reasons, it is imperative that clinician planners of IMRS treatment plans familiarize themselves with the subtle nuances of this capable but extremely complex delivery approach.

11.2.3 TREATMENT-PLANNING CONSIDERATIONS FOR CYBERKNIFE DELIVERY APPROACHES

The CyberKnife's articulated arm affords an opportunity for very high degrees of freedom for pencil-beam delivery geometry. This extremely sophisticated delivery approach arguably affords the greatest possible degrees of freedom from a beam geometry perspective. With this strength, however, also comes the obvious challenge of managing the degrees of freedom from a collision and beam-access perspective. Because the robotic arm is capable of assuming so many locations and orientations about the patient, it must be managed for both collisional and appropriateness of beam-direction criteria. Not only must the node locations for a particular treatment delivery be carefully considered, the order and trajectory of the transitions between nodes must be confirmed to be safe and appropriate. Additionally, the beamlet definition collimator for the device is interchangeable and must be selected for appropriateness of beamlet dimension for the particular treatment. Clearly, for radiosurgical applications, the selected beamlet size will often be chosen as the smallest 5 mm circular beamlet size. The manufacturer has recently released a variable aperture MLC design (Iris), which affords the opportunity to produce noncircular-shaped beams and improve conformity. The complexity and flexibility of the delivery scheme afforded by the CyberKnife delivery approach can allow for exceptionally conformal treatment delivery but also requires that the clinician treatment planner be very familiar with the intricacies of this very capable delivery approach.

11.2.4 TREATMENT-PLANNING CONSIDERATIONS FOR ARC-BASED DELIVERY APPROACHES

Whether using a cone-based, conformal arc, or intensity-modulated arc-delivery approach, all of the arc-based delivery approaches share a common treatment-planning challenge. Just as the enhanced degrees of freedom of the CyberKnife require a careful management of node locations and trajectories, the arc-based approaches must also thoughtfully manage the physical movement of the gantry from a safety and collision standpoint and from an appropriateness of beam access perspective. Treatment plans containing arcs that result in collision of the moving gantry with either other equipment, such as the treatment couch or, worse yet, the patient, must be identified in advance and avoided. Often, it is only the experience and awareness of the treatment planner that can consistently identify such situations. Additionally, it may be inappropriate

(a)

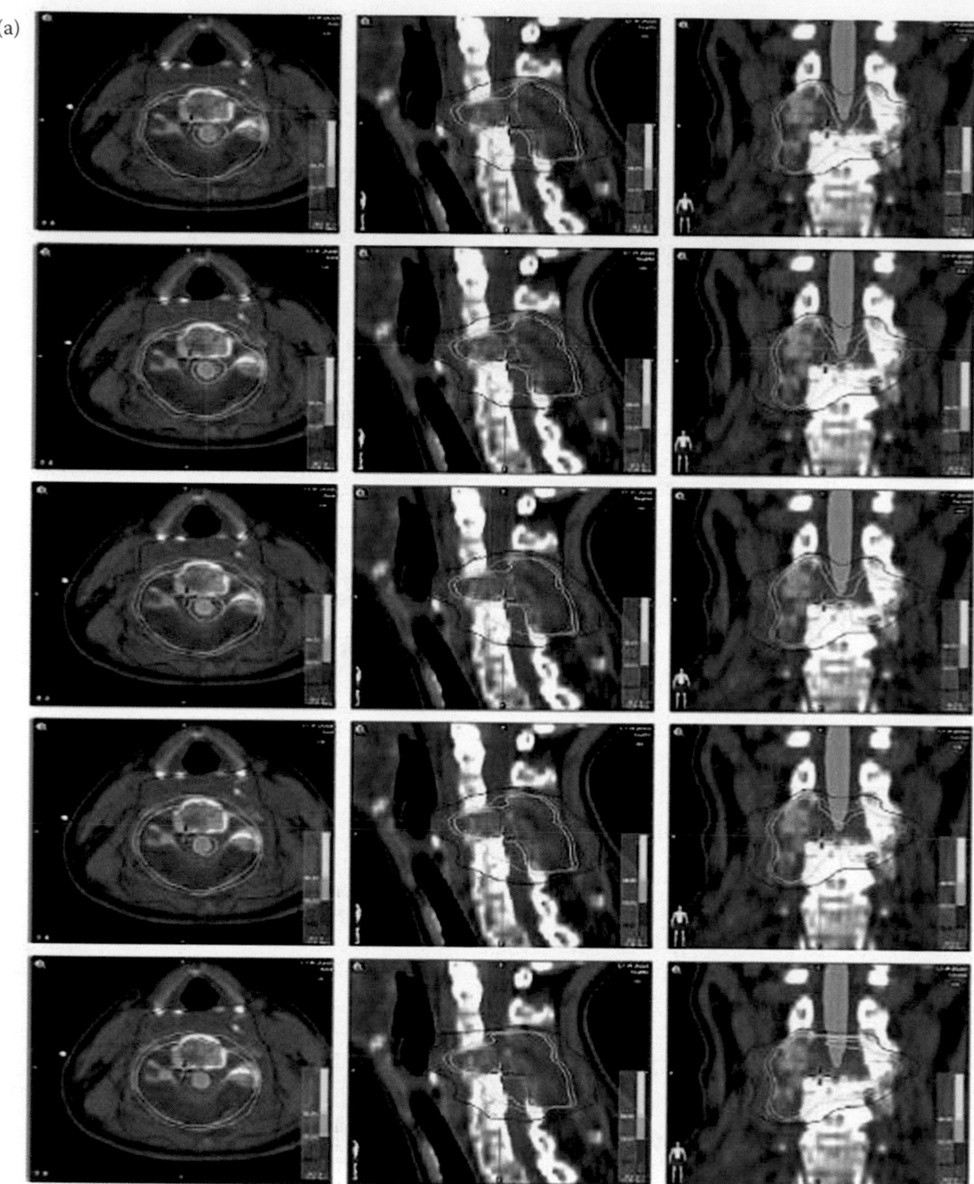

Figure 11.20 (a) Transverse (left), sagittal (middle), and coronal (right) CT images by beamlet size (from the top, CT images for beamlet sizes 1.2, 3.5, and 10 mm). The three images in each line show dose distributions around the planning target volume (red) and the spinal cord (yellow). This shows that as beamlet size was increased, the doses around the spinal cord also increased. Three isodose lines that were normalized to the prescription dose are also shown (90%: yellow, 80%: green, 50%: blue).

to allow beams to enter the patient from certain directions, and this must also be carefully identified and managed by the treatment planner. Arcs must sometimes be split into two sub-arcs with a gap placed in between to respect such regions where arcs must not be delivered or to modify weighting of dose delivery (Figure 11.21).

Interesting and valuable work has also been published describing the minimum number of arcing degrees that should be used to avoid excessive peripheral dose (Yu and Shepard 2003; Kooy et al. 1991).

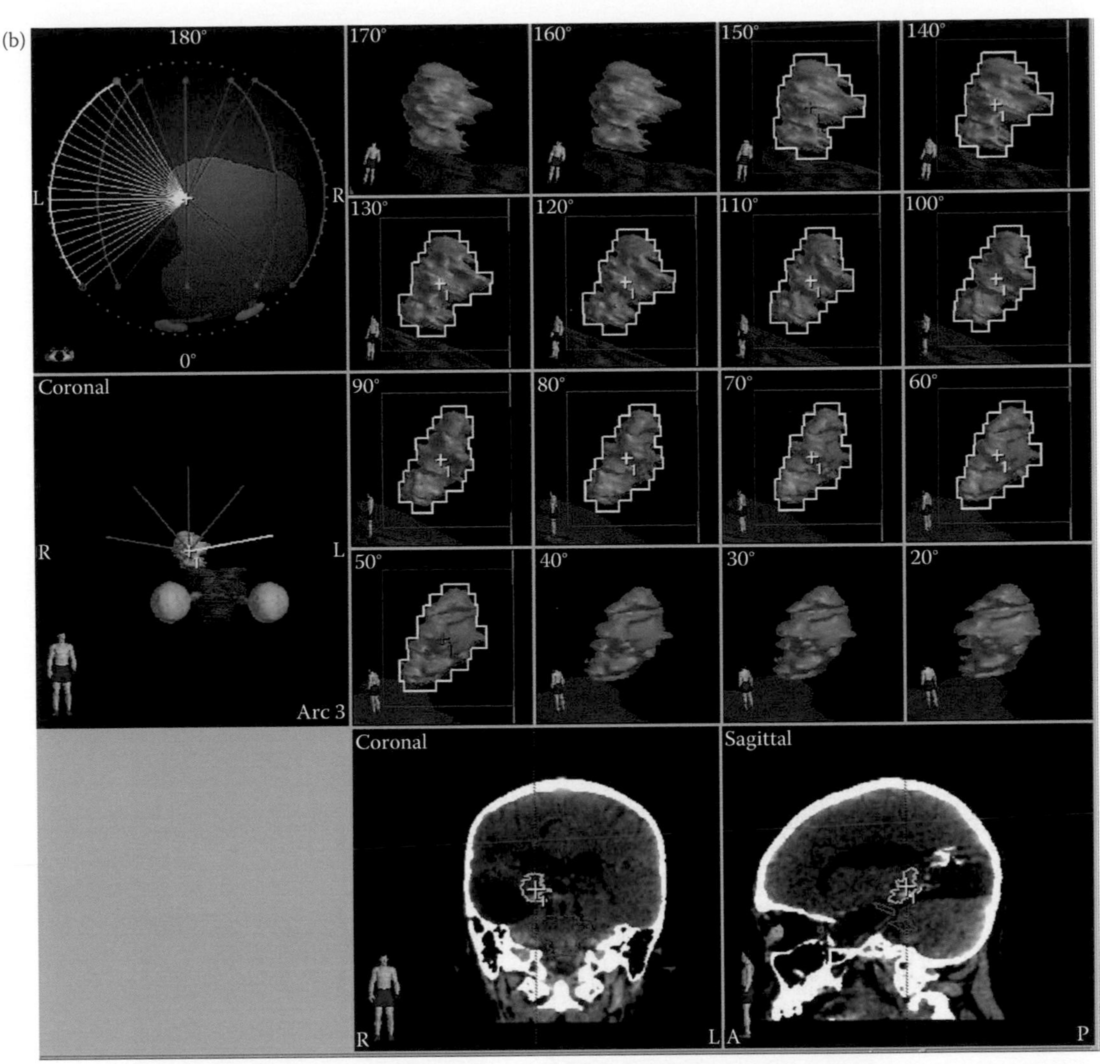

Figure 11.20 (Continued) (b) Collimator angle optimization (BEV for one arc show). IPCI was 0.585 (unoptimized) versus 0.616 (optimized), a 5% improvement. The treated (optimized) collimator angles were 120, 115, 140, 70, and 35. The default unoptimized collimator angles were 90°. The DVHs for tumor coverage are similar. All the normal structures are identical. (Image courtesy of Bill J. Salter, University of Utah.)

The recently evolved IMAT delivery approach has facilitated the delivery of treatments with many more degrees of freedom and the subsequent highly conformal dose distributions that can result from such a delivery scheme. But just as the added degrees of freedom of previously discussed delivery approaches, such as CyberKnife and arc-based approaches, afford enhanced treatment quality, they also include enhanced delivery challenges that must be carefully planned for and managed. The IMAT delivery approach facilitates high-quality, efficiently delivered treatments but is known to sometimes lack sufficient modulation from certain high-value delivery angles because the moving gantry may not be afforded sufficient opportunity to deliver enough dose from such angles. This situation must be recognized by the clinician planner, and either a multiple-pass delivery scheme must be utilized (Figure 11.22) (if this is possible for the particular vendor-supplied version of IMAT that is being used) or a more fully modulated static gantry approach must be opted for. This is yet another example of how the clinician performing and/or approving the treatment planning must be very familiar with both the capabilities and limitations of the respective delivery modality being planned for.

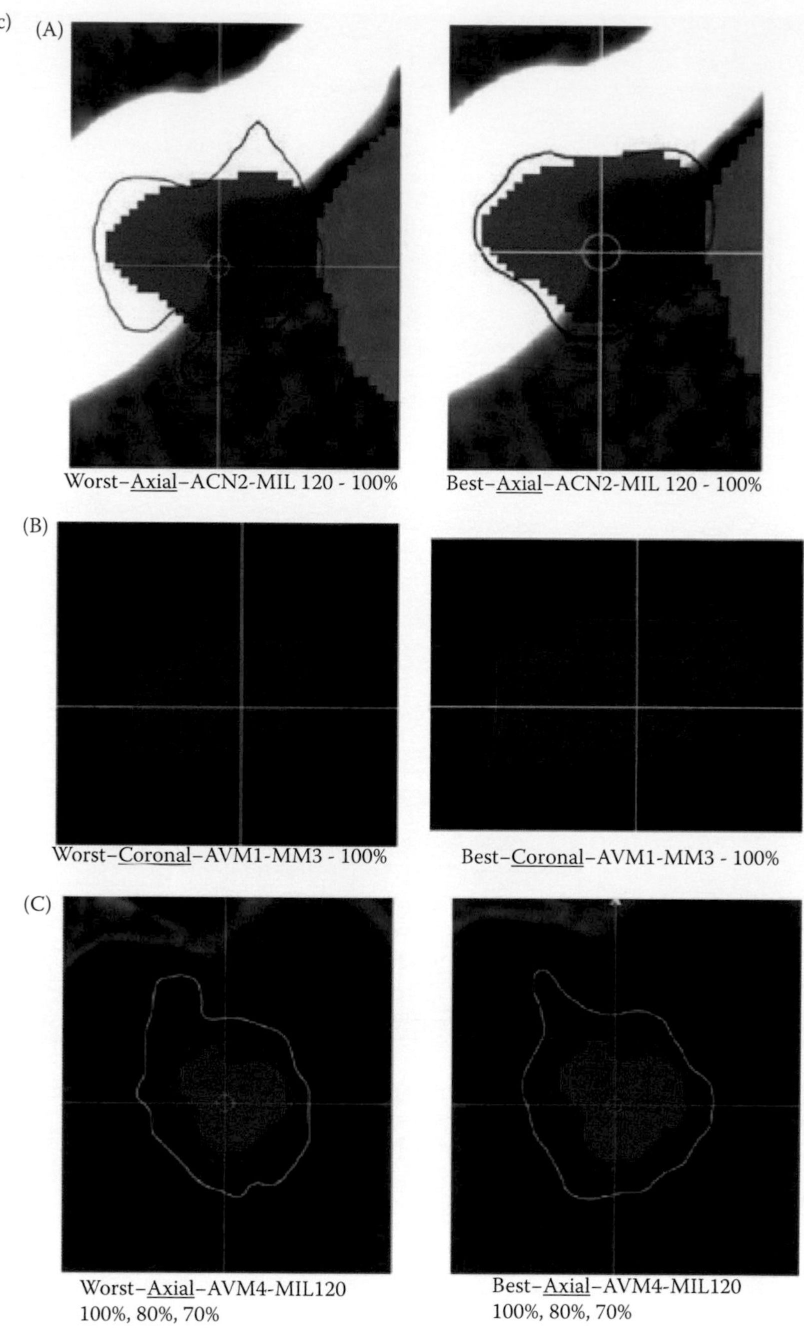

Figure 11.20 (Continued) (c) Example of advantage of optimizing isocenter location for stereotactic treatment of brain lesions. Images on left side present results of selecting an unoptimized isocenter location whereas the right side images show significantly improved conformity achieved by simply relocating isocenter location slightly (~1 mm) so that intensity-modulated pencil beams align better to the edges of the target. (From Salter, B. J. et al., *Int. J. Radiat. Oncol. Biol. Phys.*, 73, 2, 546–55, 2009.)

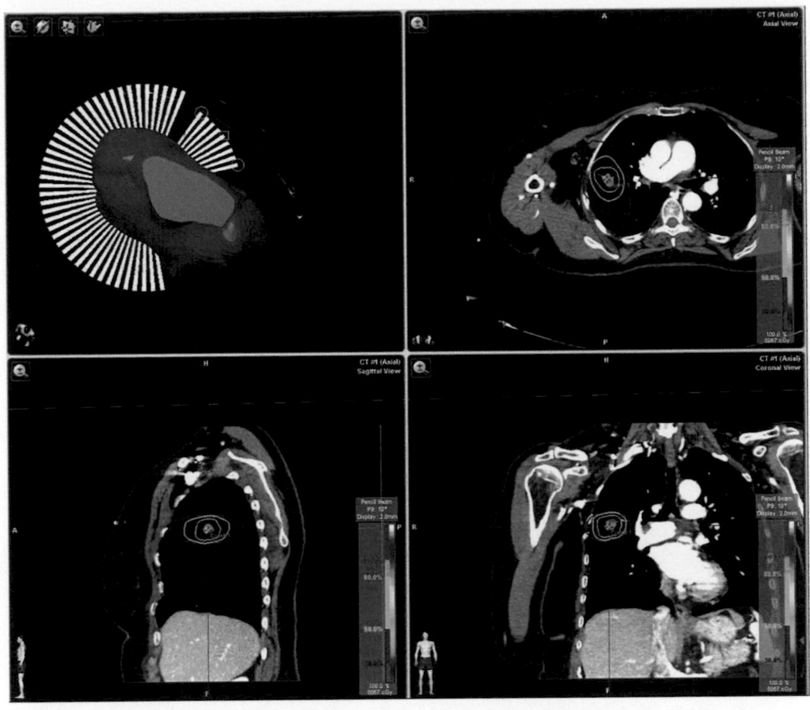

Figure 11.21 Example of split arc delivery. (Image courtesy of Bill J. Salter, University of Utah.)

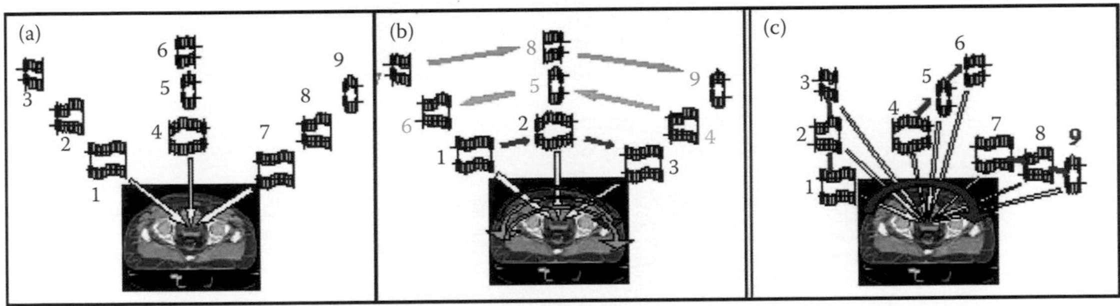

Figure 11.22 "Stacked" segments from multiple-pass intensity-modulated arcs. (a) Planned segments. (b) 'Stacked' delivery approach. (c) 'Bouquet' delivery approach. (Images courtesy of Roberto Pellegrini—Elekta.)

11.2.5 TREATMENT-PLANNING CONSIDERATIONS FOR SBRT TREATMENT APPROACHES

While SBRT represents more of a dose, fractionation, and targeting approach than a specific delivery approach, there exist numerous, subtle challenges to the treatment planning of such treatments, and these should certainly be discussed here. As a stereotactic delivery approach, many of the previously mentioned considerations, for many of the previously mentioned delivery approaches, must be considered and brought to bear for SBRT treatment planning as well. Of course, the highly specialized Gamma Knife would never be used for extracranial deliveries, such as SBRT, and even the cone-based approach would likely not be widely employed. But previously discussed considerations for static gantry IMRT, conformal arc, and IMAT approaches must certainly all be considered and appropriately managed. Additionally, while static or fixed-field delivery approaches would not be commonly encountered for intracranial stereotactic deliveries, such treatments are frequently employed for extracranial SBRT applications, owing in large part to their relative insensitivity and consistent, predictable dose delivery in the presence of moving targets. Peer-reviewed literature has reported that between 13 and 15 fields are optimal in such situations in order to prevent skin erythema or desquamation at highly weighted delivery port locations (Figure 11.23) (Kooy et al. 1991; Liu et al. 2006).

Additionally, for lung SBRT treatments, the limitations placed on both V20 and V5 by routinely employed protocols, such as radiation therapy oncology group (RTOG) 0236, 0618, and 0813, can be particularly challenging. In such situations, it can be valuable to borrow from lessons learned from intracranial treatment planning and make careful use of non-coplanar delivery approaches (Figure 11.24).

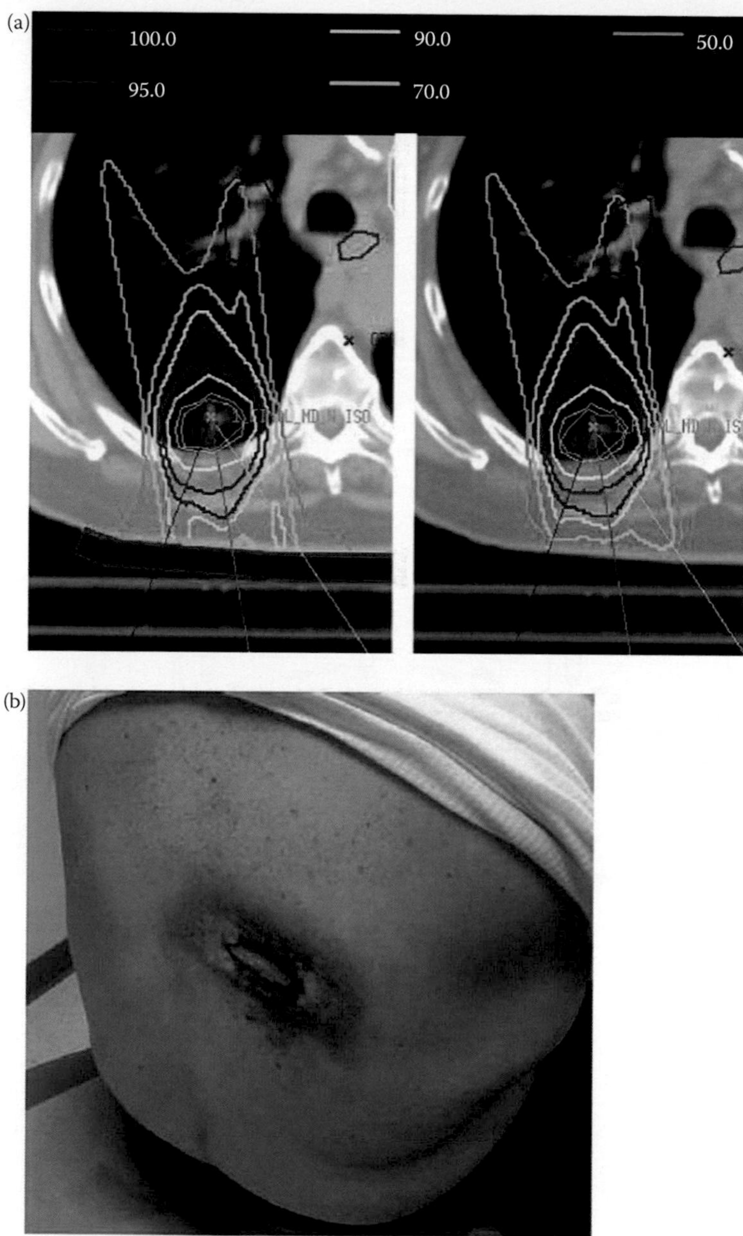

Figure 11.23 (a) Treatment plan for a patient that developed Grade 4 skin toxicity without any corrections for treating through the couch and mobilization device (right) and with 1 cm of bolus to account for the couch and mobilization device (left); (b) patient who developed Grade 4 skin necrosis from stereotactic body radiation therapy. (From Hoppe, B. S. et al., *Int. J. Radiat. Oncol. Biol. Phys.*, 72, 5, 1283–86, 2008.)

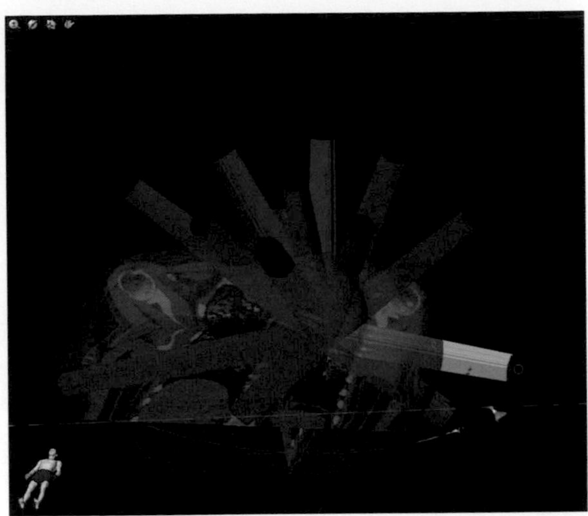

Figure 11.24 Example of SBRT treatment plan utilizing 13 non-coplanar beams. (Image courtesy of Bill J. Salter, University of Utah.)

Such non-coplanar approaches can be more challenging in the regions of the torso, but are still achievable and, sometimes, absolutely essential to achieving an optimal treatment plan (Liu et al. 2006; Tobler, Leavitt, and Watson 2004).

11.3 DIFFERENCES BETWEEN "CONVENTIONAL" AND "STEREOTACTIC" TREATMENT-PLANNING SYSTEMS

Thus far, we have discussed the overall challenges of stereotactic treatment planning to include a brief overview of some individual treatment approaches and also some of the subtle challenges to treatment planning associated with each. In discussing treatment-planning challenges, we have, obviously, assumed the availability of a suitable stereotactic treatment-planning system. The question we now explore is: "What constitutes an 'acceptable' stereotactic treatment-planning system?" It does not seem foolish to at least ask, "Are all commercial treatment-planning systems suitable for stereotactic treatment planning?" Let's begin by characterizing the typical stereotactic planning problem and requirements and then, perhaps, the answers to the above questions will become more clear.

The typical intracranial stereotactic treatment-planning problem would likely entail treatment of a relatively small brain lesion, say 1.5 cm, that could very possibly reside in close proximity to a sensitive structure, such as the brain stem or optic apparatus (Figure 11.25).

The lesion would likely be scanned with "high-resolution" CT (i.e., thin slice, smallest possible field of view) and, because of the reduced contrast-resolution of thin-slice CT images, the lesion might not be well visualized on the CT data set. This would make a fusion of the CT with an MRI image set necessary in order to fully appreciate and delineate the volume to be targeted. An effectively designed treatment plan would be highly conformal with steep dose gradients to facilitate sparing of the nearby OARs. Dose and dose-volume constraints for the OARs would be satisfied, and the target would be treated to the prescribed dose with carefully considered dose inhomogeneity within the target. Normal, nonspecific, healthy tissue surrounding the lesion would be spared of undesirable hot spots. Such a plan, if delivered correctly and accurately, might reasonably lead us to expect a high probability of local control of the lesion without the manifestation of undesirable complications. While an "accurate" delivery might reasonably suggest a *spatially* accurate treatment, it should also suggest the notion of a *dosimetrically* accurate treatment as well. In other words, the dose distribution predicted by the treatment-planning system should be delivered to the patient with high fidelity. Predicted dose levels should be accurate with regard to spatial location *and* magnitude, and dose and dose-volume metrics reported by the planning system should be achieved in the patient. The question naturally arises, "What system features must exist to ensure such a delivery?"

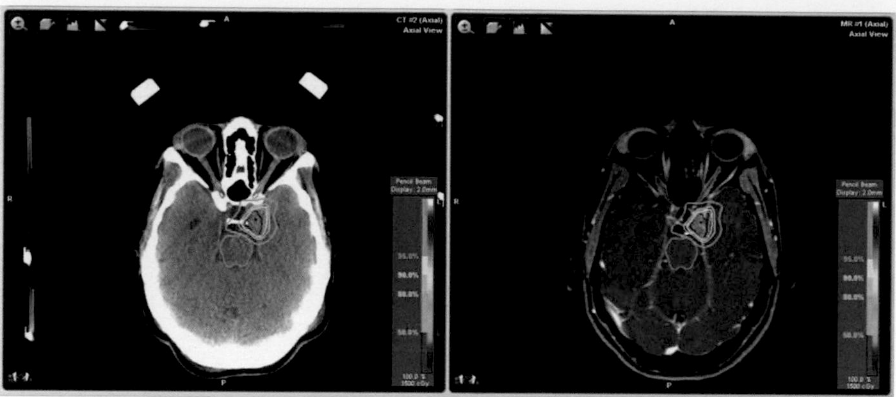

Figure 11.25 Registered CT (left) and MR (right) images of a patient showing a plan for treating a left cavernous sinus lesion (in red) that is very close to the brainstem and optic nerves. Doses are normalized using a prescription of 15 Gy to the 100% line. (Image courtesy of Bill J. Salter, University of Utah.)

11.3.1 STRUCTURE AND DOSE GRID VOXEL SIZE

The previously stated need for high target conformality implies the ability of the planning system to characterize conformality. Multiple conformality, or conformity, indices (CIs) have been proposed throughout the years, and some have been implemented into existing stereotactic treatment-planning systems. The originally popular CI method was derived from a RTOG definition and did not characterize undertreated regions (i.e., it *assumed* that 100% of the target was always covered by the prescription isodose line).

$$\mathrm{Cl}_1 = 1 + \frac{V_{\mathrm{Normal}}}{V_{\mathrm{PTV}}}; \mathrm{Cl}_1 = 1 - \infty$$

Equation 1: Originally popular RTOG-based CI; ideal value = 1.

The more recently embraced index, for intracranial lesions at least, appears to be the Ian Paddick conformity index (IPCI), named after the author of the peer-reviewed manuscript in which the index was described (Paddick 2000). The equation for the IPCI is

$$\mathrm{OTR} = \frac{\text{Target Volume (TV)} \cap \text{Isodose Volume (IV)}}{\text{Isodose Volume(IV)}}$$

$$\mathrm{UTR} = \frac{\text{Target Volume (TV)} \cap \text{Isodose Volume (IV)}}{\text{Target Volume (TV)}}$$

$$\mathrm{CI}_2 = \mathrm{IPCI} = \mathrm{OTR} \times \mathrm{UTR} = \frac{(\mathrm{TV} \cap \mathrm{IV})^2}{\mathrm{TV} \times \mathrm{IV}}$$

Equation 2: IPCI; ideal value = 1.

With an ideal value = 1 or 100%, the popularity of this index has evolved in large part due to its ability to characterize both overtreated *and* undertreated regions (Figure 11.26).

It is clear from the equation that in order to calculate the IPCI we must first accurately determine the total amount of tissue (of any kind, target or healthy) that is within the isodose volume of interest. And, likewise, we must also accurately determine the target volume (TV). The challenge lies in the fact that the volumes that are to be calculated are typically quite small. For example, if 3-mm-square structure voxels are utilized, then a spherical version of a 1.5-cm TV would be covered by only 5 voxels across its widest point. The calculated volume for our spherical lesion would be ($V = 4/3\ \pi r^3 = 1767.1\ \mathrm{mm}^3$). But if margin for error in representing this structure is ±1 voxel on each side, then the volume could be misrepresented to be of diameter = 7 voxels = 21 mm and $V = 4849\ \mathrm{mm}^3$ for an error in volume representation of 174%. It is important then, if the CI is to be used as a metric of an acceptable stereotactic treatment plan, that

	60% IDL	89% IDL	95% IDL	100% IDL
	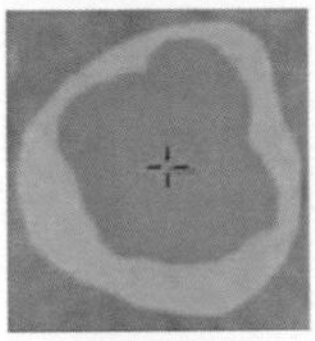	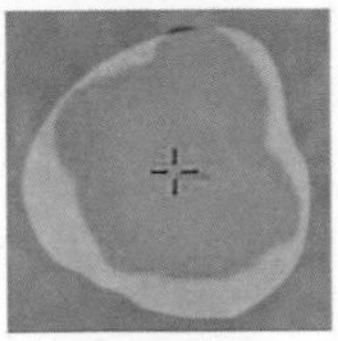	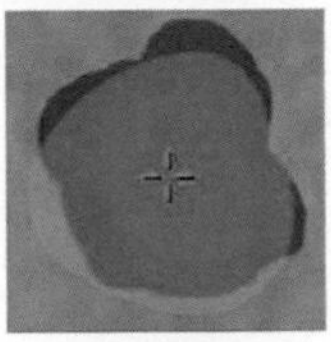	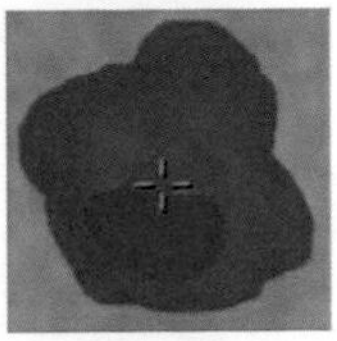
CI_1	4.08	1.64	1.15	1
CI_2	24.5%	60.8%	79.2%	10%

Figure 11.26 Results from calculating CI_1 and CI_2 for four different isodose line coverages. Target is in purple with iso-surfaces in green, yellow, orange and red (from left to right). Notice that as more and more of the target is covered by the isosurface of interest, the CI_1 continuously approaches an ideal value of 1 even though more and more healthy (nontarget) tissue is being irradiated. The CI_2 (IPCI), however, penalizes for the healthy tissue that is covered via the overtreatment component of the index. (Image courtesy of Brainlab.)

Table 11.1 Results of calculating the IPCI for a small brain lesion with the only difference being the dose voxel grid size

DVH GRID	1 MM	2 MM	3 MM	4 MM
TV_{PIV}	2.159	2.136	2.214	2.304
TV	2.264	2.216	2.403	2.432
Normal tissue	0.913	0.553	0.875	0.000
IPCI	0.670	0.766	0.660	0.947

Note: It is clear from the table that the use of dose voxels larger than 1 mm for this small lesion would lead to significant changes in the calculated CI. Note that the planning system's understanding of TV changes significantly, and its understanding of normal tissue irradiated ultimately goes to zero for 4 mm dose voxels.

small structures be represented within the system with high volumetric accuracy through the use of small voxel dimensions. Table 11.1 presents the results of calculating the IPCI for a small brain lesion with the only difference being the calculated dose voxel grid size. It is clear from the table that the use of dose voxels larger than 1 mm for this small lesion would lead to significant changes in the calculated CI.

The challenge to the use of very small voxel dimensions is, of course, in the amount of memory and processing power that must be present in such a system, and this is why, no doubt, not all treatment-planning systems include such fine structure representation. It also stands to reason that for dose volume histogram (DVH) and isodose distribution data for such treatments to be accurately represented, the structure *and* dose grid dimensions must also be sufficiently small.

Non-stereotactic treatment-planning systems often utilize 3-mm or even 5-mm-square dose grid voxels. This, again, is because of the extensive memory and processing requirements associated with such fine 3-D grid representations of the patient anatomy, and for many non-stereotactic treatment-planning situations, such voxel resolution can be perfectly acceptable. In an effort to improve on dose grid resolution while respecting the bounds of system resources, some planning systems may attempt to reduce the total number of dose grid voxels by placing a high resolution grid over the targeted region and then using a reduced resolution dose grid elsewhere in "normal" tissue (Figure 11.27). This approach can, in theory, work well but can be challenged by high-intensity "fingers" of dose that may occur for IMRS treatments in normal tissue (Figure 11.28) (see previous section for discussion). If the grid of dose calculation points is too coarse

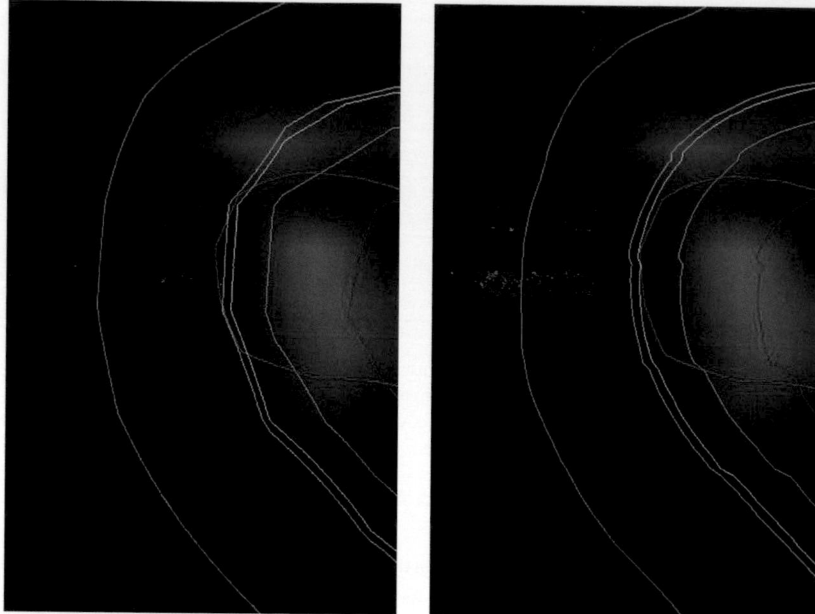

Figure 11.27 Differences in isodose distribution display when using fixed 2 mm dose calculation grid (left) versus an adaptively sized grid (right) as employed in the Brainlab treatment-planning system. Note the improved accuracy of isodose coverage representation at the purple OAR in the adaptive grid dose calculation. (Image courtesy of Brainlab.)

in normal tissue, these high-dose fingers may inadvertently pass between tissue dose calculation grid points, going undetected and, therefore, unpenalized by the inverse planning objective function.

It also stands to reason that in order to accurately characterize and represent the very steep dose gradients employed in stereotactic treatment approaches, dose grid voxels must be quite small. A typical 80%–20% gradient dimension for a stereotactic delivery modality is on the order of 3 mm or 20%/mm. This suggests that in order to accurately represent gradient differences of 10%/mm, we would need dose voxels of ~0.5 mm. Furthermore, the physically delivered beamlets employed in many stereotactic

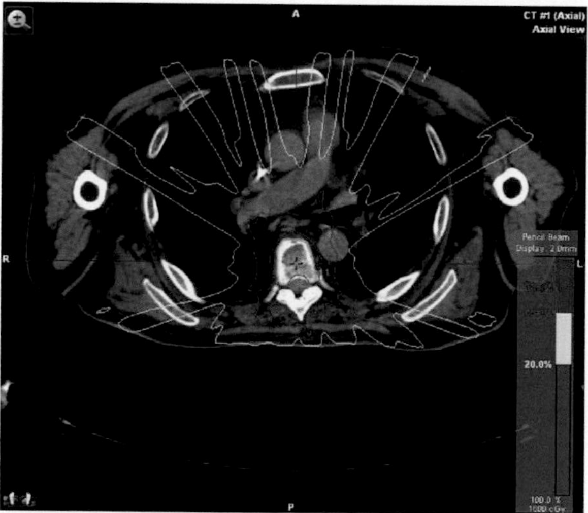

Figure 11.28 High-intensity "fingers" of dose extending into healthy tissue, thus necessitating a higher-density dose calculation grid in normal tissue to prevent under appreciation of dose to normal tissue. (Image courtesy of Bill J. Salter, University of Utah.)

delivery approaches are on the order of millimeters as opposed to the centimeter scale of conventional, fractionated radiation-therapy beams. Such small, physically delivered beams must be accurately modeled and, ultimately, represented by sufficiently small dose grid voxels. In any case, it is clear that an effective stereotactic treatment-planning system must provide for highly accurate characterization of small structure and dose volumes and small physically delivered beams through the use of small structure and dose grid voxels.

11.3.2 SMALL-FIELD DOSIMETRY

It is obvious that the small targets typically treated by stereotactic approaches require the use of very small treatment beams. Beams as small as 4–5 mm in diameter are not unusual in stereotactic treatments, and such small beams can pose nontrivial challenges to the dose calculation component of stereotactic treatment-planning systems. We have previously discussed the importance of the treatment-planning system having sufficiently small structure and dose grid voxels for accurately representing the steep gradients produced by stereotactic approaches, but it is equally important that the dose calculation algorithm employed by the treatment-planning system be capable of accurately modeling small-field dose distributions. Typical, non-stereotactic treatment-planning systems normally require the collection of beam data for validation and modeling of the dose calculation accuracy down to approximately 3-cm-square beams. This is not surprising because non-stereotactic treatment of targets smaller than this is very unusual in so-called "conventional" radiotherapy. Arguably, the lion's share of lesions treated by radiosurgical approaches occurs at, or even below, this lower limit threshold for conventional radiotherapy. Thus, it is imperative that a stereotactic treatment-planning system utilize methods for commissioning and validating dose calculation performance that are proven to be highly accurate for very small treatment fields.

One example of "tailoring" the dose-calculation algorithm to stereotactic applications can occur as the manufacturer optimizes the accuracy performance of the algorithm. Manufacturers of treatment-planning systems intended to treat large conventional fields up to 40 cm in size will be required to optimize performance across the range of field sizes up to 40 cm. Manufacturers of stereotactic-focused systems, on the other hand, have the luxury of fine-tuning the performance of their dose-calculation engine across the smaller range of stereotactic field sizes (e.g., ≤10 cm). One clear indication that a treatment-planning system is specifically designed for treatment of small stereotactic targets will be the requirement to measure beam parameters for very small fields during the commissioning process. Most manufacturers of dedicated stereotactic treatment-planning systems will also provide guidance for the medical physicist regarding what types and resolution of measurement devices to use. Very small diode or ionization detectors, possibly coupled with film measurement of cross profiles, are standard operating procedure for such situations. Additionally, some manufacturers may also require the measurement of small field sizes, using multiple combinations of MLC leaf and secondary jaw configurations in order to better model the small-field dosimetry. Failure of the manufacturer to pay appropriate attention to such details will manifest as nontrivial discrepancies between treatment planning–system predicted versus measured dose distributions for very small treatment targets (Murphy et al. 2003; Solberg et al. 2008; Wurm et al. 2008).

11.3.3 IMAGE REGISTRATION AND STEREOTACTIC FRAME OF REFERENCE

The ability to perform accurate image registration is important for all modern treatment-planning systems, but it is important to remember that in the stereotactic planning environment, lesions can be as small as a few millimeters, dose gradients are extremely steep, and GTV/CTV to PTV margins are often zero.

This, coupled with the fact that lesions are routinely delineated only on the MRI (i.e., secondary) data set, makes clear the fact that image registration routines must be as accurate as possible in the stereotactic planning environment (Figure 11.29). Highly accurate image registration can be facilitated by the identification of a precisely defined coordinate reference frame, and whether or not the frame is invasively attached (Figure 11.30) (Hitchcock et al. 1989; Phillips et al. 1990; Saunders et al. 1988) or a so-called frameless approach (see images next page) (Murphy et al. 2003; Solberg et al. 2008; Wurm et al. 2008), an accurately defined coordinate system must still be represented and accepted by the treatment-planning system. It is very important to remember that even an accurately implemented image fusion will still introduce some degree of error into the planning and delivery process, if only due to the difference

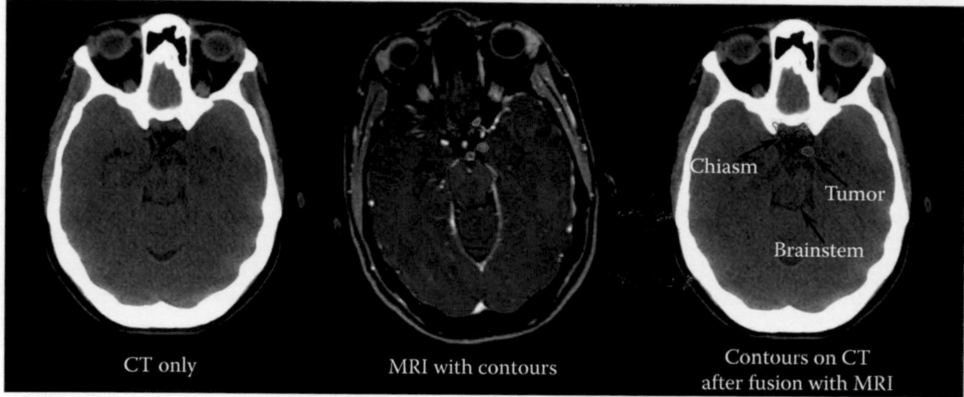

Figure 11.29 Improved visibility of small structures on MRI and why such structures are often delineated only on the MRI image set. In such situations, the accuracy of the CT to MRI fusion is extremely important to accurate targeting. (Image courtesy of Bill J. Salter, University of Utah.)

(a)

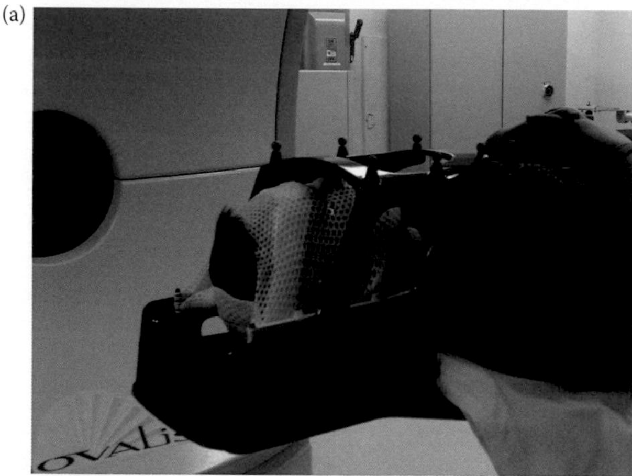

(b)

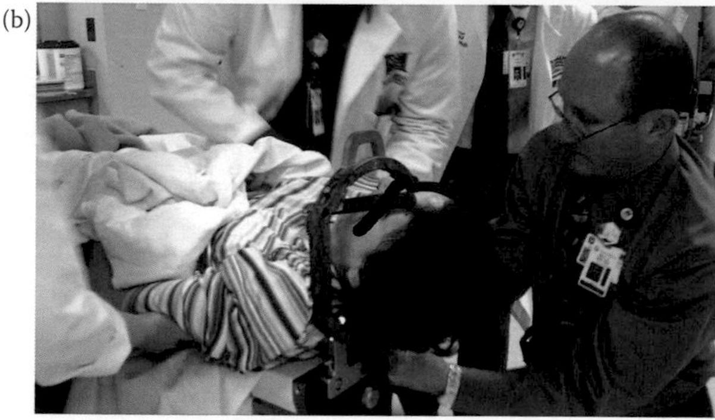

Figure 11.30 Patient setup prior to undergoing SRS with (a) frameless approach and (b) invasive halo frame. (Image courtesy of Bill J. Salter, University of Utah.)

in discretization between the two registered data sets. Any error associated with less than perfect user-approved alignment of the two data sets will only add to the total fusion-related error (Figure 11.31).

Because such a high percentage of cases are planned based on fused data sets and because the consequences of small spatial errors are amplified in the stereotactic environment, the image registration algorithm of any stereotactic treatment planning system should be carefully tested for accuracy prior to

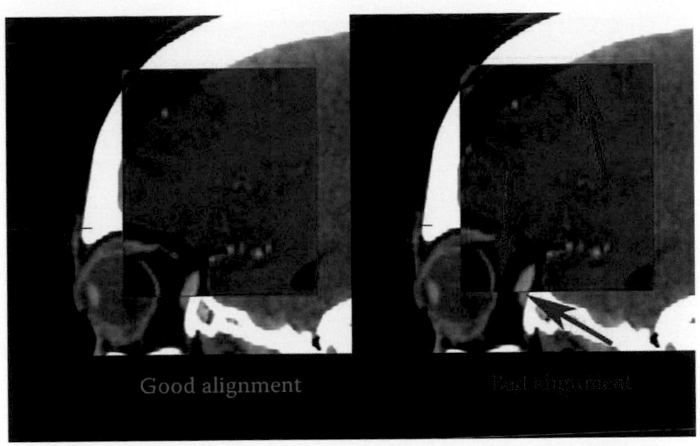

Figure 11.31 Good and bad CT to MRI fusion. Small "spyglass" window shows data from MRI, and data outside the spyglass window shows CT data. Note how globe of the eye, and the brain to bone interface is disjointed in the bad alignment (see arrows). (Image courtesy of Bill J. Salter, University of Utah.)

use on patients and, ideally, the measured accuracy of image registration at one's own institution will be factored into margin design at time of treatment planning.

11.3.4 STEREOTACTIC BODY RADIATION THERAPY: DOSE CALCULATION ACCURACY

As mentioned in a previous section of this chapter, SBRT treatment of lung, liver, and spine targets have recently evolved to represent capable and frequently employed treatment techniques. Because such targets can sometimes be very small, it is very important that the previously discussed considerations regarding small-field dosimetry and targets be heeded. Additionally, however, it is important to recognize that many such treatments may entail delivery of dosages to nontrivial amounts of lung tissue (Figure 11.32), and it in such cases, it is important to recognize the deficiencies of pencil-beam dose-calculation algorithms in such heterogeneous density environments.

AAPM Task Group 101 has stated that use of pencil beam–based dose-calculation algorithms are not recommended for lung SBRT applications because significant discrepancies between predicted versus delivered dose distributions may result (Benedict et al. 2010). Convolution- or, better still, Monte

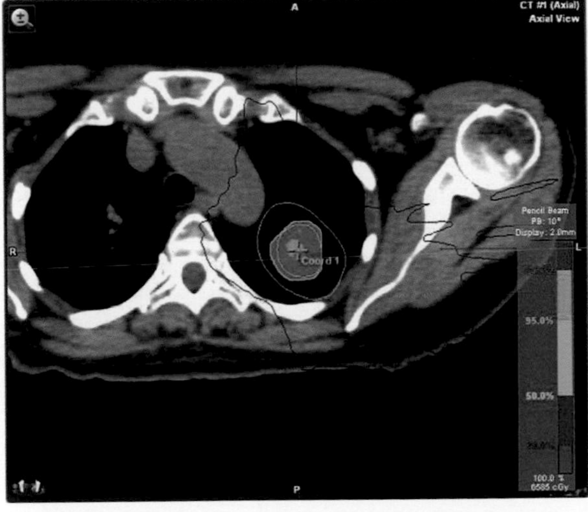

Figure 11.32 Axial isodose distribution for SBRT of lung case and nontrivial volume of lung irradiated to a relatively high dose level. Prescribed dose = 54 Gy; V_{20} = 4%; V_5 = 12%. (Image courtesy of Bill J. Salter, University of Utah.)

(a)

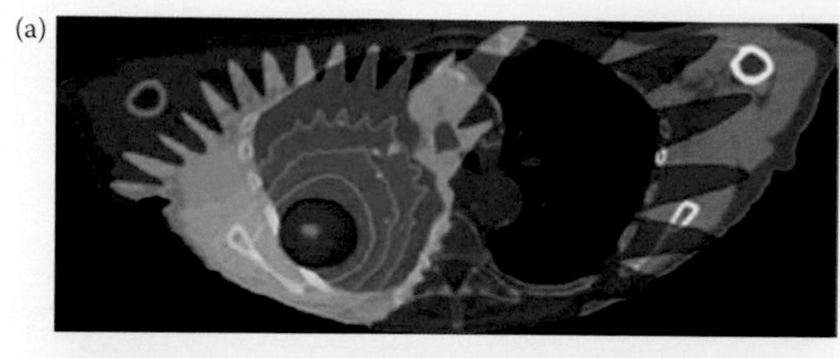

(b)

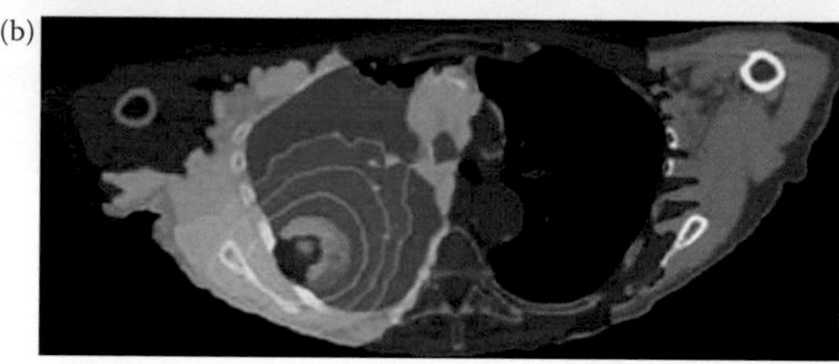

Figure 11.33 Pencil beam–calculated isodose distribution (a) versus Monte Carlo isodose distribution (b). Note the significant overestimation of red 100% isovolume region by the pencil beam calculation, relative to the more accurate prediction by Monte Carlo dose calculation. (Images courtesy of Brainlab, Inc.)

Carlo–based dose calculations are considered ideal. Figure 11.33 demonstrates the significant difference between an inaccurate pencil beam dose calculation in the lung (a) versus a highly accurate Monte Carlo dose calculation (b).

It is also important to recognize that because SBRT treatment plans frequently employ many fields, delivered from many different non-coplanar directions, it is imperative that a sufficiently large volume of the patient be scanned by CT so as to facilitate calculation of dosages at all regions of the patient through which beams may be delivered. Last, with regard to SBRT treatment, it is also important to recognize that due to the recent evolution of SBRT as a treatment modality, dose and fractionation schemes are still evolving. AAPM SBRT Task Group 101 has recommended that SBRT patients be planned according to multi-institutional protocol standards whenever feasible.

11.3.5 MARGIN DEFINITION

As mentioned previously, it is not unusual in intracranial stereotactic radiosurgery applications for a CTV-to-PTV margin of zero to be utilized. This is possibly due to historical reasons and the evolution of the Gamma Knife being initiated by a neurosurgeon, but also likely due to the early use of highly accurate invasive head frames, which have been reported to have near zero error introduction in phantom. Accuracy of invasive forms of head frames in patient treatment is likely closer to 1 mm, but, in any case, the high degree of accuracy afforded by use of such immobilization and reference frames is widely acknowledged. This is evidenced, for example, by the fact that in this era of widely employed mask-based frameless immobilization approaches, invasive head rings continue to be the immobilization method of choice of many clinicians for treatment of trigeminal neuralgia treatments, in which extreme precision of delivery is required for typical treatment of the root entry zone of the fifth cranial nerve immediately adjacent to the brain stem. More recently, image guidance–based, frameless approaches have evolved to deliver accuracy that closely approaches that of invasive head frame approaches (Solberg et al. 2008). Regardless, however, of which immobilization and stereotactic targeting approach is utilized, it is important that treatment planning margins for treatment delivery be assigned based on careful consideration of the accuracy of the immobilization methodology being used and, in the case of SBRT treatment, in which nontrivial target motion may occur, that the motion-management capability of the system being employed be carefully considered. Furthermore, it is important that the accuracy of the immobilization system be considered not only based on reported literature, but based on the measured accuracy of implementation at one's own institution. Additionally, as mentioned in a previous section, it is important that the accuracy of other planning-related processes, such image registration, be quantified at one's own institution and that this information be included in margin-definition decisions at the time of treatment planning.

11.4 CONCLUSION

Stereotactic delivery approaches have continued to evolve over the past several decades, and delivery techniques can now allow for unprecedented conformity of delivered dose distributions. The process of developing a plan of attack on a targeted lesion entails the thoughtful, intelligent use of a treatment-planning system specifically designed for the task. Dedicated stereotactic treatment-planning systems incorporate carefully considered features that are specifically designed to accurately predict the dose delivered to the small, conformally treated lesions routinely encountered in stereotactic treatment. While it is essential that a bona fide stereotactic planning system be employed for calculation of dose distributions for very small stereotactically targeted lesions, it is equally important that the subtle nuances of the specific delivery technique be factored into the treatment design. While myriad effective and capable treatment delivery options currently exist, it is imperative that both the strengths and limitations of the specific delivery modality be carefully considered when designing a stereotactic treatment plan. When all relevant factors are carefully considered, the potential for successful treatment of stereotactically targeted inter- and intracranial lesions has never been greater than it is today.

REFERENCES

Andrews, D. W., G. Bednarz, J. J. Evans, et al. 2006. A review of 3 current radiosurgery systems. *Surg. Neurol.* 66(6): 559–64.

Benedict, S. H., K. M. Yenice, D. Followill, et al. 2010. Stereotactic body radiation therapy: The report of AAPM Task Group 101. *Med. Phys.* 37(8): 4078–101.

Clark, G. M., R. A. Popple, P. E. Young, and J. B. Fiveash. 2010. Feasibility of single-isocenter volumetric modulated arc radiosurgery for treatment of multiple brain metastases. *Int. J. Radiat. Oncol. Biol. Phys.* 76(1): 296–302.

Dvorak, P., D. Georg, J. Bogner, B. Kroupa, K. Dieckmann, and R. Pötter. 2005. Impact of IMRT and leaf width on stereotactic body radiotherapy of liver and lung lesions. *Int. J. Radiat. Oncol. Biol. Phys.* 61(5): 1572–81.

Hitchcock, E., G. Kitchen, E. Dalton, and B. Pope. 1989. Stereotactic linac radiosurgery. *Br. J. Neurosurg.* 3(3): 305–12.

Hoppe, B. S., B. Laser, A. V. Kowalski, et al. 2008. Acute skin toxicity following stereotactic body radiation therapy for stage I non-small-cell lung cancer: Who's at risk? *Int. J. Radiat. Oncol. Biol. Phys.* 72(5): 1283–6.

Jin, J. Y., F. F. Yin, S. Ryu, M. Ajlouni, and J. H. Kim. 2005. Dosimetric study using different leaf-width MLCs for treatment planning of dynamic conformal arcs and intensity-modulated radiosurgery. *Med. Phys.* 32(2): 405–11.

Kooy, H. M., L. A. Nedzi, J. S. Loeffler, et al. 1991. Treatment planning for stereotactic radiosurgery of intracranial lesions. *Int. J. Radiat. Oncol. Biol. Phys.* 21(3): 683–93.

Lee, N., C. Chuang, J. M. Quivey, et al. 2002. Skin toxicity due to intensity-modulated radiotherapy for head-and-neck carcinoma. *Int. J. Radiat. Oncol. Biol. Phys.* 53(3): 630–7.

Li, K., and L. Ma. 2005. A constrained tracking algorithm to optimize plug patterns in multiple isocenter Gamma Knife radiosurgery planning. *Med. Phys.* 32(10): 3132–5.

Lindquist, C., and I. Paddick. 2007. The Leksell Gamma Knife Perfexion and comparisons with its predecessors. *Neurosurgery* 61(3 Suppl.): 130–40; discussion 140–1.

Liu, R., J. M. Buatti, T. L. Howes, J. Dill, J. M. Modrick, and S. L. Meeks. 2006. Optimal number of beams for stereotactic body radiotherapy of lung and liver lesions. *Int. J. Radiat. Oncol. Biol. Phys.* 66(3): 906–12.

Ma, L., D. Larson, P. Petti, C. Chuang, and L. Verhey. 2007. Boosting central target dose by optimizing embedded dose hot spots for Gamma Knife radiosurgery. *Stereotact. Funct. Neurosurg.* 85(6): 259–63.

Maciunas, R. J., R. L. Galloway Jr., and J. W. Latimer. 1994. The application accuracy of stereotactic frames. *Neurosurgery* 35(4): 682–94; discussion 694–5.

Monk, J. E., J. R. Perks, D. Doughty, and P. N. Plowman. 2003. Comparison of a micro-multileaf collimator with a 5-mm-leaf-width collimator for intracranial stereotactic radiotherapy. *Int. J. Radiat. Oncol. Biol. Phys.* 57(5): 1443–9.

Murphy, M. J., S. D. Chang, I. C. Gibbs, et al. 2003. Patterns of patient movement during frameless image-guided radiosurgery. *Int. J. Radiat. Oncol. Biol. Phys.* 55(5): 1400–8.

Nakamura, J. L., A. Pirzkall, M. P. Carol, et al. 2003. Comparison of intensity-modulated radiosurgery with Gamma Knife radiosurgery for challenging skull base lesions. *Int. J. Radiat. Oncol. Biol. Phys.* 55(1): 99–109.

Paddick, I. 2000. A simple scoring ratio to index the conformity of radiosurgical treatment plans. Technical note. *J. Neurosurg.* 93(Suppl. 3): 219–22.

Phillips, M. H., K. A. Frankel, J. T. Lyman, et al. 1990. Comparison of different radiation types and irradiation geometries in stereotactic radiosurgery. *Int. J. Radiat. Oncol. Biol. Phys.* 18(1): 211–20.

Rosenthal, D. I., M. S. Chambers, C. D. Fuller, et al. 2008. Beam path toxicities to non-target structures during intensity-modulated radiation therapy for head and neck cancer. *Int. J. Radiat. Oncol. Biol. Phys.* 72(3): 747–55.

Salter, B. J., M. Fuss, V. Sarkar, et al. 2009. Optimization of isocenter location for intensity modulated stereotactic treatment of small intracranial targets. *Int. J. Radiat. Oncol. Biol. Phys.* 73(2): 546–55.

Saunders, W. M., K. R. Winston, R. L. Siddon, et al. 1988. Radiosurgery for arteriovenous malformations of the brain using a standard linear accelerator: Rationale and technique. *Int. J. Radiat. Oncol. Biol. Phys.* 15(2): 441–7.

Schell, M. C., F. Bova, D. Larson, et al. 1995. Stereotactic radiosurgery: The report of AAPM Task Group 42. AAPM Report No. 54, p. 88. American Institute of Physics.

Solberg, T. D., P. M. Medin, J. Mullins, and S. Li. 2008. Quality assurance of immobilization and target localization systems for frameless stereotactic cranial and extracranial hypofractionated radiotherapy. *Int. J. Radiat. Oncol. Biol. Phys.* 71(1 Suppl.): S131–5.

Tobler, M., D. D. Leavitt, and G. Watson. 2004. Optimization of the primary collimator settings for fractionated IMRT stereotactic radiotherapy. *Med. Dosim.* 29(2): 72–9.

Wu, Q. J., Z. Wang, J. P. Kirkpatrick, et al. 2009. Impact of collimator leaf width and treatment technique on stereotactic radiosurgery and radiotherapy plans for intra- and extracranial lesions. *Radiat. Oncol.* 4: 3.

Wurm, R. E., S. Erbel, I. Schwenkert, et al. 2008. Novalis frameless image-guided noninvasive radiosurgery: Initial experience. *Neurosurgery* 62(5 Suppl.): A11–7; discussion A17–8.

Yu, C., and D. Shepard. 2003. Treatment planning for stereotactic radiosurgery with photon beams. *Technol. Cancer Res. Treat.* 2(2): 93–104.

12 Small-field dosimetry for stereotactic radiosurgery and radiotherapy

Kamil M. Yenice, Yevgeniy Vinogradskiy, Moyed Miften, Sonja Dieterich, and Indra J. Das

Contents

12.1 INTRODUCTION

Stereotactic radiosurgery (SRS), a method for high-dose irradiation of cranial tumors in a single fraction, has become a standard of care in the treatment of brain tumors, vascular malformations, functional disorders, and pain since its inception in the fifties (Leksell 1951). Modern radiosurgery can be performed noninvasively and on an outpatient basis yet with an extremely high degree of accuracy. Within the past 10 years, the field of radiosurgery has seen numerous technological enhancements, including the development of dedicated devices for stereotactic delivery, the use of relocatable frames to facilitate fractionated delivery, the development of image-guided and "frameless" approaches, and the application to extracranial tumor sites (Benedict et al. 2008). Each of these developments is accompanied by its own challenges in assuring targeting and dosimetric accuracy. These technical developments have permitted the scope of what is regarded as SRS to expand up to five fractions (Barnett et al. 2007).

SRS involves treatment of small, well-defined targets up to approximately 4 cm in diameter and delivers a much higher dose than in conventional fractionated radiation therapy, which results in a high biologically effective dose. The dose is typically prescribed to the 50%–85% isodose line covering at least 95% of the target, resulting in a pronounced central dose "hot spot" of 25%–50%. Precise target definition and target localization is, therefore, very important to assure the safety and accuracy of the radiosurgery procedure. Early development of SRS generally focused on localization accuracy of a hidden target (Lutz, Winston, and Maleki 1988) to <1 mm. For larger lesions, stereotactic radiotherapy (SRT) uses the accuracy of stereotactic localization and precision of dose delivery in a fractionated approach to take advantage of the biological differences between the normal brain tissue and tumor response. Stereotactic body radiotherapy (SBRT) was the next logical extension to treat small targets in the body by employing advances from radiosurgery, intensity modulated radiation therapy (IMRT), and imaging. These new elements allow precise treatment of extracranial targets to high fractional doses even if the target is moving. Although radiosurgery can effectively be delivered by charged-particle therapy, such as protons and carbon ions, photons are by far the most common ionizing radiation modality used for this purpose. We therefore focus our attention in this chapter on small-field dosimetry issues related to megavoltage photon beams used in SRS, SRT, or SBRT.

All stereotactic applications are characterized by the use of small radiation fields which may cause nonequilibrium conditions for electron transport. Small beams employed by various devices capable of delivering radiosurgery are created by specialized collimator assemblies in the form of precisely machined circular collimators or multileaf-collimator (MLC) systems. The older Gamma Knife models (U, B, and C) use 201 individual Co-60 sources, which are collimated by an internal (primary) collimator and a removable external semispherical assembly called a helmet, which has circular apertures of 4, 8, 14, and 18 mm. The collimating system focuses the individual beams of gamma radiation to a very precise focal point (the unit center point), and the superposition of all the beams at the focus delivers a tightly conformal, high-dose volume to the target with a rapid dose falloff away from the target edge. The new Gamma Knife Perfexion model has only one internal collimating system with 192 Co-60 sources distributed over eight movable sectors, and the final beam collimation is achieved with the appropriate alignment of sources over 4-, 8-, and 16-mm apertures to deliver focused radiation to the target.

In addition to the original set of 12 fixed collimators of 5–60 mm diameter, CyberKnife also comes with a variable-aperture collimator (Echner et al. 2009) (IRIS) to create circular apertures for beam delivery. More recently, a MLC for CyberKnife Model M6 has been released, and it consists of 41 leaf pairs of 2.5-mm thickness at 80 cm SAD and maximum nominal field size of 10 × 12 cm.

Linear accelerator–based systems, on the other hand, generate the small beams necessary for radiosurgery by either tertiary collimating systems using circular cone attachments or small leaf-width (e.g., "micro") MLC systems. Conventional linacs with fine MLC leafs of 4–2.5 mm can further offer conformal beam shaping, dynamic beam delivery (changing of beam aperture according to target shape within an arc), or intensity modulation for computer-optimized treatment plan delivery. Unlike the stationary beam arrangement of the Gamma Knife, conventional linac systems rely on positioning of the gantry and table to achieve non-coplanar beam angles and typically use a smaller number of beams (in the order of tens) or one to two volumetric modulated arcs. Both the Gamma Knife and CyberKnife usually deliver on the order of hundreds of beams per treated target.

The International Commission on Radiation Units and Measurements Report 24 set a standard, still followed today, that the delivered dose for radiation treatments should be within ±5% of the prescribed dose (International Commission on Radiation Units and Measurements 1976). This puts stringent requirements on the measurement of basic physical parameters of stationary beams collimated for radiosurgery as well as the overall dose distributions calculated by the 3-D dose calculation algorithms. The basic parameters for the 3-D treatment-planning software are essentially the same as those used in the characterization of conventional large-field radiotherapy beams: relative output factors, percentage depth doses (PDD) or tissue phantom ratios (TPR), and off-axis ratios (OAR) or dose profiles. However, as the field dimensions become small, various effects, including lack of electronic equilibrium, limited beam focal spot point of view, spectral changes within the field, and detector perturbations, confound the choice of detector and significantly affect the measurement accuracy of these parameters. Differences in the output factor measurements for collimated radiosurgery beams of diameters less than 2 cm have been reported to be up to 12% by Das et al. (2000). Inappropriate selection of a detector in the commissioning of small-field output factors resulted in mistreatment of 145 patients in Toulouse, France, in 2006–2007 (Derreumaux et al. 2008) and 152 patients in Springfield, Missouri, from 2004 to 2009 (Solberg and Medin 2011). These incidents have indicated the continuing challenges in the determination of small-field parameters for high-energy photon beams used in radiosurgery delivery. As the radiosurgery technology evolves to precisely deliver even smaller and higher-dose fields to targets, the need for more explicit guidelines in assessing the accuracy of experimental data and their modeling in the clinical environment becomes more significant.

12.2 SMALL-FIELD DOSIMETRY PROBLEM

Small-field dosimetry is challenging, both in terms of measurements of dosimetric data and modeling of small fields within the treatment-planning system. Both improper measurement and improper modeling techniques can introduce significant errors in stereotactic treatment delivery. The relative output factors, PDDs or TPRs, and OARs in traditional fields of 4×4 cm^2 to 40×40 cm^2 can be easily measured and quantified. However, for small fields, the detectors used in measuring radiation beams start violating the requirements of the Bragg-Gray cavity theory (Attix 1986). A quantitative definition of a small field is not simple and depends on the beam energy, the focal spot of the x-ray source, the collimating system, and the density of the medium. The American Association of Physicists in Medicine (AAPM) TG-106 (Das et al. 2008) defines a field size of $<4 \times 4$ cm^2 as "small." More specifically, there are three "equilibrium conditions" that determine the scale of a radiation field to be considered as small: (i) the size of viewable parts of the beam focal spot as projected from the detector location through the beam aperture, (ii) the secondary electron range in the irradiated medium, and (iii) the size of the detector needed for accurate measurements. These factors are discussed below and are elaborated in the Institute of Physics and Engineering in Medicine (IPEM) Report 103 (Aspradakis 2010) and the AAPM Task Group-155 report (Das et al. 2014).

12.2.1 EFFECTS OF THE RADIATION SOURCE SIZE

When collimating a beam from a source of finite width to a very small size, only a part of the source area can be viewed from the detector's point of view (Figure 12.1). The field size at which focal spot occlusion starts to occur is extremely variable by machine type as described by Jaffray et al. (1993). Occlusion of the focal spot results in lower output than the output for field sizes at which the entire source can be viewed from the detector's field of view (Munro, Rawlinson, and Fenster 1988; Sharpe et al. 1995; Zhu and Bjarngard 1994; Zhu, Bjarngard, and Shackford 1995; Zhu and Manbeck 1994; Zhu et al. 2000). It has been demonstrated by Ding, Duggan, and Coffey (2006) that the beam output (planar fluence profile) can be significantly influenced by the collimation geometry used to achieve the small field sizes. Once the field size becomes so small that the entire source cannot be viewed from the center of the field, the geometrical penumbra is extended all over the field cross-section (Sharpe et al. 1995; Zhu and Bjarngard 1994; Zhu and Manbeck 1994), yielding a blurred and widened profile (Figure 12.2).

12.2.2 ELECTRON RANGE AND LOSS OF CHARGED-PARTICLE EQUILIBRIUM

Radiation dose from a photon beam is delivered by secondary electrons that are produced by photon interactions with tissue, including photoelectric, Compton, and pair-production interactions. These electrons have a finite range in which they deposit energy as they travel through the medium. Typically, the range of these particles is the depth of maximum dose (d_{max}) in the forward direction for photon beams (Khan 2012). For large fields, the lateral equilibrium in the central portion of the field is maintained, giving a uniform dose; however, close to the beam edge, there is no equilibrium, and the dose is reduced. When the field size is sufficiently small, the maximum range of secondary electrons becomes greater than the beam radius, thus causing a loss of charged-particle equilibrium (CPE). Therefore, the lateral range of the electrons is, in general, a more significant factor than the forward range of the electrons for the establishment of CPE for a given field dimension (Li et al. 1995).

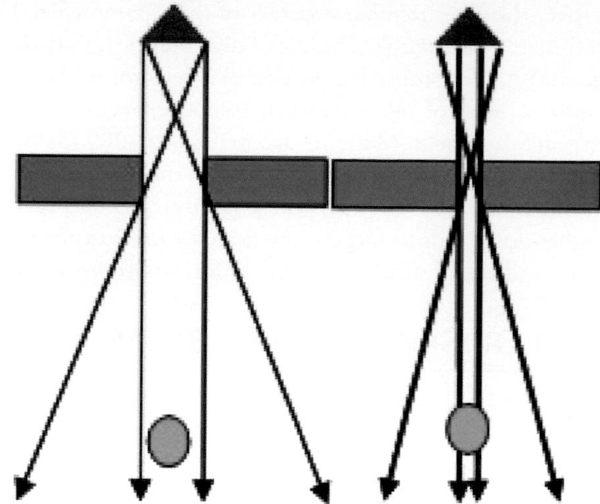

Figure 12.1 Schematic diagram of the effect of collimation setting as viewed by a detector: At very small collimator settings, the direct beam source as seen from the position of measurement (detector) is occluded by the collimating device.

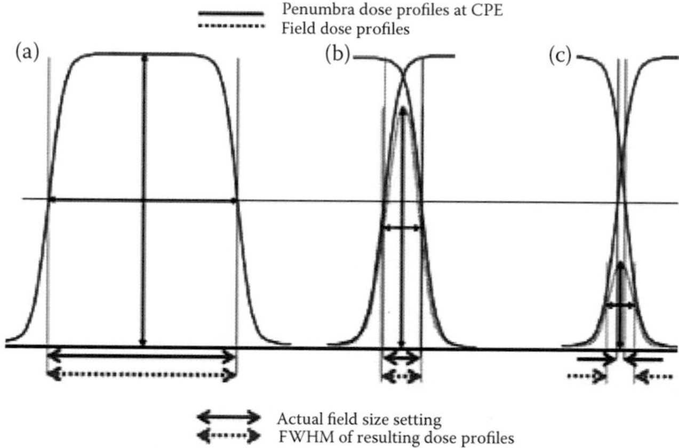

Figure 12.2 Diagram illustrating overlapping penumbrae leading to decreased output with decreasing field size. The concept of using full width at half maximum (FWHM) to define field size that is used for conventional beams (a) breaks down as the field gets smaller (b and c). (From Das, I. J. et al., *Med. Phys.*, 35, 206–15, 2008. With permission.)

Because the maximum range of secondary electrons is dependent on beam energy and the composition of the interaction material, the charged-particle disequilibrium is exacerbated in higher beam energies and by the presence of tissue heterogeneities in the treatment sites. Li et al. (1995) described the lateral range of electrons under different conditions. Studying primary dose profiles in water across a collimating edge provides information for penumbra ranges in unit density media that set the dimensions of when small field conditions apply based on overlapping electron distribution zones from different field edges (Nyholm et al. 2006). Figure 12.3 shows the primary dose profiles in water across a collimating edge for different beam energies, specified by their quality index defined as the ratio of tissue phantom ratios at 20 and 10 cm depth: ($TPR_{20/10}$). (AAPM Task Group 21, 1983; IAEA Report No. 398, 2000).

12.2.3 DETECTOR SIZE

Issues relating to the size, composition, and design of the detector arise primarily due to the finite size of the detectors used in acquiring beam data. Partial volume effects occur because the signal is averaged

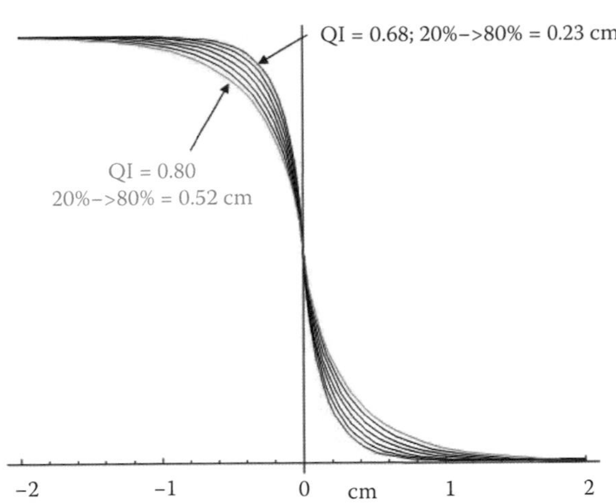

Figure 12.3 Electron transport giving rise to the dose spread in water across a collimated edge as a function of beam energy, specified by their quality index ($TPR_{20/10}$). (From Das, I. J. et al., *Med. Phys.*, 35, 206–15, 2008. With permission.)

over the finite detector active volume. Furthermore, because field sizes are relatively small, the detector itself can perturb the beam in an unpredictable manner (Ding et al. 2007). When a finite-sized detector is used to measure dose in a photon beam, the photon fluence is perturbed if the detector is not perfectly tissue equivalent. This perturbation can be interpreted as a departure from the Bragg-Gray cavity or ideal detector behavior. When the size of the cavity is smaller than the range of charged particles originated in the surrounding medium, the cavity is treated as nonperturbing, and the dose to the medium is simply related to the dose to the detector by the stopping-power ratios of medium to detector material. However, with decreasing field size, neither the CPE nor the concept of cavity theory holds well due to the changes in lateral scatter conditions. The disturbance of particle fluence becomes the major source of detector perturbation. This effect is difficult to quantify through the use of standard correction methods, such as the replacement correction factor in small fields, because the perturbation is highly dependent on the detector geometry and medium properties (Das, Ding, and Ahnesjo 2008). To address the above challenges, comprehensive reports have been compiled aimed at discussing and making recommendations regarding small-field dosimetric measurements (Alfonso et al. 2008; Aspradakis 2010; Das, Ding, and Ahnesjo 2008). A detailed discussion on the perturbation due to high-density detectors (solid state) being proposed for use in small-field dosimetry is elaborated by Fenwick et al. (2013).

12.3 DETECTORS FOR STEREOTACTIC RADIOSURGERY

Perturbation of the beam, loss of CPE, and partial volume effects greatly complicate the choice of a detector when making dosimetric measurements for small fields. IPEM Report 103 (Aspradakis 2010) and the AAPM reports TG-120 (Low et al. 2011) and TG-155 (TG-155 is not published yet. The report is under review by *Medical Physics*.) (Das et al. 2014) each compiled a comprehensive list of radiation detectors and their advantages and disadvantages with regard to small-field dosimetric measurements. Desirable detector characteristics include excellent spatial resolution, a linear response with dose and dose rate, tissue equivalence, stable short-term readings, an isotropic response, low energy dependency and minimal background signal. Particular emphasis should be placed on using a detector that has a small active volume and high spatial resolution when making small field measurements. A wide range of detector types have been used for small field measurements, including ion chambers, diodes, solid-state detectors, thermoluminescent dosimeters (TLD), film, and gel dosimeters.

The unshielded stereotactic diode, micro-ionization chambers; bis acrylamide nitrogen gelatin (BANG) polymer gels; and radiochromic film are considered the most dependable detectors to use in small field measurements. Correction factors are always required to account for detector perturbation as discussed above. Because no detector possesses the ideal combination of characteristics, measurements should be cross-referenced between at least two different detector types and compared against values published in the literature when possible (Dieterich et al. 2011).

12.3.1 IONIZATION CHAMBERS

Ionization chambers are the most widely used type of dosimeter in radiation therapy. They provide reproducible measurements and an absorbed dose-rate estimate traceable to primary national standards. Ionization chambers generally have high sensitivity and are nearly energy, dose, and dose-rate independent (Das, Ding, and Ahnesjo 2008). However, typical ionization chambers are not suitable for small-field dosimetry because of their large active volume. Microionization chambers with active volumes ranging from 2 to 7 mm^3 have been reported as used for small field measurements. Martens, De Wagter, and De Neve (2000) investigated the value of the PinPoint ion chamber for characterization of small field segments used in IMRT. Laub and Wong (2003) studied the volume effect of detectors in the dosimetry of small fields used in IMRT. Stasi et al. (2004) studied the behavior of several microionization chambers in small IMRT fields. Rice et al. (1987) measured dose distributions and output factors in 6-MV small beams. Bjarngard, Tsai, and Rice (1990) characterized the central axis doses in narrow 6-MV x-ray beams. Le Roy et al. (2011) assessed 24 ionization chambers for their performance as reference dosimeters in high-energy photon beams.

Generally, ionization chambers have a high sensitivity compared to other detectors; however, microionization chambers have decreased sensitivity because of their smaller active volumes. Therefore, leakage can be significant if not corrected for, especially in low-dose portions of the beam (Leybovich, Sethi, and Dogan 2003). In addition, stem and cable irradiation can provide a spurious signal (Le Roy et al. 2011; Lee et al. 2002). An additional difficulty for microionization chambers is that the displacement to determine the effective point of measurement is not well known for small fields. The rule of moving the chamber 0.6 times the chamber radius only applies to chambers with larger active volumes. Several publications have noted that the effective point of measurement can depend on the dimension of the detector as well as field size (Kawrakow 2006; Looe, Harder, and Poppe 2011). Some microionization chambers have a metal central electrode, which may affect the response of the detector in a radiation field, especially when the photon spectra vary under narrow field conditions (Ma and Nahum 1993). Accredited Dosimetry Calibration Laboratories advise caution when attempting to use a microionization chamber as a primary calibration standard due to their lack of long-term stability. Comparison with other detectors and modeling the ion chamber response for different energy spectra using Monte Carlo modeling may be useful in this case (Ding, Duggan, and Coffey 2006).

12.3.2 SOLID-STATE DETECTORS

Diodes are solid-state detectors that are an attractive option for small-field dosimetry because of their high sensitivity and inherently small active volumes (diameters of 0.6–1.13 mm, 0.01–0.06 mm thickness) as well as their enhanced response per unit volume due to both their density (1000 times that of air) and their lower W/e values (Dieterich and Sherouse 2011; Rikner and Grusell 1983). Multiple groups have reported on the use of diodes for small-field dosimetry (Beddar, Mason, and Obrien 1994; Cheng et al. 2007; Gotoh et al. 1996; Mack et al. 2002; McKerracher and Thwaites 1999). There are two primary types of diodes used for small field measurements: unshielded and shielded. Shielded diodes contain a layer of high atomic material, usually tungsten, behind the active volume to account for silicon over-response to low-energy photons (Cranmer-Sargison et al. 2012). The literature is inconclusive regarding the use of shielded diodes for small-field dosimetry. Several studies suggest that the added tungsten introduces a directional dependence and additional scatter (Mobit and Sandison 2002) for electron beams, and other publications conclude that shielded diodes require only minor corrections for small field measurements (Saini and Zhu 2004).

Diodes have inherent characteristics that can limit their use in small field measurements if not taken into account. Variations in measurement of up to 10% have been noted with accumulated dose (Eveling, Morgan, and Pitchford 1999). The composition of diodes is not water equivalent and can introduce a perturbation of the incident beam (Mobit and Sandison 2002). Finally, diodes exhibit an energy dependence that can be addressed with energy-based correction factors (Eklund and Ahnesjo 2009; Francescon, Cora, and Cavedon 2008).

Diamond detectors are intriguing detectors due to their small active volumes, high sensitivity, and near tissue equivalence. Their use in small-field dosimetry has been experimentally evaluated (De Angelis et al. 2002; Laub, Kaulich, and Nusslin 1999; Rustgi 1995). Diamond detectors exhibit a dose-rate dependence that can be addressed with correction factors (Laub, Kaulich, and Nusslin 1999). The main limitation for diamond detectors is that they are harder to manufacture, making them more expensive than other dosimeters. Recently, synthetic diamonds have replaced natural diamonds, making them more affordable (PTW, New York, NY).

Metal-oxide silicon semiconductor field-effect transistor (MOSFET) dosimeters have a very small active volume (0.2×0.2 mm^2), meeting the requirements of the necessary spatial resolution for small-field dosimetry. MOSFET dosimeters are relatively independent of energy (in MV beams), dose rate, and temperature (Das, Ding, and Ahnesjo 2008). MOSFET dosimeters are predominantly used for specialized *in vivo* point dose measurement. However, the detector readout process is not instantaneous for MOSFETs and would not be practical for the repeat measurements required to achieve precise positioning. Furthermore, MOSFET detectors have a limited lifetime span because they lose signal with accumulated dose. Because of their practical limitations, MOSFET detectors are currently not recommended for small-field dosimetry.

12.3.3 FILM

Both radiographic and radiochromic films have been used extensively for two-dimensional dosimetry. Radiographic film inherently has excellent spatial resolution and provides good profile measurements in small fields (Paskalev et al. 2003; Zhu et al. 2000). The biggest disadvantage of radiographic film is that processing is required to develop the film. Many hospitals are going filmless, and film processors are vanishing. Consistent dosimetric film measurements require a program for checking the integrity of the processor (Pai et al. 2007). Radiochromic films are an attractive option for small-field dosimetry because they have high spatial resolution, are nearly water-equivalent, and do not require any processing (Niroomand-Rad et al. 1998). Several authors have used films for small-field dosimetry and report dose accuracy of 2%–3% (Devic et al. 2005; Mack et al. 2003; McLaughlin et al. 1994; Pantelis et al. 2008).

12.3.4 THERMOLUMINESCENT DOSIMETER (TLD) AND OPTICALLY STIMULATED LUMINESCENCE DOSIMETER (OSLD)

TLDs are most frequently used for point dose measurements. They have a wide dynamic range and are nearly water equivalent. The biggest drawback of TLDs is that they have a large uncertainty of about ±3% (Kirby, Hanson, and Johnston 1992) and are labor intensive to read out. TLDs have been used for measurements in small fields (Francescon et al. 1998) and can provide a good cross-reference for output factor measurements.

Optically stimulated luminescence dosimeters (OSLDs) have been introduced in recent years for dosimetry measurements in megavoltage photon beams and recently for proton beams as well (Jursinic 2007). These dosimeters have an energy response, which must be evaluated and corrected for if they are to be used under nonreference conditions (Scarboro et al. 2012; Scarboro and Kry 2013). OSLDs have been widely implemented for peer review of beam calibration by the University of Texas Radiological Physics Center. In this application, they have been successfully used for reference dosimetry checks for small fields, such as the CyberKnife 60 mm collimator. This chapter defines small fields as less than 4 cm × 4 cm. Using the equivalent circle method, the 60 mm collimator is a 5.3 cm square field and therefore not small. OSLDs are commercially available as InLight/OSL Nanodots (Landauer, Inc., Glenwood, IL) (Jursinic and Yahnke 2011).

12.3.5 PLASTIC SCINTILLATING DETECTOR (PSD)

PSDs have many advantages over other types of radiation detectors, including excellent water equivalency, high spatial resolution, photon-beam quality independence, dose rate independence, dose linearity, and instantaneous reading (Beddar 1994; Beddar, Mackie, and Attix 1992b, 1992c). The major disadvantage of PSD dosimetry systems is their generation of a Cerenkov signal in the optical fiber that transports the scintillation photon to the photodetector (Archambault et al. 2005). Cerenkov radiation occurs when charged particles in a transparent medium move at a speed greater than the speed of light in that material. Because the optical fiber could be long, the contribution of Cerenkov emission to the total signal may be significant. However, significant improvements have recently been achieved in appropriate handling of background radiation in plastic scintillation dosimetry (Archambault et al. 2005; Archambault et al. 2006; Beddar, Mackie, and Attix 1992a; Deboer, Beddar, and Rawlinson 1993; Fontbonne et al. 2002). A recent report described the use of plastic scintillators for measuring total scatter factors for the CyberKnife (Morin et al. 2013).

A recently developed PSD system called Exradin W1 by Standard Imaging, Inc. (Middleton, WI) uses the principle of chromatic separation of the Cerenkov radiation from the total signal in a two-channel dosimetry processing system using a calibration procedure for accurate dose measurements in small fields. The Exradin W1 has a 1.0-mm-diameter × 3.0-mm-long scintillating material region (composed of polystyrene with an ABS plastic enclosure and polymide stem) in a 2.8-mm-diameter × 42-mm-long housing attached to a 1-m-long acrylic fiber. There is limited clinical data available in the literature with this detector at this time. However, preliminary work on using similar PSDs has shown that the scintillation detector's small detecting volume, high spatial resolution, and near tissue equivalence allow for greater confidence in the determination of absorbed dose distributions and stereotactic collimator output factors compared to other detectors, including micro ion chambers and diodes (Beddar 2006; Westermark et al. 2000).

12.3.6 OTHER DETECTORS

Several other detectors have desirable characteristics for small-field dosimetry, but they are still a work in progress in terms of development and clinical implementation. These include radiophotoluminescent glass rods (Araki et al. 2003); three-dimensional dosimeters, such as BANG dosimeters (McJury et al. 2000; Zeidan et al. 2010); and radiochromic plastic dosimeters (Clift et al. 2010). Polymer gels have been shown to perform well for dosimetry in small fields down to 5 mm full width at half maximum (FWHM) (Pantelis et al. 2008).

12.4 MEASUREMENT TECHNIQUES FOR LINAC RADIOSURGERY

Measurement techniques for depth-dose curves, profiles, and output factors for conventional fields have been addressed in AAPM TG reports 106 (Das et al. 2008) and TG 155 (Das et al. 2014) and IPEM report 103 (Aspradakis 2010). These guidelines should be followed with additional precautions necessary for small fields. This section pertains only to linear accelerator–based SRS and SRT measurements. Gamma SRS calibration and field verification require completely different techniques and equipment.

Because the partial occlusion of the focal spot causes an apparent widening of the FWHM field size (Aspradakis 2010), it is important to verify that there is no ambiguity in field size definition between measurements and the treatment-planning system. For planning systems dedicated to calculating dose using stereotactic cones, the field size of a cone is typically defined by the FWHM the cone generates at the reference SAD. Treatment-planning systems designed for a broader use, including large field sizes, tend to define the field size based on the collimator setting. For those systems, the collimator-defined field size may differ from the measured FWHM field size for small beams. It is an important safety issue during commissioning to establish which convention the treatment-planning system uses and design the small field commissioning process accordingly.

Prior to making small-field dosimetric measurement, the mechanical alignment and calibration of the collimation system (standard collimators, MLCs, and specialized stereotactic collimators) used to define the small field need to be carefully assessed. Inaccuracies in the alignment and calibration of the collimation system can have a significant effect on relative dose measurements. The beam collimation can be verified using measured dose profiles with film or a water-scanning system.

When making relative measurements in small fields, alignment of the detector on the central-axis is critical. Figure 12.4 shows the OAR of the five smallest CyberKnife cones with the size of a typical

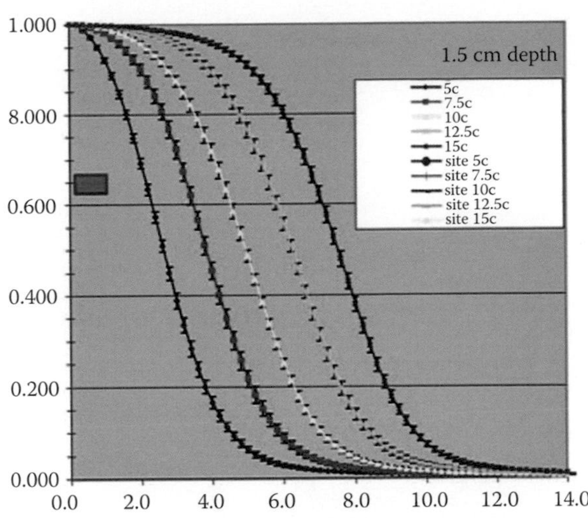

Figure 12.4 OAR of the five smallest CyberKnife cones compared to the size of a typical stereotactic diode detector (blue square).

stereotactic diode for comparison. A 1-mm detector misalignment can cause local errors on the order of 5% or higher (Ding, Duggan, and Coffey 2006). However, detector alignment uncertainty can be reduced to 0.2 mm by verifying the symmetry of orthogonal profile measurements at various depths using small field sizes (Cheng et al. 2007; Francescon, Cora, and Cavedon 2008).

12.4.1 REFERENCE DOSIMETRY

Reference dosimetry for conventional modalities is most often performed using the AAPM TG-51 protocol (Almond et al. 1999) in North America and the IAEA-398 (Mujsolino 2001) or similar protocols in other countries. Measurements are made in reference conditions by an ionization chamber whose calibration is traceable to a national primary standard. Certain treatment systems with "nonstandard geometries" used in SBRT and SRS (Gamma Knife, CyberKnife® without the M6 MLC collimator, and TomoTherapy Hi-Art®) cannot achieve the conventional reference conditions (e.g., 10 × 10 cm^2 field at 100 cm from a single source) (Das, Ding, and Ahnesjo 2008; Das et al. 2014). Additionally, ionization chambers may not be suitable for reference dose measurements in extremely small fields because of volume averaging effects and a lack of CPE. Kawachi et al. (2008) demonstrated the effect of chamber length in an unflattened stereotactic beam to be as high as 1.3% for a Farmer chamber. Established codes of practice for performing absolute dosimetry for nonstandard beams do not yet exist. There are, however, proposed methodologies that provide an extension of the conventional protocols into nonconventional fields (Alfonso et al. 2008; Dieterich et al. 2011; Ding, Duggan, and Coffey 2008; Langen et al. 2010). These suggested methodologies propose to incorporate a factor that accounts for a nonstandard field and nonreference conditions (Figure 12.5). The proposed factor can be written as $k_{Q_{msr},Q}^{f_{msr},f_{ref}}$, where Q is the beam quality of the conventional reference field f_{ref}, and Q_{msr} is the beam quality of the machine-specific reference field f_{msr}. This factor accounts for the differences between the conditions of field size, geometry, phantom material, and beam quality between a reference field and the machine-specific reference field (Alfonso et al. 2008). Some examples of machine-specific reference fields include the 60-mm-diameter collimator field for the CyberKnife, the 5 × 20 cm^2 static field in TomoTherapy, and the 16-mm- or 18-mm-diameter maximum field size for the Gamma Knife. If the field size and all other geometric and phantom conditions are the same for the reference field and the machine-specific reference field, the factor reduces to the conventional beam-quality correction factor. The correction factor can be obtained by direct calibration of a detector in different fields compared against traceable primary standards or using Monte Carlo simulation techniques (Chung, Bouchard, and Seunjens 2010; Sterpin et al. 2010).

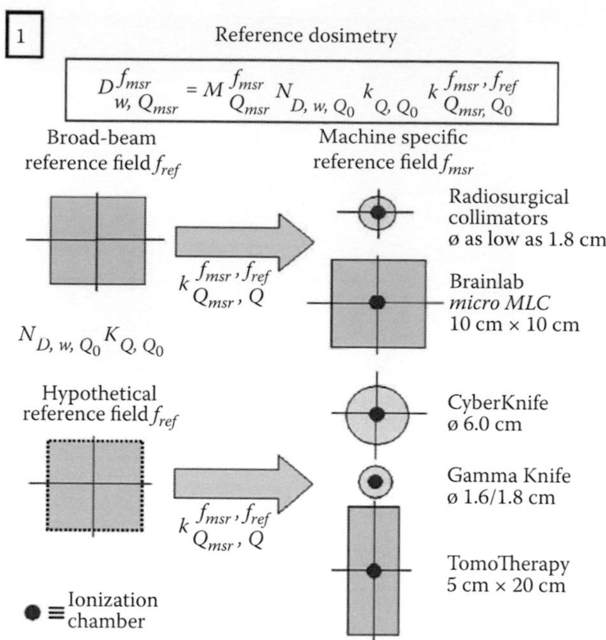

Figure 12.5 Proposed formalism for reference and relative dosimetry for treatment units when conventional reference conditions cannot be established. (From Alfonso, R. et al., *Med. Phys.*, 35, 5179–86, 2008. With permission.)

To address relative dosimetry in nonconventional beams, a clinical field factor $\Omega_{Q_{clin},Q_{msr}}^{f_{clin},f_{msr}}$ is introduced, where Q_{clin} is the beam quality of the clinical field f_{clin} (Figure 12.5) (Alfonso et al. 2008). The factor converts from absorbed dose in water in the machine-specific reference field to absorbed dose in water in the clinically used field, a concept analogous to an output factor in conventional modalities. The field factor is defined as the ratio of absorbed doses in the clinical field and machine-specific reference field. It can be calculated using Monte Carlo techniques or determined using a combination of measurements and detector correction factors $(k_\perp(Q_\perp clin, Q_\perp msr)\uparrow(f_\perp clin, f_\perp msr))$ that account for the differences in detector response for the two fields. These correction factors have been calculated for various detectors and linear-accelerator combinations (Cranmer-Sargison et al. 2011; Fenwick et al. 2013; Francescon, Cora, and Satariano 2011) (Table 12.1). For any detector for which the response is close to unity in the different geometries, the ratio of detector readings will provide a good approximation to the field factor $\Omega_{Q_{clin},Q_{msr}}^{f_{clin},f_{msr}}$.

12.4.2 DEPTH-DOSE MEASUREMENTS

The basic data to carry out dosimetric calculation requires depth-dose functions, such as PDDs, TPRs, and tissue maximum ratios. Measurement of PDD curves in standard conditions has been addressed in AAPM reports TG-106 (Das et al. 2008) and TG-155 (Das et al. 2014). Measurement of PDDs in small fields requires an appropriate detector with a small active volume and excellent spatial resolution. Using a detector that has a large active volume can result in underestimation of the dose and a steeper falloff for the PDD curve. Suitable detectors for measuring depth-dose curves are mini ionization chambers, diodes, and radiochromic film (Aspradakis 2010). Proper setup of the detector should be verified. This includes confirming that the detector is aligned with the central axis, that the scan arm travels parallel with the central axis, and that a minimal amount of stem and cable is exposed to the beam throughout scanning. Whenever possible, PDD curves should be measured and cross-referenced using two different detector types.

Dose-calculation systems often require TPR data instead of PDDs. Measurement of small-field TPRs is challenging. Therefore, TPRs are often calculated from measured PDD curves. The conversion of PDDs to TPRs is complicated in small fields because traditional methods rely on equivalent field-size definitions

Table 12.1 Example of calculated correction factors ($k_\perp(Q_\perp\ clin, Q_\perp\ 10 \times 10)^\dagger(f_\perp\ clin, f_\perp\ 10 \times 10)$) for the synergy and primus linear accelerators using the PTW diode 60012

PTW DIODE 60012						
X/Y (CM)	0.5	0.75	1.0	1.25	1.5	3.0
Primus						
0.5	0.968	0.975	0.980	0.983	0.985	0.989
0.75	0.977	0.984	0.989	0.992	0.994	0.998
1.0	0.982	0.990	0.995	0.998	1.000	1.004
1.25	0.986	0.993	0.998	1.001	1.003	1.007
1.5	0.988	0.995	1.000	1.003	1.005	1.009
3.0	0.991	0.998	1.003	1.006	1.008	1.012
Synergy						
0.5	0.964	0.974	0.980	0.983	0.986	0.989
0.75	0.972	0.982	0.988	0.992	0.994	0.998
1.0	0.977	0.987	0.994	0.997	1.000	1.003
1.25	0.981	0.991	0.997	1.001	1.003	1.007
1.5	0.983	0.993	1.000	1.004	1.006	1.009
3.0	0.987	0.998	1.004	1.008	1.010	1.014

Source: Francescon, P. et al., *Med. Phys.*, 38, 6513–27, 2011. With permission.

and scatter ratios under the condition of full CPE, which may not exist for small fields. TPRs have been calculated for small fields from measured PDDs and output factors using conversion factors (Cheng et al. 2007; McKerracher and Thwaites 1999) and small-field extrapolation techniques (Cheng et al. 2007). Because of the challenge of keeping the detector on the beam central axis for very small fields and the absence of CPE, a calculated PDD-to-TPR conversion should not be used in clinical practice for fields less than 15 mm FWHM without careful verification with directly measured TPR data. Water phantoms designed specifically for TPR measurements in small fields have become commercially available and provide a more time-efficient way to directly measure TPR.

12.4.3 PROFILES

Complete commissioning of small fields requires cross-profile measurements. These are usually performed for conventional linacs in the gun-target and left-right directions. Additional profiles are required for the CyberKnife variable-aperture collimator (Echner et al. 2009). As is the case with other small-field beam measurements, precise alignment of the measuring phantom axes with the central axis is required. The detector should be oriented such that the smallest dimension of the detector housing is facing the beam (e.g., the long axis of a diode detector parallel to the beam axis) unless the specific detector construction requires a different placement (e.g., for the Sun Nuclear Edge detector). A detector with a small active volume and high spatial resolution is critical for small-field profile measurements. Detectors with a large active volume can distort the shape of the profile in the penumbral region because of volume-averaging effects (Wurfel 2013) (Figure 12.6). One possible way to address the volume-averaging effects is to deconvolve the detector response from the profile (Sahoo, Kazi, and Hoffman 2008). However, deconvolution techniques are not straightforward, and significant expertise is needed to implement the corrections properly (Aspradakis 2010). Currently, the detectors best suited for small-field profile measurements are stereotactic diodes, microionization chambers, polymer gel, and radiochromic film. Stereotactic diodes provide the best spatial resolution when measuring profiles. If using diodes for profile measurements, it should be verified that the dose rate and energy dependence of the detector does not alter

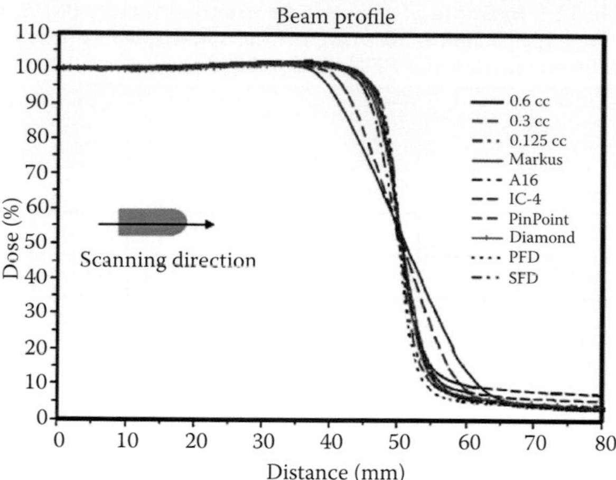

Figure 12.6 Profile measurements made with different detectors. The measured profiles illustrate that as the detector's active volume increases, the width of the penumbra increases due to volume averaging effects. (From Das, I. J. et al., *Med. Phys.*, 35, 4186–4215, 2008. With permission.)

the measurement as distance from the central axis increases. Microionization chambers have also been used for profile measurements although they generally provide worse spatial resolution than a stereotactic diode. If radiographic or radiochromic film is used for profile measurements, proper film processing and scanning should be performed, and the film energy and temporal response should be considered (Lindsay et al. 2010; Low et al. 2011). Good agreement has been reported between profile measurements made with film and deconvolved measurements made with ionization chambers (Chen, Graeff, and Baffa 2005). Whenever possible, profiles should be measured and compared using different detectors, and verified against "golden" beam data sets.

To remove any spikes and high-frequency noise in the measurements, it is acceptable to apply data processing and smoothing to the raw measured profiles. Using a sufficiently slow scan speed will eliminate noise caused by water ripples (Das et al. 2008), especially for scans taken at shallow depth. Care should be taken to smooth the data in such a manner as to not exacerbate the effect of penumbral blurring and to not change the shape of the profile. *Median filters* are preferable for smoothing of profiles because the original data points are preserved. It is always prudent to keep the original measurements so reprocessing can be performed if data smoothing provides unacceptable results.

12.4.4 OUTPUT FACTORS

Output factors relate the absorbed dose to water in a clinical field to the absorbed dose to water measured under reference conditions. In-air output factors are used in some treatment-planning systems for Monte Carlo–based beam modeling (Deng et al. 2003). Output factors are unitless and are generally measured at 5 cm or 10 cm depth to minimize effects of electron contamination near the surface of the phantom. Methods for measuring output factors have been discussed in the AAPM TG 106 (Das et al. 2008) and TG-155 reports (Das et al. 2014). Measuring output factors in small fields is complicated by a lack of CPE, source occlusion, and volume averaging, causing a reduced signal in the central part of the beam (Losasso 2008). Small-field output factors can vary by as much as 14% with choice of detector (Figure 12.7). Monte Carlo modeling of the detector response to small-field beam conditions has been shown to resolve output factor measurement discrepancies for a diverse set of detectors (Francescon, Cora, and Cavedon 2008). The detector spatial resolution, perturbation of the beam, and tissue-equivalence are aspects that should be considered when deciding on a detector. Small-field output factors should be measured in a water tank, so the detector alignment against the central axis can be verified with orthogonal profiles. An alternative method used with the CyberKnife system is to mount the detector on a cage connected to the linac head, which ensures accurate alignment of the detector with respect to SAD and the central beam axis.

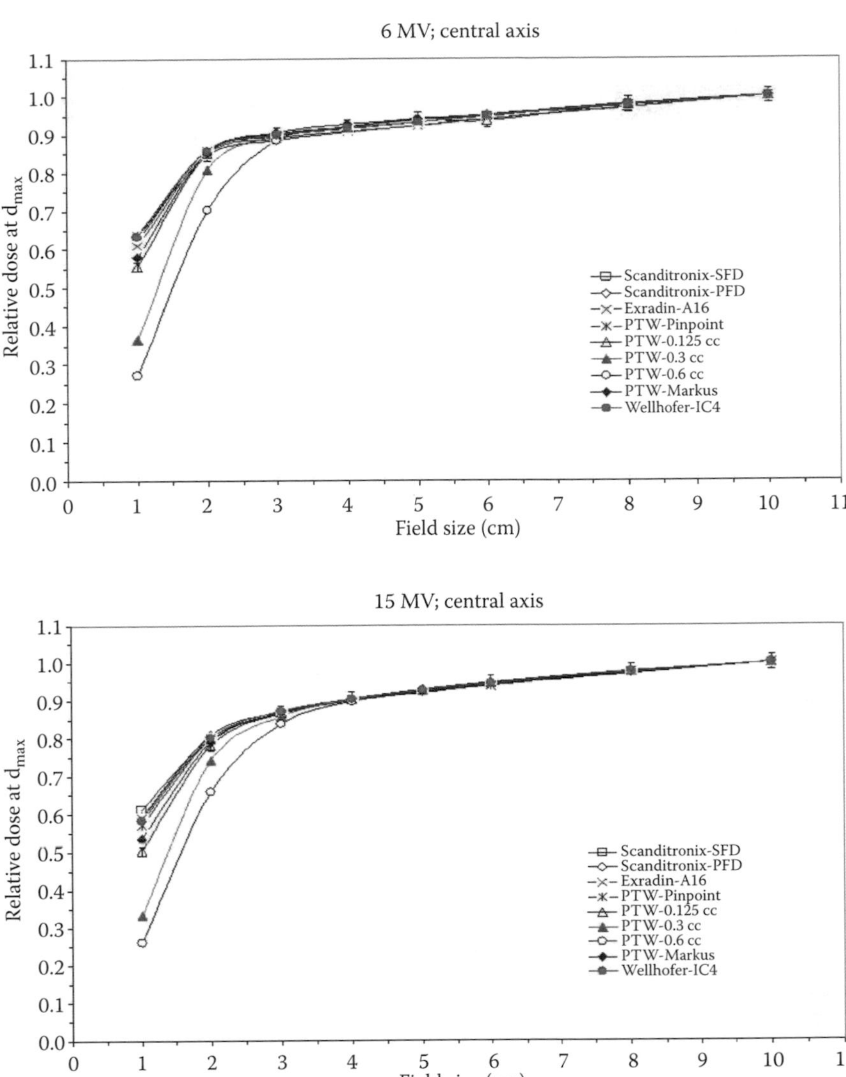

Figure 12.7 Output factor measurements using different detectors and cone sizes for 6- and 15-MV beams. The plot illustrates detector-dependent variability in output factors measurements. (From Das, I. J. et al., *Med. Phys.*, 35, 206–15, 2008. With permission.)

The total output factor, also called total scatter factor S_{cp}, is defined as the product of the phantom scatter factor (S_p) and the head scatter factor (S_c) normalized to a reference field size. For small fields created by tertiary collimation devices, S_{cp} is measured with photon jaws fixed and the fields defined by attached cones or MLCs. Although S_{cp} can be measured directly and used on its own, many but not all treatment-planning systems require the individual scatter components for use in beam modeling. If S_C and S_P are needed separately, the most common method is to measure S_{cp} and S_c and determine S_p by taking the ratio of S_{cp} and S_p.

The head scatter factor S_c accounts for variation in beam output with field size caused by changes in radiation originating from the linear accelerator. S_c values are generally measured using a water-equivalent mini phantom with dimensions smaller than the radiation field (Zhu et al. 2009). The water-equivalent mini phantom becomes impractical to provide complete CPE for small field sizes. Different combinations of detector and mini phantom design have been investigated for S_c measurements (McKerracher and Thwaites 2007a, 2007b). Certain studies have used materials, such as brass, for mini phantom design (Weber, Nilsson,

Table 12.2 Radiological Physics Center measured and institution treatment planning system calculated small field output factors for Varian (12.2A), Elekta (12.2B), and Siemens (12.2C) linear accelerators

PART A								
FIELD SIZE, CM²	VARIAN 6 MV		VARIAN 10 MV		VARIAN 15 MV		VARIAN 18 MV	
	RPC	Institution	RPC	Institution	RPC	Institution	RPC	Institution
10 × 10	1.000	1.000	1.000	1.000	1.000	1.000	1.000	1.000
6 × 6	0.921	0.929	0.946	0.953	0.951	0.950	0.949	0.950
	(0.013)	(0.004)	(0.017)	(0.016)	(0.008)	(0.008)	(0.011)	(0.014)
	[0.9%]	n = 64	[0.7%]	n = 9	[0.5%]	n = 14	[0.5%]	n = 16
4 × 4	0.865	0.874	0.900	0.912	0.909	0.909	0.902	0.900
	(0.018)	(0.021)	(0.024)	(0.030)	(0.013)	(0.017)	(0.014)	(0.024)
	[1.3%]	n = 64	[1.3%]	n = 9	[1.1%]	n = 14	[1.1%]	n = 16
3 × 3	0.828	0.841	0.867	0.875	0.874	0.877	0.861	0.856
	(0.017)	(0.025)	(0.020)	(0.025)	(0.014)	(0.019)	(0.014)	(0.027)
	[1.7%]	n = 62	[1.2%]	n = 9	[1.3%]	n = 12	[1.7%]	n = 16
2 × 2	0.786	0.796	0.817	0.828	0.803	0.813	0.784	0.782
	(0.019)	(0.031)	(0.015)	(0.019)	(0.016)	(0.038)	(0.015)	(0.034)
	[2.3%]	n = 55	[1.8%]	n = 11	[2.8%]	n = 10	[3.5%]	n = 15

PART B						
FIELD SIZE, CM²	ELEKTA 6 MV		ELEKTA 10 MV		ELEKTA 18 MV	
	RPC	Institution	RPC	Institution	RPC	Institution
10 × 10	1.000	1.000	1.000	1.000	1.000	1.000
6 × 6	0.930	0.934	0.937	0.940	0.945	0.947
	(0.010)	(0.009)	(0.004)	(0.005)	(0.002)	(0.003)
	[0.5%]	n = 18	[0.7%]	n = 6	[0.3%]	n = 5
4 × 4	0.878	0.888	0.890	0.891	0.901	0.918
	(0.015)	(0.027)	(0.009)	(0.010)	(0.002)	(0.039)
	[1.3%]	n = 22	[0.6%]	n = 8	[0.4%]	n = 6
3 × 3	0.842	0.848	0.857	0.862	0.861	0.863
	(0.012)	(0.009)	(0.003)	(0.005)	(0.003)	(0.004)
	[0.9%]	n = 17	[0.6%]	n = 6	[0.6%]	n = 4
2 × 2	0.790	0.796	0.796	0.802	0.786	0.798
	(0.007)	(0.010)	(0.009)	(0.008)	(0.006)	(0.019)
	[1.6%]	n = 17	[1.3%]	n = 6	[2.4%]	n = 4

(*continued*)

Table 12.2 (Continued) Radiological Physics Center measured and institution treatment planning system calculated small field output factors for Varian (12.2A), Elekta (12.2B), and Siemens (12.2C) linear accelerators

PART C						
FIELD SIZE, CM²	SIEMENS 6 MV		SIEMENS 10 MV		SIEMENS 18 MV	
	RPC	Institution	RPC	Institution	RPC	Institution
10 × 10	1.000	1.000	1.000	1.000	1.000	1.000
6 × 6	0.914	0.920	0.927	0.935	0.940	0.946
	(0.008)	(0.008)	(0.003)	(0.010)	(0.005)	(0.003)
	[0.7%]	n = 13	[0.9%]	n = 4	[0.6%]	n = 4
4 × 4	0.855	0.863	0.877	0.884	0.891	0.0.896
	(0.010)	(0.009)	(0.001)	(0.012)	(0.004)	(0.003)
	[1.1%]	n = 13	[1.2%]	n = 4	[0.6%]	n = 4
3 × 3	0.820	0.825	0.841	0.850	0.849	0.855
	(0.008)	(0.011)	(0.001)	(0.007)	(0.003)	(0.003)
	[1.3%]	n = 13	[1.1%]	n = 4	[0.7%]	n = 4
2 × 2	0.764	0.757	0.777	0.742	0.795	0.779
	(0.010)	(0.042)	(0.005)	(0.079)	(0.004)	(0.015)
	[2.8%]	n = 12	[5.8%]	n = 4	[1.9%]	n = 4

Source: Followill, D. et al., *J. Appl. Clin. Med. Phys.*, 13, 282–89, 2012. With permission.

and Ahnesjo 1997; Zhu et al. 2009), and other reports cite that lateral CPE does not need to be preserved as long as the electron contamination is eliminated from the mini phantom (Cranmer-Sargison et al. 2011).

Numerous groups have published output factors for various linac and detector combinations (Das, Ding, and Ahnesjo 2008; Fan et al. 2009; Fenwick et al. 2013; Followill et al. 2012; Francescon, Cora, and Cavedon 2008; Francescon et al. 1998; Mack et al. 2002; Sauer and Wilbert 2007; Zhu et al. 2000). The published output factors were measured using a combination of microionization chambers, solid-state detectors, film, and TLD. The most prevalent detectors chosen for S_{cp} measurements are the stereotactic diode (Dieterich and Sherouse 2011) and microionization chamber. Measurements should be compared using at least two different types of dosimeters. Furthermore, measured total scatter factors should be compared against data published in the literature for various linear accelerators (Table 12.2a–12.2c) (Followill et al. 2012). Whenever available, the correction factor $k_{Q_{clin},Q_{msr}}^{f_{clin},f_{msr}}$ that accounts for the differences in the detector response between the machine-specific reference field and the clinical field should be used for the pertinent combination of detector and linear accelerator. (Note that, at the time of this writing, this factor is not widely available but should be in the near future. Care should be applied to choose factors that have been independently confirmed through multiple peer-reviewed publications or disseminated through a national or international society recommendation.)

12.5 DOSE CALCULATIONS FOR SMALL FIELDS

Accurate dose calculations are essential to small-field dosimetry. Using inappropriate dose calculation algorithms can result in serious errors in dose prediction for small fields (Chetty 2007; Papanikolaou et al. 2004). There are two main challenges to achieving accurate small-field dose calculations. Calculation algorithms require measured beam data, and any inaccuracies in the beam data collection can propagate to the dose calculations. All the challenges and uncertainties associated with performing small field measurements have the potential to translate to inaccurate dose calculations. The other challenge is that the lack of CPE is further exacerbated by the presence of inhomogeneities found in common SBRT and

SRT treatment sites, such as the lung and nasal cavities. The lateral range of electrons significantly increases in lung material compared to water, causing a loss of CPE at much larger field sizes. AAPM Task Group 65 (Papanikolaou et al. 2004) and Task Group 105 (Chetty 2007) provide reviews of dose-calculation algorithms and complications associated with calculating dose in a heterogeneous medium.

Calculation of radiation dose is divided into a primary component and a scatter component. The primary component can be accurately modeled by exponential attenuation, and most dose calculations appropriately account for this factor. The differences in dose-calculation algorithms arise in how they handle scatter. There are several types of dose-calculation algorithms, including factor based, convolution techniques, and Monte Carlo. Factor-based dose-calculation algorithms model the primary fluence of the beam. They use a limited set of measured beam data and interpolation to reconstruct the entire beam data set. Tissue heterogeneities are addressed by applied scaling factors. Factor-based algorithms are limited in how they account for scatter, particularly in conditions without established CPE and are therefore not recommended for stereotactic treatment dose calculations.

The convolution superposition method convolves the total energy released per unit mass (TERMA) with a dose-deposition kernel (Mackie, Scrimger, and Battista 1985). The dose-deposition kernel is precomputed for different energies using Monte Carlo techniques. The superposition method has been shown to predict doses within 2% in heterogeneous media (Mackie, Scrimger, and Battista 1985, Ahnesjo 1989) and performs well in small-field conditions without CPE (Jones and Das 2005). The superposition method uses density-based scaling for both the primary beam and the dose spread arrays. Incorporating density-based scaling for both components improves the performance of the superposition algorithm in heterogeneous materials (Papanikolaou et al. 2004). When modeling the primary fluence for small field sizes, it is important to insure that the algorithm properly models the direct beam source size and shape (Jaffray et al. 1993), collimator geometry (Ding, Duggan, and Coffey 2006), and the leakage through any secondary collimation. These components are of primary importance for accurate small-field dose calculations.

Monte Carlo techniques track the individual particle trajectories through the linear accelerator to the patient. Interactions are based on probabilities stemming from fundamental physical principles. Monte Carlo dose calculations provide the greatest accuracy in heterogeneous media and conditions without established CPE. With recent advances in computing power and calculation efficiency (Fippel 1999), Monte Carlo dose calculations have come into routine clinical use for SBRT treatment planning (Huq et al. 2008; Wilcox et al. 2010).

Linear Boltzman transport equation solvers are similar to classical Monte Carlo methods in allowing for accurate modeling of dose in heterogeneous tissue. The Acuros XB algorithm (Vassiliev et al. 2010) is a grid-based Boltzman equation solver, which has been evaluated for small fields in Rapidarc stereotactic treatments (Fogliata et al. 2011). The authors conclude that for adequately tuned configuration parameters, Acuros was able to achieve agreement of <1.7% in monitor unit (MU) difference between calculated and measured dose for fields as small as 0.8×0.8 mm^2.

Measurement-based algorithms do not perform well in conditions in which CPE does not exist and should not be used for small-field dose calculations. Most current treatment-planning systems provide dose calculation algorithms that have good scatter models and properly deal with a lack of CPE in heterogeneous media. These algorithms should provide adequate dose calculation accuracy in most clinical situations. Recent publications have reported substantial disagreements for heterogeneous SBRT calculations (especially for lung tumors) between pencil beam algorithms and Monte Carlo treatment planning (Kry et al. 2013). Whenever available, Monte Carlo calculations are ideal to use for the most complex geometries in SBRT (Benedict et al. 2010). The dose calculation algorithms should be verified using established guidelines, and whenever possible, calculations should be compared to measurements made in heterogeneous phantoms.

12.6 SUMMARY AND RECOMMENDATIONS

Commissioning and modeling of small photon fields and quality assurance (QA) of equipment used for SRS involve additional care and expertise in these areas not easily recognized in routine radiotherapy applications using broad beams and standard dose fractionations. Accurate delivery of stereotactic treatments with small photon fields requires high accuracy of measured beam data, accurate modeling of small fields, and a stringent QA program. The QA program must include end-to-end tests and check the integrity of all system components in beam data acquisition, treatment-planning systems, precise machine functioning, patient positioning, and the image-guidance process. Various recommendations, including those in AAPM Task Groups 101 (Benedict et al. 2010), 106 (Das et al. 2008), 142 (Klein et al. 2009), and 155 (Das et al. 2014) and IPEM report 103 (Aspradakis 2010), address issues pertinent to accurate delivery of small fields, which must be carefully implemented in the clinical setting for safe and accurate patient treatments with small photon beams and equipment using such fields. Medical physicists should be specifically trained in SRS/SBRT at established programs before independently working in a small-field environment (Benedict et al. 2010).

The following is a summary of some key descriptions and recommendations regarding small photon field measurements.

12.6.1 SMALL FIELD

A photon field is considered to be small when the field size becomes comparable to the lateral range of secondary electrons for the energy and medium under consideration and/or when the beam collimation system blocks part of the direct source from the point of measurement.

12.6.2 FIELD SIZE DEFINITION

In small fields, nonequilibrium conditions prevail with overlapping penumbrae within the field and the relationship between the collimator setting and the FWHM of the dose profile changes compared to large fields. In order to use a consistent field size definition between the treatment machine and the treatment planning system, it is recommended that the user needs to do the following:

1. Understand which field size definition is used in the treatment-planning system
2. Precisely calibrate individual collimator jaws at the central axis and small field settings
3. Establish the MLC leaf positions for rounded leaf ends
4. Determine and verify the field size and penumbra width in the TPS

12.6.3 DETECTOR CHOICE FOR DOSE MEASUREMENTS

Measurement of dosimetric parameters, such as output factors, PDDs, and beam profiles for small photon fields require careful selection of the dosimeter and water tank for the task at hand. Important detector properties for small field measurements are the detector size, detector composition or water equivalency, and energy and dose rate independencies. The choice of detector may depend on how small the field size is. A detector dimension comparable to the field size causes the volume effect on the measured dose. As a consequence, the dose in the field is underestimated, and the width of the penumbra is overestimated due to volume averaging of the measured signal when the dose changes noticeably across the detector. Also the nonwater equivalence of an air ion chamber measures a blurred profile because the higher range of electrons in air than in water results in a broadening of the measured penumbra.

It is recommended that

1. For a given field, an appropriate detector is selected so that no more than 1% variation over the detector diameter is observed when scanned through the field center (Das et al. 2008). (Note that this is only appropriate for flattened beams. For flattening filter–free beams, the variation will be greater.)
2. For fields less than 4 cm^2, tissue maximum ratios and output factors should be acquired with diodes or microchambers (Aspradakis 2010).
3. Mini-ionization cylindrical and parallel plane chambers with a sensitive volume equal to or less than 0.01 cm^3 without a steel central electrode and shielded or unshielded diodes be used for scanned relative depth dose or PDD measurements in small photon beams (Aspradakis 2010).
4. Tissue-equivalent radiochromic film and diodes (stereotactic, shielded or unshielded) positioned parallel to beam central axis CAX should be used to resolve penumbra in small photon fields.
5. The detector alignment with the CAX should be well within 1 mm (Aspradakis 2010) and can likely be within 0.3 mm if care is used in measurement.

6. For output measurements, the CAX detector position could be optimized at the maximum point of two orthogonal profiles crossing each other taken at the depth of measurement (5 cm or 10 cm).
7. For depth dose measurements, the CAX detector position could be optimized at the maximum point of two orthogonal profiles crossing each other taken at two different depths (one at the beginning and one at the end of the scan)
8. The direction of scanning in a water tank should be from the larger to the shallower depths to reduce the disturbance at small depths.

REFERENCES

AAPM Task Group 21. 1983. A protocol for the determination of absorbed dose from high-energy photon and electron beams. *Med. Phys.* 10:741–71.

Ahnesjo, A. 1989. Collapsed cone convolution of radiant energy for photon dose calculation in heterogeneous media. *Med. Phys.* 16:577–92.

Alfonso, R., P. Andreo, R. Capote et al. 2008. A new formalism for reference dosimetry of small and nonstandard fields. *Med. Phys.* 35:5179–86.

Almond, P. R., P. J. Biggs, B. M. Coursey et al. 1999. AAP's TG-51 protocol for clinical reference dosimetry of high-energy photon and electron beams. *Med. Phys.* 26:1847–70.

Araki, F., T. Ikegami, T. Ishidoya, and H. D. Kubo. 2003. Measurements of Gamma-Knife helmet output factors using a radiophotoluminescent glass rod dosimeter and a diode detector. *Med. Phys.* 30:1976–81.

Archambault, L., J. Arsenault, L. Gingras, A. S. Beddar, R. Roy, and L. Beaulieu. 2005. Plastic scintillation dosimetry: Optimal selection of scintillating fibers and scintillators. *Med. Phys.* 32:2271–8.

Archambault, L., A. S. Beddar, L. Gingras, R. Roy, and L. Beaulieu. 2006. Measurement accuracy and Cerenkov removal for high performance, high spatial resolution scintillation dosimetry. *Med. Phys.* 33:128–35.

Aspradakis, M. M. (ed.). 2010. *Small Field MV Photon Dosimetry.* New York: Institute of Physics and Engineering in Medicine.

Attix, F. H. 1986. *Introduction to Radiological Physics and Radiation Dosimetry.* New York: John Wiley & Sons.

Barnett, G. H., M. E. Linskey, J. R. Adler et al. American Association of Neurological Surgeons, and Congress of Neurological Surgeons Washington Committee Stereotactic Radiosurgery Task Force. 2007. Stereotactic radiosurgery—an organized neurosurgery-sanctioned definition. *J. Neurosurg.* 106(1):1–5.

Beddar, A. S. 1994. A new scintillator detector system for the quality assurance of Co-60 and high-energy therapy machines. *Phys. Med. Biol.* 39:253–63.

Beddar, A. S. 2006. Plastic scintillation dosimetry and its application to radiotherapy. *Radiat. Meas.* 41:S124–33.

Beddar, A. S., T. R. Mackie, and F. H. Attix. 1992a. Cerenkov light generated in optical fibers and other light pipes irradiated by electron-beams. *Phys. Med. Biol.* 37:925–35.

Beddar, A. S., T. R. Mackie, and F. H. Attix. 1992b. Water-equivalent plastic scintillation detectors for high-energy beam dosimetry: I. Physical characteristics and theoretical considerations. *Phys. Med. Biol.* 37:1883–900.

Beddar, A. S., T. R. Mackie, and F. H. Attix. 1992c. Water-equivalent plastic scintillation detectors for high-energy beam dosimetry: II. Properties and measurements. *Phys. Med. and Biol.* 37:1901–13.

Beddar, A. S., D. J. Mason, and P. F. Obrien. 1994. Absorbed dose perturbation caused by diodes for small-field photon dosimetry. *Med. Phys.* 21:1075–9.

Benedict, S. H., F. J. Bova, B. Clark et al. 2008. Anniversary paper: The role of medical physicists in developing stereotactic radiosurgery. *Med. Phys.* 35:4262–77.

Benedict, S. H., K. M. Yenice, D. Followill et al. 2010. Stereotactic body radiation therapy: The report of AAPM Task Group 101. *Med. Phys.* 37:4078–101.

Bjarngard, B. E., J. S. Tsai, and R. K. Rice. 1990. Doses on the central axes of narrow 6-Mv x-ray beams. *Med. Phys.* 17(5):794–9.

Chen, F., C. F. O. Graeff, and O. Baffa. 2005. K-band EPR dosimetry: Small-field beam profile determination with miniature alanine dosimeter. *Appl. Radiat. Isot.* 62(2):267–71.

Cheng, C. W., S. H. Cho, M. Taylor, and I. J. Das. 2007. Determination of zero-field size percent depth doses and tissue maximum ratios for stereotactic radiosurgery and IMRT dosimetry: Comparison between experimental measurements and Monte Carlo simulation. *Med. Phys.* 34:3149–57.

Chetty, I. 2007. Issues associated with clinical implementation of Monte Carlo–based treatment planning: Summary of the AAPM task group report No.105. *Radiother. Oncol.* 84:S6.

Chung, E., H. Bouchard, and J. Seuntjens. 2010. Investigation of three radiation detectors for accurate measurement of absorbed dose in nonstandard fields. *Med. Phys.* 37:2404–13.

Clift, C., A. Thomas, J. Adamovics, Z. Chang, I. Das, and M. Oldham. 2010. Toward acquiring comprehensive radiosurgery field commissioning data using the PRESAGE (R)/optical-CT 3D dosimetry system. *Phys. Med. Biol.* 55:1279–93.

Cranmer-Sargison, G., S. Weston, J. A. Evans, N. P. Sidhu, and D. I. Thwaites. 2012. Monte Carlo modelling of diode detectors for small field MV photon dosimetry: Detector model simplification and the sensitivity of correction factors to source parameterization. *Phys. Med. Biol.* 57:5141–53.

Cranmer-Sargison, G., S. Weston, N. P. Sidhu, and D. I. Thwaites. 2011. Experimental small field 6 MV output ratio analysis for various diode detector and accelerator combinations. *Radiother. Oncol.* 100:429–35.

Das, I. J., C. W. Cheng, R. J. Watts et al. 2008. Accelerator beam data commissioning equipment and procedures: Report of the TG-106 of the Therapy Physics Committee of the AAPM. *Med. Phys.* 35:4186–215.

Das, I. J., G. X. Ding, and A. Ahnesjo. 2008. Small fields: Nonequilibrium radiation dosimetry. *Med. Phys.* 35:206–15.

Das, I. J., M. B. Downes, A. Kassaee, and Z. Tochner. 2000. Choice of radiation detector in dosimetry of stereotactic radiousurgery-radiotherapy. *J. Radiosurg.* 3:177–85.

Das, I. J., P. Francescon, A. Ahnesjo et al. 2014. Small fields and non-equilibrium condition photon beam dosimetry: Report of the TG-155 of the Therapy Physics Committee of the AAP. *Med Phys.* (in press).

De Angelis, C., S. Onori, M. Pacilio et al. 2002. An investigation of the operating characteristics of two PTW diamond detectors in photon and electron beams. *Med. Phys.* 29:248–54.

Deboer, S. F., A. S. Beddar, and J. A. Rawlinson. 1993. Optical filtering and spectral measurements of radiation-induced light in plastic scintillation dosimetry. *Phys. Med. Biol.* 38:945–58.

Deng, J., C. M. Ma, J. Hai, and R. Nath. 2003. Commissioning 6 MV photon beams of a stereotactic radiosurgery system for Monte Carlo treatment planning. *Med. Phys.* 30:3124–34.

Derreumaux, S., C. Etard, C. Huet et al. 2008. Lessons from recent accidents in radiation therapy in France. *Radiat. Prot. Dosim.* 131:130–5.

Devic, S., J. Seuntjens, E. Sham et al. 2005. Precise radiochromic film dosimetry using a flat-bed document scanner. *Med. Phys.* 32:2245–53.

Dieterich, S., C. Cavedon, C. F. Chuang et al. 2011. Report of AAPM TG 135: Quality assurance for robotic radiosurgery. *Med. Phys.* 38:2914–36.

Dieterich, S., and G. W. Sherouse. 2011. Experimental comparison of seven commercial dosimetry diodes for measurement of stereotactic radiosurgery cone factors. *Med. Phys.* 38:4166–73.

Ding, G. X., D. M. Duggan, and C. W. Coffey. 2006. Commissioning stereotactic radiosurgery beams using both experimental and theoretical methods. *Phys. Med. Biol.* 51:2549–66.

Ding, G. X., D. M. Duggan, and C. W. Coffey. 2008. A theoretical approach for non-equilibrium radiation dosimetry. *Phys. Med. Biol.* 53:3493–9.

Ding, G. X., D. M. Duggan, B. Lu et al. 2007. Impact of inhomogeneity corrections on dose coverage in the treatment of lung cancer using stereotactic body radiation therapy. *Med. Phys.* 34:2985–94.

Echner, G. G., W. Kilby, M. Lee et al. 2009. The design, physical properties and clinical utility of an iris collimator for robotic radiosurgery. *Phys. Med. Biol.* 54:5359–80.

Eklund, K., and A. Ahnesjo. 2009. Modeling silicon diode energy response factors for use in therapeutic photon beams. *Phys. Med. Biol.* 54:6135–50.

Eveling, J. N., A. M. Morgan, and W. G. Pitchford. 1999. Commissioning a p-type silicon diode for use in clinical electron beams. *Med. Phys.* 26:100–7.

Fan, J., K. Paskalev, L. Wang et al. 2009. Determination of output factors for stereotactic radiosurgery beams. *Med. Phys.* 36:5292–300.

Fenwick, J. D., S. Kumar, A. J. D. Scott, and A. E. Nahum. 2013. Using cavity theory to describe the dependence on detector density of dosimeter response in non-equilibrium small fields. *Phys. Med. Biol.* 58:2901–23.

Fippel, M. 1999. Fast Monte Carlo dose calculation for photon beams based on the VMC electron algorithm. *Med. Phys.* 26:1466–75.

Fogliata, A., G. Nicolini, A. Clivio, E. Vanetti, and L. Cozzi. 2011. Accuracy of Acuros XB and AAA dose calculation for small fields with reference to RapidArc (®) stereotactic treatments. *Med. Phys.* 38(11):6228–37.

Followill, D., S. Kry, L. Qin et al. 2012. The Radiological Physics Center's standard data set for small field size output factors. *J. Appl. Clin. Med. Phys.* 13:282–9.

Fontbonne, J. M., G. Iltis, G. Ban et al. 2002. Scintillating fiber dosimeter for radiation therapy accelerator. *IEEE Trans. Nucl. Sci.* 49:2223–7.

Francescon, P., S. Cora, and C. Cavedon. 2008. Total scatter factors of small beams: A multidetector and Monte Carlo study. *Med. Phys.* 35:504–13.

Francescon, P., S. Cora, C. Cavedon, P. Scalchi, S. Reccanello, and F. Colombo. 1998. Use of a new type of radiochromic film, a new parallel-plate micro-chamber, MOSFETs, and TLD 800 microcubes in the dosimetry of small beams. *Med. Phys.* 25(4):503–11.

Francescon, P., S. Cora, and N. Satariano. 2011. Calculation of k(Qclin), Q(msr) (fclin,fmsr) for several small detectors and for two linear accelerators using Monte Carlo simulations. *Med. Phys.* 38:6513–27.

Gotoh, S., M. Ochi, N. Hayashi et al. 1996. Narrow photon beam dosimetry for linear accelerator radiosurgery. *Radiother. Oncol.* 41:221–4.

Huq, M. S., B. A. Fraass, P. B. Dunscombe et al. 2008. A method for evaluating quality assurance needs in radiation therapy. *Int. J. Radiat. Oncol. Biol. Phys.* 71(1 Suppl):S170–3.

IAEA Report No. 398. 2000. Absorbed dose determination in external beam radiotherapy: An international code of practice for dosimetry on standards of absorbed dose to water. Vienna: International Atomic Energy Agency.

International Commission on Radiation Units and Measurements. 1976. Determination of Absorbed Dose in a Patient Irradiated by Beams of X or Gamma Rays, Report 24. Bethesda, MD: ICRU.

Jaffray, D. A., J. J. Battista, A. Fenster, and P. Munro. 1993. X-ray sources of medical linear accelerators: Focal and extra-focal radiation. *Med. Phys.* 20:1417–27.

Jones, A. O., and I. J. Das. 2005. Comparison of inhomogeneity correction algorithms in small photon fields. *Med. Phys.* 32:766–76.

Jursinic, P. A. 2007. Characterization of optically stimulated luminescent dosimeters, OSLDs, for clinical measurements. *Med. Phys.* 34(12):4594–604.

Jursinic, P. A., and C. J. Yahnke. 2011. *In vivo* dosimetry with optically stimulated luminescent dosimeters, OSLDs, compared to diodes: The effects of buildup cap thickness and fabrication material. *Med. Phys.* 38(10):5432–40.

Kawachi, T., H. Saitoh, M. Inoue, T. Katayose, A. Myojoyama, and K. Hatano. 2008. Reference dosimetry condition and beam quality correction factor for CyberKnife beam. *Med. Phys.* 35:4591–8.

Kawrakow, I. 2006. On the effective point of measurement in megavoltage photon beams. *Med. Phys.* 33:1829–39.

Khan, F. M. 2012. *The Physics of Radiation Therapy*. Wolters Kluwer Health.

Kirby, T. H., W. F. Hanson, and D. A. Johnston. 1992. Uncertainty analysis of absorbed dose calculations from thermoluminescense dosimeters. *Med. Phys.* 19:1427–33.

Klein, E. E., J. Hanley, J. Bayouth et al. 2009. Task Group 142 report: Quality assurance of medical accelerators. *Med. Phys.* 36:4197–212.

Kry, S. F., P. Alvarez, A. Molineau, C. Amador, J. Galvin, and D. S. Followill. 2013. Algorithms used in heterogeneous dose calculations show systematic differences as measured with the Radiological Physics Center's anthropomorphic thorax phantom used for RTOG credentialing. *Int. J. Radiat. Oncol. Biol. Phys.* 85(1):e95–100.

Langen, K. M., N. Papanikolaou, J. Balog et al. 2010. QA for helical tomotherapy: Report of the AAPM Task Group 148. *Med. Phys.* 37:4817–53.

Laub, W. U., T. W. Kaulich, and F. Nusslin. 1999. A diamond detector in the dosimetry of high-energy electron and photon beams. *Phys. Med. Biol.* 44:2183–92.

Laub, W. U., and T. Wong. 2003. The volume effect of detectors in the dosimetry of small fields used in IMRT. *Med. Phys.* 30:341–7.

Lee, H. R., M. Pankuch, J. C. Chu, and J. J. Spokas. 2002. Evaluation and characterization of parallel plate microchamber's functionalities in small beam dosimetry. *Med. Phys.* 29:2489–96.

Leksell, L. 1951. The stereotaxic method and radiosurgery of the brain. *Acta Chir. Scand.* 102:316–9.

Le Roy, M., L. de Carlan, F. Delaunay et al. 2011. Assessment of small volume ionization chambers as reference dosimeters in high-energy photon beams. *Phys. Med. Biol.* 56:5637–50.

Leybovich, L. B., A. Sethi, and N. Dogan. 2003. Comparison of ionization chambers of various volumes for IMRT absolute dose verification. *Med. Phys.* 30:119–23.

Li, X. A., M. Soubra, J. Szanto, and L. H. Gerig. 1995. Lateral electron equilibrium and electron contamination in measurements of head-scatter factors using miniphantoms and brass caps. *Med. Phys.* 22:1167–70.

Lindsay, P., A. Rink, M. Ruschin, and D. Jaffray. 2010. Investigation of energy dependence of EBT and EBT-2 Gafchromic film. *Med. Phys.* 37:571–6.

Looe, H. K., D. Harder, and B. Poppe. 2011. Experimental determination of the effective point of measurement for various detectors used in photon and electron beam dosimetry. *Phys. Med. Biol.* 56:4267–90.

Losasso, T. 2008. IMRT delivery performance with a varian multileaf collimator. *Int. J. Radiat. Oncol. Biol. Phys.* 71(1 Suppl):S85–8.

Low, D. A., J. M. Moran, J. F. Dempsey, L. Dong, and M. Oldham. 2011. Dosimetry tools and techniques for IMRT. *Med. Phys.* 38:1313–38.

Lutz, W., K. R. Winston, and N. Maleki. 1988. A system for stereotactic radiosurgery with a linear accelerator. *Int. J. Radiat. Oncol. Biol. Phys.* 14:373–81.

Ma, C. M., and A. E. Nahum. 1993. Effect of size and composition of the central electrode on the response of cylindrical ionization chambers in high-energy photon and electron-beams. *Phys. Med. Biol.* 38:267–90.

Mack, A., G. Mack, D. Weltz, S. G. Scheib, H. D. Bottcher, and V. Seifert. 2003. High precision film dosimetry with GAFCHROMIC (R) films for quality assurance especially when using small fields. *Med. Phys.* 30:2399–409.

Mack, A., S. G. Scheib, J. Major et al. 2002. Precision dosimetry for narrow photon beams used in radiosurgery-determination of Gamma Knife output factors. *Med. Phys.* 29(9):2080–9.

Mackie, T. R., J. W. Scrimger, and J. J. Battista. 1985. A convolution method of calculating dose for 15-MV x-rays. *Med. Phys.* 12(2):188–96.

Martens, C., C. De Wagter, and W. De Neve. 2000. The value of the PinPoint ion chamber for characterization of small field segments used in intensity-modulated radiotherapy. *Phys. Med. Biol.* 45(9):2519–30.
McJury, M., M. Oldham, V. P. Cosgrove et al. 2000. Radiation dosimetry using polymer gels: Methods and applications. *Br. J. Radiol.* 73:919–29.
McKerracher, C., and D. I. Thwaites. 1999. Assessment of new small-field detectors against standard-field detectors for practical stereotactic beam data acquisition. *Phys. Med. Biol.* 44:2143–60.
McKerracher, C., and D. I. Thwaites. 2007a. Head scatter factors for small MV photon fields. Part I: A comparison of phantom types and methodologies. *Radiother. Oncol.* 85:277–85.
McKerracher, C., and D. I. Thwaites. 2007b. Head scatter factors for small MV photon fields. Part II: The effects of source size and detector. *Radiother. Oncol.* 85:286–91.
McLaughlin, W. L., C. G. Soares, J. A. Sayeg et al. 1994. The use of a radiochromic detector for the determination of stereotaxic radiosurgery dose characteristics. *Med. Phys.* 21:379–88.
Mobit, P., and G. Sandison. 2002. A Monte Carlo based development of a cavity theory for solid state detectors irradiated in electron beams. *Radiat. Prot. Dosim.* 101:427–9.
Morin, J., D. Beliveau-Nadeau, E. Chung et al. 2013. A comparative study of small field total scatter factors and dose profiles using plastic scintillation detectors and other stereotactic dosimeters: The case of the CyberKnife. *Med. Phys.* 40(1):011719.
Mujsolino, S. V. 2001. Absorbed dose determination in external beam radiotherapy: An international code of practice for dosimetry based on standards of absorbed dose to water; technical reports series No. 398. *Health Phys.* 81:592–3.
Munro, P., J. A. Rawlinson, and A. Fenster. 1988. Therapy imaging: Source sizes of radiotherapy beams. *Med. Phys.* 15:517–24.
Niroomand-Rad, A., C. R. Blackwell, B. M. Coursey et al. 1998. Radiochromic film dosimetry: Recommendations of AAPM Radiation Therapy Committee Task Group 55. American Association of Physicists in Medicine. *Med. Phys.* 25:2093–115.
Nyholm, T., J. Olofsson, A. Ahnesjo, and M. Karlsson. 2006. Modeling lateral beam quality variations in pencil kernel based photon dose calculations. *Phy. Med. Biol.* 51:4111–8.
Pai, S., I. J. Das, J. F. Dempsey et al. 2007. TG-69: Radiographic film for megavoltage beam dosimetry. American Association of Physicists in Medicine. *Med. Phys.* 34(6):2228–58.
Pantelis, E., C. Antypas, L. Petrokokkinos et al. 2008. Dosimetric characterization of CyberKnife radiosurgical photon beams using polymer gels. *Med. Phys.* 35:2312–20.
Papanikolaou, N., B. Jerry, B. Arthur et al. 2004. Tissue Inhomogeneity Corrections for Megavoltage Photon Beams. Medical Physics Publishing. AAPM Report No. 85, College Park, MD: American Association of Physicists in Medicine.
Paskalev, K. A., J. P. Seuntjens, H. J. Patrocinio, and Podgorsak, E. B. 2003. Physical aspects of dynamic stereotactic radiosurgery with very small photon beams (1.5 and 3 mm in diameter). *Med. Phys.* 30:111–8.
Rice, R. K., J. L. Hansen, G. K. Svensson, and R. L. Siddon. 1987. Measurements of dose distributions in small beams of 6 Mv x-rays. *Phys. Med. Biol.* 32:1087–99.
Rikner, G., and E. Grusell. 1983. Effects of radiation damage on p-type silicon detectors. *Phys. Med. Biol.* 28(11):1261–7.
Rustgi, S. N. 1995. Evaluation of the dosimetric characteristics of a diamond detector for photon-beam measurements. *Med. Phys.* 22:567–70.
Sahoo, N., A. M. Kazi, and M. Hoffman. 2008. Semi-empirical procedures for correcting detector size effect on clinical MV x-ray beam profiles. *Med. Phys.* 35:5124–33.
Saini, A. S., and T. C. Zhu. 2004. Dose rate and SDD dependence of commercially available diode detectors. *Med. Phys.* 31:914–24.
Sauer, O. A., and J. Wilbert. 2007. Measurement of output factors for small photon beams. *Med. Phys.* 34:1983–8.
Scarboro, S. B., D. S. Followill, J. R. Kerns, R. A. White, and S. F. Kry. 2012. Energy response of optically stimulated luminescent dosimeters for non-reference measurement locations in a 6 MV photon beam. *Phys. Med. Biol.* 57:2505–15.
Scarboro, S. B., and S. F. Kry. 2013. Characterisation of energy response of Al(2)O(3):C optically stimulated luminescent dosemeters (OSLDs) using cavity theory. *Radiat. Prot. Dosimet.* 153:23–31.
Sharpe, M. B., D. A. Jaffray, J. J. Battista, and P. Munro. 1995. Extrafocal radiation: A unified approach to the prediction of beam penumbra and output factors for megavoltage x-ray beams. *Med. Phys.* 22:2065–74.
Solberg, T. D., and P. M. Medin. 2011. Quality and safety in stereotactic radiosurgery and stereotactic body radiation therapy: Can more be done? *J. Radiosurg. and SBRT* 1:13–9.
Stasi, M., B. Baiotto, G. Barboni, and G. Scielzo. 2004. The behavior of several microionization chambers in small intensity modulated radiotherapy fields. *Med. Phys.* 31:2792–5.
Sterpin, E., B. T. Hundertmark, T. R. Mackie, W. G. Lu, G. H. Olivera, and S. Vynckier. 2010. Monte Carlo–based analytical model for small and variable fields delivered by TomoTherapy. *Radiother. Oncol.* 94:229–34.

Vassiliev, O. N., T. A. Wareing, J. McGhee, G. Failla, M. R. Salehpour, and F. Mourtada. 2010. Validation of a new grid-based Boltzmann equation solver for dose calculation in radiotherapy with photon beams. *Phys. Med. Biol.* 55:581–98.

Weber, L., P. Nilsson, and A. Ahnesjo. 1997. Build-up cap materials for measurement of photon head-scatter factors. *Phys. Med. Biol.* 42:1875–86.

Westermark, M., J. Arndt, B. Nilsson, and A. Brahme. 2000. Comparative dosimetry in narrow high-energy photon beams. *Phys. Med. Biol.* 45:685–702.

Wilcox, E. E., G. M. Daskalov, H. Lincoln, R. C. Shumway, B. M. Kaplan, and J. M. Colasanto. 2010. Comparison of planned dose distributions calculated by Monte Carlo and ray-trace algorithms for the treatment of lung tumors with CyberKnife: A preliminary study in 33 patients. *Int. J. Radiat. Oncol. Biol. Phys.* 77(1):277–84.

Wurfel, J. U. 2013. Dose measurements in small fields. *Med. Phys. Int.* 1(1):81–90.

Zeidan, O. A., S. I. Sriprisan, O. Lopatiuk-Tirpak, P. A. Kupelian, and S. Meeks. 2010. Dosimetric evaluation of a novel polymer gel dosimeter for proton therapy. *Med. Phys.* 37(5):2145–52.

Zhu, T. C., A. Ahnesjo, K. L. Lam et al. 2009. Report of AAPM Therapy Physics Committee Task Group 74: In-air output ratio, S-c, for megavoltage photon beams. *Med. Phys.* 36:5261–91.

Zhu, T. C., and B. E. Bjarngard. 1994. The head-scatter factor for small field sizes. *Med. Phys.* 21:65–8.

Zhu, T. C., B. E. Bjarngard, and H. Shackford. 1995. X-ray source and the output factor. *Med. Phys.* 22:793–8.

Zhu, T. C., and K. Manbeck. 1994. CT reconstruction of the x-ray source profile of a medical accelerator. *Proc. SPIE* 2132:242–53.

Zhu, X. R., J. J. Allen, J. Shi, and W. E. Simon. 2000. Total scatter factors and tissue maximum ratios for small radiosurgery fields: Comparison of diode detectors, a parallel-plate ion chamber, and radiographic film. *Med. Phys.* 27:472–7.

Part V

New directions in stereotactic radiation therapy

Part V: New directions concludes this volume with a look at radiation biology and clinical outcomes over the last six decades and then looks at future developments.

Lars Leksell's idea of performing "bloodless" cranial surgery was the dawn of a new age in surgery. Just as brain surgery is the oldest known form of surgery, stereotactic radiosurgery ushered in a new era of minimally invasive surgery, which now includes robotic assistance. After a long, slow start, due to immature imaging technology, the twin fields of stereotactic radiosurgery and stereotactic radiotherapy have literally exploded. No major teaching center or regional medical center can be taken seriously if they do not have stereotactic radiosurgery (SRS) and stereotactic body radiation therapy (SBRT) capabilities.

Radiation biologists have also contributed enormously to this field. Pioneers such as Jack Fowler, Harold Rossi, and Rod Withers have provided good predictive models for the effects of time, dose, and fractionation in radiation therapy. These ideas have been transformative in prediction of response and in making suggestions for new clinical protocols. A new generation of highly sophisticated animal imaging and irradiation devices is now being created at major research centers that should provide more insight into SRT and SBRT.

Many years after the term "radiosurgery" was coined and the first patient treated, it is apparent that intracranial radiosurgery is now a completely accepted part of neurosurgery (and radiation oncology). New residents in these fields are expected by their prospective employers to be well versed in this field. Skull-base surgery, the most high-risk procedure in neurosurgery, is much less risky when a conservative neurosurgeon does *not* "go for broke" and attempt to resect every last tumor cell, which can lead to devastating consequences. Radiosurgery can very easily follow subtotal resections and "clean up" what is left safely and effectively.

The newer field of SBRT has had remarkable achievements in its short practice history. Patients with nonresectable lung tumors, liver tumors, and spinal tumors are often living productive lives years later when, in previous years, they may not have been treated at all or perhaps only with palliative therapy.

Finally, we look into the future of the technology of SRS and SBRT. New devices and techniques are being created continuously. Many new linear accelerators are now being marketed as "total solutions" or "hybrids," both for conventional nonstereotactic treatments as well as SRS and SBRT applications and each with completely integrated treatment and imaging suites. Other novel dedicated stereotactic radiation therapy (SRT) devices in development and presented here include a dedicated radiosurgery treatment device for breast cancer and hybrid Cobalt-60 and linacs built within an MRI for simultaneous imaging and treatment capabilities.

13 Radiation biology of stereotactic radiotherapy

Igor Barani, Zachary Seymour, Ruben Fragoso, and Andrew Vaughan

Contents

13.1 INTRODUCTION

The successful application of highly focused large radiation doses to tumor control can be traced back to the 1950s and the pioneering implementation of cranial stereotactic radiosurgery (SRS) by Lars Leksell (1983). The aim of SRS is to deposit a large, ablative dose of radiation within a defined tumor volume or selected nonmalignant target. The use of multiple low-intensity beams, which intersect at the target location, reduces normal tissue doses to safe levels. More recently, the SRS concept has been extended to extracranial body sites (stereotactic body radiotherapy [SBRT]), initially by Timmerman to medically inoperable lung cancer and later by others to primary and metastatic tumors of other extracranial organs (Leksell 1983; Timmerman et al. 2006, 2011). The ability to deliver large radiation doses to the tumor target safely in an abbreviated, "hypofractionated" schedule of treatments was facilitated by technological advances in imaging, motion management, and radiation delivery.

From a radiobiological perspective, the rationale for hypofractionation does not appear to follow the basic tenets of radiobiology that are traditionally invoked to explain the tumor responses achieved by conventional fractionation schemes (Withers 1975). During the 20th century, the radiation treatment schedule for most tumors evolved into a regimen of multiple, small fractions of irradiation. The governing scientific principles involved in this approach are widely accepted to be the classic four "Rs" of radiobiology (Withers 1975):

- Repair. Multiple small fractions of radiation allow for the repair of sublethal DNA damage in slowly dividing normal tissues, but less repair occurs in more rapidly cycling tumor cells (Time scale: hours-days).
- Reoxygenation. The availability of molecular oxygen is a major determinant of tumor cell radiosensitivity. Tumors can outgrow their blood supply and develop regions of poorly vascularized or necrotic tissue that are relatively hypoxic and up to three times more resistant than when fully oxygenated. The tumor burden reduction in the interfraction interval relieves interstitial pressure and allows some of these hypoxic cells/regions to reoxygenate, increasing cell kill in later fractions (Time scale: hours to days).
- Redistribution. Radiation sensitivity varies by cell cycle phase. After treatment, cells are usually preferentially depleted within the radiosensitive regions of the cell cycle (e.g., G2/M phase). Surviving cells, largely within less-sensitive phases of the cell cycle during first treatment, can redistribute to more radiosensitive phases between fractions and thereby be more susceptible to subsequent radiation injury (Time scale: hours).
- Repopulation. Tumor cells may divide and proliferate in the interfraction interval. Additionally, injury and cell death during treatment might induce or select an increased proportion of more rapidly dividing tumor cells that cause an accelerated repopulation. With lengthy treatment courses, there is a potential risk that these more rapidly dividing cells accumulate and then persist, leading to recurrence after treatment. Concern about accelerated repopulation is one of the motivations for the exploration of hypofractionated schedules for some cancers, notably head and neck cancer (Time scale: weeks).

These findings of classical radiobiology with a strong focus on DNA damage–dependent phenomena within tumor cells remain largely unchanged today and continue to form the basis of various radiobiological and mathematical models of disease control and normal tissue toxicity observed after conventionally fractionated radiotherapy. It is less clear whether these concepts are as relevant in the hypofractionated setting or at what fractional doses they become less relevant. The use of a small number of large fractions will offer less opportunity to exploit differential repair capacities between normal and tumor tissues with respect to sublethal damage and reduce the time frame for reoxygenation to occur (Carlson et al. 2011). On the other hand, repopulation will be of lesser importance as the time scale for treatment decreases. Similarly, redistribution of cells within the cell cycle will likely be different, but the clinical significance of this in a hypofractionated setting is unclear.

One conceptual distinction sometimes applied to the use of high-dose-per-fraction SRS and SBRT is the notion of its capacity to cause an ablative effect whereby the normal and/or tumor tissue is effectively converted into nonfunctional scar tissue. Examples include functional applications of SRS intended to shut down blood flow into an arteriovenous malformation or render a trigeminal nerve insensitive. In the SBRT setting, treatment of an isolated lung or liver tumor is given with an ablative intent when there is no clinical need to preserve the normally functioning parenchyma within or immediately adjacent to the target lesion. The "threshold" dose intensity when a treatment becomes ablative likely depends on the dose/fractionation schedule and the tumor/tissue histology being treated and might be largely mediated by vascular or other stromal effects.

13.2 RADIATION-INDUCED VASCULAR INJURY

13.2.1 VASCULAR ARCHITECTURE OF TUMORS

The vascular bed structure and physiology in tumors is markedly different than normal vasculature (Carmeliet and Jain 2000; Konerding, Miodonski, and Lametschwandtner 1995). The vessels formed from the tumors' angiogenic factors are often composed of a single-layer endothelium without innervations or the smooth muscle for autoregulation or basement membrane to prevent excessive extravasation. Gaps between endothelial cells make the vascular beds of aggressive tumors inherently leaky. The vessels are tortuous and form irregular patterns that vary in diameter and create sluggish flow and ultimately help account for the elevated interstitial pressure as well as the acidic, hypoxic, and nutritionally deprived environment within tumors. These hastily arranged vessels may mimic the more radiosensitive immature blood vessels of developing vascular beds, which have long been suspected, but only recently confirmed, to be more radiosensitive (Grabham et al. 2011; Park et al. 2012b). However, the vasculature is not uniform. Some normal vessels are incorporated into the tumor volume as tumors grow, and the architecture of benign and slow-growing tumors may ultimately resemble more normal vascular beds (Carmeliet and Jain 2000; Jain 2003).

13.2.2 CHARACTERISTICS OF RADIATION-INDUCED VASCULAR INJURY

Of the proposed "threshold" effects possibly driving tumor and normal tissue ablation, radiation-induced vascular change has the greatest support from experimental evidence with more than 40 studies on tumor vasculature and radiation having been reported in the last 60 years (Park et al. 2012a). As early as 1936, the interaction between blood supply and tumor radiosensitivity was implied when Mottram (1936) reported cancer cells at the periphery of tumors, where there is better blood supply, to be more responsive to radiation than cells in the necrotic center. With reoxygenation between fractions, the viable outer rim of the tumor is exposed to more oxygenated blood, and this interface increases in relative size as cytoreduction occurs from cancer cell death (Ng et al. 2007). Competing vascular changes, such as endothelial cell death and vessel obliteration, result in a more complicated picture of oxygenation. Overall, blood volume tends to increase at the beginning of a radiation course but decreases toward the end. This is particularly true for radiation courses longer than two weeks, which may suggest that reoxygenation effects may be more limited during later phases of more prolonged fractionation schemes (Mayr et al. 1996).

Traditionally, the impact of radiation on tumor vasculature was broadly viewed as a potential modulating effect of fractionated radiotherapy. Corroborating studies have shown that reduction in tumor vasculature correlates with improved treatment outcomes (Pirhonen et al. 1995). Vascular effects, particularly with high-dose treatments, are thought to be increasingly important and relevant as the body of supporting research in both animal and human xenograft models demonstrates dramatic vascular changes with fraction sizes as low as 8 Gy (Park et al. 2012a).

Levitt, Wong, and Song in the 1970s laid the groundwork for our current understanding of trends in extravasation and changes in vascular volume after single doses of high-dose irradiation. The irradiation of Walker 256 carcinoma in rats with a single dose of 30 Gy resulted in transient increase in the weight and size of the tumor 7–8 days after irradiation. This was then followed by a rapid decrease by day 15 (Song and Levitt 1971). In these tumors, the extravasation rate significantly increased almost immediately after irradiation before quickly declining after day 12, when the rate was well below baseline, and ultimately recovering 24 days after irradiation. The intravascular density and volume also decreased within 1 day after irradiation and continued to decline for 12 days. While smaller doses of radiation caused only a slight decrease in vascular volume, the decline in functional vascularity appears to be increasingly dose dependent, with doses between 5–20 Gy, and is substantially higher than that seen with fractionated radiotherapy (Chen et al. 2009; Kobayashi et al. 2004; Song, Payne, and Levitt 1972; Wong, Song, and Levitt 1973).

The leakiness of tumor blood vessels increases with increasing vascular damage and widening of gaps between endothelial cells. This effect is compounded by cytokine release from endothelial and surrounding stromal cells. However, cytokine-mediated injury seems to be mitigated by anti-inflammatory medications (Janssen et al. 2010; Kobayashi et al. 2004; Ng et al. 2007; Park et al. 2012a; Song, Drescher, and Tabachnick 1968; Song et al. 1974). Several mechanisms have been proposed to explain the increased leakiness of tumor vasculature: (1) Tumor vasculature may obliterate in an irregular fashion as already irregular vascular beds vasoconstrict, causing stasis (Song et al. 1974). (2) A decline in functional vascularity may develop as a consequence of transient vasoconstriction resulting from increased interstitial pressure. (3) Endothelial cell death and tumor shrinkage can variably change the configuration, volume, and hemodynamics of tumor vasculature (Kolesnick and Fuks 2003; Song et al. 1974).

Studies linking vascular change to radiation doses have used a variety of experimental techniques, including Doppler ultrasonography, MR, CT, angiography, ^{133}Xe clearance, colpophotography, and histopathology as well as window chambers (Park et al. 2012a). Regardless of the technique used, similar changes in vascularity are observed and do not seem to be dependent on the tumor type or its perceived relative radioresistance. For example, Solesvik, Rofstad, and Brustad (1984) revealed a 35% reduction in 5–15 micron diameter vessels in human melanoma xenographs one week after irradiation with a single dose of 10–15. Similar reductions in vascular density have been noted with fractional doses ≥4 Gy in human xenografts of colon and rectal carcinoma, nonsmall cell lung cancer, laryngeal squamous cell carcinoma, and ovarian cancer as well as in melanoma and glioblastoma rodent xenografts (Chen et al. 2009; Fenton, Lord, and Paoni 2001; Song and Levitt 1971; Song, Payne, and Levitt 1972; Song et al. 1974; Wong, Song, and Levitt 1973).

On direct visualization through window chambers, 5-Gy irradiation leads to increased vascular density and perfusion for 24–72 hr in R3230 mammary adenocarcinoma, but doses of 20–50 Gy result in progressive narrowing of the blood vessels and decreased vascular density (Dewhirst et al. 1990; Merwin, Algire, and Kaplan 1950). In Lewis lung carcinoma grown subcutaneously in mice, a single dose of 20 Gy causes substantial reduction in tumor blood flow, but return to baseline is seen by 4 days following irradiation. This recovery is suppressed if the tumors are again irradiated with 20 Gy on the second day (Kim et al. 2006). Interestingly, delay in delivery of a second fraction (on day four) is not as effective as treatment on day two in terms of tumor perfusion, suggesting that minor changes in the fractionation schedule may result in potentially significant outcome differences.

In aggregate, these studies (even though heterogeneous in terms of tumor type, radiation dose, and schedule as well as the choice of model system and choice of experimental techniques) tend to support the notion that increasing vascular damage occurs with increasing dose and likely influences tumor and potentially normal tissue response to irradiations. A few generalizations emerge from these data: (1) Doses of 5–10 Gy (per fraction) lead to transient increase in tumor blood flow before returning to (or close to) baseline levels 2–3 days following irradiation. (2) Increasing fractional dose in the 5–10 Gy (per fraction) range produces a more persistent (but variable) vascular perfusion decrease that often returns to baseline levels over time. (3) Between 15 and 25 Gy (per fraction), a fairly rapid, widespread, and likely persistent functional (perfusion) and structural vascular loss ("vascular dropout") occurs (Park et al. 2012a).

13.2.3 CELLULAR MECHANISMS OF RADIATION-INDUCED VASCULAR INJURY

Classical radiation biology has focused almost exclusively on radiation-induced DNA damage as a direct mediator of cell death. However, it has been suggested that dose-dependent vascular effects from high doses of radiation (≥8 Gy) may be mediated in part by the acidic-sphingomyelinase (ASMase) pathway and lead to generation of ceramide within tumor vascular endothelium (Kolesnick and Fuks 2003). ASMase hydrolyzes sphingomyelin to generate pro-apoptotic ceramide, which can act as a secondary messenger and cause a mitochondrial mediated apoptotic response or create membrane rafts to alter extracellular and intracellular signaling (Gajate and Mollinedo 2001; Hueber et al. 2002; Kolesnick, Goni, and Alonso 2000; Zundel and Giaccia 1998). The secretory form of ASMase is found in endothelial cells at 20-fold higher concentrations than any other cell in the body and rapidly translocates from the cytosol to cholesterol-enriched rafts in the cell membrane after high doses of radiation in both normal and tumor-related vasculature (Fuks, Haimovitz-Friedman, and Kolesnick 1995; Garcia-Barros et al. 2003; Paris et al. 2001; Santana et al. 1996). Ceramide stimulates the Bax pathway of pro-apoptotic signals with resultant cytochrome C release from the mitochondria (Kolesnick and Fuks 2003). Garcia-Barros et al. (2003) reported ceramide-mediated apoptosis in tumor endothelial cells with a single dose of 15–20 Gy 1–6 hr after exposure; the effect that was blocked in ASMase and Bax knockout mice. Tumor cells remained intact until 48–72 hr after initial exposure and then began to show signs of treatment effect and suggested that early-phase microvascular apoptosis was necessary to achieve tumor control in these models. The apparent threshold to induce the ASMase pathway is 8–10 Gy and appears to have a dose response up to 20–25 Gy (Fuks and Kolesnick 2005). This is consistent with experimental data showing increasing degrees of vascular injury and durability of this effect with increasing doses of radiation (as discussed above). More work is needed before these pathways are fully understood and universally accepted although these new insights offer promising glimpses into mechanisms responsible for the dramatic biological effects of SBRT.

By extension, ceramide-mediated apoptosis observed after high-dose hypofractionated radiotherapy may lead to better understanding of relative radioresistance to conventionally fractionated radiation in which induction of HIF-1 may lead to endothelial preservation and thereby, by extension, preservation of vasculogenesis and angiogenesis (Moeller et al. 2004, 2005). Each radiation treatment creates a wave of HIF-1 activation from preformed mRNA transcripts stimulated by reactive oxygen species. HIF-1 engages more than 60 target genes with a hypoxia response element, including, vascular epithelial growth factor (VEGF), which promotes radioresistance of vascular epithelium as well as angiogenesis (Semenza 2003). Although HIF-1 can also promote apoptosis, it requires p53, which is mutated in more than half of tumors and therefore would favor radioresistance rather than apoptosis in most tumors. HIF-1 induced by fractionated radiation appears in many respects a perfect foil to the vascular effects of radiation although further data is required (Dewhirst et al. 2007). Observations by Kioi et al. (2010) also suggest that vasculogenesis, implying the process of recruitment of bone marrow–derived vascular precursor cells and not angiogenesis alone plays a role in tumor recurrence. In a series of experiments, the investigators inhibited vasculogenesis by blocking bone marrow–derived cells from infiltrating the irradiated regions of glioblastoma xenografts by either HIF-1 activation or HIF-1–induced CXCR4/SDR interactions, which are necessary for cells to infiltrate damaged tissues (Figures 13.1 and 13.2).

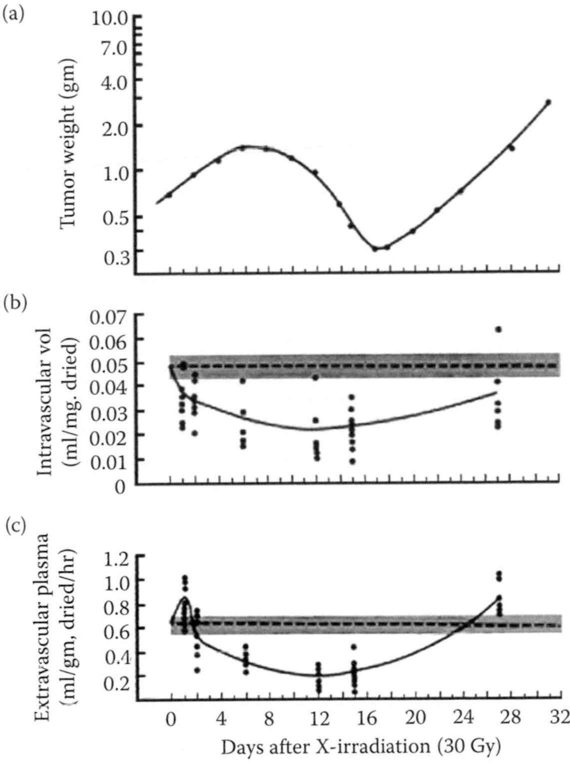

Figure 13.1 Effects of 30-Gy radiation given in a single dose on the tumor size and vascular functions in Walker 256 tumors (s.c.) grown in the legs of Sprague-Dawley rats. (a) Dried tumor weight. (b) Intravascular volume. (c) Extravasation rate of plasma protein. The solid lines in each panel indicate the means of 6–10 tumors used at the different times indicated. The dotted lines in each panel are the mean values of 15 control tumors weighing 0.3–2.0 g. The shaded areas show the range of standard error of the mean. (From Park, H. J. et al., *Radiat. Res.*, 177, 3, 311–27, 2012.)

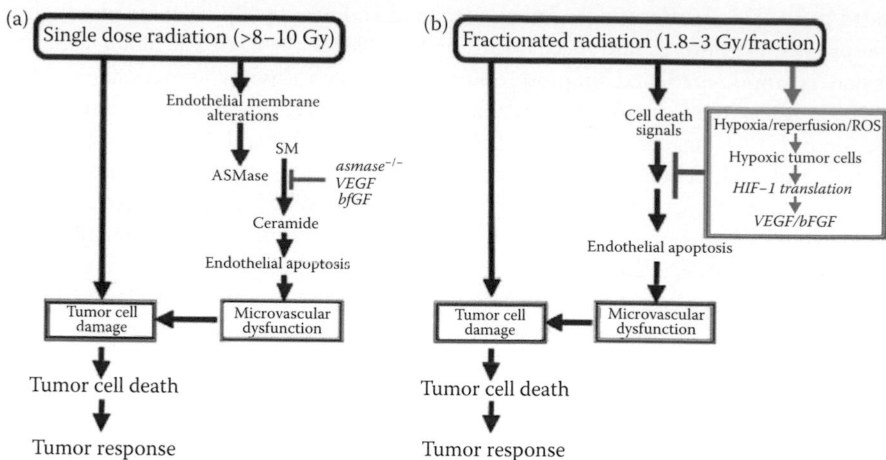

Figure 13.2 Models of microvascular endothelial engagement in tumor response to single-dose or fractionated radiotheraphy. Endothelial damage appears to be induced by both the high treatment doses (>8–10 Gy) of single-dose radiotherapy (a) and the low-dose (1.8–3 Gy) exposures of fractionated radiotherapy (b). The resulting microvascular dysfunction confers conversion of sublethal radiation lesions in tumor cells into lethal lesions via as yet unknown mechanism. (From Fuks Cancer Cell.)

13.3 TUMOR MICROENVIRONMENT: STROMAL, PARACRINE, VASCULAR, AND IMMUNE RESPONSES

13.3.1 VASCULARITY AND MICROENVIRONMENT

The understanding of the tumor microenvironment and its impact on tumor response to treatment as well as tumor-stromal interactions is still incomplete. The vasculature is a key component of the microenvironment with a single endothelial cell having the ability to influence as many as 2000 surrounding tumor cells (Denekamp 1984). Given the serial arrangement of most vascular beds, obliteration of a single vessel can potentially have major downstream effects on the tumor and its microenvironment, including adjacent normal tissues.

Radiation-induced vascular changes can cause rapid changes in oxygen tension and perfusion, resulting in formation of relative regions of hypoxia. Such hypoxic regions create a less hospitable environment for cell growth and also diminish local effects of low-dose irradiation through well-understood processes. For example, Kioi et al. (2010) described glioblastoma xenografts in nude mice irradiated with a single dose of 15 Gy showing not only a decrease in functional vascularity, but also an increase in hypoxic regions at 17–18 days post treatment. Similarly, in prostate cancer, Chen et al. (2009) noted substantial reduction in vascular density (25%) at three weeks after a single 20-Gy dose. The surrounding perivascular areas were also notable for increased hypoxia (Chen et al. 2009). However, despite these promising early reports, robust data on the impact of the tumor microenvironment, and hypoxia in particular, after high doses of irradiation is sparse and requires further corroboration.

Oxygen tension can vary widely within a tumor, and hypoxia can occur on a microscopic scale. The classic work of Thomlinson and Gray (1955) established that oxygen diffusion may be compromised with distances between tumor cells and a vascular supply as small as 100 μm.

As a result, variable regions of hypoxia might be partially to blame for inconsistent experimental data on oxygen tension within tumors. Most studies assessing oxygen tension were conducted prior to key technological advancements required for accurate data collection (Molls et al. 1998). Modern studies assessing hypoxia in tumors have highly variable results with tumor oxygen tensions either increased, stable, or decreased even among tumors of the same histological subtype (Brizel et al. 1999; Cooper et al. 1999; Dunst et al. 1999; Fyles et al. 1998; Lyng et al. 2001; Stadler et al. 1998). The only discernible trend is that overall pO2 in human tumors tends to decrease toward the end of a prolonged (>2 weeks) fractionated radiation course (Molls et al. 1998).

Beyond the issues directly related to oxygenation, the role of blood vessels in development of cancer continues to be defined. The perivascular niche is comprised of the cellular connections and paracrine interactions of endothelial cells; pericytes; lymphocytes; organ-dependent supportive cells, such as glial cells in the brain; and normal stem cells. There is an increasing body of work that suggests some cancer stem cells (CSCs) exploit a perivascular niche (Al-Hajj et al. 2003; Lapidot et al. 1994; Singh et al. 2004). The CSCs are self-renewing, multipotent, and tumor-initiating (Calabrese et al. 2006). They may not be conserved across all malignancies but have been found in hematologic malignancies, gliomas, head and neck, lung, pancreatic, prostate, hepatic, and colon cancers (Dalerba et al. 2007; Lapidot et al. 1994; Li et al. 2007; Ritchie and Nör 2012). Not only can they exploit the same niche, but they appear, as in the example of CD133 and Nestin expressing CSC in the brain, to mimic normal stem cells in gene expression. The greatest preponderance of research regarding the perivascular niche is in the brain, but similar findings have been found in head and neck cancer and neuroblastoma, suggesting that exploitation of the perivascular niche is necessary for tumor growth and that vascular changes play an important role in recurrence for many tumor types (Calabrese et al. 2006; Pietras et al. 2008; Prince et al. 2007; Ritchie and Nör 2012). Disturbance of the perivascular niche by large doses of irradiation may therefore impact the role of CSCs in tumor recurrence.

13.3.2 IMMUNE RESPONSE

Traditionally, radiotherapy has been thought of as being immunosuppressive. Lymphocytes and myeloid progenitors are notoriously radiosensitive although it is unknown if by varying delivery (e.g., dose and fractionation), irradiation can independently activate T-cells or regulatory molecules to reliably stimulate an immune response. Even within the brain, single-dose irradiation has been found to cause an intense inflammatory reaction characterized by infiltration of predominantly CD68-positive macrophages and CD3-positive T lymphocytes. Interestingly, the magnitude or intensity of tumor tissue infiltration seems to correlate with improved local tumor control (Szeifert et al. 2005, 2012). There are also multiple reports of an abscopal effect in the literature, a phenomenon with which limited irradiation causes a T-cell–mediated tumor regression at distant sites, often far away from the treated site (Demaria et al. 2004). It is unclear how or if this phenomenon can be better understood and/or exploited for therapeutic purposes given its relative rarity.

Immune modulation can occur through a variety of radiation-induced mechanisms, such as stimulation of toll-like receptor 4 dendritic cells with downstream CD8+ T-cell formation leading to a modified tumor cell phenotype that is more susceptible to T-cell–mediated cell killing. Alternatively, the focal alteration of the tumor microenvironment can cause systemic stimulation of effector immune cells (Apetoh et al. 2007; Chakraborty et al. 2004; Lee et al. 2009). A single dose of 8 Gy of focal irradiation to the tumor can also lead to induction and upregulation of the Fas receptor, a so-called "death" receptor, in situ for up to 11 days (Chakraborty et al. 2004). Ablative doses of 15–25 Gy (single fraction) generated a CD8+ T-cell–driven immune response within the draining lymphatic sites and resultant reduction of the primary site tumor as well as eradication of some disease at distant sites (Lee et al. 2009).

Not all data supports immune modulation effects with a single-fraction of high-dose irradiation. In breast cancer xenografts, primary site irradiation to a high-dose (20 Gy, single fraction) caused significant cytoreduction of the primary tumor but secondary tumor sites appeared to be unaffected without the presence of CTLA-4 antibodies, and then this phenomenon was only seen for fractionated radiotherapy schemes of three to five fractions of 8 or 6 Gy, respectively (Dewan et al. 2009). This response corresponded to the degree of tumor-specific CD8+ T-cells produced and presence of CTLA-4 antibodies (Masucci et al. 2012).

13.3.3 PARACRINE AND STROMAL EFFECTS

Autocrine and paracrine growth factors have an important influence on the tumor microenvironment and directly impact cancer growth and proliferation. Autocrine-paracrine loops have been described in non-small cell lung cancer with hepatocellular growth factor, and in theory, a small volume of cancer cells outside of the planned target volume could be controlled if the paracrine loops could be substantially reduced (Siegfried et al. 1997). In addition, niches other than perivascular are starting to be recognized during investigations of autocrine and paracrine effects in tumors (Malanchi et al. 2011).

There are no consistent biologically derived standards that justify and establish treatment margins for benign or malignant lesions; most treatment margins and fields are derived clinically from historical observations and patterns of failure studies (Story, Kodym, and Saha 2008). Many such studies assume direct and focal radiation effects on the treated tissue but rarely consider effects at or beyond the treatment margin. It is plausible to consider that margin size can modulate the stromal effects that impact both tumor and normal tissue response to radiation therapy. Tumor cells often resemble and exploit their tissues of origin and organize niches, perivascular and stromal. Normal tissue irradiation adjacent to tumor cells may have effects that both promote recurrence, as with HIF-1 activation, as well as inhibit growth and recurrence. For example, co-culturing normal fibroblast cells increases the radiosensitivity of mammary carcinoma and illustrates how little is known about tumor-stromal effects (Rossi et al. 2000). Similarly, radiation is believed to alter the immediate tumor microenvironment such that it inhibits perineural invasion and reduces the likelihood of recurrence (Bakst et al. 2012; Baumann et al. 1994; Milas et al. 1988).

Recurrence in preirradiated tissue may occur in part through HIF-1 upregulation and/or possibly through other factors, such as VEGF-induced motility of glioma cells post-radiation (Kil, Tofilon, and Camphausen 2012; Moeller et al. 2004). Tumors that recur in the setting of previous irradiation may behave very differently due to an altered microenvironment (Rofstad et al. 2005). Further research is required to better understand the impact of radiation on normal stroma as well as differential radiation effects observed among patients. The expanding body of research on the complex and interrelated components that govern radiation response of tissues is leading to identification of factors that can modulate both normal and tumor tissue responses. These can potentially be exploited in therapeutic applications to either (or both) improve tumor control and/or mitigate normal tissue toxicities.

13.4 RADIOBIOLOGICAL MODELING OF SRT: NORMAL TISSUE TOLERANCES

Fractionation was implemented to decrease complications related to the inclusion of large volumes of normal tissue. Prolonged fractionation courses allow repair of normal tissues but also repopulation of the cancer cells and alterations in the target volume that may increase toxicity and decrease control. Emami et al. (1991) first estimated partial volume tolerance limits for most body organs assuming conventional fractionation (1.8 to 2 Gy/fraction) based on viewing the organ at risk as one third, two thirds, or the whole organ exposed to each treatment.

With the advent of 3-D imaging and image-processing technologies, it was possible to further reduce treatment fields and to accurately calculate absorbed doses based on tissue-density tables. These developments, coupled with improved immobilization and treatment verification systems, allow complex, small-field treatments to be carried out. SRT is the ultimate extension of this paradigm and requires knowledge of dose limits to small or very small tissue volumes. This is further compounded by the use of high doses over one or a few treatments. Early attempts to extend classical radiobiologic models of normal tissue toxicity to SRT treatment regimens dramatically overestimated complication rates. For example, rates of radiation pneumonitis or myelopathy with SBRT are well below expectations of our models for conventionally fractionated radiation (Daly et al. 2012; Kavanagh 2008; McGarry et al. 2005).

Increasing the dose per fraction increases the theoretical risk of long-term complications, and robust and long-term clinical follow up is necessary to better inform our models. To date, no existing model has been universally accepted for high-dose hypofractionated, small-field radiotherapy. Because practical needs necessitate estimation of risk, various modifications and/or new models were introduced to help estimate such risk. These are briefly summarized below.

13.4.1 CLINICAL ESTIMATES OF NORMAL TISSUE COMPLICATIONS

Current estimates of normal tissue tolerances for most SBRT treatments are based on clinical data. In some cases, estimates of tolerance doses remain speculative. The American Association of Physicists in Medicine (AAPM) Task Group 101 publication offers suggested dose constraints to apply in SBRT regimens of one, three, or five fractions (Table 13.1) (Benedict et al. 2010).

Table 13.1 Summary of suggested dose constraints for various critical organs

		ONE FRACTION		THREE FRACTIONS		FIVE FRACTIONS		
SERIAL TISSUE	MAX CRITICAL VOLUME ABOVE THRESHOLD	THRESHOLD DOSE (Gy)	MAX POINT DOSE (Gy)**	THRESHOLD DOSE (Gy)	MAX POINT DOSE (Gy)**	THRESHOLD DOSE (Gy)	MAX POINT DOSE (Gy)**	ENDPOINT (≥GRADE 3)
Optic pathway	<0.2 cc	8 Gy	10 Gy	15.3 Gy (5.1 Gy/fx)	17.4 Gy (5.8 Gy/fx)	23 Gy (4.6 Gy/fx)	25 Gy (5 Gy/fx)	Neuritis
Cochlea			9 Gy		17.1 Gy (5.7 Gy/fx)		25 Gy (5 Gy/fx)	Hearing loss
Brainstem (not medulla)	<0.5 cc	10 Gy	15 Gy	18 Gy (6 Gy/fx)	23.1 Gy (7.7 Gy/fx)	23 Gy (4.6 Gy/fx)	31 Gy (6.2 Gy/fx)	Cranial neuropathy
Spinal cord and medulla	<0.35 cc <1.2 cc	10 Gy 7 Gy	14 Gy	18 Gy (6 Gy/fx) 12.3 Gy (4.1 Gy/fx)	21.9 Gy (7.3 Gy/fx)	23 Gy (4.6 Gy/fx) 14.5 Gy (2.9 Gy/fx)	30 Gy (6 Gy/fx)	Myelitis
Spinal cord subvolume (5–6 mm above and below level treated per Ryu)	<10% of subvolume	10 Gy	14 Gy	18 Gy (6 Gy/fx)	21.9 Gy (7.3 Gy/fx)	23 Gy (4.6 Gy/fx)	30 Gy (6 Gy/fx)	Myelitis
Cauda equina	<5 cc	14 Gy	16 Gy	21.9 Gy (7.3 Gy/fx)	24 Gy (8 Gy/fx)	30 Gy (6 Gy/fx)	32 Gy (6.4 Gy/fx)	Neuritis
Sacral plexus	<5 cc	14.4 Gy	16 Gy	22.5 Gy (7.5 Gy/fx)	24 Gy (8 Gy/fx)	30 Gy (6 Gy/fx)	32 Gy (6.4 Gy/fx)	Neuropathy
Esophagus*	<5 cc	11.9 Gy	15.4 Gy	17.7 Gy (5.9 Gy/fx)	25.2 Gy (8.4 Gy/fx)	19.5 Gy (3.9 Gy/fx)	35 Gy (7 Gy/fx)	Stenosis/fistula
Brachial plexus	<3 cc	14 Gy	17.5 Gy	20.4 Gy (6.8 Gy/fx)	24 Gy (8 Gy/fx)	27 Gy (5.4 Gy/fx)	30.5 Gy (6.1 Gy/fx)	Neuropathy

(continued)

Table 13.1 (Continued) Summary of suggested dose constraints for various critical organs

		ONE FRACTION		THREE FRACTIONS		FIVE FRACTIONS		
SERIAL TISSUE	MAX CRITICAL VOLUME ABOVE THRESHOLD	THRESHOLD DOSE (Gy)	MAX POINT DOSE (Gy)**	THRESHOLD DOSE (Gy)	MAX POINT DOSE (Gy)**	THRESHOLD DOSE (Gy)	MAX POINT DOSE (Gy)**	ENDPOINT (≥GRADE 3)
Heart/pericardium	<15 cc	16 Gy	22 Gy	24 Gy (8 Gy/fx)	30 Gy (10 Gy/fx)	32 Gy (6.4 Gy/fx)	38 Gy (7.6 Gy/fx)	Pericarditis
Great vessels	<10 cc	31 Gy	37 Gy	39 Gy (13 Gy/fx)	45 Gy (15 Gy/fx)	47 Gy (9.4 Gy/fx)	53 Gy (10.6 Gy/fx)	Aneurysm
Trachea and large bronchus*	<4 cc	10.5 Gy	20.2 Gy	15 Gy (5 Gy/fx)	30 Gy (10 Gy/fx)	16.5 Gy (3.3 Gy/fx)	40 Gy (8 Gy/fx)	Stenosis/fistula
Bronchus-smaller airways	<0.5 cc	12.4 Gy	13.3 Gy	18.9 Gy (6.3 Gy/fx)	23.1 Gy (7.7 Gy/fx)	21 Gy (4.2 Gy/fx)	33 Gy (6.6 Gy/fx)	Stenosis with atelectasis
Rib	<1 cc <30 cc	22 Gy	30 Gy	28.8 Gy (9.6 Gy/fx) 30.0 Gy (10.0 Gy/fx)	36.9 Gy (12.3 Gy/fx)	35 Gy (7 Gy/fx)	43 Gy (8.6 Gy/fx)	Pain or fracture
Skin	<10 cc	23 Gy	26 Gy	30 Gy (10 Gy/fx)	33 Gy (11 Gy/fx)	36.5 Gy (7.3 Gy/fx)	39.5 Gy (7.9 Gy/fx)	Ulceration
Stomach	<10 cc	11.2 Gy	12.4 Gy	16.5 Gy (5.5 Gy/fx)	22.2 Gy (7.4 Gy/fx)	18 Gy (3.6 Gy/fx)	32 Gy (6.4 Gy/fx)	Ulceration/ fistula
Duodenum*	<5 cc <10 cc	11.2 Gy 9 Gy	12.4 Gy	16.5 Gy (5.5 Gy/fx) 11.4 Gy (3.8 Gy/fx)	22.2 Gy (7.4 Gy/fx)	18 Gy (3.6 Gy/fx) 12.5 Gy (2.5 Gy/fx)	32 Gy (6.4 Gy/fx)	Ulceration
Jejunum/ileum*	<5 cc	11.9 Gy	15.4 Gy	17.7 Gy (5.9 Gy/fx)	25.2 Gy (8.4 Gy/fx)	19.5 Gy (3.9 Gy/fx)	35 Gy (7 Gy/fx)	Enteritis/ obstruction

Colon*	<20 cc	14.3 Gy	18.4 Gy	24 Gy (8 Gy/fx)	28.2 Gy (9.4 Gy/fx)	25 Gy (5 Gy/fx)	38 Gy (7.6 Gy/fx)	Colitis/fistula
Rectum*	<20 cc	14.3 Gy	18.4 Gy	24 Gy (8 Gy/fx)	28.2 Gy (9.4 Gy/fx)	25 Gy (5 Gy/fx)	38 Gy (7.6 Gy/fx)	Proctitis/fistula
Bladder wall	<15 cc	11.4 Gy	18.4 Gy	16.8 Gy (5.6 Gy/fx)	28.2 Gy (9.4 Gy/fx)	18.3 Gy (3.65 Gy/fx)	38 Gy (7.6 Gy/fx)	Cystitis/fistula
Penile bulb	<3 cc	14 Gy	34 Gy	21.9 Gy (7.3 Gy/fx)	42 Gy (14 Gy/fx)	30 Gy (6 Gy/fx)	50 Gy (10 Gy/fx)	Impotence
Femoral heads (right & left)	<10 cc	14 Gy		21.9 Gy (7.3 Gy/fx)		30 Gy (6 Gy/fx)		Necrosis
Renal hilum/ vascular trunk	<2/3 volume	10.6 Gy		18.6 Gy (6.2 Gy/fx)		23 Gy (4.6 Gy/fx)		Malignant hypertension
		ONE FRACTION		THREE FRACTIONS		FIVE FRACTIONS		
PARALLEL TISSUE	MINIMUM CRITICAL VOLUME BELOW THRESHOLD	THRESHOLD DOSE (Gy)	MAX POINT DOSE (Gy)**	THRESHOLD DOSE (Gy)	MAX POINT DOSE (Gy)**	THRESHOLD DOSE (Gy)	MAX POINT DOSE (Gy)**	ENDPOINT (≥GRADE 3)
Lung (right & left)	1500 cc	7 Gy	NA–parallel tissue	10.5 Gy (3.5 Gy/fx)	NA–parallel tissue	12.5 Gy (2.5 Gy/fx)	NA–parallel tissue	Basic lung function
Lung (right & left)	1000 cc	7.4 Gy	NA–parallel tissue	11.4 Gy (3.8 Gy/fx)	NA–parallel tissue	13.5 Gy (2.7 Gy/fx)	NA–parallel tissue	Pneumonitis
Liver	700 cc	9.1 Gy	NA–parallel tissue	17.1 Gy (5.7 Gy/fx)	NA–parallel tissue	21 Gy (4.2 Gy/fx)	NA–parallel tissue	Basic liver function
Renal cortex (right & left)	200 cc	8.4 Gy	NA–parallel tissue	14.4 Gy (4.8 Gy/fx)	NA–parallel tissue	17.5 Gy (3.5 Gy/fx)	NA–parallel tissue	Basic renal function

13.4.2 MODELS OF NORMAL TISSUE COMPLICATION PROBABILITY

Dose per fraction has always been tailored to provide a "reasonable" outcome with regard to risk-benefit analysis of tumor control and toxicity. One model for normal tissue complication probability (NTCP) for a given organ at risk (OAR) with nonuniform, partial organ irradiation is the Lyman-Kutcher-Burman model with standard fractionation (Kutcher and Burman 1989; Kutcher et al. 1991; Lyman 1985).

$$NTCP = 1 - e^{R}$$

where R is related to dose and volume.

$$R = -\ (d/d_0)^k$$

where k is a constant correlating volume and d_0 determines the slope of the sigmoidal curve for a given clinical endpoint. As volume and dose increase, toxicity should also increase. Application of the similar dose-volume models have been applied to estimate various toxicity endpoints, including cerebral radionecrosis, cranial neuropathy, myelopathy, pneumonitis, and duodenal toxicity at individual institutions with mixed results (Daly et al. 2012; Meeks et al. 2000; Murphy et al. 2010; Wennberg et al. 2011).

13.4.3 COMPARING DOSES FOR NORMAL TISSUE EFFECTS

The linear-quadratic (LQ) model of cell survival after radiotherapy can also be adapted to compare the efficacy of differing dose and fractionation. Calculation of a biological equivalent dose (BED) is commonly used in clinical practice despite its well-acknowledged limitations at fractional doses (>6 Gy).

$$\frac{D_1}{D_2} = \left(\frac{\alpha}{\beta} + d_2\right) \Big/ \left(\frac{\alpha}{\beta} + d_1\right)$$

where D is the isoeffective total dose for a given tissue with a defined α/β ratio (empirically estimated), and d is the prescribed fractional dose. The α/β of a tissue can vary but is generally thought to be approximately three for late-responding tissues (e.g., brain, spinal cord, prostate); for early-responding tissues, including tumors, it is commonly estimated to be ~10. It is also worth noting that the α/β ratio can vary considerably for different histologies, depending on whether they are thought to behave more like early- or late-responding tissues. The BED concept is used to estimate the likely effects on both tumor and normal tissues and is most relevant for conventionally fractionated RT. Inherent in this formulation is the well-established LQ model (see below) of radiation toxicity.

13.4.4 DOSE-DEPENDENT CELL KILL CONCEPT

Analysis of clonogenic cell survival after irradiation formed the basis of classical radiobiology research for more than half a century (Puck and Marcus 1956). For human cells, the dose response curves commonly have a curvilinear appearance as shown in Figure 13.3. For sparsely ionizing radiation at low doses (≤5 Gy), a shoulder is apparent which terminates in a straight line on a log-linear plot. Without invoking any mechanistic analysis of the dose response, it is clear that the dose region relevant to conventional fractionation has a different dose-response curve than higher fractional doses of radiation (>5 Gy). The survival of cells after conventional irradiation (1.8–2 Gy/fraction) has been interpreted using the theory of dual radiation action (Fowler 1989; Kellerer and Rossi 1978). According to this theory and supporting experimental data, radiation fragments individual chromosomes such that it increases the possibility of incorrect rejoining, creating aberrant chromosomal structures that are not compatible with cell survival. These include dicentrics, chromosome chimeras that contain two centromeres (Figure 13.4). These aberrations interrupt normal mitotic processes by attaching to the spindle of each potential daughter cell,

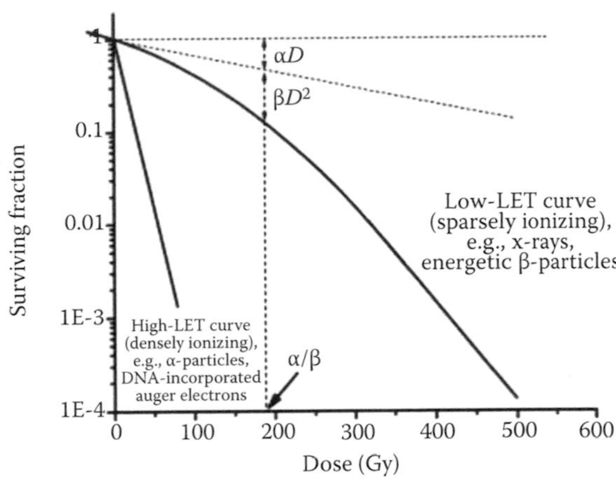

Figure 13.3 Mammalian cell survival curves after high- and low-LET irradiation. With high-LET radiation (alpha particles and non-energetic electrons), curve shows exponential decrease in cell survival; with low-LET radiation (energetic electrons), curve exhibits a shoulder. (From Kassiss, A., *Semin. Nucl. Med.*, 38, 5, 358–66, 2008.)

physically restricting mitosis and killing the cell as it attempts to divide. This mechanism has been linked to a mathematic interpretation that is encapsulated in the LQ formalism as

$$E = n\,(\alpha d + \beta d^2)$$

where E is the biological radiation effect, n is the number of fractions, and d is the fractional dose. In some cases, it is useful to use E/α or BED to rewrite the equation and add factors to account for repopulation effects in prolonged treatments:

$$BED = nd\left(1 + \frac{d}{\frac{\alpha}{\beta}}\right) - \ln 2\left(\frac{T - T_k}{\alpha T_p}\right)$$

where T is the overall treatment time in days, T_k is the time at which proliferation begins, and T_p is the potential doubling time. Here, the two coefficients used, α and β, represent chromosome damage that is

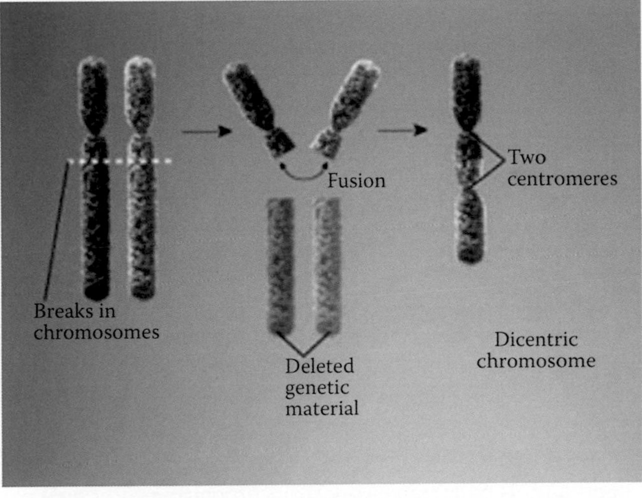

Figure 13.4 Generation of a dicentric by simultaneous fragmentation of two chromosomes. (From National Library of Medicine.)

formed by a single ionization event (αD) or by two separate such events (βd^2) that result in lethal lesions. In the latter case, individual lesions may be repaired independently and, therefore, they do not immediately result in cell death, in effect, reducing the number of lethal lesions. Because two independent lesions are involved, a power term is used in the equation above and determines the "bend" in the cell survival curve. This mechanistic interpretation underlies many of the current radiobiological models but grossly oversimplifies clinical observations (Fowler 2010).

The application of the LQ model in the high dose region (large fraction sizes) is limited by the mathematically predicted continued bending of the survival curve due to the contribution of the β component (Figure 13.3). All studies, however, tend to show that human survival curves do not exhibit such bending but approach a linear response on the normal log-linear plots (Fertil and Malaise 1985). Thus, the LQ model is not readily extrapolated to high-dose SBRT treatments without modifications. The utility of the LQ equation for modeling the response to high-dose irradiation is contentious and discussed further below (Brenner 2008; Kirkpatrick, Meyer, and Marks 2008).

Although these equations provide a useful guide for comparing effects of different treatment schedules, they assume that observed clinical responses can be explained solely by tumor cell kill that adheres closely to that predicted by the LQ equation. Although this may be true for some tumor types that are exquisitely sensitive to irradiation, this assumption does not always hold and appears to fail when applied to short-course hypofractionated regimens in which a threshold trigger may apply as discussed above.

13.5 PREDICTING HYPOFRACTIONATED RADIATION RESPONSES

Predicting the response of a tumor to irradiation is central to the therapeutic application of radiation. Below, we review several radiobiological and mathematical models attempting to predict tissue responses to hypofractionated treatment regimens.

13.5.1 TUMOR CONTROL PROBABILITY (TCP)

Ever-expanding clinical applications for SBRT have generated considerable interest in tools or models that would predict the biological effect of such aggressive treatment schemes on both the tumor and normal tissue. As discussed above, central to most models is the implicit assumption that tumor control requires eradication of all clonogens that can potentially repopulate the tumor. Various TCP models have been developed with this assumption in mind. Parallel models also exist for prediction of NTCPs. Both model types generally result in sigmoidal survival curves (Figure 13.5). The therapeutic ratio is governed by the difference (e.g., separation) of the TCP and NTCP curves and varies for different tumor subtypes.

Early attempts by Munro and Gilbert (1961) applied Poisson-like distributions to model TCP. These models are based on the LQ assumptions of cell kill and typically involve the linear component ($e^{-\alpha D}$) to represent the surviving fraction. The linear component dominates in the low-dose regions of the curve that are characteristic of many conventionally fractionated radiation regimens. If the goal is to completely eradicate all the tumor clonogens, then the probability of obtaining tumor control for N_x clonogens exposed to total dose D is approximated by the equation:

$$TCP = e^{-N_x SF}$$

where SF is the surviving fraction of cells.

These initial models used simple assumptions and did not involve more complex parameters later found to be important. They are time independent and do not consider the influence of tumor repopulation although the latter may not be as relevant to short-course hypofractionation schemes. Interestingly, most of these models, when applied to various clinical scenarios, tend to generally underestimate the observed TCP (Withers, Thames, and Peters 1983).

More recent models build on this basic framework by incorporating additional parameters to better approximate clinical observations. Tucker and Taylor (1996) proposed models that address clonogen repopulation. Webb and Nahum (1993) proposed a modified formula that incorporated nonuniform clonogen

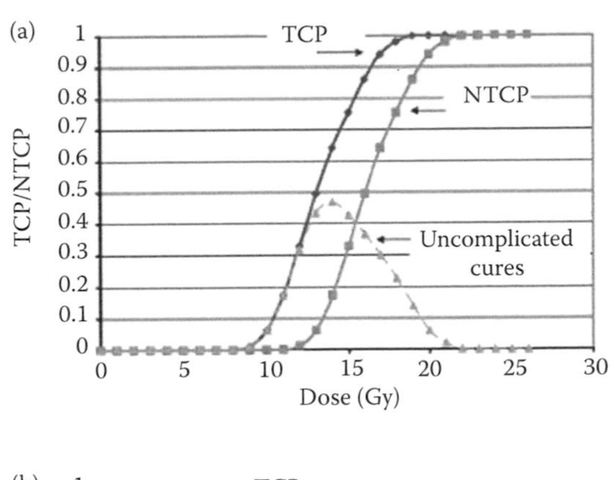

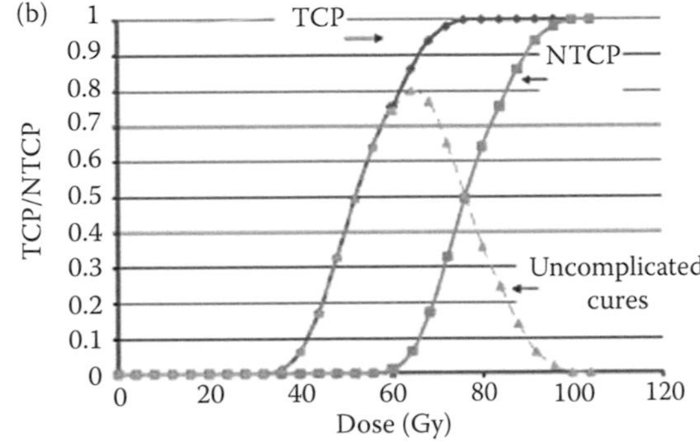

Figure 13.5 Curves comparing the TCP with the NTCP. (a) The curves are positioned close to one another. Normal tissue complications may be avoided only by minimizing the dose to the critical normal structure. Such a situation may occur when a normal structure, such as the optic nerve, lies adjacent to a benign tumor being treated with single-dose radiosurgery. (b) Dose fractionation separates the TCP and NTCP curves, allowing for a higher probability of tumor control without significant risk of normal tissue complication. The uncomplicated cure curve is TCP–NTCP. (Adapted from Anker, C. J. and Shrieve, D. C., *Otolaryngol. Clin. N. Am.*, 42, 601–21, 2009.)

cell density and nonuniform dose distribution factors. A non-Poisson model was proposed by Zaider and Minerbo (2000) that incorporates temporal effects of dose distribution and repopulation based on birth-and-death stochastic processes. These various models have been further extended and others generated to account for cell cycle parameters (O'Rourke, McAneney, and Hilen 2009). However, models that address TCP in the range of radiation doses that one would typically use in SRT are limited. Strigari et al. (2012) proposed a model that introduces a parameter "B" in a TCP model, creating a so-called "linear-quadratic-linear" (LQ-L) model. Their model takes into account the presence of hypoxic cells and their reoxygenation in the setting of hypofractionated regimens for early stage non-small cell lung cancer in which the B factor represents the fraction of hypoxic cells that survive and become oxygenated after each irradiation session. Their parameters were calculated from clinical data obtained prior to 2009 and then validated based on post-2009 data. The authors also suggest that their modified model, by means of the B-factor, takes into account both the direct cytotoxic effect as well as indirect vascular and stromal effects of radiation.

The normalized total dose (NTD), also known as the equivalent total dose, relates total doses from nonstandard fractionated schemes to the "equivalent total dose" of more standard fractionated schemes used clinically. It is occasionally referred to as "biologically equivalent dose," but it should not be confused with the more commonly encountered biologically effective dose (BED) that was previously described. The idea of relating different radiation regimens is not new and dates back to the early 1980s when Barendsen introduced the concept of extrapolated tolerance dose, discussed below (Barendsen 1982, Withers, Thames, and Peters 1983). It relayed an isoeffective outcome in terms of a total dose based on a standard fractionated regimen and not as an abstract unit as obtained from other models at the time. The idea was that most, if not all, clinical radiation oncologists would be familiar with the total dose in reference to a common clinical fractionation dose, typically 2 Gy. The terminology of NTD was introduced by Maciejewski,

Taylor, and Withers (1986). His concept is not limited to the LQ formulation, but it can be used with any type of model as reviewed by Flickinger and Kalend (1990), who also provide formula derivations and examples.

The LQ model remains widely used today with one of its most common applications being the calculation of the BED (see above). The BED formulation can be extended to determine a dose/fractionation regimen with equivalent efficacy (or biological effect). The concept was put forward by Barendsen (1982) and was first known as the extrapolated tolerance dose or extrapolated response dose. It was the tolerance dose for an infinite number of very small fractions. The model also allowed for the summation of different fractionated schedules and low dose rate treatments. The model was put forward as an alternative to the existing nominal standard dose model that was thought to be inadequate. The current term "biologically effective dose" was then introduced by Fowler who further expanded the concept by adding a time factor to allow for proliferation (Fowler 1989, 2010) when the time factor is zero for late-reacting tissues (see Section 13.4.4).

Although the LQ model has been helpful in comparing various conventional radiation schedules, particularly for those in head and neck cancer, there is considerable debate regarding its relevance in the SRT/SBRT setting. The primary objections include the limitations of the underlying LQ model and the fact that hypofractionated doses of radiation are likely to induce additional mechanisms that lead to tumor destruction.

As discussed in detail above, the LQ model has a firm grounding in classical radiobiology in that it describes the generation of chromosome rearrangements that lead to a mitotic catastrophe-type cell death at least within the range of conventional fractionation. In terms of its application to large radiation doses, such as those used in SBRT, Brenner contends that the application of the LQ formula is appropriate (Brenner 2008). In vitro data is cited that shows the LQ model is a good fit for cell survival data up to doses of 15 Gy. Also, various endpoints for normal tissue effects are cited as examples that the model holds up to 20 Gy. It is argued that the LQ model is reasonably well validated, experimentally and theoretically, up to about 10 Gy/fraction and would be reasonable for use up to about 18 Gy per fraction. It is also said that there is no evidence of problems when the LQ model has been applied in the clinic.

In contrast, those that contend that the LQ model is not appropriate in the hypofractionated setting note various inconsistencies at commonly used SRT/SBRT doses (Kirkpatrick, Meyer, and Marks 2008). The LQ model predicts a continuously bending curve at these higher doses, but experimentally it appears to be more linear in nature, thus resulting in an overestimation of cell kill by the LQ model (Webb and Nahum 1993). Clinically, the LQ model has underestimated the biological effect of higher doses, which appears at odds with a simple application of the LQ equation. A partial explanation may reflect the fact that the LQ model does not properly reflect the tumor complexity and the heterogeneity of cell types within the tumor and does not consider tissue-level effects (e.g., stromal and vascular interactions). Also, it does not consider other potentially important mechanisms of cell death, other than mitotic catastrophe (e.g., ceramide-mediated apoptosis of endothelial cells).

Others have used a range of mathematical tools with the goal of modeling cell survival after hypofractionated doses. The majority of these models continue to be based on the LQ formula and aim to correct the "overestimate" of the classical LQ model in the high-dose region as in the universal survival model (Figure 13.6) (Park et al. 2008). Such approaches must be applied with caution in that (relatively) simple mathematics are used to describe a complex and still little-understood combination of biological effects induced by the radiation.

13.5.2 UNIVERSAL SURVIVAL CURVE (USC)

The USC proposed by Park et al. (2008) is a hybrid model. The LQ component for the linear and shoulder portions of the survival curve is maintained when the classical LQ model provides a good approximation to clinical or experimental data. However, for larger doses beyond the shoulder region where a linear component is expected to dominate, the historic multitarget model is used (Elkind and Whitmore 1968). In this particular model, dose D_T, is the transition point at which the linear component of the multitarget model is tangential to the curved component of the LQ component. Thus, at doses of D_T or below, the curve is identical to the LQ curve, and at doses of D_T or greater, it approximates the multitarget model.

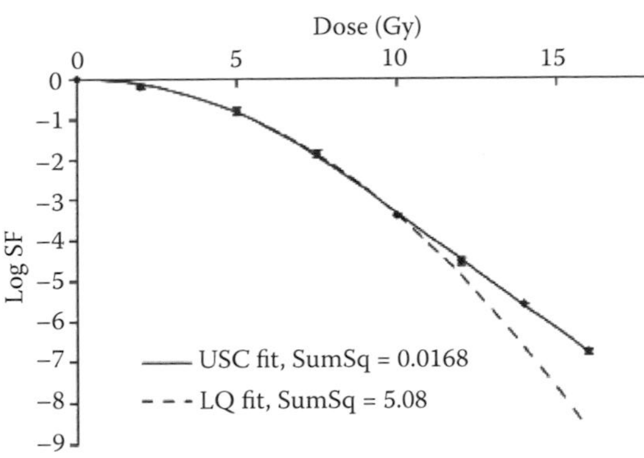

Figure 13.6 Survival curve of H460 fitted with LQ model (using points ≤8 Gy) and with USC model. Fit over entire range drastically improved with USC model fit. If LQ model fit over the entire dose range, sum of square improves compared with low-dose LQ model fit (sum of squares = 0.285) but still much inferior to USC model (data not shown). Details of generating this survival curve will be published separately. (From Park, C. et al., *Int. J. Radiat. Oncol. Biol. Phys.*, 70, 3, 847–52, 2008.)

Although such a combination may have predictive ability, it is worth noting that the LQ formalism appears to be founded on actual radiobiologic observations whereas the USC modification utilizes a mathematical extrapolation that fits empiric, macro-level observations.

13.5.3 LINEAR QUADRATIC LINEAR (LQ-L) MODEL

The LQ-L model proposed by Astrahan (2008) was intended as a more manageable model than the USC and avoided the somewhat arbitrary fusion of the LQ and the multitarget principles. Whereas in the USC model various parameters had to be extrapolated and involved multiple mathematical manipulations, the LQ-L model eliminated the multitarget aspect and simply specified the $\log_e$ cell kill per Gy in the final linear portion of the survival curve, where dose D_T was the start of the linear portion. This composite approach also introduced an additional factor, γ, which represents the $\log_e$ cell kill per Gy in the final linear portion of the survival curve (in the high-dose region). Additional mathematical calculations are therefore required to solve for γ in order to estimate the BED.

13.5.4 GENERALIZED LINEAR-QUADRATIC MODEL

Wang et al. (2010) tried to address inconsistencies for large dose fractions by extending the LQ mechanism of DNA damage at higher doses. He suggested that at higher radiation doses, sublethal damage is decreased because it is converted to lethal damage due to the intensity of the higher dose irradiation. The proportional reduction of sublethal damage becomes large at ablative doses. By introducing this concept into the formulation, it was possible to mechanistically extend the LQ concepts into the high-dose region while preserving some form of underlying biological rationale.

13.5.5 MODIFIED LINEAR QUADRATIC

Guerrero and Li (2004) used a different approach to extend the LQ model in the high dose region. They utilized the surviving fraction formula and added a dose protraction factor, G, to account for dose-rate effects:

$$SF = e^{(-\alpha D - \beta G(\lambda T) D^2)}$$

where

$$G(\lambda T) = \frac{2(\lambda T + e^{-\lambda T} - 1)}{(\lambda T)^2}$$

and λ is the repair rate, T is the delivery time, D is the dose, and α and β are LQ coefficients. They introduced a new parameter, δ, to cause a shift in the dose protraction factor and thereby modified the high-dose portion of the curve to give it a linear component:

$$G(\lambda T) = G(\lambda T + \delta D)$$

One possible interpretation of this formulation is that repair rate is better modeled; thereby no overestimation of cell killing at the high-dose region occurs.

13.5.6 LETHAL–POTENTIALLY LETHAL (LPL) MODEL

In 1986, Curtis introduced the δ parameter to augment the classical LQ formalism. The LPL model assumes that two different kinds of radiation lesions lead to cell death: (1) irreparable (lethal) and (2) repairable (potentially lethal) lesions. The LPL model differed from the LQ model in that at higher doses the LPL model becomes an exponential with constant slope as opposed to continuous curvature. This model, however, did not gain much acceptance.

In general, modification of the LQ formalism may offer some utility when predicting biological effects of large fractional doses used in SRT/SBRT regimens; however, the models vary considerably in the amount of biological relevance and underlying experiment/clinical support. It is increasingly likely that tumor control is a discontinuous phenomenon, related to multiple tumor and host factors that cannot be adequately described by simple formalisms grounded in basic cell kill/survival concepts alone. Furthermore, many such biological mechanisms exhibit threshold effects (e.g., vascular "drop out" described above) that can produce dramatic changes even after small perturbations in the system. As such, it is likely that the quality and relevance of our radiobiological models will have to evolve concurrently with increasing understanding of the mechanisms that underlie hypofractionated, high-dose effects observed clinically at the macroscopic level.

13.6 HISTOPATHOLOGIC OUTCOMES OF HIGH-DOSE RT

To provide an interpretive framework for the above discussion of biological and mathematical concepts, it is instructive to review select studies in which large fractional (or single-session) doses were applied clinically. Here, we use SRS of brain tumor and AVMs as examples of the response of both tumor and normal tissue to high does per fraction treatment.

There have been scant publications reporting on the radiobiological effect of SRS on brain tumors. During the initial years of radiosurgery, several autopsy reports and studies were reported in which extremely high doses of radiation (exceeding 180–200 Gy) were applied to limited target volumes (Szeifert et al. 2007, 2012). The reports described necrotic tissue finely demarcated from surrounding normal parenchyma in the shape of the treatment portal. Immediately surrounding the necrotic region was a submillimeter area of reactive tissue beyond which only normal parenchyma remained. Within the necrotic tissue, vessels were thrombosed and, in some cases, had necrotic walls. In subsequent years, a few reports were published describing histopathological effects of high-dose RT despite increasing application of these treatments.

Human tissue examples have included autopsy and surgery data, primarily in the setting of brain tumor metastases (Hirato et al. 1996; Jagannathan et al. 2010; Koike et al. 1994; Tago et al. 1996), primary brain tumors (Shinoda et al. 2002), and AVMs (Schneider, Eberhard, and Steiner 1997; Yamamoto et al. 1995). These cases have employed more commonly used and accepted dosing regimens, typically in the range of 12–25 Gy. For metastatic tumors, radionecrosis was seen as early as 3 weeks after treatment with later assessments showing a well-demarcated necrotic region surrounded by reactive and fibrotic tissue. The study by Jagannathan and colleagues (2010) looked at 15 cases in which patients required surgery for either presumed tumor progression or worsening neurological symptoms associated with increased mass effect, some of whom had received prior radiation. They found viable tumor, necrotic tumor, vascular hyalinization, hemosiderin-laden macrophages, reactive gliosis in the surrounding brain tissue, and an elevated MIB-1 proliferation index in cases in which a viable tumor was identified (10 cases). In all irradiated tumors, densely necrotic changes that correlated well with the 50% isodose line were observed. Seven of these cases also showed areas of reactive gliosis and hemorrhage. Demarcation between viable tumor and necrotic tissue did not fit a particular pattern but the tumor was found to be adjacent to necrotic tissue. MIB-1 was found to be elevated for tissues that had received prior radiation. In the five cases in which no viable tumor was identified, they reported finding necrotic tumor, vascular hyalinization, hemosiderin-laden macrophages, reactive gliosis, and edema. Significant for this study of SRS/SBRT, endothelial damage was a consistent finding with vascular endothelium appearing hypertrophic with an abundant fibroblastic proliferation and deposition of collagen within the perivascular areas. Vasculature that was obliterated was completely replaced by degenerated hyaline scar tissue (Figure 13.7). This was most pronounced for those cases with prior radiation history. Changes in the surrounding brain tissue were also reported and included vascular hyalinization, hemosiderin-laden macrophages, reactive gliosis, and edema. Within 12 months of radiosurgery, damage was characterized by inflammatory cell proliferation, predominantly by polymorphonuclear cells. After 12 months, much of the inflammatory infiltrate was replaced by gliotic, hypocellular tissue that showed varying degrees of hyaline degeneration. Inflammatory reactions, if present, appeared to be dominated by lymphocytes.

In 2007, Szeifert et al. reported on 38 patients with a total of 47 lesions who went on to have a craniotomy for tumor resection after radiosurgery for radiological and/or clinical progression. The treated histologies included brain metastases, anaplastic astrocytomas, low-grade astrocytoma, meningiomas, atypical meningiomas, sporadic vestibular schwannomas, NF-2 related vestibular schwannomas, a jugular bulb schwannoma, and a hemangioblastoma. They report that the morphological appearance of the various

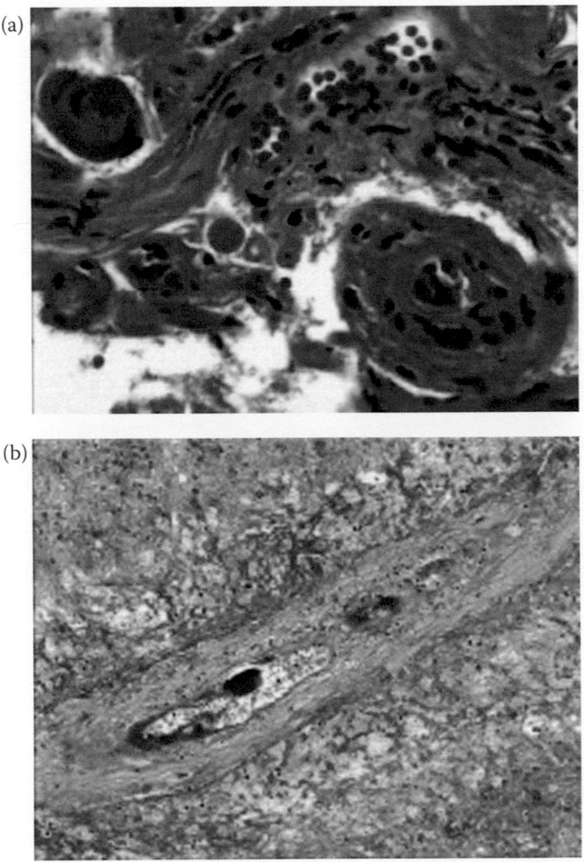

Figure 13.7 Photomicrographs showing vascular changes seen in cases of radiation necrosis. (a) Small arterioles from within the prior tumor cavity showing partial to complete luminal obliteration by hyaline change. Extravasated erythrocytes and perivascular inflammation are also evident (hematoxylin and eosin; original magnification, ×200). (b) Medium-sized arteriole showing hyalinization of the vessel wall with surrounding tumor necrosis (hematoxylin and eosin; original magnification, ×100), showing vascular changes. (Adapted from Jagannathan, J. et al., *Neurosurgery*, 66, 1, 208–17, 2010.)

lesions was suggestive of radiation effect and not significantly related to the histopathological type of the irradiated tumor, likely because ablative SRS/SBRT doses were used. It was also not related to the time interval between radiosurgery and craniotomy. The histopathological changes due to radiation appeared to occur within the tumor parenchyma, adjacent connective tissue stroma, and surrounding vessels. The typical changes were degenerative, primarily localized to the tumor parenchyma, and proliferative in the adjacent connective tissue stroma and surrounding area. The radiation-induced lesion was well circumscribed and showed sharp demarcation from the surrounding tissues, consistent with the steep radiation gradient that is characteristic of radiosurgery. Three types of lesions were characterized over time: acute, subacute, and chronic. The acute type, which can occur in an early or early-delayed time frame, consists of sharply demarcated coagulative necrosis. There was no prominent gliosis or immune reaction noted. Stromal vessels around the necrotic core demonstrated endothelial destruction, fibrinoid necrosis, undulation of the internal elastic membrane, and vacuolation and accumulation of eosinophilic material with the vessel wall. The subacute type, which can occur in early or delayed fashion as well, showed well-circumscribed coagulative necrosis within the tumor parenchyma, but stromal alterations now also included evidence of an inflammatory response. A rim of macrophages surrounded the necrotic core, and beyond the rim was a zone of granulation tissue. Vascular lumen narrowing was seen at the periphery of the necrotic region. The chronic type, occurring in delayed fashion, is notable hypocellular scar tissue with sharp demarcation with the tumor target region. Observed stromal alterations included focal lymphocytic infiltration, hyaline degenerated scar tissue, and calcifications. Vascular changes included subendothelial spindle-shaped cell

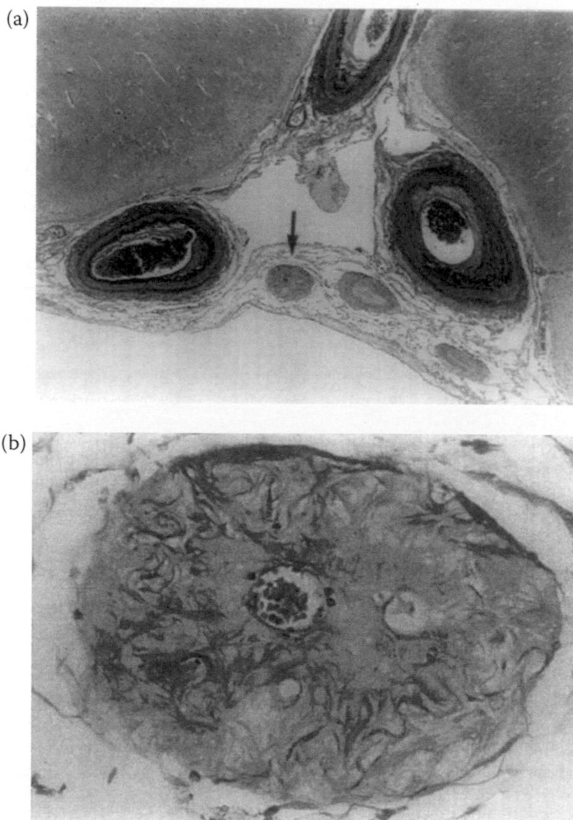

Figure 13.8 Pial arteries (*arrows* in Figure 13.8a), considered to have been exposed to an irradiation dose of about 25 Gy. (a) Significant intimal hypertrophy and marked fragmentation of the elastic laminae were observed in the larger-sized vessels. In the smaller-sized vessels, the lumina was completely obstructed by intimal hypertrophy and no elastic laminae were discernible. Recanalizing vessels were observed within the obstructed vessel (b, *arrow* in Figure 13.8a) (elastica van Gieson, original magnification: a, ×10; b, ×100). Vascular changes after radiosurgery for an AVM. (Adapted from Yamamoto, M. et al., *Surg. Neurol.*, 44, 5, 421–27, 1995.)

proliferation and hyaline degeneration leading to subtotal or complete lumen obliteration. Acute, subacute, and chronic-type lesions were seen at 2–17 months, 4–59 months, and 18–82 months, respectively.

In 1995, Yamamoto et al. reported on an autopsy for a treated AVM. The vascular malformation had been treated 2 years before the autopsy. Immediately prior to the patient's death, angiography confirmed complete obliteration of the AVM. Histopathological assessment showed that the vasculature within the nidus contained obstructed lumens consequent to intimal hypertrophy. Interestingly, some of the vessels (even within the treated nidus) retained lumen patency with normal surrounding vascular endothelium. The majority of the remaining vessels were thin walled and contained elastic fibers; these were presumed to be veins. Normal pial arteries, which were thought to have received 10 Gy or greater, showed intimal hypertrophy and fragmentation of the elastic lamina but not to the extent seen in pial arteries that received the full dose of 25 Gy (Figure 13.8). Vessels outside the 5-Gy isodose gradient did not exhibit these changes. A later study by Schneider, Eberhard, and Steiner (1997) showed similar findings. In this later report, AVMs from nine patients were assessed after surgery or autopsy 10 months to 5 years after radiosurgery. Most patients experienced a hemorrhage prior to complete obliteration; one patient did have a complete obliteration. The earliest observed histopathological event was endothelial cell injury. The next observed event along the temporal continuum appeared to be progressive thickening of the intimal layer of the vessels, likely secondary to smooth muscle cell proliferation resulting in an extracellular matrix deposition, including of the type IV collagen. The next event was cellular degeneration and hyaline transformation of the wall matrix. The final appearance was total obliteration of the vessel wall structure and nidus vascular architecture. Muscularized vessels throughout the AVM were noted to exhibit these changes with preferential and more severe effects being seen in smaller vessels.

These limited reports do indeed appear to support vascular endothelium (and tumor vasculature in general) as a significant target of high-dose irradiation as well as documenting the presence of an active inflammatory response. Thus tumor destruction in these specific settings is clearly a composite of direct tumor toxicity coupled with vascular destruction and immune reactivity.

13.7 FUTURE APPROACHES TO SBRT RISK MANAGEMENT AND PREDICTION

Multiple aspects of high-dose radiation need to be addressed clinically and modeled before we can understand and predict toxicity in all OARs. It is unlikely that a simple and reliable model across all OARs will be found. Even for conventionally fractionated radiation, different endpoints have been evaluated using different radiobiological or mathematical models (e.g., neuret and optic ret for radiation necrosis and optic neuropathy for conventional fractionation) (Goldsmith et al. 1992; Sheline, Wara, and Smith 1980). It is likely that what is considered conventional radiation biology (repair, reoxygenation, redistribution, and repopulation) may be less relevant at large doses per fraction and be uninformative when suprathreshold events, such as vascular or immune system changes, are triggered.

Historically, organs were broadly classified based on perceived functional structure into two categories: (1) parallel (lung, liver, kidney, and parotid gland) and (2) serial (spinal cord, cranial nerves, GI tract, and urethra). SBRT treatments may reduce complications in both types of OARs—in serial structures, by reducing the overall dose and, in parallel structures, by reducing the number of units affected. Our understanding of organ structure continues to evolve as we learn more about the microanatomy and the biology of normal tissue effects and as new imaging modalities are introduced and allow a different choice of endpoints. As imaging resolution improves and treatment delivery is optimized, it may be possible to adapt RT plans to specific organ and/or tumor microanatomy and to exploit potential (even small) differences in radiosensitivity.

It is possible, and indeed likely, that as we learn more about tumor and tissue structures and functional relationships, simple dose-volume relationships will become less relevant with an increased emphasis being placed on organ-specific metrics critical for reliable predictions of biological effect with SRT/SBRT. High-dose irradiation of small targets adjacent to critical structures often relies on the use of non-coplanar (and sometimes nonisocentric techniques). These techniques result in dose distributions that are inherently inhomogenous. This inhomogeneity alone may help explain observed toxicities as the in case of spinal cord myelopathy in the setting of SBRT (Sahgal et al. 2012). Additionally, small changes in the irradiated volume due to planning and/or set-up errors or organ motion could have clinically significant effects. Image-guided radiation therapy follows either the target directly or fiducial markers within the target but does not necessarily reflect or account for changes in the surrounding normal anatomy. The associated changes in the tumor–normal tissue interface could be significant, particularly in cases in which significant dose heterogeneity is applied to a small tumor target. These issues coupled with the use of highly variable treatment schedules and inherent differences in patient radiation tolerances will not only necessitate the need to develop better radiobiological models for SRT/SBRT, but should also spur development of radioprotectors (Greve et al. 2012; Story, Kodym, and Saha 2008).

Note that for serial tissues, the volume-dose constraints are given in terms of the critical maximum tissue volume that should receive a dose equal to or greater than the indicated threshold dose for the given number of fractions used. For parallel tissue, the volume-dose constraints are based on a critical minimum volume of tissue that should receive a dose equal to or less than the indicated threshold dose for the given number of fractions used.

REFERENCES

Al-Hajj, M., M. S. Wicha, A. Benito-Hernandez, S. J. Morrison, and M. F. Clarke. 2003. Prospective identification of tumorigenic breast cancer cells. *Proc. Natl. Acad. Sci. USA* 100(7): 3983–8.

Apetoh, L., F. Ghiringhelli, A. Tesniere et al. 2007. Toll-like receptor 4-dependent contribution of the immune system to anticancer chemotherapy and radiotherapy. *Nat. Med.* 13(9): 1050–9.

Astrahan, M. 2008. Some implications of linear-quadratic-linear radiation dose-response with regard to hypofractionation. *Med. Phys.* 35(9): 4161–72.

Bakst, R., N. Lee, S. He et al. 2012. Radiation impairs perineural invasion by modulating the nerve microenvironment. *PLoS One* 7(6): e39925.

Barendsen, G. W. 1982. Dose fractionation, dose rate and iso-effect relationships for normal tissue responses. *Int. J. Radiat. Oncol. Biol. Phys.* 8(11): 1981–97.

Baumann, M., F. Würschmidt, A. Twardy, and H. P. Beck-Bornholdt. 1994. Impact of tumor stroma on expression of the tumor bed effect in R1H rat rhabdomyosarcoma. *Radiat. Res.* 140(3): 432–6.

Benedict, S. H., K. M. Yenice, D. Followill et al. 2010. Stereotactic body radiation therapy: The report of AAPM Task Group 101. *Med. Phys.* 37(8): 4078–101.

Brenner, D. J. 2008. The linear-quadratic model is an appropriate methodology for determining isoeffective doses at large doses per fraction. *Seminars in Radiat. Oncol.* 18(4): 234–9.

Brizel, D. M., R. K. Dodge, R. W. Clough, and M. W. Dewhirst. 1999. Oxygenation of head and neck cancer: Changes during radiotherapy and impact on treatment outcome. *Radiother. Oncol.* 53(2): 113–7.

Calabrese, C., H. Poppleton, M. Kocak et al. 2006. A perivascular niche for brain tumor stem cells. *Cancer Cell* 11(1): 69–82.

Carlson, D. J., P. J. Keall, B. W. Loo Jr., Z. J. Chen, and J. M. Brown. 2011. Hypofractionation results in reduced tumor cell kill compared to conventional fractionation for tumors with regions of hypoxia. *Int. J. Radiat. Oncol. Biol. Phys.* 79(4): 1188–95.

Carmeliet, P., and R. K. Jain. 2000. Angiogenesis in cancer and other diseases. *Nature* 407(6801): 249–57.

Chakraborty, M., S. I. Abrams, C. N. Coleman, K. Camphausen, J. Schlom, and J. W. Hodge. 2004. External beam radiation of tumors alters phenotype of tumor cells to render them susceptible to vaccine-mediated T-cell killing. *Cancer Res.* 64(12): 4328–37.

Chen, F. H., C. S. Chiang, C. C. Wang et al. 2009. Radiotherapy decreases vascular density and causes hypoxia with macrophage aggregation in TRAMP-C1 prostate tumors. *Clin. Cancer Res.* 15(5): 1721–9.

Cooper, R. A., C. M. West, J. P. Logue et al. 1999. Changes in oxygenation during radiotherapy in carcinoma of the cervix. *Int. J. Radiat. Oncol. Biol. Phys.* 45(1): 119–26.

Curtis, S. B. 1986. Lethal and potentially lethal lesions induced by radiation—a unified repair model. *Radiat. Res.* 106(2): 252–70.

Dalerba, P., S. J. Dylla, I. K. Park et al. 2007. Phenotypic characterization of human colorectal cancer stem cells. *Proc. Natl. Acad. Sci. USA* 104(24): 10158–63.

Daly, M. E., G. Luxton, C. Y. Choi et al. 2012. Normal tissue complication probability estimation by the Lyman-Kutcher-Burman method does not accurately predict spinal cord tolerance to stereotactic radiosurgery. *Int. J. Radiat. Oncol. Biol. Phys.* 82(5): 2025–32.

Demaria, S., B. Ng, M. L. Devitt et al. 2004. Ionizing radiation inhibition of distant untreated tumors (abscopal effect) is immune mediated. *Int. J. Radiat. Oncol. Biol. Phys.* 58(3): 862–70.

Denekamp, J. 1984. Vascular endothelium as the vulnerable element in tumours. *Acta Radiol. Oncol.* 23(4): 217–25.
Dewan, M. Z., A. E. Galloway, N. Kawashima et al. 2009. Fractionated but not single-dose radiotherapy induces an immune-mediated abscopal effect when combined with anti-CTLA-4 antibody. *Clin. Cancer Res.* 15(17): 5379–88.
Dewhirst, M. W., Y. Cao, C. Y. Li, and B. Moeller. 2007. Exploring the role of HIF-1 in early angiogenesis and response to radiotherapy. *Radiother. Oncol.* 83(3): 249–55.
Dewhirst, M. W., R. Oliver, C. Y. Tso, C. Gustafson, T. Secomb, and J. F. Gross. 1990. Heterogeneity in tumor microvascular response to radiation. *Int. J. Radiat. Oncol. Biol. Phys.* 18(3): 559–68.
Dunst, J., G. Hansgen, C. Lautenschlager, G. Fuchsel, and A. Becker. 1999. Oxygenation of cervical cancers during radiotherapy and radiotherapy + cis-retinoic acid/interferon. *Int. J. Radiat. Oncol. Biol. Phys.* 43(2): 367–73.
Elkind, M. M., and G. F. Whitmore. 1968. Single-cell technique (book reviews: *The Radiobiology of Cultured Mammalian Cells*). *Science* 161(3837): 152.
Emami, B., J. Lyman, A. Brown et al. 1991. Tolerance of normal tissue to therapeutic irradiation. *Int. J. Radiat. Oncol. Biol. Phys.* 21(1): 109–22.
Fenton, B. M., E. M. Lord, and S. F. Paoni. 2001. Effects of radiation on tumor intravascular oxygenation, vascular configuration, development of hypoxia, and clonogenic survival. *Radiat. Res.* 155(2): 360–8.
Fertil, B., and E. P. Malaise. 1985. Intrinsic radiosensitivity of human cell lines is correlated with radioresponsiveness of human tumors: Analysis of 101 published survival curves. *Int. J. Radiat. Oncol. Biol. Phys.* 11(9): 1699–707.
Flickinger, J. C., and A. Kalend. 1990. Use of normalized total dose to represent the biological effect of fractionated radiotherapy. *Radiother. Oncol.* 17(4): 339–47.
Fowler, J. F. 1989. The linear-quadratic formula and progress in fractionated radiotherapy. *Br. J. Radiol.* 62(740): 679–94.
Fowler, J. F. 2010. 21 years of biologically effective dose. *Br. J. Radiol.* 83(991): 554–68.
Fuks, Z., A. Haimovitz-Friedman, and R. N. Kolesnick. 1995. The role of the sphingomyelin pathway and protein kinase C in radiation-induced cell kill. *Important Adv. Oncol.* 19–31.
Fuks, Z., and R. Kolesnick. 2005. Engaging the vascular component of the tumor response. *Cancer Cell* 8(2): 89–91.
Fyles, A. W., M. Milosevic, M. Pintilie, and R. P. Hill. 1998. Cervix cancer oxygenation measured following external radiation therapy. *Int. J. Radiat. Oncol. Biol. Phys.* 42(4): 751–3.
Gajate, C., and F. Mollinedo. 2001. The antitumor ether lipid ET-18-OCH(3) induces apoptosis through translocation and capping of Fas/CD95 into membrane rafts in human leukemic cells. *Blood* 98(13): 3860–3.
Garcia-Barros, M., F. Paris, C. Cordon-Cardo et al. 2003. Tumor response to radiotherapy regulated by endothelial cell apoptosis. *Science* 300(5622): 1155–9.
Goldsmith, B. J., S. A. Rosenthal, W. M. Wara, and D. A. Larson. 1992. Optic neuropathy after irradiation of meningioma. *Radiology* 185(1): 71–6.
Grabham, P., B. Hu, P. Sharma, and C. Geard. 2011. Effects of ionizing radiation on three-dimension human vessel models: Differential effects according to radiation quality and cellular development. *Radiat. Res.* 175(1): 21–8.
Greve, B., T. Bölling, S. Amler et al. 2012. Evaluation of different biomarkers to predict individual radiosensitivity in an inter-laboratory comparison—lessons for future studies. *PLoS One* 7(10): e47185.
Guerrero, M., and X. A. Li. 2004. Extending the linear–quadratic model for large fraction doses pertinent to stereotactic radiotherapy. *Phys. Med. Biol.* 49(20): 4825–35.
Hirato, M., J. Hirato, A. Zama et al. 1996. Radiobiological effects of Gamma Knife radiosurgery on brain tumors studied in autopsy and surgical specimens. *Stereotact. Funct. Neurosurg.* 66(1): 4–16.
Hueber, A. O., A. Bernard, Z. Hérincs, A. Couzinet, and H. T. He. 2002. An essential role for membrane rafts in the initiation of Fas/CD95-triggered cell death in mouse thymocytes. *EMBO Rep.* 3(2): 190–6.
Jagannathan, J., T. D. Bourne, D. Schlesinger et al. 2010. Clinical and pathological characteristics of brain metastasis resected after failed radiosurgery. *Neurosurgery* 66(1): 208–17.
Jain, R. K. 2003. Molecular regulation of vessel maturation. *Nat. Med.* 9(6): 685–93.
Janssen, M. H., H. J. Aerts, R. G. Kierkels et al. 2010. Tumor perfusion increases during hypofractionated short-course radiotherapy in rectal cancer: Sequential perfusion-CT findings. *Radiother. Oncol.* 94(2): 156–60.
Kavanagh, B. 2008. Clinical experience shows that catastrophic late effects associated with ablative fractionation can be avoided by technological innovation. *Semin. Radiat. Oncol.* 18(4): 223–8.
Kellerer, A. M., and H. H. Rossi. 1978. A generalized formulation of dual radiation action. *Radiat. Res.* 75(3): 471–88.
Kil, W. J., P. J. Tofilon, and K. Camphausen. 2012. Post-radiation increase in VEGF enhances glioma cell motility in vitro. *Radiat. Oncol.* 7: 25.
Kim, D. W., J. Huamani, K. J. Niermann et al. 2006. Noninvasive assessment of tumor vasculature response to radiation-mediated, vasculature-targeted therapy using quantified power Doppler sonography: Implications for improvement of therapy schedules. *J. Ultrasound Med.* 25(12): 1507–17.
Kioi, M., H. Vogel, G. Schultz, R. M. Hoffman, G. R. Harsh, J. M. Brown. 2010. Inhibition of vasculogenesis, but not angiogenesis, prevents the recurrence of glioblastoma after irradiation in mice. *J. Clin. Invest.* 120(3): 694–705.

Kirkpatrick, J. P., J. J. Meyer, and L. B. Marks. 2008. The linear-quadratic model is inappropriate to model high dose per fraction effects in radiosurgery. *Semin. Radiat. Oncol.* 18(4): 240–3.

Kobayashi, H., K. Reijnders, S. English et al. 2004. Application of a macromolecular contrast agent for detection of alterations of tumor vessel permeability induced by radiation. *Clin. Cancer Res.* 10(22): 7712–20.

Koike, Y., H. Hosoda, Y. Ishiwata, K. Sakata, and K. Hidaka. 1994. Effect of radiosurgery using Leksell gamma unit on metastatic brain tumor: Autopsy case report. *Neur. Med. Chir.* 34(8): 534–7.

Kolesnick, R., and Z. Fuks. 2003. Radiation and ceramide-induced apoptosis. *Oncogene* 22(37): 5897–906.

Kolesnick, R. N., F. M. Goni, and A. Alonso. 2000. Compartmentalization of ceramide signaling: Physical foundations and biological effects. *J. Cell Physiol.* 184(3): 285–300.

Konerding, M. A., A. J. Miodonski, and A. Lametschwandtner. 1995. Microvascular corrosion casting in the study of tumor vascularity: A review. *Scanning Microsc.* 9(4): 1233–44.

Kutcher, G. J., and C. Burman. 1989. Calculation of complication probability factors for non-uniform normal tissue irradiation: The effective volume method. *Int. J. Radiat. Oncol. Biol. Phys.* 16(6): 1623–30.

Kutcher, G. J., C. Burman, L. Brewster, M. Goitein, and R. Mohan. 1991. Histogram reduction method for calculating complication probabilities for three-dimensional treatment planning evaluations. *Int. J. Radiat. Oncol. Biol. Phys.* 21(1): 137–46.

Lapidot, T., C. Sirard, J. Vormoor et al. 1994. A cell initiating human acute myeloid leukaemia after transplantation into SCID mice. *Nature* 367(6464): 645–8.

Lee, Y., S. L. Auh, Y. Wang et al. 2009. Therapeutic effects of ablative radiation on local tumor require CD8+ T cells: Changing strategies for cancer treatment. *Blood* 114(3): 589–95.

Leksell, L. 1983. Stereotactic radiosurgery. *J. Neurol. Neurosurg. Psychiatry* 46(9): 797–803.

Li, C., D. G. Heidt, P. Dalerba et al. 2007. Identification of pancreatic cancer stem cells. *Cancer Res.* 67(3): 1030–7.

Lyman, J. T. 1985. Complication probability as assessed from dose-volume histograms. *Radiat. Res. Suppl.* 8: S13–9.

Lyng, H., A. O. Vorren, K. Sunfor et al. 2001. Assessment of tumor oxygenation in human cervical carcinoma by use of dynamic Gd-DTPA-enhanced MR imaging. *J. Magn. Reson. Imaging* 14(6): 750–6.

Maciejewski, B., J. M. Taylor, and H. R. Withers. 1986. Alpha/beta value and the importance of size of dose per fraction for late complications in the supraglottic larynx. *Radiother. Oncol.* 7(4): 323–6.

Malanchi, I., A. Santamaria-Martinez, E. Susanto et al. 2011. Interactions between cancer stem cells and their niche govern metastatic colonization. *Nature* 481(7379): 85–9.

Masucci, G. V., P. Wersäll, R. Kiessling, A. Lundqvist, and R. Lewensohn. 2012. Stereotactic ablative radio therapy (SABR) followed by immunotherapy a challenge for individualized treatment of metastatic solid tumours. *J. Transl. Med.* 10: 104.

Mayr, N. A., W. T. Yuh, V. A. Magnotta et al. 1996. Tumor perfusion studies using fast magnetic resonance imaging technique in advanced cervical cancer: A new noninvasive predictive assay. *Int. J. Radiat. Oncol. Biol. Phys.* 36(3): 623–33.

McGarry, R. C., L. Papiez, M. Williams, T. Whitford, and R. D. Timmerman. 2005. Stereotactic body radiation therapy of early-stage non-small-cell lung carcinoma: Phase I study. *Int. J. Radiat. Oncol. Biol. Phys.* 63(4): 1010–5.

Meeks, S. L., J. M. Buatti, K. D. Foote, W. A. Friedman, and F. J. Bova. 2000. Calculation of cranial nerve complication probability for acoustic neuroma radiosurgery. *Int. J. Radiat. Oncol. Biol. Phys.* 47(3): 597–602.

Merwin, R., G. H. Algire, and H. S. Kaplan. 1950. Transparent-chamber observations of the response of a transplantable mouse mammary tumor to local roentgen irradiation. *J. Natl. Cancer Inst.* 11(3): 593–627.

Milas, L., H. Hirata, N. Hunter, and L. J. Peters. 1988. Effect of radiation-induced injury of tumor bed stroma on metastatic spread of murine sarcomas and carcinomas. *Cancer Res.* 48(8): 2116–20.

Moeller, B. J., Y. Cao, C. Y. Li, and M. W. Dewhirst. 2004. Radiation activates HIF-1 to regulate vascular radiosensitivity in tumors: Role of reoxygenation, free radicals, and stress granules. *Cancer Cell* 5(5): 429–41.

Moeller, B. J., M. R. Dreher, Z. N. Rabbani et al. 2005. Pleiotropic effects of HIF-1 blockade on tumor radiosensitivity. *Cancer Cell* 8(2): 99–110.

Molls, M., H. J. Feldman, P. Stadler, and R. Jund. 1998. Changes in tumor oxygenation during radiation therapy. In *Blood Perfusion and Microenvironment of Human Tumors*, ed. M. Molls, and P. Vaupel, 81–87. Berlin, Heidelberg: Springer-Verlag.

Mottram, J. C. 1936. A factor of importance in the radiosensitivity of tumors. *Br. J. Radiol.* 9: 606–14.

Munro, T. R., and C. W. Gilbert. 1961. The relation between tumour lethal doses and the radiosensitivity of tumour cells. *Br. J. Radiol.* 34(400): 246–51.

Murphy, J. D., C. Christman-Skieller, J. Kim, S. Dieterich, D. T. Chang, and A. C. Koong. 2010. A dosimetric model of duodenal toxicity after stereotactic body radiotherapy for pancreatic cancer. *Int. J. Radiat. Oncol. Biol. Phys.* 78(5): 1420–6.

Ng, Q. S., V. Goh, J. Milner, A. R. Padhani, M. I. Saunders, and P. J. Hoskin. 2007. Acute tumor vascular effects following fractionated radiotherapy in human lung cancer: in vivo whole tumor assessment using volumetric perfusion computed tomography. *Int. J. Radiat. Oncol. Biol. Phys.* 67(2): 417–24.

O'Rourke, S. F., H. McAneney, and T. Hillen. 2009. Linear quadratic and tumour control probability modelling in external beam radiotherapy. *J. Math. Biol.* 58(4): 799–817.
Paris, F., Z. Fuks, A. Kang et al. 2001. Endothelial apoptosis as the primary lesion initiating intestinal radiation damage in mice. *Science* 293(5528): 293–7.
Park, C., L. Papiez, S. Zhang, M. Story, and R. D. Timmerman. 2008. Universal survival curve and single fraction equivalent dose: Useful tools in understanding potency of ablative radiotherapy. *Int. J. Radiat. Oncol. Biol. Phys.* 70(3): 847–52.
Park, H. J., R. J. Griffin, S. Hui, S. H. Levitt, and C. W. Song. 2012a. Radiation-induced vascular damage in tumors: Implications of vascular damage in ablative hypofractionated radiotherapy (SBRT and SRS). *Radiat. Res.* 177(3): 311–27.
Park, M. T., E. T. Oh, M. J. Song et al. 2012b. The radiosensitivity of endothelial cells isolated from human breast cancer and normal tissue in vitro. *Microvasc. Res.* 84(2): 140–8.
Pietras, A., D. Gisselsson, I. Ora et al. 2008. High levels of HIF-2alpha highlight an immature neural crest-like neuroblastoma cell cohort located in a perivascular niche. *J. Pathol.* 214(4): 482–8.
Pirhonen, J. P., S. A. Grenman, A. B. Bredbacka, R. O. Bahado-Singh, and T. A. Salmi. 1995. Effects of external radiotherapy on uterine blood flow in patients with advanced cervical carcinoma assessed by color Doppler ultrasonography. *Cancer* 76(1): 67–71.
Prince, M. E., R. Sivanandan, A. Kaczorowski et al. 2007. Identification of a subpopulation of cells with cancer stem cell properties in head and neck squamous cell carcinoma. *Proc. Natl. Acad. Sci. USA* 104(3): 973–8.
Puck, T. T., and P. I. Marcus. 1956. Action of X-rays on mammalian cells. *J. Exp. Med.* 103(5): 653–66.
Ritchie, K. E., and J. E. Nör. 2012. Perivascular stem cell niche in head and neck cancer. *Cancer Lett.* 338(1): 41–6.
Rofstad, E. K., B. Mathiesen, K. Henriksen, K. Kindem, and K. Galappathi. 2005. The tumor bed effect: Increased metastatic dissemination from hypoxia-induced up-regulation of metastasis-promoting gene products. *Cancer Res.* 65(6): 2387–96.
Rossi, L., D. Reverberi, G. Podestá, S. Lastraioli, and R. Corvó. 2000. Co-culture with human fibroblasts increases the radiosensitivity of MCF-7 mammary carcinoma cells in collagen gels. *Int. J. Cancer* 85(5): 667–73.
Sahgal, A., L. Ma, J. Fowler et al. 2012. Impact of dose hot spots on spinal cord tolerance following stereotactic body radiotherapy: A generalized biological effective dose analysis. *Technol. Cancer Res. Treat.* 11(1): 35–40.
Santana, P., L. A. Peña, A. Haimovitz-Friedman et al. 1996. Acid sphingomyelinase-deficient human lymphoblasts and mice are defective in radiation-induced apoptosis. *Cell* 86(2): 189–99.
Schneider, B. F., D. A. Eberhard, and L. E. Steiner. 1997. Histopathology of arteriovenous malformations after Gamma Knife radiosurgery. *J. Neurosurg.* 87(3): 352–7.
Semenza, G. L. 2003. Targeting HIF-1 for cancer therapy. *Nat. Rev. Cancer* 3(10): 721–32.
Sheline, G. E., W. M. Wara, and V. Smith. 1980. Therapeutic irradiation and brain injury. *Int. J. Radiat. Oncol. Biol. Phys.* 6(9): 1215–28.
Shinoda, J., H. Yano, H. Ando et al. 2002. Radiological response and histological changes in malignant astrocytic tumors after stereotactic radiosurgery. *Brain Tumor Pathol.* 19(2): 83–92.
Siegfried, J. M., L. A. Weissfeld, P. Singh-Kaw, R. J. Weyant, J. R. Testa, and R. J. Landreneau. 1997. Association of immunoreactive hepatocyte growth factor with poor survival in resectable non-small cell lung cancer. *Cancer Res.* 57(3): 433–9.
Singh, S. K., C. Hawkins, I. D. Clarke et al. 2004. Identification of human brain tumour initiating cells. *Nature* 432(7015): 396–401.
Solesvik, O. V., E. K. Rofstad, and T. Brustad. 1984. Vascular changes in a human malignant melanoma xenograft following single-dose irradiation. *Radiat. Res.* 98(1): 115–28.
Song, C. W., J. J. Drescher, and J. Tabachnick. 1968. Effect of anti-inflammatory compounds on beta-irradiation-induced increase in vascular permeability. *Radiat. Res.* 34: 616–25.
Song, C. W., and S. H. Levitt. 1971. Vascular changes in Walker 256 carcinoma of rats following X irradiation. *Radiology* 100(2): 397–407.
Song, C. W., J. T. Payne, and S. H. Levitt. 1972. Vascularity and blood flow in x-irradiated Walker carcinoma 256 of rats. *Radiology* 104(3): 693–7.
Song, C. W., J. G. Sung, J. J. Clement, and S. H. Levitt. 1974. Vascular changes in neuroblastoma of mice following x-irradiation. *Cancer Res.* 34(9): 2344–50.
Stadler, P., H. J. Feldmann, C. Creighton, R. Kau, and M. Molls. 1998. Changes in tumor oxygenation during combined treatment with split-course radiotherapy and chemotherapy in patients with head and neck cancer. *Radiother. Oncol.* 48(2): 157–64.
Story, M., R. Kodym, and D. Saha. 2008. Exploring the possibility of unique molecular, biological, and tissue effects with hypofractionated radiotherapy. *Semin. Radiat. Oncol.* 18(4): 244–8.
Strigari, L., M. Benassi, A. Sarnelli, R. Polico, and M. D'Andrea. 2012. A modified hypoxia-based TCP model to investigate the clinical outcome of stereotactic hypofractionated regimes for early stage non-small-cell lung cancer (NSCLC). *Med. Phys.* 39(7): 4502–14.

Szeifert, G. T., D. Kondziolka, M. Levivier, and L. D. Lunsford, eds. 2007. *Radiosurgery and Pathological Fundamentals.* Basel: Karger.

Szeifert, G. T., D. Kondziolka, M. Levivier, and L. D. Lunsford. 2012. Histopathology of brain metastases after radiosurgery. *Prog. Neurol. Surg.* 25: 30–8.

Szeifert, G. T., I. Salmon, S. Rorive et al. 2005. Does Gamma Knife surgery stimulate cellular immune response to metastatic brain tumors? A histopathological and immunohistochemical study. *J. Neurosurg.* 102(Suppl): 180–4.

Tago, M., Y. Aoki, A. Terahara et al. 1996. Gamma knife radiosurgery for brain stem metastases: Two autopsy cases. *Stereotact. Funct. Neurosurg.* 66(Suppl 1): 225–30.

Thomlinson, R. H., and L. H. Gray. 1955. The histological structure of some human lung cancers and the possible implications for radiotherapy. *Br. J. Cancer* 9(4): 539–49.

Timmerman, R., M. Bastasch, D. Sara, R. Abdulrahman, W. Hittson, and M. Story. 2011. Stereotactic body radiation therapy: Normal tissue and tumor control effects with large dose per fraction. *Front. Radiat. Ther. Oncol.* 43: 383–94.

Timmerman, R., J. Galvin, J. Michalski et al. 2006. Accreditation and quality assurance for Radiation Therapy Oncology Group: Multicenter clinical trials using stereotactic body radiation therapy in lung cancer. *Acta Oncol.* 45(7): 779–86.

Tucker, S. L., and J. M. Taylor. 1996. Improved models of tumour cure. *Int. J. Radiat. Biol.* 70(5): 539–53.

Wang, J. Z., Z. Huang, S. S. Lo, W. T. C. Yuh, and N. A. Mayr. 2010. A generalized linear-quadratic model for radiosurgery, stereotactic body radiation therapy, and high–dose rate brachytherapy. *Sci. Transl. Med.* 2(39): 39ra48.

Webb, S., and A. E. Nahum. 1993. A model for calculating tumour control probability in radiotherapy including the effects of inhomogeneous distributions of dose and clonogenic cell density. *Phys. Med. Biol.* 38(6): 653–66.

Wennberg, B. M., P. Baumann, G. Gagliardi et al. 2011. NTCP modelling of lung toxicity after SBRT comparing the universal survival curve and the linear quadratic model for fractionation correction. *Acta Oncol.* 50(4): 518–27.

Withers, H. R. 1975. Four R's of radiotherapy. *Adv. Radiat. Biol.* 5: 241–7.

Withers, H. R., H. D. Thames Jr., and L. J. Peters. 1983. A new isoeffect curve for change in dose per fraction. *Radiother. Oncol.* 1(2): 187–91.

Wong, H. H., C. W. Song, and S. H. Levitt. 1973. Early changes in the functional vasculature of Walker carcinoma 256 following irradiation. *Radiology* 108(2): 429–34.

Yamamoto, M., M. Jimbo, M. Ide et al. 1995. Gamma knife radiosurgery for cerebral arteriovenous malformations: An autopsy report focusing on irradiation-induced changes observed in nidus-unrelated arteries. Commentary. *Surg. Neurol.* 44(5): 421–7.

Zaider, M., and G. N. Minerbo. 2000. Tumour control probability: A formulation applicable to any temporal protocol of dose delivery. *Phys. Med. Biol.* 45(2): 279–93.

Zundel, W., and A. Giaccia. 1998. Inhibition of the anti-apoptotic PI(3)K/Akt/Bad pathway by stress. *Genes Dev.* 12(13): 1941–6.

14 Clinical outcomes using stereotactic body radiation therapy and stereotactic radiosurgery

Megan E. Daly, Anthony L. Michaud, and Kyle E. Rusthoven

Contents

14.1 INTRODUCTION

Stereotactic body radiation therapy (SBRT) and stereotactic radiosurgery (SRS) are precise, tightly focused, highly accurate techniques for the delivery of ablative external beam radiotherapy in an abbreviated course of treatment completed in five or fewer sessions. These techniques have been used extensively for primary and metastatic tumors in the brain, spine, lung, and liver as well as for recurrent lesions in the head and neck and selected other sites. In this chapter, the clinical outcomes of SBRT and SRS will be reviewed. The concept of oligometastases and the application of SBRT and SRS for patients with limited-burden metastatic disease will also be discussed.

14.2 BRAIN

14.2.1 BRAIN METASTASES

The use of stereotactic radiosurgery for lesions in the brain traces its origins to the pioneering work of Lars Leksell in the mid-20th century (Leksell 1951). Thousands of clinical studies since published have refined the technical aspects of SRS and patient selection criteria. Although SRS is an available treatment for certain functional disorders and selected primary brain tumors that recur after conventional therapy, the single most common application is in the treatment of brain metastases from solid tumors.

Major randomized clinical trials have addressed the question of whether SRS adds clinical benefit to the use of whole brain radiotherapy (WBRT) for brain metastases and whether, conversely, WBRT adds clinical benefit to SRS or surgical resection. To address the first question, the Radiation Therapy Oncology Group (RTOG) randomized 333 patients with one to three newly diagnosed brain metastases to receive either WBRT alone (37.5 Gy in 15 fractions) or WBRT followed by an SRS boost (15–24 Gy). Although baseline clinical status as determined by RTOG recursive partitioning analysis criteria was the major determinant of overall survival (OS) on multivariate analysis, in the subgroup of patients with a solitary metastasis, there was a survival advantage with the addition of SRS that approached clinical significance ($p = 0.0533$). Additionally, patients in the SRS group were more likely to maintain a stable or improved Karnofsky Performance Status (KPS) score at 6 months follow-up than were patients who received WBRT alone (43% vs. 27%, respectively; $p = 0.03$) (Andrews et al. 2004).

Three randomized studies published in the last decade have evaluated whether WBRT adds clinical benefit to the use of surgery or SRS for brain metastases. First, in a multi-institutional study of 132 patients, Aoyama and colleagues (2006) observed no survival benefit with the addition of WBRT following SRS for patients with one to four brain metastases. While the brain recurrence rate was higher without WBRT, salvage therapy apparently compensated adequately (Aoyama et al. 2006). Similar findings were observed by Kocher and colleagues (2011) in a randomized study of 359 patients conducted by the European Organization for Research and Treatment of Cancer (EORTC), which studied clinical outcomes after either surgery or SRS for patient with one to three metastases from a variety of solid tumors. In the EORTC study, median OS among all patients was approximately 11 months with or without WBRT (Kocher et al. 2011). In a smaller single institutional randomized study, Chang E. L. and colleagues (2009) randomized patients between SRS alone and SRS plus WBRT. Detailed neurocognitive follow-up studies were performed, and it was observed that the addition of WBRT led to a significant decline in learning and memory function at 4 months, providing further support to a strategy of SRS alone for most patients in an effort to avoid this toxicity and associated adverse effect on quality of life (Chang E. L. et al. 2009).

To evaluate whether there is an upper limit on the number of discrete brain metastases beyond which SRS is contraindicated, Serizawa and colleagues (2010) analyzed 778 patients who met the following criteria: (1) newly diagnosed brain metastases, (2) 1–10 brain lesions, (3) less than 10 cm^3 volume of the largest tumor, (4) less than 15 cm^3 total tumor volume, (5) no magnetic resonance findings of cerebrospinal fluid dissemination, and (6) no impaired activity of daily living (<70 KPS) due to extracranial disease. Most patients were male, and the most common histology was a lung cancer primary. On multivariate analysis, prognostic factors for worse (OS) were active systemic disease, poor initial KPS, and male gender, but not number of brain lesions (Serizawa et al. 2010).

Investigators from the MD Anderson Cancer Center made similar observations in a group of 251 patients who underwent SRS for the initial treatment of brain metastases from a variety of primary

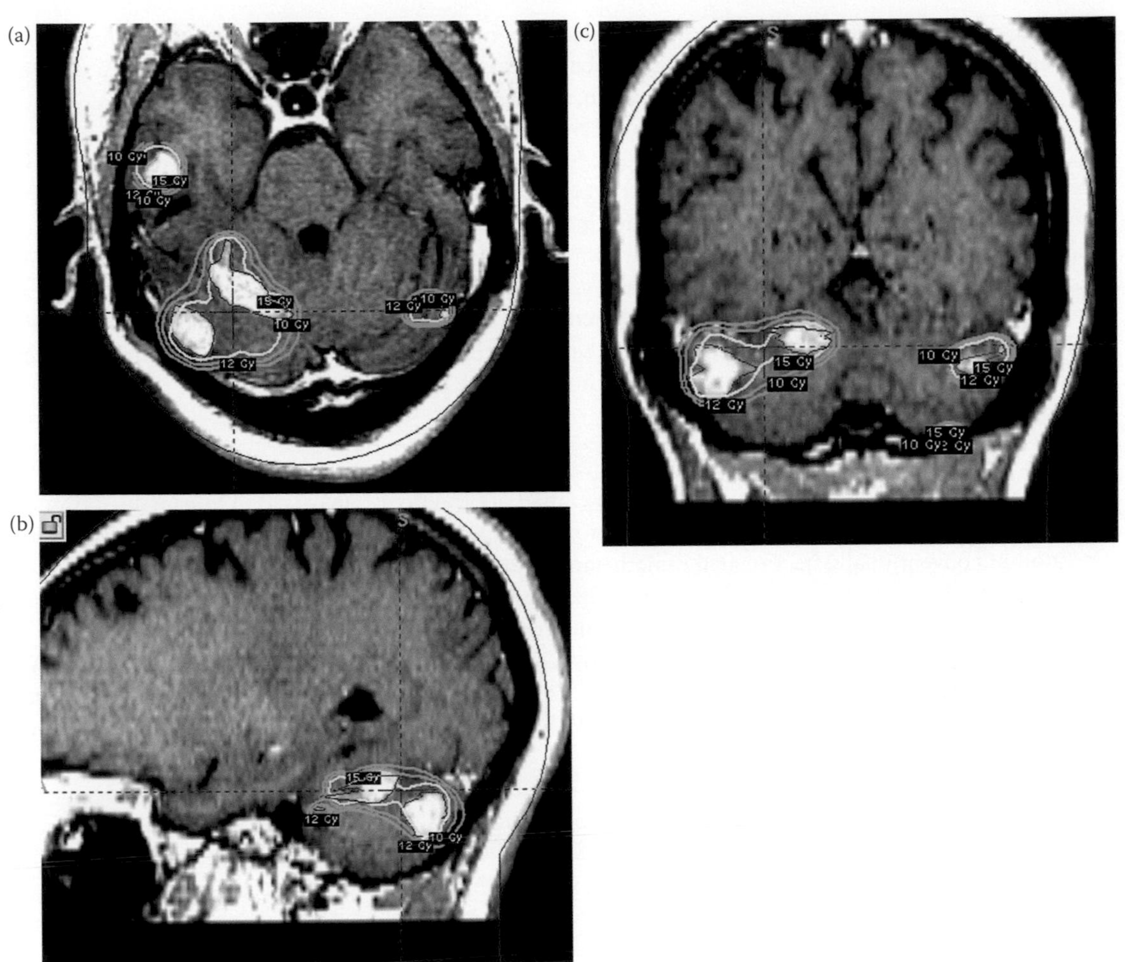

Figure 14.1 Gamma Knife SRS plan targeting four intracranial metastases from melanoma is shown in the axial (a), coronal (b), and sagittal (c) planes.

malignancies, including non-small cell lung cancer (NSCLC), melanoma, and breast carcinoma (Likhacheva et al. 2012). Some patients had subsequent salvage therapy with SRS, WBRT, or resection. Median OS in their cohort was 11.1 months, and on multivariate analysis, the only significant predictors of OS were the presence of extracranial disease, total tumor volume greater than 2 cm^3, age ≥60 years, and diagnosis-specific graded prognostic assessment (Sperduto et al. 2010). The number of brain metastases was not predictive of local or distant brain failure or OS. Figure 14.1 demonstrates axial, sagittal, and coronal slices from a representative Gamma Knife SRS plan targeting multiple brain metastases from melanoma.

14.2.2 PRIMARY INTRACRANIAL TUMORS AND OTHER INDICATIONS

SRS may also be used in the management of a wide variety of primary intracranial tumors, and dozens of institutions have published observational series on the use of SRS for meningiomas, acoustic schwannomas, glomus tumors, and recurrent malignant gliomas, among others. SRS is also used for selected patients with CNS vascular malformations and trigeminal neuralgia. A comprehensive review of each of these indications is beyond the scope of the current chapter, but an overview with representative clinical reports from each of these clinical scenarios are described.

14.2.2.1 Meningioma

Meningioma represents the most common benign intracranial tumor in adults (Dolecek et al. 2012). Radiosurgery is an established treatment approach for appropriately selected patients with unresectable or recurrent intracranial meningiomas. Most published clinical series report single fraction approaches although multisession treatments are also described, particularly for tumors in close proximity to sensitive critical structures, such as the optic tracts or brainstem. Although prospective dose finding studies are unavailable, doses of 12–14 Gy are typically used for single-session treatments.

No prospectively randomized studies compare SRS to surgical resection; however, a retrospective report from the Mayo clinic demonstrated equivalent progression-free survival for SRS and complete surgical resection (Simpson grade 1) and inferior PFS for incomplete resection (Simpson grade 2–4) (Pollock et al. 2003). In a large, recent retrospective review of 416 patients with benign intracranial meningiomas treated to a median marginal tumor dose of 16 Gy, Pollock and colleagues identified 5- and 10-year local control rates of 96% and 89%, respectively, with corresponding disease-specific survival of 97% and 94%. Male gender; previous surgery; or location of the tumor in the parasagittal, falx, or convexity regions were associated with increased hazard ratio for local failure (Pollock et al. 2012).

14.2.2.2 Acoustic neuroma

Acoustic neuromas, also known as vestibular schwannomas, are benign tumors derived from Schwann cells covering the vestibular portion of the eighth cranial nerve. Growth may lead to hearing loss, tinnitus, hydrocephalus, and cranial neuropathies. SRS has emerged as an established alternative to microsurgical resection or conventionally fractionated radiotherapy. The earliest report of SRS for acoustic neuroma, by Leksell in 1971, delivered single fraction marginal doses of 18–25 Gy. With publication of subsequent studies demonstrating excellent local control and improved serviceable hearing preservation with lower doses, single fraction regimens of 12–13 Gy are now commonly used. In a recent critical review, Murphy and Suh (2011) identify local control ranging from 92% to 100% following single-fraction SRS in modern series with corresponding rates of cranial nerve V, VII, and serviceable hearing preservation of 92%–100%, 94%–100%, and 32%–71%, respectively.

Multisession approaches designed to reduce the risk of hearing loss, cranial neuropathies, and trigeminal neuralgia are reported although prospective comparisons to a single-fraction technique have not been performed. Hansasuta et al. (2011) from Stanford University describe 383 patients treated with a multisession regimen of 18 Gy in three fractions. The authors identified 5-year actuarial local control of 96% with a crude rate of serviceable hearing preservation of 76%.

SRS for large acoustic neuromas (>3 cm diameter) remains controversial although select published series suggest an acceptable toxicity profile for patients without brainstem compression (Chung et al. 2010; Milligan et al. 2012; Sun and Liu 2012). Other authors advocate a strategy of optimal debulking followed by SRS to the residual tumor (Fuentes et al. 2008; Pan et al. 2012; van de Langenberg et al. 2011). A multi-institution observational cohort study is currently evaluating this approach (ClinicalTrials.gov. 2012).

14.2.2.3 Glomus tumors

Glomus tumors, also known as chemodectomas or nonchromaffin paragangliomas, are benign chromaffin-negative tumors of the chemoreceptor system, such as the carotid bodies. They may also be found near the jugular bulb (glomus jugulare tumor), near the tympanum, orbital cavity, larynx, and other thoracic sites. In the carotid region, the preferred treatment of these lesions is embolization followed by surgical resection. When these tumors develop in proximity to the intracranial vasculature, they may eventually compromise cranial nerve function. A recent critical review analyzing 19 published studies by Guss et al. (2011) identified a 97% crude tumor control rate across all studies. The authors conclude SRS should be considered a standard frontline management approach for these cases. Single-fraction marginal doses ranging from 13–18 Gy are reported (Guss et al. 2011).

14.2.2.4 High-grade glioma

A number of investigations have evaluated the application of SRS to both newly diagnosed and recurrent malignant gliomas. SRS is a particularly appealing option for recurrent high-grade gliomas for which no standard treatment paradigm exists. Among the largest of such studies, Cuneo and colleagues (2012) from Duke University reported a large series of 63 patients treated with SRS to a median dose of 15 Gy (range: 12.5–25 Gy) for recurrent, previously irradiated high-grade glioma. The median OS from SRS was 10 months with an 11% rate of grade 3 toxicity (Cuneo et al. 2012). The optimal dose and fractionation in the recurrent setting is poorly defined. Fractionated approaches ranging from three to 10 fractions are reportedly well tolerated. A large retrospective analysis from Thomas Jefferson University evaluated hypofractionated radiotherapy for recurrent glioma using a median dose of 35 Gy in 10 fractions and demonstrated a promising median survival of 11 months with one case of grade 3 toxicity, suggesting more protracted regimens warrant further investigation (Fogh et al. 2010).

Interest also exists in combining SRS with novel biologically targeted agents for recurrent gliomas. Two single-arm prospective trials from the University of Colorado explored this approach. In the first trial, SRS was escalated from 18 Gy to 36 Gy in three fractions with concurrent gefitinib with no dose-limiting toxicity identified in 15 treated patients (Schwer et al. 2008). In the second trial, three-fraction SRS to 36 Gy was combined with escalating doses of vandetinib in 10 recurrent glioma patients. The median OS was 6 months from SRS with significant dose-limiting toxicity from vandetinib.

The standard of care for newly diagnosed glioblastoma in good performance status patients remains conventionally fractionated radiotherapy to 60 Gy with concurrent and adjuvant temozolamide. However, given the dismal prognosis for these patients, significant interest exists in alternate treatment strategies. A recent phase II study performed by investigators at Case Western Reserve University evaluated the addition of SRS to conventionally fractionated radiotherapy targeting biologically active tumoral subregions identified by magnetic resonance spectroscopy (Einstein et al. 2012). The median OS among the entire cohort of 35 patients was 15.8 months with median OS of 20.8 months among the subgroup also completing temozolamide. These results compare favorably with the median OS of 14.6 months achieved on the EORTC trial that established temozolamide with radiotherapy as the standard of care. Further prospective studies will be needed for a conclusive role to be established for SRS in either the up-front or recurrent management of high-grade glioma.

14.2.2.5 Trigeminal neuralgia

Trigeminal neuralgia is an often excruciating, episodic pain affecting one side of the face. Although the precise etiology remains incompletely understood, vascular compression of the trigeminal nerve roots is believed to cause most cases (Thomas and Vilensky 2013). Therapeutic options for medically refractory trigeminal neuralgia include microvascular decompression, radiofrequency or percutaneous glycerol rhizotomy, and SRS. The target for SRS is the trigeminal nerve just outside the root entry zone in the brainstem.

Tuleasca et al. (2012) recently reported one of the largest published case series, including 500 patients treated with Gamma Knife to a median dose max of 85 Gy for classical trigeminal neuralgia. With median follow-up of 44 months, the authors identified a 91% complete pain response, 37% of whom experienced pain relief within 48 hr of treatment. However, 34% subsequently developed recurrent pain at a median of 24 months (Tuleasca et al. 2012). The use of repeat SRS for recurrent trigeminal pain after prior SRS has been reported from several institutions. For example, the University of Pittsburgh group described a series of 119 patients treated with a median dose of 70 Gy (range, 50–90 Gy). Initial pain relief was achieved in 87% of patients, and the actuarial rate of sustained pain relief was approximately 70% at three years (Park et al. 2012).

14.2.2.6 CNS vascular malformations

SRS is employed in the management of CNS vascular malformations, including cranial and spinal arteriovenous malformations (AVMs), cavernous malformations, and dural arteriovenous fistulas. AVMs remain the most common vascular application of SRS. AVMs are abnormal vascular shunts between the arterial and venous systems. When located within the cerebral vasculature, AVMs are associated with a significant lifetime risk of hemorrhagic stroke. Treatment options include intravascular embolization, surgery, and SRS.

The first radiosurgical treatment for a cerebral AVM was performed with Gamma Knife SRS in 1970 (Steiner et al. 1972). Following further technique refinements, SRS has emerged as an efficacious strategy for cerebral and spinal vascular malformations. SRS is particularly suited for small, deep-seated lesions located in eloquent regions of the brain for which resection poses particular risk. Single-fraction approaches are generally selected for small lesions, and anatomically staged procedures are preferred for larger (>10–15 cc) AVMs (Fogh et al. 2012; Huang et al. 2012; Kano et al. 2012). Among the largest series published, the University of Pittsburgh reported 351 patients treated with Gamma Knife SRS for AVM with a minimum 3-year follow up. The authors identified an obliteration rate of 75% with marginal dose and gender predictive of obliteration by multivariate analysis (Flickinger et al. 2002).

14.2.3 ORGANS AT RISK AND TOXICITY

Brain necrosis is a particularly feared complication of intracranial radiotherapy, particularly in light of the long natural history of some indications for intracranial SRS. The risk of brain injury is dependent total dose, fractionation, interfraction interval, volume, specific dose volume parameters, treatment site, concurrent systemic therapies, prior cranial radiotherapy, and other patient-specific factors. Symptoms of radiation necrosis of the brain include focal neurologic deficits, including speech and motor function, seizures, and cognitive sequelae among others. Minniti et al. 2011 describe outcomes among 206 patients treated with SRS for 310 intracranial metastases. A 24% rate of radionecrosis was identified with 10% of patients developing symptomatic necrosis at a median of 11 months. Symptomatic patients developed deficits in speech, cognitive and motor functions, and seizures. The volume of brain receiving 10 and 12 Gy (V10 and V12) were highly predictive of radionecrosis with a V12 > 8.5 cc predictive of a >10% risk. Yaacov Richard Lawrence and colleagues (2010) comprehensively reviewed the reported literature correlating the doses to normal brain and the risk of necrosis and likewise concluded that toxicity increases rapidly once the volume of the brain exposed to >12 Gy is >5–10 cm^3.

Radiation-induced optic neuropathy (RION) may result from high doses to the anterior visual pathways. Leber, Bergloff, and Pendl (1998) evaluated 45 patients treated with single-fraction SRS for tumors of the central skull base. The rate of RION was 0%, 26.7%, and 77.8% for patients receiving a maximum dose of <10 Gy, 10–15 Gy, and >15 Gy to the anterior visual pathways, respectively (Leber, Bergloff, and Pendl 1998). In a review of 215 patients treated with Gamma Knife SRS for tumors of the sellar or parasellar region by Stafford and colleagues (2003) found a 1.1% rate of RION when the optic apparatus was constrained to <12 Gy. The available literature consistently suggests a low rate of RION when doses are limited to <10 Gy, and the recently published *Quantitative Analyses of Normal Tissue Effects in the Clinic* suggests this constraint is appropriately conservative (Mayo et al. 2010).

Radiation-induced hearing loss most commonly occurs following SRS for acoustic neuroma. The precise mechanism of radiation-induced hearing loss is not fully understood and may involve damage to the cochlea, the vestibular nerve, and/or the auditory canal (Gephart et al. 2013; Massager et al. 2007). Retrospective data on the correlation between cochlear dose and serviceable hearing preservation suggest a relationship. Among 82 patients treated with SRS for acoustic neuroma, Massager et al. (2007) identified a cochlear dose of 3.7 Gy among patients with hearing preservation as compared to 5.3 Gy among those with audiological worsening ($p < 0.05$) (Massager et al. 2007). Tamura et al. (2009) identified increased long-term hearing preservation among 74 acoustic neuroma patients with a cochlear dose <4 Gy (90.9% versus 78.4% for all patients).

14.3 SPINE

14.3.1 PAIN RESPONSE AND IN-FIELD TUMOR CONTROL

Palliative radiation therapy for painful spinal metastases with conventional fractionation results in pain relief for approximately two thirds of patients and complete pain relief in 15%–20% of cases. RTOG 97-14 randomized patients to two palliative fractionation schemes, 30 Gy in 10 fractions versus 8 Gy in a single fraction for painful metastases (Hartsell et al. 2005). In this trial, the overall and complete pain response rates were 66% and 18% for the 30 Gy arm and 65% and 15% for the 8 Gy arm, respectively. The rate of pain relief at 3 months post treatment was 51%. A meta-analysis of 10 randomized trials comparing single-fraction and multifraction palliative regimens reported an overall response rate of 62% (single fraction) and 59% (multifraction) and a complete response rate of 33% (single fraction) and 32% (multifraction), respectively (Wu et al. 2003).

Recent studies have evaluated hypofractionated SBRT as a means of improving pain response and local control without added toxicity. Early single-institution reports first established the safety and efficacy of this approach. Ryu and colleagues (2008), from Henry Ford Hospital, reported on a series of patients treated with single-fraction SBRT to a median dose of 16 Gy for painful spinal metastases with no prior in-field radiation. They identified overall and complete response rates of 85% and 45%, respectively, and in-field local control at

Table 14.1 Single-fraction SRS for spinal metastases

STUDY	PRIOR RT	DOSE	PAIN RESPONSE	LOCAL CONTROL
Ryu et al. (2008)	0%	Median 16 Gy	CR: 45% OR: 85%	95%
Yamada et al. (2008)	0%	Median 24 Gy	NR	1 yr: 90% 95% (24 Gy) vs. 80% (18–23.9 Gy)
Garg et al. (2012)	0%	16–24 Gy	NR	18 months: 88%
Gerszten et al. (2007)	69%	Median 20 Gy	OR: 86%	90%

Note: NR, not reported.

last follow-up of 95%. Investigators from Memorial Sloan-Kettering Cancer Center identified a dose-response relationship for local control among patients treated with single-fraction SBRT to previously unirradiated spinal metastases (Yamada et al. 2008). In this series, local control as defined radiographically by MRI at 1 year was 95% for lesions treated to 24 Gy as compared to 80% for lesions prescribed 18–23 Gy (p = 0.03). Garg and colleagues (2012) from MD Anderson recently reported a single institution prospective phase I/II trial of single-fraction SRS for previously unirradiated spinal metastasis to 16–24 Gy. The 18-month actuarial local control was 88%. Two patients developed grade ≥3 toxicity (myelopathy and radiculopathy, respectively). A summary of select studies evaluating single-fraction SRS for spinal metastases is outlined in Table 14.1.

Cooperative group efforts are now evaluating spinal SBRT prospectively. RTOG 0631 is an ongoing phase II/III trial evaluating single-fraction radiosurgery for painful, previously unirradiated spinal metastases (Radiation Therapy Oncology Group: Protocol RTOG 0631). In the now-completed phase II component, patients underwent single-fraction SRS to 16 Gy. The currently accruing phase III component randomizes patients in a 2:1 fashion to single-fraction SRS (16 Gy) or single-fraction conventional RT (8 Gy). The primary endpoint is pain relief at 3 months post treatment, and the study is powered to detect a 19% improvement in this endpoint with SRS (70% vs. 51%). The target volume includes the entire vertebral body and bilateral pedicles. Epidural extension is permitted, provided there is at least a 3-mm gap between the lesion and the spinal cord on MRI. The spinal cord is contoured 5–6 mm above and below the planning target volume (PTV) using fused MRI. The investigators limit the spinal cord as contoured 5–6 mm above and below the target lesion to <10 Gy to <10% of the contoured cord and <0.35 cc cord receiving 10 Gy.

14.3.2 IMAGE-GUIDED RADIATION AFTER PRIOR IN-FIELD SPINAL RT

Multiple large studies demonstrate that stereotactic radiation with single- or multifraction regimens can safely treat painful spinal lesions arising within a previously irradiated field. In this setting, stereotactic technique is necessary to avoid exceeding cumulative tolerance doses for the spinal cord and cauda equina.

Gerszten and colleagues (2007) reported a series of 500 cases of spinal metastases treated with single fraction SRS using CyberKnife. Of these 500 cases, 344 (69%) had received prior radiation at the involved level. The mean SBRT dose was 20 Gy (range 12.5–25 Gy) prescribed to the 80% isodose line. At a median follow-up of 21 months, pain relief was observed in 86% of patients with local control of 90%.

Several investigators have evaluated multifraction SBRT regimens in patients with spinal metastases arising in a previously irradiated field. A summary of trials using three-fraction SBRT is presented in Table 14.2. Chang and colleagues (2007) performed a prospective phase I/II trial of three- to five-fraction SBRT for spinal metastases. The first 32 patients received 30 Gy in five fractions and, after a protocol amendment, the final 31 patients received 27 Gy in three fractions. Thirty-five patients (56%) had received prior in-field spinal radiation to a median dose of 33 Gy (range 30–54 Gy). Actuarial one-year local control as defined by MRI was achieved in 84% of cases. Investigators from the University of California–San Francisco reported a series of 39 patients with 60 spinal lesions and 62% in previously irradiated fields (Sahgal et al. 2009). The median dose was 24 Gy in three fractions. One-year progression-free probability (PFP) for the entire cohort was 85% with 96% 1-year PFP in previously irradiated cohort. Similarly, in a series of 32 patients

Table 14.2 Three-fraction SBRT for spinal metastases

STUDY	PRIOR RT	REGIMEN	PAIN RESPONSE	LOCAL CONTROL
Chang et al. (2007)	56%	30 Gy/5 27 Gy/3	NR	1 yr: 84%
Sahgal (Ryu et al. 2008)	62%	Median 24 Gy/3	NR	1 yr: 85%–96%
Nelson (Yamada et al. 2008)	67%	Median 21 Gy/3	CR: 40% OR: 94%	Crude: 88%

Note: NR, not reported.

with 33 spinal lesions treated at Duke University, patients underwent SBRT to a median dose of 21 Gy in three fractions with 67% in a previously irradiated field (Nelson et al. 2009). Overall and complete pain response rates were 94% and 40%, respectively, and the crude rate of MRI-evaluated local control at a median of 6 months follow up was 88%.

14.3.3 ORGANS AT RISK AND TOXICITY

The spinal cord is the dose-limiting critical structure for spinal SRS and SBRT. The incidence of spinal myelopathy has been low with presently used spinal cord constraints. Ryu and colleagues (2007) analyzed spinal cord tolerance in 177 patients with 230 spinal metastases treated with single-fraction spinal radiosurgery. No patients had received prior involved-field radiation. Eighty-six patients had follow-up of at least 1 year. Only one case of myelopathy was observed in a breast cancer patient treated to the C1 vertebrae to a single-fraction dose of 16 Gy with a spinal cord dose maximum of 14.6 Gy. Thirteen months after SRS, she developed 4/5 right lower extremity weakness, which improved with decadron therapy. No severe or persistent myelopathy was observed in any patient.

Gibbs and colleagues (2009) from Stanford University and the University of Pittsburgh published the largest series of SBRT-induced spinal myelopathy cases available. They describe six cases of myelopathy among 1075 patients treated with spinal radiosurgery for benign or malignant spinal tumors at a mean latency of 6.3 months (range 2–10 months) for a cumulative incidence of 0.6% (Gibbs et al. 2009). The authors did not identify clear dosimetric predictors of myelopathy among these patients but recommended a limit on the volume of cord receiving >8 Gy in a single fraction to <1 cc.

Sahgal and colleagues (2012) performed an analysis of dosimetric predictors of radiation myelopathy among patients treated with salvage stereotactic radiation after prior in-field radiation. The authors evaluated thecal sac dosimetry in five patients developing radiation myelopathy. Dosimetry from these cases was compared to a matched control group of 14 patients without radiation myelopathy treated at the University of California–San Francisco. The median prior radiation dose was 38 Gy for those with myelopathy and 39.8 Gy for those without myelopathy. The maximum point dose and the maximum dose delivered to 0.5, 1.0, and 2.0 cc were each significantly higher among patients with radiation myelopathy compared to those without myelopathy. Notably, the target volume definition among the control group who did not develop radiation myelopathy included the gross tumor volume with no margin added for microscopic disease (clinical target volume) or setup inaccuracy (PTV). No attempt was made to electively cover the entire vertebral body or bilateral pedicles, which likely permitted better sparing of the cal sac and spinal cord in these patients.

A high incidence of vertebral fracture has also been reported after single-fraction radiosurgery for spinal lesions. In a series of 62 patients treated with spinal radiosurgery (median dose 24 Gy) to 71 vertebrae at Memorial Sloan-Kettering Cancer Center, vertebral fracture was noted in 27 (39%) vertebrae at a median follow-up of 13 months (Rose et al. 2009). Several predictors of post-treatment fracture were identified. Spinal location, lytic (versus blastic) metastases, and the percentage of the vertebrae involved were independent predictors of fracture. Lesions caudal to T10 were 4.6 times more likely to fracture compared to lesions at other sites. Similarly, lytic lesions were 6.8 times more likely to fracture than blastic lesions. The incidence of fracture also increased with the percentage of vertebral body involvement with a plateau in incidence for percentage involvement above 40%. The authors of this series suggest prophylactic kyphoplasty or vertebroplasty for high-risk patients.

Acute skin toxicity has also been reported after spinal radiosurgery. Investigators from Memorial Sloan-Kettering prospectively analyzed the incidence of radiation dermatitis after single-fraction spine radiosurgery (Murphy and Suh 2011). Analysis was performed in 24 patients treated to 29 spinal or paraspinal sites. The incidence of grade 0, 1, 2, and 3 acute skin toxicity was 28%, 41%, 28%, and 3%. Skin dose was lower among patients with no skin toxicity compared to those with grade 1–3 toxicity, but no statistical analysis was performed. Grade 3 skin toxicity occurred in only one patient with a dosimeter-measured skin dose of 13.5 Gy.

Other toxicities reported after spinal SRS and SBRT include esophagitis and nausea/vomiting (Gomez et al. 2009). The risk of these toxicities is location-dependent. Both acute and late esophagitis have been reported following spinal SRS. Cox and colleagues (2012a) identified the dose to 2.5 cc (D2.5) as highly predictive of esophageal toxicity with a 2% rate below 14 Gy an 12% above 14 Gy. Prophylactic antiemetics should be considered prior to treatment for lesions in the lower thoracic or upper lumbar vertebrae.

14.3.4 PATTERNS OF FAILURE AND TARGET DELINEATION

Some variation exists regarding SRS and SBRT target delineation for spinal metastases. These lesions often lack a discrete border, making precise target delineation challenging. Two approaches to target delineation

(a)

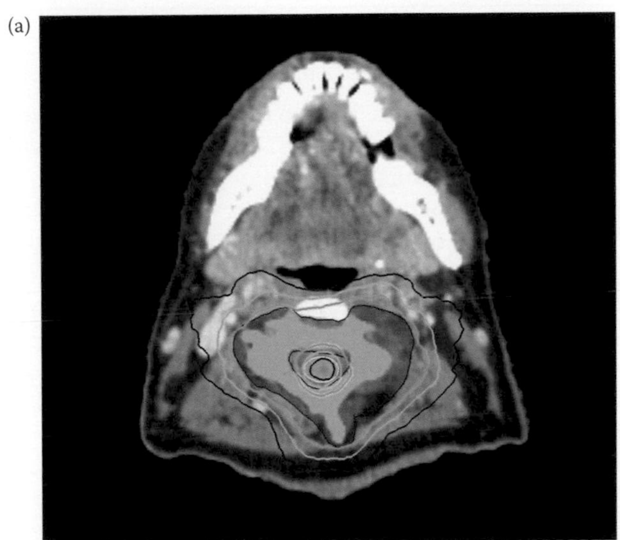

(b)

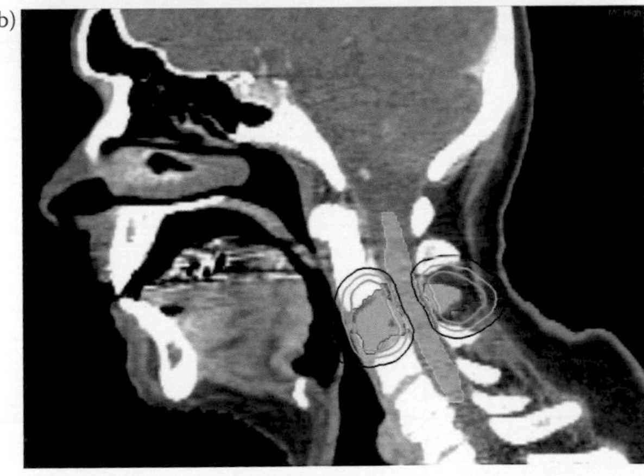

Figure 14.2 Axial (a) and sagittal (b) slices from a representative spinal SRS treatment plan targeting a circumferential C3 metastasis with mild epidural disease. The treatment plan met RTOG 0631 constraints. 18 Gy prescribed to the 74% isodose line (green line) covered 95% of the C3 PTV (red contour). Spinal cord (yellow contour) was limited to a Dmax of 14 Gy with a V10 of 0.3 cc. Also displayed are the 14 Gy (cyan) and 10 Gy (blue) isodose lines. (Images courtesy of Scott Soltys, MD.)

have been described. In the first approach, the target volume includes both the gross tumor volume as well as a surrounding area without macroscopic disease or a clinical target volume. This approach is supported by an analysis of the patterns of failure. In a study by Chang and colleagues (2007), two patterns of failure were identified: (1) the osseous margin posterior to the vertebral body treated and (2) the epidural space. Several investigators have advocated for the treatment of the entire vertebral body and proximal pedicles for lesions arising in the vertebral body (Chang et al. 2007; Ryu et al. 2007). These investigators recommend more extensive elective coverage of the vertebrae for lesions involving the posterior elements. A spinal SBRT plan with comprehensive coverage of the vertebral body and posterior elements is shown in Figure 14.2.

In the second approach, the target volume includes both the gross tumor volume and a PTV expansion to account for setup error and intrafraction motion. This approach is more consistent with the target delineation approach used in SRS and SBRT of other sites. Although this approach opens the door for marginal failures, it also allows for superior spinal cord sparing, particularly in previously irradiated patients. Recently published consensus guidelines advocate inclusion of adjacent normal bony expansion based on tumor location and known patterns of spread (Cox et al. 2012b).

14.4 LUNG

14.4.1 STAGE I NONSMALL CELL LUNG CANCER

Many patients with early stage NSCLC are technically resectable but medically inoperable secondary to severe cardiopulmonary or other comorbidities. Prior to the widespread implementation of thoracic SBRT, patients with medically inoperable stage I NSCLC (MI-NSCLC) had poor survival and low rates of disease control. Despite a high risk of death from comorbid conditions, single institution and population-based registries demonstrate that the majority of untreated patients with stage I MI-NSCLC die of lung cancer (McGarry et al. 2002), and the median survival of untreated patients is only 13–14 months (Rose et al. 2009; Sahgal et al. 2012). Disease-control rates are slightly improved with treatment using conventionally fractionated external beam radiation, but local progression with this approach is frequent and long-term survival rates are low, particularly among patients with T2N0 disease (Powell et al. 2009). Because of the radiobiologic architecture of the lung, dose escalation is tolerable provided an adequate volume of normal lung tissue is spared. SBRT allows for ablative doses of radiation to be delivered to macroscopic disease without exceeding lung tolerances in patients with medically inoperable disease. A summary of select prospective trials using SBRT for stage I MI-NSCLC is presented in Table 14.3.

Table 14.3 Select prospective trials using SBRT for stage I MI-NSCLC

STUDY	INCLUSION	REGIMEN	PRESCRIPTION IDL*	LOCAL CONTROL
Fakiris et al. (2009)	T1–2N0, <7 cm	T1: 60 Gy/3 T2: 66 Gy/3	80%	3 yr: 88.1%
Baumann et al. (2009)	T1–2N0, ≤5 cm	45 Gy/3	67%	3 yr: 92%
Nagata et al. (2005)	T1–2N0, <4 cm	48 Gy/4	Isocenter	5 yr: 96%
Timmerman et al. (2010)	T1–2 (≤5 cm), peripheral	60 Gy/3	NR	3 yr: 97.6%
Lagerwaard et al. (2008)	T1–T2 (≤5 cm)	60 Gy/3 60 Gy/5 60 Gy/8	80%	2 yr: 93%
Dunlap et al. (2010b)	T1–T2, N0, peripheral	42–60 Gy in 3–5 fxns	D_{95}[a]	2 yr: 90% (T1) and 70% (T2)
Ricardi et al. (2010)	T1–2 (≤5 cm), peripheral	45 Gy/3	80%	3 yr: 87.8%

Note: NR, not reported.
[a] Using tomotherapy.

Investigators from Indiana University performed sequential single-institution phase I and phase II clinical trials evaluating three-fraction SBRT in the treatment of stage I MI-NSCLC. In the phase I component, dose escalation was performed beginning at 24 Gy in three fractions (8 Gy per fraction). Cohorts were escalated by 6 Gy total (2 Gy per fraction) in three patient cohorts (McGarry et al. 2005). Dose-limiting toxicity was defined as any acute or late-grade 3–4 skin, soft tissue, esophageal, cardiac or pulmonary toxicity or any grade 4–5 toxicity potentially related to SBRT treatment. For patients with T1N0M0 disease, dose was safely escalated to 60 Gy in three fractions without dose-limiting toxicity. For the T2N0M0 strata, dose-limiting pulmonary toxicity was realized at a dose of 72 Gy in three fractions for tumors >5 cm. For T2N0M0, the maximum tolerated dose was 66 Gy. A phase II trial, including 70 patients (34 T1 and 36 T2) with tumors <7 cm in greatest dimension, was performed (Fakiris et al. 2009). The prescription dose for T1 and T2 tumors was 60 Gy and 66 Gy, respectively, delivered in three fractions prescribed to the 80% isodose volume. At a median follow-up of 50.2 months, the Kaplan-Meier 3-year estimates of local control and OS were 88.1% and 42.7%, respectively. Median survival was 38.7 months for patients with T1 tumors and was 24.5 months for patients with T2 tumors. A subsequent report suggested excessive toxicity among tumors located within 2 cm of the proximal bronchial tree (Timmerman et al. 2006) although in the 4-year follow-up report the difference in toxicity rates between peripheral and central tumors lost statistical significance (p = 0.08) (Fakiris et al. 2009). In aggregate, these data have led to caution using three-fraction regimens for centrally located tumors.

Nagata and colleagues (2005) reported a phase I/II trial of 45 patients with stage I NSCLC treated with SBRT to a dose of 48 Gy in four fractions. Patients either had medically inoperable disease or refused surgery. Seventy-one percent presented with T1N0M0 disease. Treatment was prescribed to isocenter. Five-year local relapse-free survival rates were 95% and 100% and 3-year OS was 83% and 72% for stage IA and IB disease, respectively.

Lagerwatrd et al. (2008) reported a large retrospective series including 206 patients using risk-adapted SBRT for patients with stage I disease from VU University in Amsterdam. Treatment was delivered to a total dose of 60 Gy using one of three fractionation schemes, selected based upon tumor location and estimated risk of SBRT-associated toxicity. Three fractions (20 Gy per fraction) were used for T1 tumors; five fractions (12 Gy) were used for T1 with broad contact with the chest wall or for T2 tumors; and eight fractions (7.5 Gy) were used for central tumors abutting the heart, hilum, or mediastinum. The 2-year actuarial local progression-free survival was 93%. Treatment was well tolerated even in patients with central tumors.

In a multi-institutional trial from Sweden, Norway, and Denmark, Baumann and colleagues (2009) evaluated SBRT using a dose of 45 Gy delivered in three fractions for patients with stage I MI-NSCLC. Fifty-seven patients were enrolled, 40 with T1 and 17 with T2 disease. Treatment was prescribed to the periphery of the PTV, corresponding to roughly the 67% isodose line. Three-year overall and cancer-specific survival rates were 60% and 88%, respectively. Local relapse was observed in four patients (7%), and the Kaplan-Meier estimate of 3-year local control was 92%. Distant failure was the predominant pattern of relapse and was more common in patients with T2 disease.

Building on these favorable results, the RTOG initiated the first cooperative group trial evaluating SBRT in the treatment of MI-NSCLC. RTOG 0236 accrued 59 patients (55 evaluable) with T1-T2N0M0 MI-NSCLC with a maximum diameter less than 5 cm in greatest dimension (Timmerman et al. 2010). Patients were treated with 60 Gy in three fractions without heterogeneity correction, corresponding to 54 Gy in three fractions (18 Gy per fraction) with heterogeneity correction. Eighty percent of patients presented with T1 primary tumors. Fifty-one percent of patients had a radiographic complete response at a median of 6.5 months after SBRT. At a median follow-up of 34.4 months, 3-year actuarial in-field and in-lobe tumor control was 97.6% and 90.6%, respectively. Regional nodal failures occurred in only two patients despite reliance on noninvasive methods of mediastinal staging. The primary pattern of failure was distant with a 3-year rate of distant recurrence of 14.7% for T1 tumors and 47% for T2 tumors. Three-year OS was 55.8%. By histology, the 3-year rate of distant recurrence was 5.9% for squamous and 30.7% for nonsquamous histology. Figure 14.3 illustrates a representative SBRT plan for a patient with a peripherally located T1N0 NSCLC of the right upper lobe of the lung.

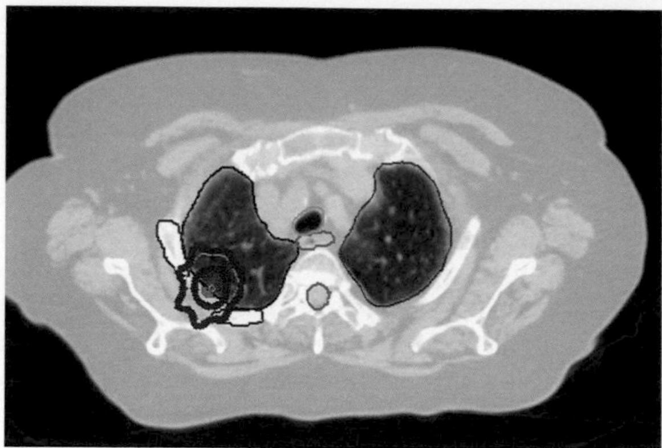

Figure 14.3 A representative axial slice from an SBRT plan targeting a peripheral T1aN0 NSCLC abutting the chest wall. 50 Gy was prescribed in four fractions to the 79.9% isodose line. 95% of the PTV (red colorwash) is covered by the prescription isodose line (blue). The 50% (purple) isodose line is also shown.

RTOG 0915 accrued 94 patients with peripherally located, T1-2N0 MI-NSCLC measuring less than 5 cm to a randomized phase II trial comparing 48 Gy in four fractions to 34 Gy in a single fraction (Radiation Therapy Oncology Group: Protocol RTOG 0915). Now closed to accrual, results are pending and should help delineate the effectiveness of these alternative fractionation schedules.

The recently completed RTOG 0813 is a phase I/II trial evaluating five-fraction SBRT in patients with centrally located stage I tumors (Radiation Therapy Oncology Group: Protocol RTOG 0813). This trial enrolled 120 patients with T1-2N0 MI-NSCLC arising within 2 cm of the proximal bronchial tree or abutting the mediastinal or pericardial pleura. Dose escalation started at 50 Gy in five fractions, escalating in 2.5 Gy increments to the final dose level of 60 Gy in five fractions. A representative centrally located T2N0 NSCLC SBRT case meeting RTOG 0813 dose constraints is shown in Figure 14.4.

Several independent predictors of local control for patients treated with SBRT for stage I NSCLC have been identified. SBRT dose, histology, and primary tumor size/stage have each been shown to predict for local control. Grills and colleagues (2012) report a collaborative multi-institutional analysis of 483 patients with 505 lung tumors treated with SBRT. At a median follow-up of 1.3 years, independent predictors of improved local control were smaller tumor size, nonsquamous histology, and higher biologic equivalent prescription dose. Two-year local recurrence by maximum cross-sectional GTV dimension was 3% for tumors <2.7 cm and 9% for tumors >2.7 cm ($p = 0.03$). Local recurrence based on prescribed biologic equivalent dose (BED) was 15% for $BED_{10} < 105$ and 4% for $BED_{10} \geq 105$ Gy ($p < 0.001$). Local control has also been shown to vary according to T-stage for peripheral NSCLC treated with SBRT. Dunlap and colleagues (2010b) evaluated the records of 40 consecutive patients treated with three- or five-fraction SBRT to a median dose of 60 Gy. Two-year local control was 90% and 70% for T1 and T2 tumors, respectively ($p = 0.03$). The difference in local control comparing T1 versus T2 primaries may be more pronounced in patients treated to lower SBRT doses (Koto et al. 2007).

Based on the success of SBRT for patients with medically inoperable disease, interest is growing in SBRT as an alternative to surgery for medically operable patients. At the 2010 ASTRO annual meeting, Nagata and associates (2010) reported the initial results of JCOG 0403, a multi-institutional phase II trial evaluating SBRT in patients with operable T1N0M0 (stage IA) NSCLC. Patients were required to have PaO2 of >60 torr and FEV1 > 700 mL. Sixty-five patients were enrolled at 15 institutions. Median age was 79 years. Treatment was 48 Gy in four fractions prescribed to isocenter. Three-year OS, progression-free survival, local progression-free survival, and local control were 76%, 54.5%, 68.5%, and 84%, respectively. Only six grade 3 toxicities and no grade 4–5 toxicities were observed. RTOG 0618, a phase II trial evaluating SBRT for patients with peripherally located, medically operable stage I/II NSCLC completed accrual of 33 patients as of May 2010, and results are pending (Radiation Therapy Oncology Group: Protocol RTOG 0618). Ultimately, randomized comparisons with surgery are needed to establish the appropriate role of SBRT for medically operable disease. The American College of Surgeons Oncology

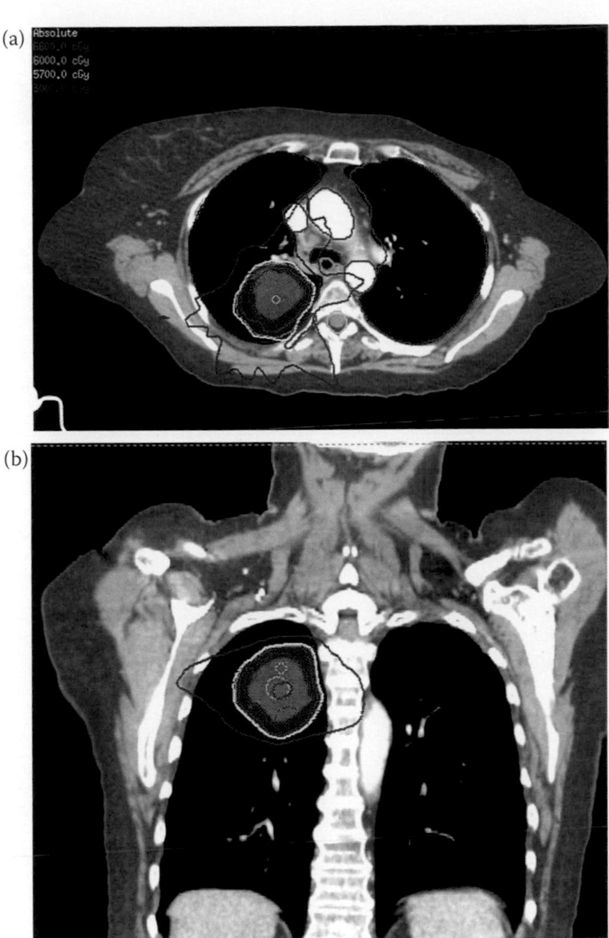

Figures 14.4 Representative axial (a) and coronal (b) slices are shown from an SBRT plan targeting a T2aN0 centrally located NSCLC. A prescription dose of 60 Gy was delivered in five fractions using a 10-field IMRT plan with all normal tissue constraints from RTOG 0813 achieved. 95% of the PTV is covered by the 100% isodose line (green). The 110% (purple), 95% (yellow), and 50% (blue) isodose lines are also shown.

Group (ACOSOG) Z4099/RTOG 1021 trial attempted to enroll high surgical–risk patients with early stage NSCLC to a phase III randomized comparison of sublobar resection and SBRT, but recently closed secondary to poor accrual (ACOSOG Protocol Z4099/RTOG Protocol 1021). Similarly, the lung cancer STARS trial was an international collaborative trial enrolling medically operable early stage NSCLC patients to a phase III comparison of lobectomy and SBRT with the CyberKnife system (Lung Cancer STARS Trial, CriticalTrials.gov.) The trial also recently closed due to slow accrual. The ROSEL trial, a Dutch multi-institution randomized phase III comparison of lobectomy and SBRT for medically operable patients, closed early secondary to poor accrual (ROSEL Lung Cancer Trial, CriticalTrials.gov.). Table 14.4 outlines current trials evaluating thoracic SBRT in the medically operable population.

Published prospective trials have not yet incorporated chemotherapy or targeted therapies with SBRT, and the role of systemic therapy for patients treated with SBRT is largely unexplored. As the failure pattern after SBRT remains predominantly distant (Chi et al. 2010), improved systemic control remains an important goal. The comorbidities that render patients medically inoperable will often limit tolerance to systemic agents. However, as SBRT is explored in the medically operable population, the appropriate integration of systemic agents will be increasingly crucial. Furthermore, as oral targeted agents are increasingly used in molecularly defined patient subsets for metastatic disease, integration of such agents for selected patients early stage disease is a logical next step.

Table 14.4 Recent and currently accruing trials evaluating thoracic SBRT in the medically operable population

STUDY	INCLUSION CRITERIA	STUDY DESIGN
RTOG 0618[a]	T1, T2, or T3 (chest wall) <5 cm, N0M0, peripheral only	Phase II non-randomized: 60 Gy in three fractions
ACOSOG Z4099/RTOG 1021[b]	Ia (<3 cm) or Ib (if visceral pleura involved), N0M0, peripheral only	Phase III Randomized: 54 Gy in three fractions versus sublobar resection
STARS Lung Cancer Trial[c] JCOG 0403[d]	T1 or T2 (<4 cm), N0M0 T1N0M0	Phase III Randomized: 60 Gy in three fractions (peripheral) or 60 Gy in four fractions (central) versus lobectomy Phase II non-randomized: 48 Gy in four fractions

[a] Radiation Therapy Oncology Group: Protocol RTOG 0618. A phase II trial of stereotactic body radiation therapy (SBRT) in the treatment of patients with operable stage I/II non-small cell lung cancer. Available at http://www.rtog.org/ClinicalTrials/ProtocolTable/StudyDetails.aspx?study = 0618.
[b] American College of Surgeons Oncology Group (Protocol ACOSOG Z4099)/Radiation Therapy Oncology Group (Protocol RTOG 1021). A randomized phase III study of sublobar resection (±brachytherapy) versus stereotactic body radiation therapy in high-risk patients with stage I NSCLC. Available at http://www.rtog.org/ClinicalTrials/ProtocolTable/StudyDetails.aspx?study = 1021.
[c] Lung Cancer STARS Trial. Available at http://clinicaltrials.gov/ct2/show/NCT00840749.
[d] Nagata, Y., Hiraoka, M., Shibata, T. et al. *Int. J. Radiat. Oncol. Biol. Phys.*, 78, 3 Suppl., S27–S28, 2010.

14.4.2 LUNG METASTASES

Several prospective trials have evaluated the role of SBRT for lung metastases. Similarly to SBRT for MI-NSCLC, SBRT for metastatic lesions is associated with high rates of in-field local control and a comparatively low incidence of high-grade toxicity. A summary of prospective trials using SBRT for lung metastases is presented in Table 14.5.

In a phase I/II trial of three-fraction SBRT for lung metastases, investigators from the University of Colorado evaluated three-fraction SBRT in the treatment of three or fewer pulmonary metastases with maximum cumulative diameter of <7 cm. In phase I, the dose of SBRT was safely escalated from 48 Gy to 60 Gy in three fractions without dose-limiting toxicity (Schefter et al. 2006). Subsequently, 24 patients were treated on a phase II trial to 60 Gy in three fractions. In total, nine patients received doses less than 60 Gy, and 29 patients received 60 Gy in three fractions. Only one in-field local failure was observed among 63 treated lesions, correlating to an actuarial local control rate of 96% at two years (Rusthoven et al. 2009a). No cases of grade 4–5 toxicity were observed. The crude rate of grade 3 toxicity was 7.6% (one case of each of pneumonitis, dermatitis, and rib fracture). At last follow-up, 18 of 38 patients were alive for a 2-year estimate of OS of 39%.

Table 14.5 Prospective trials of SRS/SBRT for lung metastases

STUDY	INCLUSION	REGIMEN	PRESCRIPTION IDL[a]	LOCAL CONTROL
Hof et al. (2007)	1–2 mets, ≤4 cm	12–30 Gyx1, Med 24 Gy	Isocenter, 80% IDL covers PTV	3 yr: 63.1%
Okunieff et al. (2006)	1–5 mets, No size constraints	50–55 Gy/10, Med 5.4 Gy/fx	Isocenter, 80% IDL covers PTV	3 yr: 91%
Hara et al. (2006)	≤3 mets, <4 cm	20–25 Gy/1 30–34 Gy/1	Prescribed as minimum dose to PTV	2 yr: 83% (30–34 Gy), 52% (20–25 Gy)
Rusthoven et al. (2009a)	1–3 mets, cumulative diameter <7 cm	60 Gy/3	80%–90%	2 yr: 96%

[a] Isodose line.

Okunieff and colleagues (2006) reported the outcomes for patients with lung metastases treated on a trial of hypofractionated imaged-guided radiotherapy (HIGRT) for oligometastases, using a more protracted regimen of 50–55 Gy in 10 fractions. The dose was prescribed to the isocenter, and the 80% isodose line was required to cover the PTV. Actuarial 3-year local control on a lesion-based analysis (n = 125) was 91% and on a patient-based analysis (n = 49) was 82.5%. One patient (2%) experienced grade 3 toxicity, which was a nonmalignant pleural effusion requiring pleurocentesis and sclerosis.

Only limited data is available using single-fraction radiosurgery for lung metastases. Hof and colleagues (2007) reported a series of 61 patients with 71 pulmonary metastases treated with single-fraction SBRT. The first 20 patients were treated as part of a prospective phase I–II trial. Treatment was to a median of 24 Gy (range 12–30 Gy). Dose was prescribed to isocenter with the 80% isodose line covering the PTV. Local PFP was 74% at 2 years and 63% at 3 years. Three patients (4.9%) experienced grade 3 pulmonary toxicity requiring oxygen supplementation. Hara and associates (2006) reported a series of 59 lung lesions (11 primary and 48 metastatic) treated with single-fraction radiosurgery using a modified microtron device. Nine lesions were treated to doses less than 30 Gy, and 50 lesions were treated to doses ≥30 Gy. Locoregional progression-free survival was increased among patients treated to doses of 30 Gy or higher. Two-year locoregional progression-free survival was 83% in patients treated to single-fraction doses of 30–34 Gy.

A dose response for SBRT in the treatment of lung metastases has been demonstrated. In a retrospective analysis of 246 lung and liver lesions treated with three-fraction SBRT at the University of Colorado, local control was analyzed according to nominal dose and equivalent uniform dose (EUD) (McCammon et al. 2009). In this study, both nominal dose and EUD were predictive of in-field local control of SBRT-treated lesions. Three-year actuarial local control was 8.1%, 56%, and 89.3% for nominal doses of <36 Gy, 36–53.9 Gy, and 54–60 Gy, respectively. Similarly, 3-year local control according to EUD was 12%, 61.4%, and 89.9% for EUD values of <45 Gy, 45–65.3 Gy, and >65.3 Gy, respectively.

14.4.3 ORGANS AT RISK AND TOXICITY

Radiation pneumonitis rates in prospective studies of SBRT and HIGRT for lung metastases have been low. Lung parenchyma obeys the parallel architecture model of radiation biology and, as such, ablative doses of radiation can be delivered to small volumes provided an adequate volume of normal lung is spared. Grade 2–3 pneumonitis, defined as symptoms requiring steroids or supplemental oxygen, has been observed in only 5%–10% of patients receiving SBRT for lung metastases (Hara et al. 2006; Hof et al. 2007; Okunieff et al. 2006; Rusthoven et al. 2009b; Schefter et al. 2006). In the multi-institutional trial led by the University of Colorado, the volume of normal lung (both lungs minus the cumulative gross tumor volume) receiving 15 Gy in three fractions was limited to <35% and only one patient (2.8%) experienced radiation pneumonitis (Rusthoven et al. 2009b). Several recent analyses have identified DVH predictors of radiation pneumonitis after SBRT. Guckenberger and colleagues (2010) reported an increased mean lung dose (MLD) in patients with grade II radiation pneumonitis after SBRT compared to those without pneumonitis. Ipsilateral MLD in patients with pneumonitis was 12.5 Gy compared to 9.9 Gy in those without pneumonitis. In an analysis of 251 patients treated at Indiana University, bilateral MLD >4 Gy and lung V20 (volume of total lung minus GTV receiving 20 Gy) >4% were both significant predictors of grade 2 or greater pneumonitis (Barriger et al. 2012).

Chang and colleagues (2012) likewise analyzed a single institution experience of 130 patients with stage I NSCLC treated with SBRT in four fractions and observed a significantly higher rate of radiation pneumonitis when the mean ipsilateral lung dose exceeded 9.1 Gy.

In the setting of stage I MI-NSCLC, increased toxicity has been reported for patients with centrally located lesions. As previously discussed, Timmerman and colleagues (2006) reported an 11-fold increased incidence of grade 3–5 toxicity occurring in SBRT-treated lesions within 2 cm of the proximal bronchial tree on the Indiana phase II trial. However, in the 4-year update of this trial, the cumulative incidence of grade 3–5 toxicity was 27% for centrally located lesions compared to 11% for peripheral lesions (Timmerman et al. 2010). This difference was no longer statistically significant (p = 0.088). Other studies in MI-NSCLC have not reported an increase in high-grade toxicity for central lesions albeit using lower doses or more protracted SBRT regimens (Baumann et al. 2009; Lagerwaard et al. 2008; Nagata et al. 2005). Moreover, in the reported trials using SBRT for metastatic disease, an increased incidence of severe toxicity in patients with central lesions has not been observed (Hara et al. 2006; Hof et al. 2007; Okunieff et al. 2006; Rusthoven et al. 2009b; Schefter et al. 2005). To date, no dosimetric constraints for the proximal airways have been clinically identified to help inform SBRT planning for centrally located lesions. Caution should be taken when administering SBRT for lesions occurring within 2 cm of the proximal bronchial tree.

For peripheral lesions treated with SBRT, the most frequent adverse event is toxicity involving the chest wall. Chronic pain and rib fracture have both occurred after lung SBRT. Both critical volume and maximum dose parameters are predictive of chest wall toxicity. A combined analysis of patients treated with three- or five-fraction SBRT from the University of Virginia and the University of Colorado revealed that the volume of chest wall receiving 30 Gy (V30) was the strongest predictor of chest wall pain requiring narcotics or rib fracture (Dunlap et al. 2010a). A volume threshold of 30 mL was identified. Moreover, Pettersson, Nyman, and Johansson (2009) identified serial dose parameters associated with rib fracture in a cohort of patients treated with a regimen of 45 Gy in three fractions for MI-NSCLC. In this study, a total dose of 27.3 Gy and 50 Gy to at least 2 mL of rib were associated with a 5% and 50% incidence, respectively, of radiation-induced rib fracture.

Other toxicities associated with lung SBRT include radiation dermatitis (Hoppe et al. 2008) and brachial plexopathy (Forquer et al. 2009). Radiation myelopathy has not been reported to date in setting of lung SBRT. In each of these studies, conservative spinal cord constraints have been utilized to mitigate this risk. Excluding the spinal cord from the high isodose distributions is recommended given the limited data addressing spinal cord tolerance in the setting of high fractional doses.

14.4.4 TREATMENT OF MULTIPLE LUNG LESIONS AND SEQUENTIAL COURSES OF SBRT

Questions remain regarding the number of concurrent or sequential lung lesions that can safely be treated by SBRT. Because lung parenchyma is arranged in parallel functional subunits, the risk of symptomatic radiation pneumonitis and radiation fibrosis increases as a function of treatment volume and volume of irradiated lung parenchyma. The University of Colorado trial included patients with up to three lung lesions with a maximum cumulative diameter less than 7 cm (Rusthoven 2009a). The University of Rochester oligometastases study allowed enrollment of patients with up to five lesions (Milano et al. 2008). Twenty-four of the 121 patients (20%) in this trial had four or five discernible metastases treated with SBRT; however, the number of patients with four or five lung metastases was not reported. Without further published data, SBRT for ≥4 lung metastases should be limited to the setting of a clinical trial.

There is also uncertainty regarding the safety of multiple sequential courses of SBRT or the safety of SBRT administered after prior conventionally fractionated external beam radiation therapy (EBRT). With conventionally fractionated radiation and chemoradiation, radiation pneumonitis typically occurs within the first 12 months after treatment with a peak incidence approximately 4–6 months after treatment (Marks et al. 2010). By contrast, several studies suggest toxicity is significantly delayed following hypofractionated regimens. As a result, SBRT administered sequentially after prior EBRT or SBRT may confer an additive risk of radiation-related lung toxicity regardless of the interval between treatment courses. Kelly and colleagues (2010) recently reported a series of 36 patients treated with SBRT after prior thoracic EBRT. The median dose of EBRT was 61.5 Gy, and the median interval between EBRT and SBRT was 22 months. The median SBRT dose delivered was 50 Gy in four fractions. The incidence of any symptomatic pneumonitis was 50%, and dyspnea requiring oxygen supplementation (grade 3 pneumonitis) was 19%. Further analysis revealed an association between SBRT for an out-of-field relapse and grade 3 pneumonitis ($p = 0.03$).

14.5 LIVER AND PANCREAS

14.5.1 LIVER METASTASES

Several recent studies have evaluated the use of SBRT for the treatment of limited hepatic metastases from a variety of primary tumors. Similar to the lung, the liver is an attractive organ for SBRT due to its parallel arrangement in functional subunits. This architecture allows for safe delivery of ablative radiation doses to a small volume provided that an adequate proportion of normal functioning liver is successfully spared.

In-field local control rates are similarly high to those observed with lung SBRT provided adequate doses are administered. A summary of select prospective trials using fractionated SBRT for liver metastases is presented in Table 14.6.

In a phase I/II trial, investigators from the University of Colorado demonstrated a high rate of local control with SBRT for patients with three or fewer liver metastases, each measuring less than 6 cm in greatest dimension. This trial was similar in design to the previously referenced study for lung metastases. In phase I, the dose of SBRT was safely escalated from 36 Gy to 60 Gy in three fractions without dose-limiting toxicity (Schefter et al. 2005). In phase II, the dose was 60 Gy in three fractions. In total, 13 patients received doses less than 60 Gy, and 36 patients received 60 Gy in three fractions. Only three in-field local failures among 47 lesions evaluable for local control (patients with at least 6 months radiographic follow-up) after SBRT. The actuarial local control of all SBRT-treated lesions was 92% at 2 years (Rusthoven et al. 2009b). Moreover, among lesions measuring <3 cm in greatest dimension, the 2-year actuarial local control was 100%. Only one case of grade 3 toxicity was observed, and no patients experienced grade 4–5 toxicity. No cases of radiation-induced liver disease (RILD) were observed. At last follow-up, 20 of 47 patients were alive, and the two-year survival after SBRT was 30%.

The favorable outcomes from the University of Colorado study were confirmed in a prospective study from the University of Texas Southwestern (UT-SW). Rule and colleagues (2011) reported a phase I clinical trial using SBRT for liver metastases. In this dose-escalation study, three dose cohorts were evaluated: 30 Gy/3 fx, 50 Gy/5 fx, and 60 Gy/5 fx. No grade 4–5 toxicity was reported, and only one grade 3 event occurred in the 50 Gy group, an asymptomatic grade 3 transaminitis. Local control at 24 months was 56%, 89%, and 100% for the 30 Gy, 50 Gy, and 60 Gy cohorts, respectively.

In an early study of liver SBRT from the University of Heidelberg, Herfarth and colleagues (2001) evaluated single-fraction SBRT at a dose of 14–26 Gy for 55 liver metastases. At 18 months, local control was achieved in 67% of patients, and no high-grade toxicity was observed. Hoyer and colleagues (2006) evaluated SBRT (45 Gy in three fractions) for 141 colorectal cancer metastases, including 44 hepatic metastases. The 2-year actuarial local control was 79%. In this series, one patient died of liver failure, one patient experienced colonic perforation, and two patients experienced duodenal ulceration (Hoyer et al. 2006). Méndez-Romero and colleagues (2006) reported local control of 82% in a phase II trial of 45 primary or metastatic hepatic lesions treated with SBRT, most treated with 37.5 Gy in three fractions. Among patients with metastases, the 2-year local control was 86%, and only three grade 3 toxicities were observed. No grade 4–5 toxicity was reported (Méndez-Romero et al. 2006).

A separate approach for liver metastases, using individualized doses derived from normal tissue complication probability (NTCP) modeling, has also been evaluated (Lee et al. 2009). Investigators from Princess Margaret

Table 14.6 Select prospective trials of SBRT for liver metastases

STUDY	INCLUSION	REGIMEN	PRESCRIPTION IDL	LOCAL CONTROL
Lee et al. (2009)	No max size or number	28–60 Gy/6, med 42 Gy	≥71.4%	1 yr: 71%[a]
Mendez-Romero et al. (2006)	1–3 lesions, largest <7 cm	37.5 Gy/3	65%	2 yr: 86%
Rule et al. (2011)	1–5 lesions, must meet liver constraints	30 Gy/3, 50 Gy/5, 60 Gy/5	70%–85%	30 Gy: 56% 50 Gy: 89% 60 Gy: 100%
Rusthoven et al. (2009b)	1–3 lesions, largest <6 cm	60 Gy/3	80%–90%	2 yr: 92% <3 cm: 100%
Herfarth et al. (2001)	1–3 lesions	14–26 Gy/1	80%	18 mo: 67%
Hoyer et al. (2006)	1–6 lesions, largest ≤6 cm	45 Gy/3	67%	2 yr: 86%

[a] Isodose line.

Hospital performed a phase I trial of SBRT for liver metastases with radiation doses chosen to maintain the same numerical risk of RILD. The NTCP-calculated risk of RILD was escalated from 5% to 10% to 20% in three cohorts. SBRT was delivered in six fractions over 2 weeks to a maximum dose of 60 Gy. The median dose delivered was 41.8 Gy. Sixty-eight patients were enrolled, most (40) with metastases from colorectal cancer. No dose-limiting toxicity was observed. One-year local control was 71%. Median survival was 17.6 months.

As with SBRT for lung metastases, a dose response has also been demonstrated using SBRT in the treatment of liver metastases. In the aforementioned study by McCammon and colleagues (2009) (lung 24.1.1), both nominal dose and EUD were predictive of local control. Additionally, in the previously mentioned phase I study of SBRT for liver metastases from UT-Southwestern, a clear dose-response relationship was also identified (30 Gy versus 60 Gy, $p = 0.009$) (Rule et al. 2011).

14.5.2 HEPATOCELLULAR CARCINOMA

SBRT has also been employed in the treatment of unresectable hepatocellular carcinoma (HCC). Similarly to MI-NSCLC, SBRT is an attractive modality for the treatment of unresectable HCC (Rusthoven and Hasselle 2010). Among patients with limited dysfunction (Child-Pugh class A cirrhosis), SBRT has been shown to be both safe and effective in the available studies.

In North America, two recent prospective trials have evaluated SBRT in patients with primary liver malignancies. Tse and colleagues (2008) enrolled 41 patients with Child-Pugh class A HCC (31 patients) or intrahepatic cholangiocarcinoma (10 patients) on a phase I trial of individualized SBRT for inoperable disease. Patients were treated in six fractions to a median dose of 36 Gy. The individual radiation dose was dependent on the volume of normal liver irradiated and the calculated NTCP. Median tumor volume was 173 cc. The 1-year in-field local control rate was 65% and, in patients with HCC, the median survival duration was 11.7 months. No patient developed RILD or grade 4–5 treatment-related toxicity. Five patients (12%) developed grade 3 elevation of liver enzymes. In another phase I trial from Indiana University, 17 patients with 25 lesions were enrolled on a trial of three-fraction SBRT for Child-Pugh class A and B primary HCC (Cardenes et al. 2010). Dose was escalated from 36 Gy to 48 Gy in 6 Gy (2 Gy per fraction) increments. No dose-limiting toxicity was observed in patients with Child-Pugh class A disease. Two patients with Child-Pugh class B disease developed grade 3 hepatotoxicity. The dose in these patients was subsequently modified to 40 Gy in five fractions, and one liver failure was observed. Actuarial 2-year local control and survival rates were 100% and 60%, respectively.

In a retrospective series from Korea, Seo and colleagues (2010) demonstrated a dose response using SBRT for inoperable HCC. In this series, 38 patients with tumors <10 cm in greatest dimension were treated with SBRT to a total dose of 33–57 Gy. Two-year local progression-free survival and OS were 66.4% and 61.4%, respectively. On multivariate analysis, OS was significantly higher among patients treated to a dose >42 Gy in three fractions ($p = 0.001$). The 2-year survival among patients treated to >42 Gy in three fractions was 81%. Local control according to dose was not reported.

The recently activated RTOG 1112 is a phase III trial randomizing HCC patients unsuitable for resection, transplant, or radiofrequency ablation (RFA) to five-fraction SBRT followed by sorafenib versus a control arm of sorafenib monotherapy (Radiation Therapy Oncology Group: Protocol RTOG 1112). As the first cooperative group trial to explore SBRT for HCC, the results will be greatly anticipated.

14.5.3 PANCREAS

Survival and disease-control rates using conventional therapies for patients with unresectable pancreatic adenocarcinoma are dismal. Median survival following conventionally fractionated chemoradiation is less than a year, and treatment duration using this approach is 5–6 weeks. In an effort to shorten treatment duration and improve in-field local control, several recent studies have evaluated single fraction or short multifraction SBRT regimens.

Investigators from Stanford prospectively evaluated the combination of single-fraction SBRT and gemcitabine chemotherapy in patients with locally advanced, nonmetastatic pancreatic cancer (Schellenberg et al. 2008). Sixteen patients received a single 25-Gy fraction using CyberKnife with implanted fiducials, administered 2 weeks after the first cycle of gemcitabine. Only three patients (19.4%) developed local disease progression, all 14 or more months after SBRT. Median survival was 11.4 months. One grade 3 duodenal stenosis and one grade 4 duodenal perforation were observed. The Stanford group also published a separate retrospective review of 77 patients treated with a single 25-Gy fraction for unresectable or metastatic pancreatic adenocarcinoma (Chang et al. 2009). Among patients with nonmetastatic disease, median survival from SBRT treatment and from initial diagnosis was 6.7 months and 11.5 months, respectively. One-year freedom from local progression (FFLP) was 84%, and isolated local recurrence occurred in only 5% of patients. The cumulative rate of grade ≥3 late toxicity was 9%, and the 12-month rate of grade ≥2 toxicity was 25%. The noted toxicity led to development of the currently accruing, multi-institution phase II trial evaluating sequential gemcitabine and fractionated SBRT to 33 Gy over five sessions for unresectable pancreatic adenocarcinoma (Clinical Trial NCT01360593). Figure 14.5 illustrates a five-fraction pancreas SBRT plan.

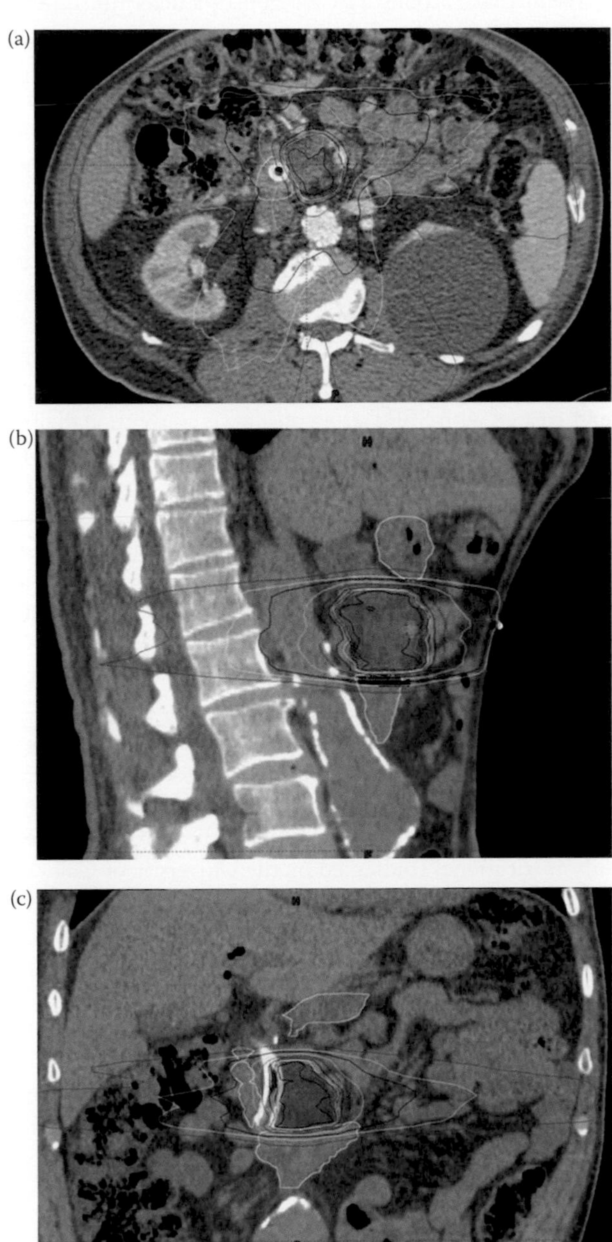

Figures 14.5 Representative axial (a), sagittal (b), and coronal (c) slices from a conformal SBRT plan targeting a locally advanced pancreatic adenocarcinoma are shown. 34 Gy was prescribed in five fractions using volume-modulated arc therapy. (Images courtesy of Daniel Chang, MD.)

Other retrospective studies have confirmed the local control efficacy of SBRT with or without chemotherapy for patients with unresectable pancreatic cancer. Rwigema and colleagues (2011) reviewed the outcomes of 71 patients treated with SBRT for pancreatic cancer. Forty-one patients (56%) had unresectable disease, and 11 (16%) had locally recurrent disease following prior surgical resection. Patients were treated to a median dose of 24 Gy (18–25 Gy). Ninety-four percent were treated with single-fraction SBRT. Median survival was 10.3 months. Overall FFLP was 64.8%. Tumor volume ($p = 0.02$) and SBRT dose ($p = 0.004$) were both associated with FFLP. Grade 3 acute toxicity was observed in three patients. In another study, investigators from Beth Israel Deaconess Medical Center published a review of 36 patients treated with three-fraction SBRT to 24–36 Gy followed by Gemcitabine for six months or until disease progression (Mahadevan et al. 2010). The median carbohydrate antigen 19-9 determined time to progression was 7.9 months, and the median computed tomography determined time to progression was 9.6 months. At a median follow up of 24 months, in-field local control was maintained in 78% of patients. Median survival duration was 14.3 months. Five grade 3 (14%) toxicities and no grade 4 toxicities were observed.

Investigators at the University of Pittsburgh have also investigated to role of SBRT in patients with close or positive margins after surgical resection (Rwigema et al. 2012). Twenty-four patients with close (33%) or positive (67%) margins were treated with SBRT to a median dose of 24 Gy in a single fraction. FFLP was achieved in 87.5% of patients with close margins and 62.5% of those with positive margins. Nineteen patients (79%) received adjuvant gemcitabine chemotherapy at a median of 18 days after SBRT. At a median follow-up for surviving patients of 16.3 months, 10 patients (41.7%) remained progression free. No grade 3–4 SBRT-related toxicity was observed.

Another potential application of SBRT for pancreatic cancer is as a boost after conventionally fractionated chemoradiation for unresectable disease. Seo and colleagues (2009) retrospectively reviewed the outcomes of 30 patients treated with SBRT as a boost after conventionally fractionated radiation. Patients were treated to a dose of 40 Gy in 20 fractions. Twenty-one received concurrent chemotherapy. SBRT boost was administered as a single fraction to a dose of 14–17 Gy. One-year survival and local progression-free survival rates were 60% and 70%, respectively.

Currently, open protocols are exploring integration of SBRT into aggressive systemic chemotherapy and immunotherapy regimens (Clinical Trial NCT01595321) and investigating the role of SBRT in converting borderline resectable and unresectable disease to resectable (Clinical Trial NCT01446458; Clinical Trial NCT01754623).

14.5.4 ORGANS AT RISK AND TOXICITY

RILD has been infrequently observed in trials using SBRT for hepatic metastases. Prevention of RILD in patients treated with SBRT hinges upon successful avoidance of an adequate volume of normal liver. Constraints designed to spare a necessary volume of normally functioning organ are known as critical volume constraints. Critical volume constraints have been empirically established for three- and five-fraction SBRT. In the multi-institutional studies led by the University of Colorado, 700 mL of normal liver was required to receive less that 15 Gy in three fractions (Rusthoven et al. 2009b). Similarly, in the UT-SW study, 700 mL of normal liver was required to receive less than 21 Gy in five fractions (Rule et al. 2011). No cases of RILD were observed in either trial. In other trials, grade 3 elevation in hepatic enzymes has been observed after SBRT for hepatic metastases, but in most cases, these episodes were asymptomatic.

Caution should be taken when considering SBRT for patients with abnormal liver function tests or with cirrhosis. Trials of SBRT for liver metastases have generally required patients to have adequate pretreatment liver function. By contrast, SBRT in the setting of unresectable HCC has been associated with high-grade toxicity and liver failure in patients with Child-Pugh class B cirrhosis (Cardenes et al. 2010; Méndez-Romero et al. 2006). SBRT for patients with Child-Pugh class B HCC should be limited to the setting of a clinical trial. SBRT should be avoided in patients with Child-Pugh class C cirrhosis (Rusthoven and Hasselle 2010).

Injury to the stomach and small bowel has been reported after hypofractionated radiation to the upper abdomen, particularly in the setting of SBRT for unresectable pancreatic adenocarcinoma. Murphy and colleagues from Stanford University analyzed dosimetric predictors of duodenal ulceration after single-fraction pancreas SBRT (Murphy et al. 2010). In the Stanford series, a volume of duodenum receiving

15 Gy (V15) greater than 9.1 cm³, a duodenal V20 greater than 3.3 cm³, and duodenal maximum dose greater than 23 Gy were each associated with a significantly increased risk of grade 2–4 duodenal toxicity. The incidence of gastric and small intestinal toxicity has been lower in trials using SBRT for hepatic metastases, which is likely attributable to greater separation between these structures and the SBRT-treated liver lesions. Liver SBRT should be avoided in patients with active peptic ulcer disease.

Similar to SBRT for lung metastases, the chest and abdominal wall are also at risk of radiation-associated toxicity in patients with peripherally located hepatic metastases treated with SBRT. Additionally, high-grade skin and soft tissue toxicity has been observed with SBRT plans using an inadequate number of exclusively coplanar fields (Rusthoven et al. 2009b). SBRT planning using a sufficient number of carefully selected beam orientations is critical to avoid giving excessive superficial doses at beam entry points. Non-coplanar beam arrangements are recommended, whenever possible, for peripherally located lesions to create a steep dose falloff to soft tissue of the chest/abdominal wall and skin. Specific constraints for the abdominal wall have not yet been identified; however, the maximum dose and partial volume constraints previously discussed for chest wall toxicity in the setting of lung SBRT may serve as reasonable dosimetric parameters (see organs at risk and toxicity in the "Lung" Section 14.4).

14.6 PROSTATE

14.6.1 CLINICAL EFFICACY

Following the widespread implementation of prostate-specific antigen screening, the majority of men with prostate cancer are diagnosed with early-stage, organ-confined disease (Lu-Yao et al. 2009). For men with low and intermediate risk disease (T1-2 tumor with Gleason score <8 and PSA <20 ng/mL), treatment with surgery or radiation therapy is associated with high rates of long-term biochemical control and prostate cancer-specific survival (Rwigema et al. 2012; Seo et al. 2009). For patients treated with EBRT, a typical course of treatment is delivered over 7 to 9 weeks. For many patients, this poses logistical challenges. In addition to the added convenience of an abbreviated course of treatment, the unique radiobiology of prostate cancer also lends itself to hypofractionation. Tumors with a low alpha-beta ratio (2–4 Gy) are considered more sensitive to changes in fractionation than are tumors with a high alpha-beta ratio (>8 Gy). The alpha-beta ratio of prostate cancer has been estimated at 1.5 Gy (95% confidence interval 0.8–2.2 Gy) (Brenner and Hall 1999; Fowler, Chappell, and Ritter 2001). As a result, interest in shortened treatment using hypofractionated radiation regimens has increased in the past decade. Trials have reported equivalent short-term biochemical control and toxicity following conformal hypofractionated radiation as compared to conventionally fractionated regimens (Lukka et al. 2005; Pollack et al. 2009). More recently, two trials have reported superior biochemical control outcomes for high-risk patients with hypofractionation (Arcangeli et al. 2010; Yeoh et al. 2011). The early success of hypofractionation has paved the way for further increases in fractional doses using SBRT. A summary of published prospective trials using SBRT for low- and intermediate-risk prostate cancer is presented in Table 14.7.

Table 14.7 Prospective trials using SBRT for low- and intermediate-risk prostate cancer

STUDY	INCLUSION	REGIMEN	PRESCRIPTION IDL[a]	BFFF[b]
Madsen et al. (2007)	NCCN low risk	33.5 Gy/5 fx	Isocenter, 90% IDL covers PTV	ASTRO: 70% Nadir + 2: 90%
King et al. (2009)	NCCN low risk	36.25 Gy/5 fx	89%–90%	100% (median FU 33 months)
Boike et al. (2011)	NCCN low & intermediate risk	45–50 Gy/5 fx	NR	100% (mean FU of 11 months)
Tang et al. (2008)	NCCN low & intermediate risk	35 Gy/5 fx	PTV	100% (median FU of 12 months)

[a] Isodose line.
[b] Biochemical freedom from failure.

Berit L. Madsen and colleagues (2007) reported a phase I/II trial of prostate SBRT from Virginia Mason Medical Center in Seattle. Forty patients with low-risk prostate cancer were enrolled. Patients had three fiducial markers placed in the prostate gland and underwent computed tomography and magnetic resonance imaging to guide treatment planning. Treatment was delivered in the prone position. Patients were treated to a dose of 33.5 Gy in five fractions (6.7 Gy per fraction). Median follow-up was 41 months. The four-year actuarial biochemical control was 70% using the ASTRO consensus definition (three consecutive rises) and was 90% using the nadir plus 2 ng/mL definition. The difference observed was due to late prostate specific antigen (PSA) bounces observed in a proportion of patients.

In a prospective phase II trial of prostate SBRT performed at Stanford University, 41 low-risk prostate cancer patients were treated to a dose of 36.25 Gy in five fractions (7.25 Gy per fraction) (King et al. 2009). Gold fiducial markers were implanted in the prostate, and real-time intrafraction tracking was performed with CyberKnife in the prone position. At a median follow-up of 33 months, no PSA recurrences have been observed. Median PSA nadir was 0.32 ng/mL, and 78% of patients with at least 12 months of follow-up achieved a PSA nadir <0.4 ng/mL. PSA bounce was observed in 29% of patients.

Investigators from the UT-SW performed a phase I dose escalation trial using SBRT for patients with low- and intermediate-risk disease and prostate volume <60 cc (Boike et al. 2011). Forty patients were enrolled and were treated using intensity-modulated SBRT in a prone position with a rectal balloon with pretreatment enema. Dose was escalated from 45 Gy to 50 Gy in five fractions in increments of 2.5 Gy (0.5 Gy/fraction). At a median follow-up of 11 months, no PSA relapses have been observed using nadir + 2 ng/mL or ASTRO consensus failure definitions.

The currently accruing RTOG 0938 randomizes men with favorable-risk prostate cancer (Gleason 206, PSA <10, T1-2a) to 36.25 Gy in five fractions or 51.6 Gy in 12 fractions (Radiation Therapy Oncology Group: Protocol RTOG 0938). A number of single institution trials further evaluating prostate SBRT are also active.

14.6.2 ORGANS AT RISK AND TOXICITY

Confirmation of the safety of SBRT for prostate cancer is critical for its adoption as a standard treatment option for low- and intermediate-risk prostate cancer. Conventionally fractionated EBRT using intensity-modulated radiation therapy is associated with high rates of biochemical control and low rates of grade 3 or 4 gastrointestinal (GI) or genitourinary (GU) morbidity (Vora et al. 2007; Zelefsky et al. 2008). Setup accuracy and precision is further improved with daily image guidance, allowing for a reduction in planning margins and further sparing of the bladder and rectum (Chung et al. 2009; Moseley et al. 2007; Serago et al. 2006). The prostate is in close proximity to the rectum, which is comprised of serially arranged functional subunits. Given this proximity, there are significant theoretical risks of increased radiation-related toxicity following hypofractionation if strict dose volume constraints are not achieved.

In the available prospective trials, high-grade GU and GI toxicity was uncommon. In the Virginia Mason trial, acute grade 1–2 GU and GI toxicity occurred in 48.5% and 39%, respectively (Madsen et al. 2007). Only one case of acute grade 3 GU toxicity was observed. Late grade 1–2 GU and GI toxicity rates were 45% and 37%, respectively, and no late grade 3 toxicities occurred. In the Stanford trial, two cases of late grade 3 urinary toxicity and no cases of late grade 3 rectal toxicity were observed (King et al. 2009). Interestingly, a difference in the rate of objectively measured rectal symptoms and rectal symptom–related quality of life was better among patients treated every other day compared to those treated on five consecutive days. The rate of any Expanded Prostate Cancer Index Composite (EPIC) rectal symptom score of 4–5 and overall EPIC QOL score of 4–5 was 38% and 24%, respectively, for patients treated daily and was 0% (p = 0.0035) and 0% (p = 0.048), respectively, for patients treated every other day. No difference was observed for International Prostate Symptom Score urinary toxicity comparing the two treatment schedules.

Robert M. Meier and colleagues (2010) reported toxicity and quality of life outcomes from a multi-institutional trial of 211 patients treated at 17 institutions with SBRT for low- and intermediate-risk disease. Patients received 40 Gy in five fractions to the prostate gland. Intermediate risk patients also concomitantly received 36.25 Gy in five fractions to the proximal seminal vesicles. All patients had implanted fiducial markers. Acute grade 2 GU and GI toxicities occurred in 20% and 8.5%, respectively.

Three patients (1.4%) required temporary catheterization for acute urinary retention. Late grade 2 GU and GI toxicities occurred in 6% and 1%, respectively. Only one grade 3 toxicity was observed. Erectile dysfunction increased from 49% at baseline to 58% at 12-month follow-up.

In the UT-SW trial, a rectal balloon with a pretreatment enema was used with each SBRT fraction (Boike et al. 2011). Theoretically, the rectal balloon allows for (1) better prostate immobilization, (2) decreased mean and intermediate radiation doses to be received by the rectal wall, and (3) the creation of a region of hypoxia in the rectal wall, thereby protecting against late radiation-associated toxicity. In a randomized trial of patients treated with conventionally fractionated three-dimensional conformal radiation, endorectal balloon use was shown to reduce the volume of rectal wall receiving 40 Gy (V40 Gy) and the rate of endoscopically visualized grade 2–3 telangiectasias at 1 and 2 years' post-treatment (van Lin et al. 2007). In the UT-SW trial, comparatively high SBRT doses were administered, but only 2.5% of treated patients experienced grade 2 rectal toxicity, and no grade 3 toxicity was observed. The mean American Urologic Association symptom score increased from a score of five prior to treatment to a score of eight to nine in follow-up.

14.7 ADRENAL AND PARA-AORTIC METASTASES

Limited clinical data evaluate SBRT for adrenal metastases. Investigators from the University of Rochester performed an open-design prospective pilot study of SBRT for patients with ≤5 metastatic lesions (Milano et al. 2008). The SBRT regimen employed was 50 Gy in 10 fractions prescribed to the 100% isodose with the 80% isodose covering the periphery of the PTV. In a separate retrospective report from these investigators, Sheema Chawla and colleagues (2009) reported the outcomes for 30 patients with adrenal metastases treated with SBRT to a median dose of 40 Gy (16–50 Gy). Actuarial local control at one year was only 55%; however, 16 of the 30 patients were treated with palliative intent, and outcomes were not separately analyzed for patients treated with definitive intent. No grade 2 or greater acute or late toxicity was observed. In another study from Hokkaido University, 10 patients received adrenal SBRT after placement of internal fiducials to a dose of 48 Gy in eight fractions (Katoh et al. 2008). These investigators reported 1-year local control of 100%, and no patient developed appreciable toxicity.

SBRT has also been applied to para-aortic (PA) and abdominal lymph nodes. In the largest series from the Korea Institute of Radiological and Medical Sciences, 30 patients with PA nodal metastases from cervix or uterine cancer were treated with three-fraction SBRT to 33–45 Gy (Choi et al. 2009). Patients were treated using a CyberKnife treatment machine with implanted fiducials and PTV margins of 2 mm around the gross tumor volume. In the 29 patients with follow-up imaging, complete or partial response was observed in 28 (97%) of cases. In this retrospective report, four-year actuarial local control was 67%. One late grade 3 ureteral stricture was observed 20 months after SBRT in a patient treated to 36 Gy in three fractions. In a separate report from this institution, seven patients with isolated PA nodal recurrence after surgical resection for gastric cancer were treated with SBRT to a dose of 45–51 Gy in three fractions (Kim et al. 2009). Patients in this series were also treated with CyberKnife and implanted fiducials using 2–3 mm PTV margins. At a median follow-up of 26 months, six of seven patients had local control in the SBRT-treated lesion. No patient developed late toxicity from SBRT. Finally, in a study from Italy, 19 patients with solitary or oligometastases to abdominal lymph nodes were treated with hypofractionated stereotactic radiation to a dose of 45 Gy in six fractions (Bignardi et al. 2011). Despite a high rate of local-control (78% at 2 years), the majority of patients experienced out-of-field progression. The 1- and 2-year rates of progression-free survival were only 29.5% and 19.7%, respectively.

14.8 HEAD AND NECK

14.8.1 REIRRADIATION

Squamous cell carcinoma of the head and neck has a high propensity for locoregional recurrence. Local and regional failures after prior in-field radiation represent a significant therapeutic challenge. Several recent trials have established the feasibility of reirradiation with concurrent chemotherapy for unresectable locoregional recurrence after prior full-dose radiation therapy, but this approach is associated with significant morbidity and median survival marginally better than historical controls using systemic therapy alone (Kramer et al. 2005; Langer et al. 2007; Salama et al. 2006; Spencer et al. 2001). Similar to stereotactic radiation for unresectable pancreatic cancer, SBRT and SRS for recurrent head and neck cancer are attractive because of the convenience of an abbreviated course of therapy in a cohort of patients with a poor prognosis. SBRT and SRS, in some circumstances, may also improve avoidance of adjacent critical structures, such as the spinal cord, brainstem, and optic pathway.

Dwight E. Heron and colleagues (2009) from the University of Pittsburgh performed a phase I clinical trial of SBRT for recurrent squamous cell carcinoma of the head and neck after prior radiation. Twenty-five patients were enrolled and treated with five-fraction SBRT. The median prior radiation dose was 64.7 Gy. The most common head and neck sites were the larynx, oral cavity, and oropharynx. Median tumor volume was 44.8 cm^3. The dose was safely escalated from 25 Gy (5 Gy per fraction) to 44 Gy (8.8 Gy per fraction) without dose-limiting toxicity. Median time to progression was 4 months and median survival was 6 months. Spinal cord and brainstem maximum doses were limited to ≤8 Gy. No radiation myelopathy or other grade 3–4 toxicities were observed.

Other retrospective series also report outcomes for fractionated SBRT for recurrent head and neck cancer. Investigators from Henry Ford review their institutional experience using stereotactic radiation for patients with primary, recurrent, or metastatic tumors of the head and neck (Siddiqui et al. 2009). Fifty-five lesions were irradiated in 44 patients. Thirty-seven lesions were treated with fractionated SBRT, and 18 received single-fraction SRS. Twenty-one patients with 29 lesions had recurrent disease arising in a previously irradiated region. The median prior radiation dose was 63.5 Gy. Doses were 13–18 Gy for SRS and 36–48 Gy in five to eight fractions for SBRT. For all patients, the complete and partial response rates were 31% and 46%, respectively. The corresponding response rates for patients with recurrent disease were 31% and 38%, respectively. One-year local control was 60.6%, and median survival was 6.7 months for patients with recurrent disease. Several high-grade toxicities were observed in the group with recurrent disease. Five patients experienced grade 3–4 late toxicity, including grade 4 cutaneous fistulas in three. Keith R. Unger and colleagues (2010) reported a series of 65 patients with recurrent head and neck cancer in a previously irradiated region. Thirty-eight patients were treated definitively with SBRT, and 27 received SBRT with palliative intent. The median initial radiation dose was 67 Gy. Patients were treated with 2–5 fractions to a total dose of 21–35 Gy. The most frequently used fractionation scheme was 30 Gy in five fractions. Thirty patients (54%) had a clinically complete response. Patients receiving higher doses (≥30 Gy) had a complete response rate of 69% compared to a rate of 29% in patients treated to lower doses (<30 Gy; $p = 0.01$). The median progression-free and OS durations were 5.7 months and 12 months, respectively. Two-year locoregional control was maintained in 30% of patients. Seven patients (11%) experienced severe late treatment-associated toxicity, including one death attributed to SBRT.

14.9 SPECIFIC APPLICATIONS: OLIGOMETASTATIC DISEASE

Several conceptual models of cancer dissemination and progression also support the aggressive treatment of discrete deposits of metastatic disease in selected patients. Perhaps the most popular is the theory of oligometastases (Hellman and Weichselbaum 1995; Withers and Lee 2006), whereby it is hypothesized that there is an intermediate state between early, localized cancer and fatally widespread disease. In this context, it is proposed that spatially targeted therapy, which eliminates all recognized disease, will achieve long-term disease control in some patients.

Clinical data also empirically supports the oligometastatic theory. Retrospective studies have documented favorable rates of long-term progression free and OS in select patients with limited hepatic metastases from colorectal cancer following surgical resection. In a large series from Memorial Sloan-Kettering Cancer Center, Fong et al. (1999) reported 5- and 10-year survival rates of 37% and 22%, respectively, in 1001 patients with limited hepatic metastases from colorectal cancer treated with surgical resection. Statistical analysis revealed that the most favorable group of patients had negative surgical margins, solitary metastases, tumor size <5 cm, carcinoembryonic antigen <200 ng/mL, disease-free interval >12 months, node-negative primary tumor and no evidence of extra-hepatic metastases. Among this select group, the 5-year survival rate was 60%. Similarly, Aloia et al. (2006) compared rates of local control, disease-free survival, and overall survival in 180 patients with solitary liver metastases from colorectal cancer treated with radiofrequency ablation versus hepatic resection. In this series, 5-year local recurrence-free survival and OS were higher with hepatic resection (92% and 71% versus 60% and 27%, respectively). No differences were observed between the two groups for distant hepatic recurrence or systemic recurrence. These findings suggest that effective local therapy is essential to achieve long-term disease-free and OS for patients with hepatic oligometastases from colorectal cancer.

Large population–based data sets also support the concept of an oligometastatic state in a proportion of patients with metastatic colorectal cancer. An analysis of the Ontario Cancer Registry identified 841 hepatic resections performed for metastatic colorectal cancer from 1996 to 2004 (Shah et al. 2007). In this study, 5-year survival was 43% and was higher when surgery was performed for solitary nodules at high-volume centers and in more recent years. Similarly, an analysis of the Surveillance, Epidemiology and End Results–Medicare registry identified 7673 patients aged ≥65 with liver metastases from colorectal cancer (Cummings, Payes, and Cooper 2007); 833 (6.1%) underwent hepatic resection, and the 5-year survival in this cohort was 32.8% as compared to 10.5% in those not undergoing hepatic resection ($p < 0.001$). Despite the selection biases inherent to population-based database studies and retrospective reviews, the observed long-term survival benefits identified in these studies strongly supports the oliogmetastatic hypothesis. The favorable outcomes observed in a subgroup of metastatic patients have encouraged evaluation of nonsurgical methods of metastatectomy, including RFA and SBRT.

Similar findings have been observed in patients with lung metastases. The International Lung Metastatases Registry (ILMR) analysis documented the survival outcomes of 5206 patients with lung metastases treated with surgical resection (Anon. 1997). In this study, the 5-, 10-, and 15-year survival rates for patients undergoing complete metastatectomy were 36%, 26%, and 22%, respectively. Incomplete metastatectomy was associated with inferior long-term survival rates of 13% at 5 years and 7% at 10 years. Concordant with the observations in studies of surgery for liver metastases, longer disease-free intervals (>36 months) and solitary metastases had improved survival. Histology was also an important predictor of survival in the ILMR study. Most patients had either an epithelial primary (43.4%) or sarcoma (41.7%). Germ cell tumors represented a minority of the metastases studied (7%), but were associated with improved survival compared to other primary tumor histologies.

Long-term survival has also been reported for select patients treated with surgery or stereotactic radiosurgery for limited brain metastases. Douglas Kondziolka et al. (2005) reviewed the records for 677 patients with brain metastases treated with radiosurgery and identified 44 (6.5%) patients with survival of at least four years. Compared with patients who had a shorter survival after radiosurgery (<3 months), patients with long-term survival had a higher initial KPS ($p = 0.01$), fewer brain metastases ($p = 0.04$), and a lower extracranial disease burden ($p < 0.00005$). In a study of patients with NSCLC presenting with synchronous, solitary brain metastases, Todd W. Flannery et al. (2008) reported the long-term outcomes for patients treated with radiosurgery. Five-year survival in this series was 21% for the entire cohort with 35% five-year survival among the subgroup of patients who received definitive local therapy (chemoradiation, surgery, or both) for their thoracic disease. In this study, the use of curative-intent therapy in the treatment of thoracic disease and KPS ≥90 was associated with improved long-term survival. These data suggest that long-term survival can be achieved in select patients with cerebral metastatic disease in the context of aggressive efforts to eradicate extracranial disease.

Further clinical evidence supporting the existence of an oligometastatic disease state is derived from the patterns of metastatic disease progression. Two studies of patients with limited-burden metastatic NSCLC have demonstrated that progression of existing metastases is the most common pattern of first disease progression, both in the *de novo* and *induced* oligometastatic states. Neil Mehta et al. (2004) analyzed the patterns of progression in 38 patients with stage IIIB or stage IV NSCLC. Among the 17 patients who were eligible for local therapy (with one to four metastases and no pleural effusion), 11 (65%) had no disease progression outside of initially involved sites at a median follow-up of nine months. Similarly, investigators from the University of Colorado analyzed patterns of disease progression in 64 patients with advanced NSCLC after first-line systemic therapy (Rusthoven et al. 2009). Among the 34 patients (53%) in this series who were eligible for SBRT (using institutional eligibility criteria) after first-line therapy, first extracranial progression was at sites of initial disease in 68%. Moreover, progression at sites initially involved with disease after first-line therapy occurred at a median of three months compared to progression at distant sites, which occurred at a median of 5.7 months. These findings suggest that in the setting of *induced* oligometastases from NSCLC, a limited window may exist during which effective local therapy could be administered and potentially prolong the interval to disease progression.

14.9.1 STUDIES USING SBRT FOR OLIGOMETASTATIC DISEASE

In the previously mentioned study performed at the University of Rochester, 121 patients with five or fewer metastatic lesions underwent SBRT. The 2-year and 4-year local control rate was 67% and 60%, respectively, with corresponding 2-year and 4-year progression-free survival of 26% and 20%. The predominant pattern of failure was outside the radiation field (Milano et al. 2008). Investigators at the University of Chicago accrued 61 patients with 113 oligometastatic lesions with a life expectancy >3 months to a dose escalation trial. Eligible patients had one to five metastatic lesions, each measuring <10 cm. Patients were assigned doses based on anatomical cohort and dose escalation was cohort specific. The five cohorts were head and neck, lung, liver, abdominal, and extremity based on the potential for normal tissue toxicity. Doses for each cohort were escalated independently with a 3 × 3 design. The starting dose was 8 Gy per fraction for three fractions with dose escalation at 2 Gy/fraction increments. At a median follow up of 20.9 months, the two-year PFS was 22% and the two-year OS was 56.7% (Salama et al. 2012).

Several studies have specifically evaluated the role of oligometastasis-directed SBRT in metastatic NSCLC. Investigators from the University of Chicago performed a retrospective review of 24 patients treated with HIGRT for metastatic NSCLC (Hasselle et al. 2012). The median total and fractional dose of HIGRT was 42 Gy and 8 Gy, respectively. One-year local control, OS, and progression-free survival were, 84%, 77%, and 48%, respectively. Only two patients experienced grade 3 toxicity. In a small phase II trial performed at the University of Colorado and UT-SW, 15 patients were treated with SBRT in combination with erlotinib for ≤6 FDG-avid, extracranial metastases from NSCLC (Kavanagh et al. 2010). SBRT was delivered in one to five fractions using conservative tumor doses and normal tissue constraints. In this trial, the 6-month PFS was 55%. The predominant pattern of relapse was outside of the SBRT field.

Based on these results, a multi-institutional phase II trial led by Coastal Carolina Radiation Oncology (Wilmington, NC) and Wake Forest University is currently enrolling patients with NSCLC with five or fewer metastatic lesions. Patients with RECIST-defined partial response or stable disease after four cycles of initial chemotherapy are treated with SBRT with the dose and fraction size tailored to the site(s) of metastatic disease (Rusthoven and Urbanic). Patients also receive thoracic stage–specific treatment to their lung primary. The target accrual is 57 patients with a primary endpoint of 6-month progression-free survival.

Similarly, encouraging results have been reported using SBRT for the treatment of oligometastases from breast cancer. Michael T. Milano and colleagues (2009) reported the preliminary results from a prospective trial evaluating SBRT for oligometastatic breast cancer. Fifty-one patients with five or fewer metastatic lesions underwent SBRT with curative intent, with a 4-year overall survival and progression-free survival of 59% and 38%, respectively. These data suggest that durable disease control can be achieved with SBRT for oligometastatic breast cancer, but additional follow-up is required given the long natural history of metastatic breast cancer in favorable-risk patients.

Future studies will continue to refine the selection of patients for aggressive local, metastasis-directed therapy for oligometastatic disease. Careful patient selection for this treatment paradigm is critical to maximize the therapeutic ratio. Tailored systemic therapy according to tumor subtype, gene expression, and mutational analysis is rapidly gaining acceptance in the oncology community. In the future, these tools might also be applied in the selection of patients with oligometastatic disease for curative-intent treatment by identifying the subgroups of patients for whom this approach is beneficial.

REFERENCES

Aloia, T. A., J.-N. Vauthey, E. M. Loyer et al. 2006. Solitary colorectal liver metastasis: Resection determines outcome. *Arch. Surg.* 141(5): 460–6.

American College of Surgeons Oncology Group (Protocol ACOSOG Z4099)/Radiation Therapy Oncology Group (Protocol RTOG 1021). A randomized phase III study of sublobar resection (±brachytherapy) versus stereotactic body radiation therapy in high risk patients with stage I NSCLC. Available at http://www.rtog.org/ClinicalTrials/ProtocolTable/StudyDetails.aspx?study = 1021.

Andrews, D. W., C. B. Scott, P. W. Sperduto et al. 2004. Whole brain radiation therapy with or without stereotactic radiosurgery boost for patients with one to three brain metastases: Phase III results of the RTOG 9508 randomised trial. *Lancet* 363(9422): 1665–72.

Anon. 1997. Long-term results of lung metastasectomy: Prognostic analyses based on 5206 cases. The International Registry of Lung Metastases. *J. Thorac. Cardiovasc. Surg.* 113(1): 37–49.

Aoyama, H., H. Shirato, M. Tago et al. 2006. Stereotactic radiosurgery plus whole-brain radiation therapy vs. stereotactic radiosurgery alone for treatment of brain metastases: A randomized controlled trial. *JAMA* 295(21): 2483–91.

Arcangeli, G., B. Saracino, S. Gomellini et al. 2010. A prospective phase III randomized trial of hypofractionation versus conventional fractionation in patients with high-risk prostate cancer. *Int. J. Radiat. Oncol. Biol. Phys.* 78: 11–18.

Barriger, R. B., J. A. Forquer, J. G. Brabham et al. 2012. A dose-volume analysis of radiation pneumonitis in non-small cell lung cancer patients treated with stereotactic body radiation therapy. *Int. J. Radiat. Oncol. Biol. Phys.* 82(1): 457–62.

Baumann, P., J. Nyman, M. Hoyer et al. 2009. Outcome in a prospective phase II trial of medically inoperable stage I non-small-cell lung cancer patients treated with stereotactic body radiotherapy. *J.Clin. Oncol.* 27: 3290–6.

Bignardi, M., P. Navarria, P. Mancosu et al. 2011. Clinical outcome of hypofractionated stereotactic radiotherapy for abdominal lymph node metastases. *Int. J. Radiat. Oncol. Biol. Phys.* 81(3): 831–8.

Boike, T. P., Y. Lotan, L. C. Cho et al. 2011. Phase I dose-escalation study of stereotactic body radiation therapy for low- and intermediate-risk prostate cancer. *J. Clin. Oncol.* 29(15): 2020–6.

Brenner, D. J., and E. J. Hall. 1999. Fractionation and protraction for radiotherapy of prostate carcinoma. *Int. J. Radiat. Oncol. Biol. Phys.* 43(5): 1095–101.

Cardenes, H. R., T. R. Price, S. M. Perkins et al. 2010. Phase I feasibility trial of stereotactic body radiation therapy for primary hepatocellular carcinoma. *Clin. Transl. Oncol.* 12(3): 218–25.

Chang, D. T., D. Schellenberg, J. Shen et al. 2009. Stereotactic radiotherapy for unresectable adenocarcinoma of the pancreas. *Cancer* 115(19117351): 665–72.

Chang, E. L., A. S. Shiu, E. Mendel et al. 2007. Phase I/II study of stereotactic body radiotherapy for spinal metastasis and its pattern of failure. *J. Neurosurg. Spine* 7(17688054): 151–60.

Chang, E. L., J. S. Wefel, K. R. Hess et al. 2009. Neurocognition in patients with brain metastases treated with radiosurgery or radiosurgery plus whole-brain irradiation: A randomised controlled trial. *Lancet Oncol.* 10(11): 1037–44.

Chang, J. Y., H. Liu, P. Balter et al. 2012. Clinical outcome and predictors of survival and pneumonitis after stereotactic ablative radiotherapy for stage I non-small cell lung cancer. *Radiat. Oncol.* 7: 152.

Chawla, S., Y. Chen, A. W. Katz et al. 2009. Stereotactic body radiotherapy for treatment of adrenal metastases. *Int. J. Radiat. Oncol. Biol. Phys.* 75(19250766): 71–5.

Chi, A., Z. Liao, N. P. Nguyen, J. Xu, B. Stea, and R. Komaki. 2010. Systemic review of the patterns of failure following stereotactic body radiation therapy in early-stage non-small-cell lung cancer: Clinical implications. *Radiother. Oncol.* 94(1): 1–11.

Choi, C. W., C. K. Cho, S. Y. Yoo et al. 2009. Image-guided stereotactic body radiation therapy in patients with isolated para-aortic lymph node metastases from uterine cervical and corpus cancer. *Int. J. Radiat. Oncol. Biol. Phys.* 74(18990511): 147–53.

Chung, H. T., P. Xia, L. W. Chan, E. Park-Somers, and M. Roach. 2009. Does image-guided radiotherapy improve toxicity profile in whole pelvic-treated high-risk prostate cancer? Comparison between IG-IMRT and IMRT. *Int. J. Radiat. Oncol. Biol. Phys.* 73(18501530): 53–60.

Chung, W. Y., D. H. Pan, C. C. Lee et al. 2010. Large vestibular schwannomas treated by Gamma Knife surgery: Long-term outcomes. *J. Neurosurg.* 113(Suppl.): 112–21.

ClinicalTrials.gov. 2012. Subtotal resection of large acoustic neuromas with possible stereotactic radiation therapy. Available at http://clinicaltrials.gov/show/NCT01129687.

Clinical Trial NCT01360593. Gemcitabine/Capecitabine followed by SBRT in pancreatic adenocarcinoma. Available at http://www.clinicaltrials.gov/ct2/show/NCT01360593.

Clinical Trial NCT01446458. Phase I study of stereotactic body radiation therapy and FOLFIRINOX in the neoadjuvant therapy of pancreatic cancer. Available at http://www.clinicaltrials.gov/ct2/show/NCT01446458.

Clinical Trial NCT01595321. Pancreatic tumor cell vaccine (GVAX), low dose cyclophosphamide, fractionated stereotactic body radiation therapy (SBRT), and FOLFIRINOX chemotherapy in patients with resected adenocarcinoma of the pancreas. Available at http://www.clinicaltrials.gov/ct2/show/NCT01595321.

Clinical Trial NCT01754623. GTX-RT in borderline resectable pancreatic cancer. Available at http://www.clinicaltrials.gov/ct2/show/NCT01754623.

Cox, B. W., A. Jackson, M. Hunt, M. Bilsky, and Y. Yamada. 2012a. Esophageal toxicity from high-dose, single-fraction paraspinal stereotactic radiosurgery. *Int. J. Radiat. Oncol. Biol. Phys.* 83(5): e661–7.

Cox, B. W., D. E. Spratt, M. Lovelock et al. 2012b. International Spine Radiosurgery Consortium consensus guidelines for target volume definition in spinal stereotactic radiosurgery. *Int. J. Radiat. Oncol. Biol. Phys.* 83(5): e597–605.

Cummings, L. C., J. D. Payes, and G. S. Cooper. 2007. Survival after hepatic resection in metastatic colorectal cancer: A population-based study. *Cancer* 109(17238180): 718–26.

Cuneo, K. C., J. J. Vredenburgh, J. H. Sampson et al. 2012. Safety and efficacy of stereotactic radiosurgery and adjuvant bevacizumab in patients with recurrent malignant gliomas. *Int. J. Radiat. Oncol. Biol. Phys.* 82(5): 2018–24.

Dolecek, T. A., J. M. Propp, N. E. Stroup, and C. Kruchko. 2012. CBTRUS statistical report: Primary brain and central nervous system tumors diagnosed in the United States in 2005–2009. *Neuro-Oncology* 14(Suppl. 5): v1–49.

Dunlap, N. E., J. Cai, G. B. Biedermann et al. 2010a. Chest wall volume receiving >30 Gy predicts risk of severe pain and/or rib fracture after lung stereotactic body radiotherapy. *Int. J. Radiat. Oncol. Biol. Phys.* 76(3): 796–801.

Dunlap, N. E., J. M. Larner, P. W. Read et al. 2010b. Size matters: A comparison of T1 and T2 peripheral non-small-cell lung cancers treated with stereotactic body radiation therapy (SBRT). *J. Thorac. Cardiovasc. Surg.* 140(20478576): 583–9.

Einstein, D. B., B. Wessels, B. Bangert et al. 2012. Phase II trial of radiosurgery to magnetic resonance spectroscopy-defined high-risk tumor volumes in patients with glioblastoma multiforme. *Int. J. Radiat. Oncol. Biol. Phys.* 84(3): 668–74.

Fakiris, A. J., R. C. McGarry, C. T. Yiannoutsos et al. 2009. Stereotactic body radiation therapy for early-stage non-small-cell lung carcinoma: Four-year results of a prospective phase II study. *Int. J. Radiat. Oncol. Biol. Phys.* 75(3): 677–82.

Flannery, T. W., M. Suntharalingam, W. F. Regine et al. 2008. Long-term survival in patients with synchronous, solitary brain metastasis from non-small-cell lung cancer treated with radiosurgery. *Int. J. Radiat. Oncol. Biol. Phys.* 72(18280058): 19–23.

Flickinger, J. C., D. Kondziolka, A. H. Maitz, and L. D. Lunsford. 2002. An analysis of the dose-response for arteriovenous malformation radiosurgery and other factors affecting obliteration. *Radiother. Oncol.* 63(3): 347–54.

Fogh, S. E., D. W. Andrews, J. Glass et al. 2010. Hypofractionated stereotactic radiation therapy: An effective therapy for recurrent high-grade gliomas. *J. Clin. Oncol.* 28(18): 3048–53.

Fogh, S., L. Ma, N. Gupta et al. 2012. High-precision volume-staged Gamma Knife surgery and equivalent hypofractionation dose schedules for treating large arteriovenous malformations. *J. Neurosurg.* 117 (Suppl.): 115–9.

Fong, Y., J. Fortner, R. L. Sun, M. F. Brennan, and L. H. Blumgart. 1999. Clinical score for predicting recurrence after hepatic resection for metastatic colorectal cancer: Analysis of 1001 consecutive cases. *Ann. Surg.* 230(10493478): 309–18.

Forquer, J. A., A. J. Fakiris, R. D. Timmerman et al. 2009. Brachial plexopathy from stereotactic body radiotherapy in early-stage NSCLC: Dose-limiting toxicity in apical tumor sites. *Radiother. Oncol.* 93(19454366): 408–13.

Fowler, J., R. Chappell, and M. Ritter. 2001. Is alpha/beta for prostate tumors really low? *Int. J. Radiat. Oncol. Biol. Phys.* 50(11429230): 1021–31.

Fuentes, S., Y. Arkha, G. Pech-Gourg, F. Grisoli, H. Dufour, and J. Regis. 2008. Management of large vestibular schwannomas by combined surgical resection and Gamma Knife radiosurgery. *Prog. Neurol. Surg.* 21: 79–82.

Garg, A. K., A. S. Shiu, J. Yang et al. 2012. Phase 1/2 trial of single-session stereotactic body radiotherapy for previously unirradiated spinal metastases. *Cancer* 118(20): 5069–77.

Gephart, M. G., A. Hansasuta, R. R. Balise et al. 2013. Cochlea radiation dose correlates with hearing loss after stereotactic radiosurgery of vestibular schwannoma. *World Neurosurg.* 80(3–4): 359–63.

Gerszten, P. C., S. A. Burton, C. Ozhasoglu, and W. C. Welch. 2007. Radiosurgery for spinal metastases: Clinical experience in 500 cases from a single institution. *Spine* 32(2): 193–9.

Gibbs, I. C., C. Patil, P. C. Gerszten, J. R. Adler Jr., and S. A. Burton. 2009. Delayed radiation-induced myelopathy after spinal radiosurgery. *Neurosurgery* 64(2 Suppl): A67–72.

Gomez, D. R., M. A. Hunt, A. Jackson et al. 2009. Low rate of thoracic toxicity in palliative paraspinal single-fraction stereotactic body radiation therapy. *Radiother. Oncol.* 93(19923027): 414–8.

Grills, I. S., A. J. Hope, M. Guckenberger et al. 2012. A collaborative analysis of stereotactic lung radiotherapy outcomes for early-stage non-small-cell lung cancer using daily online cone-beam computed tomography image-guided radiotherapy. *J. Thorac. Oncol.* 7(9): 1382–93.

Guckenberger, M., K. Baier, B. Polat et al. 2010. Dose-response relationship for radiation-induced pneumonitis after pulmonary stereotactic body radiotherapy. *Radiother. Oncol.* 97(20605245): 65–70.

Guss, Z. D., S. Batra, C. J. Limb et al. 2011. Radiosurgery of glomus jugulare tumors: A meta-analysis. *Int. J. Radiat. Oncol. Biol. Phys.* 81(4): e497–502.

Hansasuta, A., C. Y. Choi, I. C. Gibbs et al. 2011. Multisession stereotactic radiosurgery for vestibular schwannomas: Single-institution experience with 383 cases. *Neurosurgery* 69(6): 1200–9.

Hara, R., J. Itami, T. Kondo et al. 2006. Clinical outcomes of single-fraction stereotactic radiation therapy of lung tumors. *Cancer* 106(16475150): 1347–52.

Hartsell, W. F., C. B. Scott, D. W. Bruner et al. 2005. Randomized trial of short- versus long-course radiotherapy for palliation of painful bone metastases. *J. Natl. Cancer Inst.* 97(15928300): 798–804.

Hasselle, M. D., D. J. Haraf, K. E. Rusthoven et al. 2012. Hypofractionated image-guided radiation therapy for patients with limited volume metastatic non-small cell lung cancer. *J. Thorac. Oncol.* 7(2): 376–81.

Hellman, S., and R. R. Weichselbaum. 1995. Oligometastases. *J. Clin. Oncol.* 13(1): 8–10.

Herfarth, K. K., J. Debus, F. Lohr et al. 2001. Stereotactic single-dose radiation therapy of liver tumors: Results of a phase I/II trial. *J.Clin. Oncol.* 19(11134209): 164–70.

Heron, D. E., R. L. Ferris, M. Karamouzis et al. 2009. Stereotactic body radiotherapy for recurrent squamous cell carcinoma of the head and neck: Results of a phase I dose-escalation trial. *Int. J. Radiat. Oncol. Biol. Phys.* 75(19464819): 1493–500.

Hof, H., A. Hoess, D. Oetzel, J. Debus, and K. Herfarth. 2007. Stereotactic single-dose radiotherapy of lung metastases. *Strahlenther. Onkol.* 183(18040611): 673–8.

Hoppe, B. S., B. Laser, A. V. Kowalski et al. 2008. Acute skin toxicity following stereotactic body radiation therapy for stage I non-small-cell lung cancer: Who's at risk? *Int. J. Radiat. Oncol. Biol. Phys.* 72(19028267): 1283–6.

Hoyer, M., H. Roed, A. Traberg Hansen et al. 2006. Phase II study on stereotactic body radiotherapy of colorectal metastases. *Acta Oncol.* 45(7): 823–30.

Huang, P. P., S. C. Rush, B. Donahue et al. 2012. Long-term outcomes after staged-volume stereotactic radiosurgery for large arteriovenous malformations. *Neurosurgery* 71(3): 632–43; discussion 643–4.

Kano, H., D. Kondziolka, J. C. Flickinger et al. 2012. Stereotactic radiosurgery for arteriovenous malformations, Part 6: Multistaged volumetric management of large arteriovenous malformations. *J. Neurosurg.* 116(1): 54–65.

Katoh, N., R. Onimaru, Y. Sakuhara et al. 2008. Real-time tumor-tracking radiotherapy for adrenal tumors. *Radiother. Oncol.* 87(18439693): 418–24.

Kavanagh, B., R. Abdulrahman, D. R. Camidge et al. 2010. A Phase II trial of stereotactic body radiation therapy (SBRT) combined with erlotinib for patients with recurrent non-small cell lung cancer (NSCLC). *Int. J. Radiat. Oncol. Biol. Phys.* 78(3, Suppl.): S15.

Kelly, P., P. A. Balter, N. Rebueno et al. 2010. Stereotactic body radiation therapy for patients with lung cancer previously treated with thoracic radiation. *Int. J. Radiat. Oncol. Biol. Phys.* 78(20381271): 1387–93.

Kim, M. S., S. Y. Yoo, C. K. Cho et al. 2009. Stereotactic body radiotherapy for isolated para-aortic lymph node recurrence after curative resection in gastric cancer. *J. Korean Med. Sci.* 24(3): 488–92.

King, C. R., J. D. Brooks, H. Gill, T. Pawlicki, C. Cotrutz, and J. C. Presti. 2009. Stereotactic body radiotherapy for localized prostate cancer: Interim results of a prospective phase II clinical trial. *Int. J. Radiat. Oncol. Biol. Phys.* 73(18755555): 1043–8.

Kocher, M., R. Soffietti, U. Abacioglu et al. 2011. Adjuvant whole-brain radiotherapy versus observation after radiosurgery or surgical resection of one to three cerebral metastases: Results of the EORTC 22952-26001 study. *J. Clin. Oncol.* 29(2): 134–41.

Kondziolka, D., J. J. Martin, J. C. Flickinger et al. 2005. Long-term survivors after Gamma Knife radiosurgery for brain metastases. *Cancer* 104(16288488): 2784–91.

Koto, M., Y. Takai, Y. Ogawa et al. 2007. A phase II study on stereotactic body radiotherapy for stage I non-small cell lung cancer. *Radiother. Oncol.* 85(18022720): 429–34.

Kramer, N. M., E. M. Horwitz, J. Cheng et al. 2005. Toxicity and outcome analysis of patients with recurrent head and neck cancer treated with hyperfractionated split-course reirradiation and concurrent cisplatin and paclitaxel chemotherapy from two prospective phase I and II studies. *Head Neck* 27(15719391): 406–14.

Lagerwaard, F. J., C. J. A. Haasbeek, E. F. Smit, B. J. Slotman, and S. Senan. 2008. Outcomes of risk-adapted fractionated stereotactic radiotherapy for stage I non-small-cell lung cancer. *Int. J. Radiat. Oncol. Biol. Phys.* 70(18164849): 685–92.

Langer, C. J., J. Harris, E. M. Horwitz et al. 2007. Phase II study of low-dose paclitaxel and cisplatin in combination with split-course concomitant twice-daily reirradiation in recurrent squamous cell carcinoma of the head and neck: Results of Radiation Therapy Oncology Group Protocol 9911. *J. Clin. Oncol.* 25(17947728): 4800–5.

Lawrence, Y. R., X. A. Li, I. el Naqa et al. 2010. Radiation dose-volume effects in the brain. *Int. J. Radiat. Oncol. Biol. Phys.* 76(3 Suppl.): S20–7.

Leber, K. A., J. Bergloff, and G. Pendl. 1998. Dose-response tolerance of the visual pathways and cranial nerves of the cavernous sinus to stereotactic radiosurgery. *J. Neurosurg.* 88(1): 43–50.

Lee, M. T., J. J. Kim, R. Dinniwell et al. 2009. Phase I study of individualized stereotactic body radiotherapy of liver metastases. *J. Clin. Oncol.* 27(19255313): 1585–91.

Leksell, L. 1951. The stereotaxic method and radiosurgery of the brain. *Acta Chir. Scand.* 102(4): 316–9.

Leksell, L. 1971. A note on the treatment of acoustic tumours. *Acta Chir. Scand.* 137(8): 763–5.

Likhacheva, A., C. C. Pinnix, N. R. Parikh et al. 2012. Predictors of survival in contemporary practice after initial radiosurgery for brain metastases. *Int. J. Radiat. Oncol. Biol. Phys.* 85(3): 656–61.

Lukka, H., C. Hayter, J. A. Julian et al. 2005. Randomized trial comparing two fractionation schedules for patients with localized prostate cancer. *J.Clin. Oncol.* 23(16135479): 6132–8.

Lung Cancer STARS Trial. Available at http://clinicaltrials.gov/ct2/show/NCT00840749.

Lu-Yao, G. L., P. C. Albertsen, D. F. Moore et al. 2009. Outcomes of localized prostate cancer following conservative management. *JAMA* 302(19755699): 1202–9.

Madsen, B. L., R. A. Hsi, H. T. Pham, J. F. Fowler, L. Esagui, and J. Corman. 2007. Stereotactic hypofractionated accurate radiotherapy of the prostate (SHARP), 33.5 Gy in five fractions for localized disease: First clinical trial results. *Int. J. Radiat. Oncol. Biol. Phys.* 67(17336216): 1099–105.

Mahadevan, A., S. Jain, M. Goldstein et al. 2010. Stereotactic body radiotherapy and gemcitabine for locally advanced pancreatic cancer. *Int. J. Radiat. Oncol. Biol. Phys.* 78(20171803): 735–42.

Marks, L. B., S. M. Bentzen, J. O. Deasy et al. 2010. Radiation dose-volume effects in the lung. *Int. J. Radiat. Oncol. Biol. Phys.* 76(20171521): 70–6.

Massager, N., O. Nissim, C. Delbrouck et al. 2007. Irradiation of cochlear structures during vestibular schwannoma radiosurgery and associated hearing outcome. *J. Neurosurg.* 107(4): 733–9.

Mayo, C., M. K. Martel, L. B. Marks, J. Flickinger, J. Nam, and J. Kirkpatrick. 2010. Radiation dose-volume effects of optic nerves and chiasm. *Int. J. Radiat. Oncol. Biol. Phys.* 76(3, Suppl.): S28–35.

McCammon, R., T. E. Schefter, L. E. Gaspar, R. Zaemisch, D. Gravdahl, and B. Kavanagh. 2009. Observation of a dose-control relationship for lung and liver tumors after stereotactic body radiation therapy. *Int. J. Radiat. Oncol. Biol. Phys.* 73(18786780): 112–8.

McGarry, R. C., L. Papiez, M. Williams, T. Whitford, and R. D. Timmerman. 2005. Stereotactic body radiation therapy of early-stage non-small-cell lung carcinoma: Phase I study. *Int. J. Radiat. Oncol. Biol. Phys.* 63(16115740): 1010–5.

McGarry, R. C., G. Song, P. des Rosiers, and R. Timmerman. 2002. Observation-only management of early stage, medically inoperable lung cancer: Poor outcome. *Chest* 121(11948046): 1155–8.

Mehta, N., A. M. Mauer, S. Hellman et al. 2004. Analysis of further disease progression in metastatic non-small cell lung cancer: Implications for locoregional treatment. *Int. J. Oncol.* 25(15547705): 1677–83.

Meier, R., A. Beckman, I. Kaplan et al. 2010. Stereotactic radiotherapy for organ-confined prostate cancer: Early toxicity and quality of life outcomes from a multi-institutional trial. *Int. J. Radiat. Oncol. Biol. Phys.*78(3, Suppl.): S57.

Méndez-Romero, A., W. Wunderink, S. M. Hussain et al. 2006. Stereotactic body radiation therapy for primary and metastatic liver tumors: A single institution phase I-II study. *Acta Oncol.* 45(7): 831–7.

Milano, M. T., A. W. Katz, A. G. Muhs et al. 2008. A prospective pilot study of curative-intent stereotactic body radiation therapy in patients with 5 or fewer oligometastatic lesions. *Cancer* 112(18072260): 650–8.

Milano, M. T., H. Zhang, S. K. Metcalfe, A. G. Muhs, and P. Okunieff. 2009. Oligometastatic breast cancer treated with curative-intent stereotactic body radiation therapy. *Breast Cancer Res. Treat.* 115(18719992): 601–8.

Milligan, B. D., B. E. Pollock, R. L. Foote, and M. J. Link. 2012. Long-term tumor control and cranial nerve outcomes following Gamma Knife surgery for larger-volume vestibular schwannomas. *J. Neurosurg.* 116(3): 598–604.

Minniti, G., E. Clarke, G. Lanzetta et al. 2011. Stereotactic Radiosurgery for brain metastases: Analysis of outcome and risk of brain radionecrosis. *Radiation Oncology* 6: 48.

Moseley, D. J., E. A. White, K. L. Wiltshire et al. 2007. Comparison of localization performance with implanted fiducial markers and cone-beam computed tomography for on-line image-guided radiotherapy of the prostate. *Int. J. Radiat. Oncol. Biol. Phys.* 67(17293243): 942–53.

Murphy, E. S., and J. H. Suh. 2011. Radiotherapy for vestibular schwannomas: A critical review. *Int. J. Radiat. Oncol. Biol. Phys.* 79(4): 985–97.

Murphy, J. D., C. Christman-Skieller, J. Kim, S. Dieterich, D. T. Chang, and A. C. Koong. 2010. A dosimetric model of duodenal toxicity after stereotactic body radiotherapy for pancreatic cancer. *Int. J. Radiat. Oncol. Biol. Phys.* 78(20399033): 1420–6.

Nagata, Y., M. Hiraoka, T. Shibata et al. 2010. A phase II trial of stereotactic body radiation therapy for operable T1N0M0 non-small cell lung cancer: Japan Clinical Oncology Group (JCOG0403). *Int. J. Radiat. Oncol. Biol. Phys.* 78(3, Suppl.): S27–8.

Nagata, Y., K. Takayama, Y. Matsuo et al. 2005. Clinical outcomes of a phase I/II study of 48 Gy of stereotactic body radiotherapy in 4 fractions for primary lung cancer using a stereotactic body frame. *Int. J. Radiat. Oncol. Biol. Phys.* 63(16169670): 1427–31.

Nelson, J. W., D. S. Yoo, J. H. Sampson et al. 2009. Stereotactic body radiotherapy for lesions of the spine and paraspinal regions. *Int. J. Radiat. Oncol. Biol. Phys.* 73(19004569): 1369–75.

Okunieff, P., A. L. Petersen, A. Philip et al. 2006. Stereotactic body radiation therapy (SBRT) for lung metastases. *Acta Oncol.* 45(16982544): 808–17.

Pan, H. C., J. Sheehan, M. L. Sheu, W. T. Chiu, and D. Y. Yang. 2012. Intracapsular decompression or radical resection followed by Gamma Knife surgery for patients harboring a large vestibular schwannoma. *J. Neurosurg.* 117(Suppl.): 69–77.

Park, K. J., D. Kondziolka, O. Berkowitz et al. 2012. Repeat Gamma Knife radiosurgery for trigeminal neuralgia. *Neurosurgery* 70(2): 295–305; discussion 305.

Pettersson, N., J. Nyman, and K.-A. Johansson. 2009. Radiation-induced rib fractures after hypofractionated stereotactic body radiation therapy of non-small cell lung cancer: A dose- and volume-response analysis. *Radiother. Oncol.* 91(19410314): 360–8.

Pollack, A., T. Li, M. Buyyounouski et al. 2009. Hypofractionation for prostate cancer: Interim results of a randomized trial. *Int. J. Radiat. Oncol. Biol. Phys.* 75(3, Suppl.): S81–2.

Pollock, B. E., S. L. Stafford, M. J. Link, P. D. Brown, Y. I. Garces, and R. L. Foote. 2012. Single-fraction radiosurgery of benign intracranial meningiomas. *Neurosurgery* 71(3): 604–12; discussion 613.

Pollock, B. E., S. L. Stafford, A. Utter, C. Giannini, and S. A. Schreiner. 2003. Stereotactic radiosurgery provides equivalent tumor control to Simpson Grade 1 resection for patients with small- to medium-size meningiomas. *Int. J. Radiat. Oncol. Biol. Phys.* 55(4): 1000–5.

Powell, J. W., E. Dexter, E. M. Scalzetti, and J. A. Bogart. 2009. Treatment advances for medically inoperable non-small-cell lung cancer: Emphasis on prospective trials. *Lancet Oncol.* 10(19717090): 885–94.

Radiation Therapy Oncology Group: Protocol RTOG 0618. A phase II trial of stereotactic body radiation therapy (SBRT) in the treatment of patients with operable stage I/II non-small cell lung cancer. Available at http://www.rtog.org/ClinicalTrials/ProtocolTable/StudyDetails.aspx?study = 0618.

Radiation Therapy Oncology Group: Protocol RTOG 0631. Phase II/III study of image-guided radiosurgery/SBRT for localized spine metastasis. Available at http://www.rtog.org/clinicaltrials/protocoltable/studydetails.aspx?study = 0631.

Radiation Therapy Oncology Group: Protocol RTOG 0813. Seamless phase I/II study of stereotactic lung radiotherapy (SBRT) for early stage, centrally located, non-small cell lung cancer (NSCLC) in medically inoperable patients. Available at http://www.rtog.org/ClinicalTrials/ProtocolTable/StudyDetails.aspx?study = 0813.

Radiation Therapy Oncology Group: Protocol RTOG 0915. A randomized phase II study comparing 2 stereotactic body radiation therapy (SBRT) schedules for medically inoperable patients with stage I peripheral non-small cell lung cancer. Available at http://www.rtog.org/ClinicalTrials/ProtocolTable/StudyDetails.aspx?study = 0915.

Radiation Therapy Oncology Group: Protocol RTOG 0938. A randomized phase II trial of hypofractionated radiotherapy for favorable risk prostate cancer—RTOG CCOP study. Available at http://www.rtog.org/ClinicalTrials/ProtocolTable/StudyDetails.aspx?study = 0938.

Radiation Therapy Oncology Group: Protocol RTOG 1112. Randomized phase III study of sorafenib versus stereotactic body radiation therapy followed by sorafenib in hepatocellular carcinoma. Available at http://www.rtog.org/ClinicalTrials/ProtocolTable/StudyDetails.aspx?study = 1112.

Ricardi, U., A. R. Filippi, A. Guarneri et al. 2010. Stereotactic body radiation therapy for early stage non-small cell lung cancer: Results of a prospective trial. *Lung Cancer* 68(1): 72–7.

Rose, P. S., I. Laufer, P. J. Boland et al. 2009. Risk of fracture after single fraction image-guided intensity-modulated radiation therapy to spinal metastases. *J. Clin. Oncol.* 27(19738130): 5075–9.

ROSEL Lung Cancer Trial. Available at http://clinicaltrials.gov/ct2/show/NCT00687986.

Rule, W., R. Timmerman, L. Tong et al. 2011. Phase I dose-escalation study of stereotactic body radiotherapy in patients with hepatic metastases. *Ann. Surg. Oncol.* 18(4): 1081–7.

Rusthoven, K. E., S. F. Hammerman, B. D. Kavanagh, M. J. Birtwhistle, M. Stares, and D. R. Camidge. 2009. Is there a role for consolidative stereotactic body radiation therapy following first-line systemic therapy for metastatic lung cancer? A patterns-of-failure analysis. *Acta Oncol.* 48(19373699): 578–83.

Rusthoven, K. E., and M. D. Hasselle. 2010. SBRT for unresectable HCC: A familiar tune? *J. Surg. Oncol.* 102(20740575): 207–8.

Rusthoven, K. E., B. D. Kavanagh, S. H. Burri et al. 2009a. Multi-institutional phase I/II trial of stereotactic body radiation therapy for lung metastases. *J. Clin. Oncol.* 27(10): 1579–84.

Rusthoven, K. E., B. D. Kavanagh, H. Cardenes et al. 2009b. Multi-institutional phase I/II trial of stereotactic body radiation therapy for liver metastases. *J. Clin. Oncol.* 27(19255321): 1572–8.

Rusthoven, K. E., and J. Urbanic. Personal communication with co-principle investigators.

Rwigema, J. C., D. E. Heron, S. D. Parikh et al. 2012. Adjuvant stereotactic body radiotherapy for resected pancreatic adenocarcinoma with close or positive margins. *J. Gastrointest. Cancer* 43(1): 70–6.

Rwigema, J. C., S. D. Parikh, D. E. Heron et al. 2011. Stereotactic body radiotherapy in the treatment of advanced adenocarcinoma of the pancreas. *Am. J. Clin. Oncol.* 34(1): 63–9.

Ryu, S., J.-Y. Jin, R. Jin et al. 2007. Partial volume tolerance of the spinal cord and complications of single-dose radiosurgery. *Cancer* 109(17167762): 628–36.

Ryu, S., R. Jin, J.-Y. Jin et al. 2008. Pain control by image-guided radiosurgery for solitary spinal metastasis. *J. Pain Symptom Manage.* 35(3): 292–8.

Sahgal, A., C. Ames, D. Chou et al. 2009. Stereotactic body radiotherapy is effective salvage therapy for patients with prior radiation of spinal metastases. *Int. J. Radiat. Oncol. Biol. Phys.* 74(19095374): 723–31.

Sahgal, A., L. Ma, V. Weinberg et al. 2012. Reirradiation human spinal cord tolerance for stereotactic body radiotherapy. *Int. J. Radiat. Oncol. Biol. Phys.* 82(20951503): 107–16.

Salama, J. K., M. D. Hasselle, S. J. Chmura et al. 2012. Stereotactic body radiotherapy for multisite extracranial oligometastases: Final report of a dose escalation trial in patients with 1 to 5 sites of metastatic disease. *Cancer* 118(11): 2962–70.

Salama, J. K., E. E. Vokes, S. J. Chmura et al. 2006. Long-term outcome of concurrent chemotherapy and reirradiation for recurrent and second primary head-and-neck squamous cell carcinoma. *Int. J. Radiat. Oncol. Biol. Phys.* 64(16213104): 382–91.

Schefter, T. E., B. D. Kavanagh, D. Raben et al. 2006. A phase I/II trial of stereotactic body radiation therapy (SBRT) for lung metastases: Initial report of dose escalation and early toxicity. *Int. J. Radiat. Oncol. Biol. Phys.* 66(4, Suppl.): S120–7.

Schefter, T. E., B. D. Kavanagh, R. D. Timmerman, H. R. Cardenes, A. Baron, and L. E. Gaspar. 2005. A phase I trial of stereotactic body radiation therapy (SBRT) for liver metastases. *Int. J. Radiat. Oncol. Biol. Phys.* 62(16029795): 1371–8.

Schellenberg, D., K. A. Goodman, F. Lee et al. 2008. Gemcitabine chemotherapy and single-fraction stereotactic body radiotherapy for locally advanced pancreatic cancer. *Int. J. Radiat. Oncol. Biol. Phys.* 72(18395362): 678–86.

Schwer, A. L., D. M. Damek, B. D. Kavanagh et al. 2008. A phase I dose-escalation study of fractionated stereotactic radiosurgery in combination with gefitinib in patients with recurrent malignant gliomas. *Int. J. Radiat. Oncol. Biol. Phys.* 70(4): 993–1001.

Seo, Y., M.-S. Kim, S. Yoo et al. 2009. Stereotactic body radiation therapy boost in locally advanced pancreatic cancer. *Int. J. Radiat. Oncol. Biol. Phys.* 75(19783379): 1456–61.

Seo, Y. S., M.-S. Kim, S. Y. Yoo et al. 2010. Preliminary result of stereotactic body radiotherapy as a local salvage treatment for inoperable hepatocellular carcinoma. *J. Surg. Oncol.* 102(20740576): 209–14.

Serago, C. F., S. J. Buskirk, T. C. Igel, A. A. Gale, N. E. Serago, and J. D. Earle. 2006. Comparison of daily megavoltage electronic portal imaging or kilovoltage imaging with marker seeds to ultrasound imaging or skin marks for prostate localization and treatment positioning in patients with prostate cancer. *Int. J. Radiat. Oncol. Biol. Phys.* 65(16863936): 1585–92.

Serizawa, T., T. Hirai, O. Nagano et al. 2010. Gamma Knife surgery for 1–10 brain metastases without prophylactic whole-brain radiation therapy: Analysis of cases meeting the Japanese prospective multi-institute study (JLGK0901) inclusion criteria. *J. Neurooncol.* 98(2): 163–7.

Shah, S. A., R. Bromberg, A. Coates, E. Rempel, M. Simunovic, and S. Gallinger. 2007. Survival after liver resection for metastatic colorectal carcinoma in a large population. *J. Am. Coll. Surg.* 205(17964443): 676–83.

Siddiqui, F., M. Patel, M. Khan et al. 2009. Stereotactic body radiation therapy for primary, recurrent, and metastatic tumors in the head-and-neck region. *Int. J. Radiat. Oncol. Biol. Phys.* 74(19327895): 1047–53.

Spencer, S. A., J. Harris, R. H. Wheeler et al. 2001. RTOG 96-10: Reirradiation with concurrent hydroxyurea and 5-fluorouracil in patients with squamous cell cancer of the head and neck. *Int. J. Radiat. Oncol. Biol. Phys.* 51(11728690): 1299–304.

Sperduto, P. W., S. T. Chao, P. K. Sneed et al. 2010. Diagnosis-specific prognostic factors, indexes, and treatment outcomes for patients with newly diagnosed brain metastases: A multi-institutional analysis of 4,259 patients. *Int. J. Radiat. Oncol. Biol. Phys.* 77(3): 655–61.

Stafford, S. L., B. E. Pollock, J. A. Leavitt et al. 2003. A study on the radiation tolerance of the optic nerves and chiasm after stereotactic radiosurgery. *Int. J. Radiat. Oncol. Biol. Phys.* 55(5): 1177–81.

Steiner, L., L. Leksell, T. Greitz, D. M. Forster, and E. O. Backlund. 1972. Stereotaxic radiosurgery for cerebral arteriovenous malformations. Report of a case. *Acta Chir. Scand.* 138(5): 459–64.

Sun, S., and A. Liu. 2012. Long-term follow-up studies of Gamma Knife surgery with a low margin dose for vestibular schwannoma. *J. Neurosurg.* 117(Suppl.): 57–62.

Tamura, M., R. Carron, S. Yomo et al. 2009. Hearing preservation after Gamma Knife radiosurgery for vestibular schwannomas presenting with high-level hearing. *Neurosurgery* 64(2): 289–96; discussion 296.

Tang, C. I., D. A. Loblaw, P. Cheung et al. 2008. Phase I/II study of a five-fraction hypofractionated accelerated radiotherapy treatment for low-risk localised prostate cancer: Early results of pHART3. *Clin. Oncol.(R. Coll. Radiol., Gr. Br.).* 20(10): 729–37.

Thomas, K. L., and J. A. Vilensky. 2013. The anatomy of vascular compression in trigeminal neuralgia. *Clin. Anat.* doi:10.1002/ca.22157.

Timmerman, R., R. McGarry, C. Yiannoutsos et al. 2006. Excessive toxicity when treating central tumors in a phase II study of stereotactic body radiation therapy for medically inoperable early-stage lung cancer. *J. Clin. Oncol.* 24(17050868): 4833–9.

Timmerman, R., R. Paulus, J. Galvin et al. 2010. Stereotactic body radiation therapy for inoperable early stage lung cancer. *JAMA* 303(11): 1070–6.

Tse, R. V., M. Hawkins, G. Lockwood et al. 2008. Phase I study of individualized stereotactic body radiotherapy for hepatocellular carcinoma and intrahepatic cholangiocarcinoma. *J. Clin. Oncol.* 26(18172187): 657–64.

Tuleasca, C., R. Carron, N. Resseguier et al. 2012. Patterns of pain-free response in 497 cases of classic trigeminal neuralgia treated with Gamma Knife surgery and followed up for least 1 year. *J. Neurosurg.* 117 (Suppl.): 181–8.

Unger, K. R., C. E. Lominska, J. F. Deeken et al. 2010. Fractionated stereotactic radiosurgery for reirradiation of head-and-neck cancer. *Int. J. Radiat. Oncol. Biol. Phys.* 77(20056341): 1411–9.

van de Langenberg, R., P. E. Hanssens, J. J. van Overbeeke et al. 2011. Management of large vestibular schwannoma. Part I. Planned subtotal resection followed by Gamma Knife surgery: Radiological and clinical aspects. *J. Neurosurg.* 115(5): 875–84.

van Lin, E. N. J. T., J. Kristinsson, M. E. P. Philippens et al. 2007. Reduced late rectal mucosal changes after prostate three-dimensional conformal radiotherapy with endorectal balloon as observed in repeated endoscopy. *Int. J. Radiat. Oncol. Biol. Phys.* 67(17161552): 799–811.

Vora, S. A., W. W. Wong, S. E. Schild, G. A. Ezzell, and M. Y. Halyard. 2007. Analysis of biochemical control and prognostic factors in patients treated with either low-dose three-dimensional conformal radiation therapy or high-dose intensity-modulated radiotherapy for localized prostate cancer. *Int. J. Radiat. Oncol. Biol. Phys.* 68(17398023): 1053–8.

Withers, H. R., and S. P. Lee. 2006. Modeling growth kinetics and statistical distribution of oligometastases. *Semin. Radiat. Oncol.* 16(16564446): 111–9.

Wu, J. S.-Y., R. Wong, M. Johnston, A. Bezjak, and T. Whelan. 2003. Meta-analysis of dose-fractionation radiotherapy trials for the palliation of painful bone metastases. *Int. J. Radiat. Oncol. Biol. Phys.* 55(12573746): 594–605.

Yamada, Y., M. H. Bilsky, D. M. Lovelock et al. 2008. High-dose, single-fraction image-guided intensity-modulated radiotherapy for metastatic spinal lesions. *Int. J. Radiat. Oncol. Biol. Phys.* 71(2): 484–90.

Yeoh, E. E., R. J. Botten, J. Butters, A. C. Di Matteo, R. H. Holloway, and J. Fowler. 2011. Hypofractionated versus conventionally fractionated radiotherapy for prostate carcinoma: Final results of phase III randomized trial. *Int. J. Radiat. Oncol. Biol. Phys.* 81(20934277): 1271–8.

Zelefsky, M. J., E. J. Levin, M. Hunt et al. 2008. Incidence of late rectal and urinary toxicities after three-dimensional conformal radiotherapy and intensity-modulated radiotherapy for localized prostate cancer. *Int. J. Radiat. Oncol. Biol. Phys.* 70(18313526): 1124–9.

The future of stereotactic radiosurgery

Joshua Evans, Cedric Yu, Michael Wright, Edwin Crandley, and David Wilson

Contents

15.1 EMERGING TECHNOLOGY FOR STEREOTACTIC RADIOTHERAPY DELIVERY SYSTEMS

15.1.1 INTRODUCTION

Stereotactic radiotherapy (SRT) treatments aim to deliver high doses in very few fractions meaning the consequences of a geometric miss can be catastrophic. Geometric misses can arise from a number of sources, including but not limited to patient setup variability and tumor motion. Systems that support more accurate patient alignment, real-time target tracking, and adaptive radiation delivery are thus of crucial importance for enhancing the safety and effectiveness with which SRT is delivered. This chapter introduces a few emerging systems designed to address some of these issues.

15.1.2 6-D ROBOTIC COUCH SYSTEMS

Imaging systems, such as cone-beam CT or stereoscopic x-ray imaging, that can detect setup errors in six dimensions, three translational and three rotational (yaw, pitch, and roll), are becoming common features on modern linacs. While conventional linac couches allow for setup correction in four dimensions (three translational and yaw rotation), they do not offer the ability to correct for rotational pitch and roll errors. Literature shows that patients can exhibit pitch and roll rotational setup errors of up to 4° (Ahn et al. 2009; Kaiser et al. 2006). Therapists may attempt pitch and roll corrections by physically adjusting the patient, but this is inexact, requires reimaging, and is thus not an efficient solution.

Rotational setup errors, if left uncorrected, have been shown to degrade the plan quality for SRT. Schreibmann, Fox, and Crocker (2011) studied the dosimetric effects of setup errors in spinal radiosurgery using cone-beam CT imaging. They showed that four of their 10 patients had a significant decrease in tumor coverage due to the inability to correct for observed rotational errors. Potentially even worse than reduced target coverage, rotational corrections may place a critical structure into the beam path leading to higher than intended doses to the organ at risk. For example, in spinal stereotactic body radiotherapy (SBRT), characterized by close proximity to the spinal cord and high dose gradients, a rotational error that places the cord in the beam path could lead to catastrophic toxicity, such as paralysis (Gutfeld et al. 2009; Kim et al. 2009; Wang et al. 2008).

The degree to which a plan is degraded by rotational setup errors depends on the details of each particular case. Translational corrections may be used in an attempt to center the target volume in the presence of rotational setup errors (Cao et al. 2011). Murphy (2007) provides guidance on how to manage observed rotational setup errors with translational shifts. Irregularly shaped targets, volumes with a long extent, or targets in close proximity to critical structures will suffer the most from uncorrected rotational setup errors (Peng et al. 2011) and will also be the most resistant to translational-only corrections.

Several commercial robotic couch systems are currently available to address the issue of patient setup correction in six-dimensions. These 6-D robotic couch systems are gaining popularity as they allow efficient, daily correction of patient setup errors in both three translational and three rotational dimensions. The commercial systems also interface directly with a number of in-room imaging systems that calculate the required shifts with 6° of freedom and oftentimes allow the therapist to adjust the patient setup without needing to re-enter the room. Shown in Figure 15.1 is CIVCO's robotic couch system, the Protura™. The Protura system can smoothly translate and rotate (up to 5°) the couch top in six dimensions around a dynamic pivot point with a stated accuracy of 0.1 mm and 0.1°. The dynamic pivot point for the couch top is set individually for each patient to coincide with the treatment isocenter, and multiple dynamic pivot points can be defined for patients with multiple isocenters.

Linthout and colleagues (2007) studied the residual setup error following 6-D setup correction on an early robotic couch system. They performed repeat stereoscopic x-ray imaging following 6-D setup correction to assess the residual setup accuracy. They identified two groups of patients: The first, in which residual setup error after setup correction was less than 3.0 mm and 2.0° for almost all fractions, accounted for 92.3% of all patients. The second group of patients consistently showed residual setup errors of greater than 3.0 mm and 2.0°, requiring a second round of correctional shifts and verification imaging to be performed. Linthout et al. (2007) suggest performing verification imaging for the first five fractions to

(a)

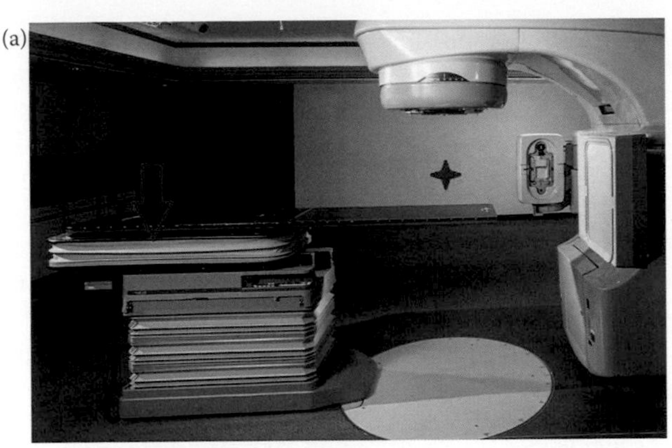

(b)

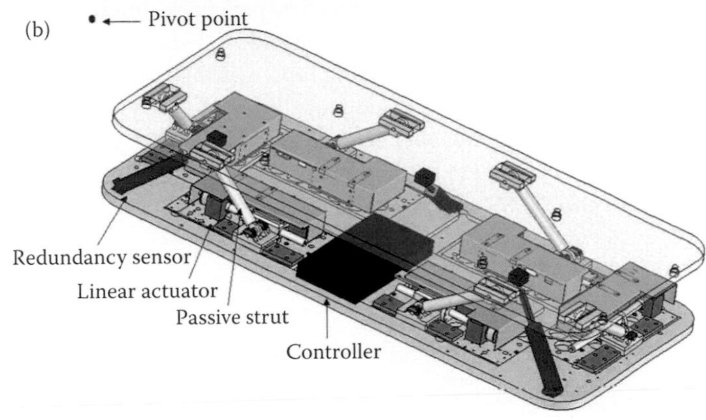

Figure 15.1 (a) The CIVCO Protura™ robotic couch system mounts on top of the vendor's couch pedestal. The active portion controlling the couch movement lies under the couch top indicted by the red arrow. (b) Schematic diagram of the active module of the CIVCO Protura™ robotic couch design. The linear actuators and struts smoothly move the couch in all six translational and rotational dimensions around the dynamic pivot point. The dynamic pivot point is set individually for each patient case to correspond to the treatment isocenter (red star in [a]). The three redundancy sensors (highlighted in red in [b]) are used to autocalibrate the system. (Images courtesy of CIVCO.)

assess which group the patient falls into. They show good agreement with the first-five-fraction method giving a false positive rate (identifying patient as a mover, when remaining fractions do not show residual errors) of 2.9% and a false negative rate (identify patient as a nonmover and remaining fractions exhibit error >3.0 mm and 2.0°) of 0.9%.

The robotic couch systems can also be used for real-time tumor tracking (D'Souza, Naqvi, and Yu 2005). For example, Buzurovic et al. (2011) evaluated an approach for 4-D active tracking and dynamic delivery and utilized a computer simulation of the ability of the HexaPOD and Elekta Precise Table robotic couch systems to simulate real-time tracking of lung tumors. Their results showed that their tracking algorithm combined with the robotic couch systems had the ability to track irregular breathing patterns to within 1 mm. Furthermore, they illustrated that this 1-mm residual tracking error resulted in insignificant changes in dosimetric coverage (Buzurovic et al. 2011).

15.1.3 INTEGRATED MRI-RADIOTHERAPY PLATFORMS

Most modern radiation therapy equipment utilizes in-room x-ray–based imaging technology, such as planar imaging or cone-beam CT, to verify daily patient setup accuracy or to perform real-time tumor tracking. These x-ray imaging systems perform well for targets with high contrast, such as bone or lung lesions. However, for identification of low contrast targets, such as liver lesions, x-ray imaging systems are often suboptimal. Radio-opaque fiducial markers may be placed for reliable target identification with x-ray imaging; however, this procedure is invasive and is not suitable for all patients and all target sites. A number of systems that integrate MRI scanners with MV radiotherapy treatment machines have been proposed (Dempsey et al. 2005; Fallone et al. 2007; Kron et al. 2006; Lagendijk et al. 2008), and one such integrated MRI-RT system, the ViewRay™, has recently hit the market and gained FDA clearance.

Compared to the x-ray– or optical-based imaging technologies, MRI has two major advantages. The first is that the superior soft tissue contrast of MRI can allow direct target visualization without the need to implant radio-opaque fiducial markers. The second advantage is that MRI is a nonionizing modality, meaning real-time imaging can be performed for each fraction without additional normal tissue dose. However, the integration of a MRI scanner, which requires highly homogeneous magnetic fields on the order of a few parts per million with a MV energy radiotherapy delivery system poses unique technical hurdles.

15.1.3.1 General MRI-RT system configurations

The orientation of the MRI's magnetic field relative to the MV treatment beam is of great importance as it impacts system engineering and design limitations as well as patient dosimetry. In a *parallel configuration*, the treatment beam is in line with the magnetic field (Constantin, Fahrig, and Keall 2011). Rotating a super-conducting MRI magnet is a major engineering task, making the use of lower field permanent biplanar magnet MRI systems more feasible to achieve in this MRI-RT system configuration. This configuration, in which a biplanar magnet is rotated around the patient is generally called a rotating biplanar (RBP) design. When the treatment beam is oriented parallel, or in line, to the magnetic field, it is called a *longitudinal rotating biplanar* (*L-RBP*) configuration (Figures 15.2a and d).

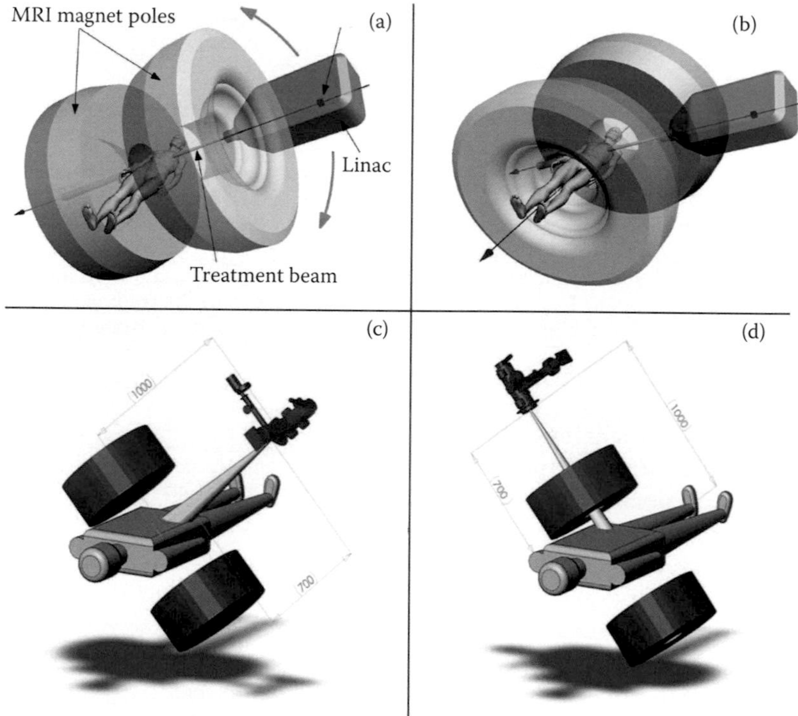

Figure 15.2 Illustration of the varying integrated MRI-RT configurations. (a) In the *L-RBP* configuration, the magnetic field and MV treatment beam are oriented parallel to one another and rotate around the patient in unison. (b) In the *FC* configuration, the magnetic field and treatment beam are perpendicular to one another. Here the magnets are fixed with respect to the patient, and the MV treatment beam rotates around the patient. (c) In the *T-RBP* configuration, the magnetic field is oriented perpendicular to the treatment beam. Here both the magnets and treatment beam are rotated around the patient in unison. (d) A second illustration of the L-RBP configuration (as in panel a), showing the magnetic field and treatment beam are parallel and rotate around the patient in unison. (Images a and b from d. Constantin et al., *Med. Phys.*, 38, 7, 4174–85, 2011, permissions granted by Medical Physics; images c and d from Kirkby et al., *Med. Phys.*, 35, 3, 1019–27, 2008, permissions granted by Medical Physics.)

Alternatively, the unit may be designed with a *perpendicular configuration*, in which the magnetic field is oriented perpendicular to the treatment beam (Constantin, Fahrig, and Keall 2011). There are two proposed perpendicular configurations. In the *transverse rotating bi-planar* (*T-RBP*) configuration, the magnet and treatment unit rotate in unison around the patient (Figure 15.2c). In the second perpendicular configuration, called *fixed cylindrical* (*FC*), the magnetic field remains fixed along the patient's cranio-caudal axis while the treatment unit rotates independently in the patient's axial plane (Figure 15.2b) (Kirkby et al. 2008).

The magnetic field of the MRI unit has important effects on the operation of a linac. St. Aubin and colleagues have demonstrated that even small fringe fields can dramatically reduce the linac output to nearly zero due to magnetic deflection in the electron gun (St. Aubin, Santos et al. 2010; St. Aubin, Steciw, and Fallone 2010). The parallel L-RBP configuration exhibits less effect from the fringe field deflection as the magnetic fringe field is in line with the electrons emitted from the linac electron gun. Furthermore, passive shielding of the electron gun has been shown to effectively recover up to >99% of the original target current in the parallel configuration (Santos et al. 2012). Work is ongoing to design MRI-RT systems that retain normal linac operating parameters, for example, active magnetic shielding for the linac; however, more complex MRI-linac systems will certainly increase system cost.

15.1.3.2 Impact on patient dosimetry

The orientation of the magnetic field with the treatment beam also has important implications for patient dosimetry. The Lorentz force from the magnetic field acts on the secondary electrons in the patient that are responsible for depositing the majority of the dose in tissue. In a L-RBP system in which the treatment beam and magnetic field remain parallel, or in line, to one another, the Lorentz force may actually improve the dosimetry in certain cases. Kirkby et al. (2010) have utilized Monte Carlo simulations to demonstrate that a MRI-RT configuration in which the magnetic field and treatment beam are parallel causes the beam penumbra to shrink as the Lorentz force helps collimate dose-depositing electrons along the beam direction. This effect, illustrated in Figure 15.3, is especially noticeable in low-density tissues, such as lung, in which electrons have a larger range. Kirkby et al. (2010) further demonstrated the effect in a clinical five-field lung plan (reproduced in Figure 15.4), showing that the parallel configuration actually increases the dose to the PTV and reduces the dose from lateral electron spread to healthy lung tissue adjacent to the field edges but at the cost of increased dose to normal tissue along the beam path.

In perpendicular MRI-RT configurations, the Lorentz force can worsen the dosimetry as the dose-depositing secondary electrons are shifted systematically away from the beam axis as shown in Figure 15.3

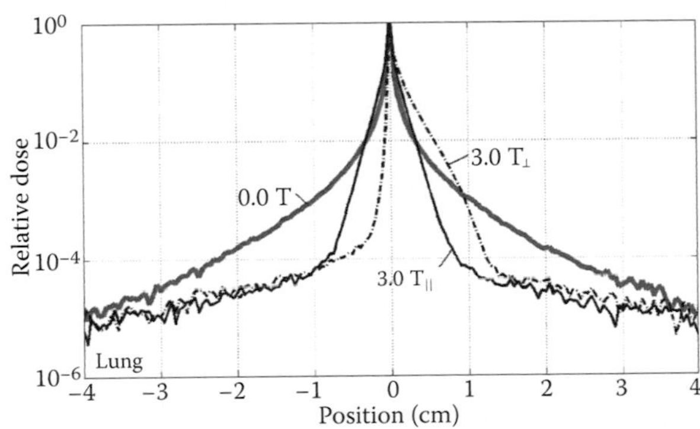

Figure 15.3 Comparison of pencil beam dose kernels at 10 cm depth in lung tissue. Shown here are the baseline 0 field case (0.0 T), the case in which a 3.0 T magnetic field is aligned parallel to the beam direction (3.0 $T_{\parallel}$), and the case in which a 3.0 T magnetic field is perpendicular to the beam direction (3.0 $T_{\perp}$). Compared to the 0 field case, a parallel, or in line, configuration of the magnetic field with respect to the MV beam leads to a collimating effect on the secondary dose-depositing electrons, leading to a sharper pencil beam dose penumbra. In the case of a perpendicular magnetic field/MV beam configuration, the secondary electrons are systematically shifted orthogonal to the beam direction leading to a lateral shift in the dose kernel. (From Kirkby et al., *Med. Phys.*, 37, 9, 4722–4732, 2010, permissions granted by Medical Physics.)

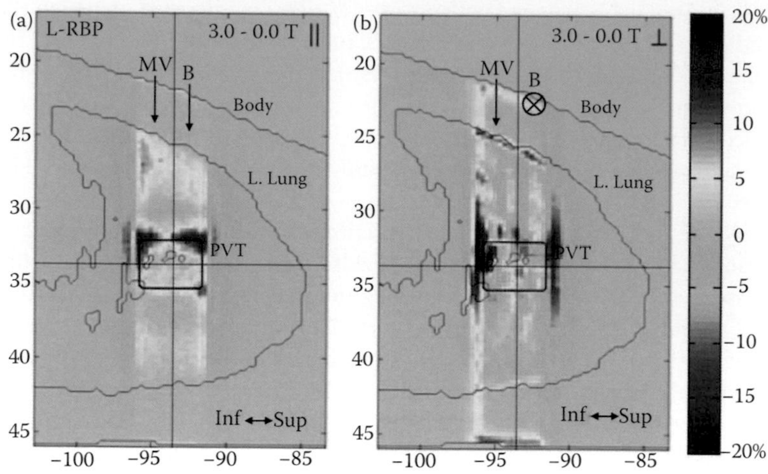

Figure 15.4 Sagittal slices of dose difference maps between 0 field and 3.0 T magnetic field for a five-field coplanar lung plan investigated by Kirkby et al. (2010) that illustrates the importance of the MRI-RT configuration in the patient dosimetry. In both RBP cases shown here, the magnetic field and treatment beam rotate around the patient in unison. (a) In the L-RBP case, in which the MV beam and MRI magnetic field, B, are parallel (or in line) with one another, secondary electrons are collimated along the beam path leading to an increase in dose along the beam path and in the PTV and a reduction of dose to the lung tissue in the beam's penumbra region. (b) In the T-RBP case, in which the MV beam and B-field are perpendicular to one another, the Lorentz force on the secondary electrons leads to a systematic dose shift in the inferior direction. (From Kirkby et al., *Med. Phys.*, 37, 9, 4722–32, 2010, permissions granted by Medical Physics.)

(Kirkby et al. 2008). At lung-tissue interfaces, the secondary electrons deflected in the magnetic field can return to the upstream lung boundary, leading to substantial dose increases at the interface, known as the "electron return effect" (Raaijmakers, Raaymakers, and Lagendijk 2005). In the T-RBP configuration, in which the magnet and gantry rotate in unison, the direction of the magnetic deflection remains constant with respect to the patient, leading to a systematic dose shift in the patient's cranio-caudal direction as demonstrated in Figure 15.4. In the FC configuration, in which the magnet remains fixed while the treatment beam rotates, the direction of the magnetic deflection changes with respect to the patient as the treatment beam is rotated (Kirkby et al. 2008).

The choice of MRI field strength will affect the magnitude of machine and dose perturbations. While typical diagnostic field strengths of 1.5 T to 3.0 T give good SNR in the MRI image, these field strengths worsen the dosimetric effects described above (Kirkby et al. 2010). Higher field strength means increased Lorentz force on the secondary electrons that are depositing dosage from the radiotherapy beam, and it has been well documented that these fields can cause significant dose differences from the zero field case. Lower field strengths on the order of 0.2 T have been shown to have minor dose variations (Raaijmakers, Raaymakers, and Lagendijk 2008). Here, the tradeoff is image quality versus dose perturbation: Lower strength magnetic fields create less dose perturbation, but image quality must be evaluated for clinical value in each scenario. In addition, the effect of the magnetic field on the dose distribution will need to be included in dose calculation algorithms while keeping the tradeoff between accuracy and calculation efficiency in mind.

15.1.3.3 MRI Cobalt-60

ViewRay is the first company to bring a commercial hybrid MRI-RT unit to the market, having gained FDA 510k clearance in May 2012. The first installed unit at Washington University in St. Louis is shown in Figure 15.5. The ViewRay system has a FC perpendicular configuration composed of a split-coil low-field super-conducting MRI system with a slip-ring gantry treatment module in the center that houses three radioactive Co-60 sources.

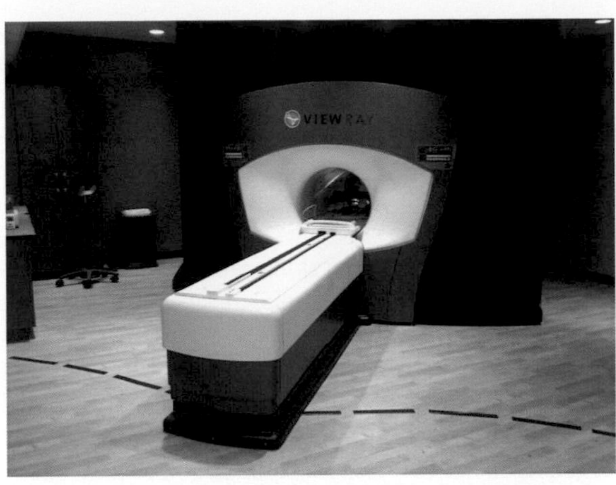

Figure 15.5 Picture of the first ViewRay™ system installed at Washington University in St. Louis in early 2012. (Image courtesy of ViewRay™.)

The ViewRay's magnetic resonance imaging system consists of a specially adapted 0.35 T Siemens Avantago MRI. The low-field magnet gives clinically acceptable image quality and reduces dosimetric effects, such as the electron return effect discussed in the previous section (Raaijmakers, Raaymakers, and Lagendijk 2008). The ViewRay system's scanner can image a single plane at four frames per second or three planes (axial, sagittal, and coronal) at two frames per second. This imaging frame rate allows real-time monitoring of tissue motion in support of beam gating or dynamic tumor tracking. With the recent installation of the first ViewRay systems, it will be exciting to see the results of real-time intrafractional tumor motion monitoring, especially for difficult-to-visualize soft-tissue tumors, such as in the liver.

The central treatment module consists of a slip-ring gantry with three Co-60 sources mounted 120° apart. Each treatment source has a volumetric MLC for intensity modulation. Figure 15.6 displays a

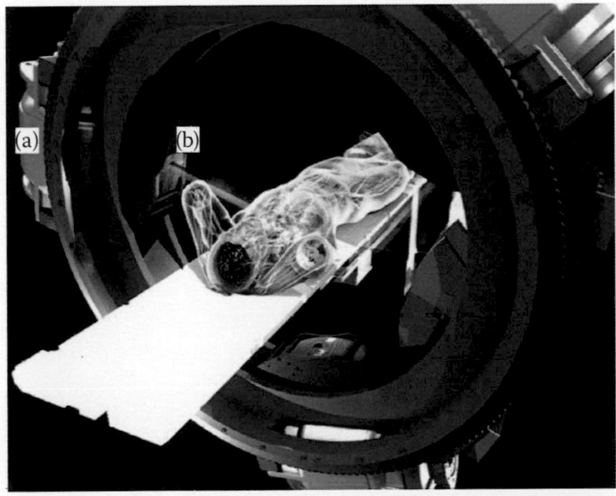

Figure 15.6 Schematic cut-away diagram of the ViewRay™ system's central Co-60 radiotherapy module. Three Co-60 sources (a), each with a volumetric multileaf collimator (b), rotate around the patient. The MLC leaves travel on curved trajectories to reduce the transmission penumbra. The 0.35 T super-conducting MRI magnets are positioned on the superior and inferior sides of the Co-60 RT module (not shown here), giving this system a fixed cylindrical geometrical configuration. Note that this image is a conceptual schematic to illustrate the source geometry and the volumetric MLC-based delivery technique and is not representative of the actual ViewRay™ system's dimensions. (Image courtesy of ViewRay™.)

schematic cut-away of the ViewRay's treatment module to illustrate the source geometry. Each source has an initial activity of 15,000 Ci. This initial activity gives a dose rate of ~200 cGy/min per source; up to 600 cGy/min if all three sources are treating simultaneously.

The choice of Cobalt-60 as the radiation source for the ViewRay hybrid MRI-RT system avoids some of the difficult technical hurdles of integrating a linear accelerator with a MRI scanner as the operation of the radioactive Co-60 source is unaffected by the MRI's magnetic field. However, Co-60 has received criticism for use in modern radiation therapy. Concerns include larger beam penumbras due to larger physical source size; lower energy (1.25 MeV), leading to decreased skin sparing and a larger volume of low-dose bath in normal tissue; lower achievable dose rates; and regulatory considerations with high-activity radioactive sources (Cadman, Paliwal, and Orton 2010).

Work is ongoing to alleviate these concerns and to help clear the stigma that Co-60 is an outdated source for use in modern radiotherapy. Intensity modulated radiotherapy (IMRT) plans using Co-60, while having slightly larger low-dose baths in surrounding normal tissue, have been shown to provide very similar plan quality as conventional MV linacs (Cadman and Bzdusek 2011; Fox et al. 2008). A head and neck treatment plan comparison between the ViewRay system and a conventional Varian linac is demonstrated in Figure 15.7. The ViewRay system's MLCs are designed to move along arc trajectories to reduce the transmission penumbra that is characteristic of MLCs with rounded leaf ends that travel on linear trajectories. The ability of double-focused MLCs to deliver highly conformal Co-60 IMRT has been demonstrated on a bench-top tomotherapy system (Schreiner et al. 2009).

The ViewRay system's design using three sources helps alleviate dose rate concerns, but because a radioactive Co-60 source is used, the dose rate will decrease over time, necessitating administrative planning and coordination for source exchanges. In addition, the presence of a very high-activity radiation source presents special security issues. These, however, are not new concerns; Gamma Knife users will be very familiar with these special considerations in using Co-60. Institutions interested in the ViewRay system should keep these logistical and regulatory issues in mind to avoid delays in bringing the system to clinical operation.

Integrating a MRI, which has superior soft-tissue contrast and is a nonionizing *modality*, with a radiotherapy treatment machine exemplifies an emerging technology for SRT. While hurdles to widespread

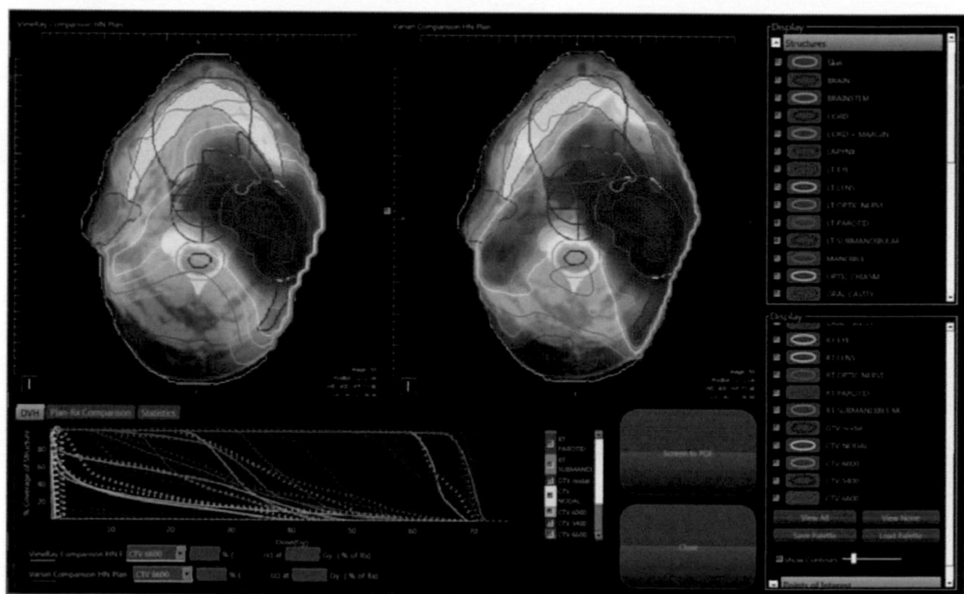

Figure 15.7 A head-and-neck plan comparison illustrating the ability of volumetric Co-60 IMRT to deliver plan quality comparable to an IMRT plan on a conventional Varian linear accelerator. (Image courtesy of ViewRay.)

clinical adoption remain, research continues to be very active in this area, and solutions to solve or minimize many of the aforementioned technical problems have been presented. The ViewRay system, in bringing the first hybrid MRI-RT system to the market, gives us confidence that the future of SRT may be very close indeed.

15.1.4 DEDICATED GAMMA STEREOTACTIC RADIOTHERAPY BREAST IRRADIATION DEVICE

A new breast SRT device, named GammaPod™, has been developed by Xcision Medical Systems and the University of Maryland to deliver highly focused and localized doses to a target in the breast under stereotactic image guidance (Yu et al. 2013). The GammaPod system uses a hemispherical source carrier containing 36 ^{60}Co sources (total activity 4380 Ci), a tungsten collimator assembly, a dynamically controlled patient support table, and the breast immobilization system, which also functions as a stereotactic frame. Highly focused radiation is achieved at the isocenter by the cross-firing of 36 nonoverlapping radiation arcs generated by rotating the 36 individual Cobalt-60 beams. Stereotactic localization of the breast is achieved by a vacuum-assisted breast immobilization cup with built-in stereotactic frame. A dedicated treatment-planning system optimizes the path of the focal spot using an algorithm borrowed from computational geometry such that the target can be covered by 90%–95% of the prescription dose, and the doses to surrounding tissues are minimized. The treatment plan is delivered with the continuous motion of the treatment couch, allowing the GammaPod to dynamically paint the desired dose distributions. Due to the highly focused dose delivery and precise stereotactic localization mechanism, it is envisioned that GammaPod technology has the potential to significantly shorten postoperative radiation treatments and even eliminate surgery by ablating the tumor and sterilizing the tumor bed simultaneously.

Figure 15.8 illustrates the structures of the GammaPod™ system. The patient wearing the vacuum-assisted breast cup is first set up on the patient couch at a near standing position with the breast cup locked onto the couch surface. The loading mechanism then lifts and rotates the treatment couch onto the bed frame with the patient in the prone position. When treatment is started, the upper shielding door opens to allow the treatment couch to be lowered, placing the focal spot at the target location. The bed frame moves dynamically during treatment, such that the focal spot paints the desired dose distribution conforming to the target. The arrangement of the sources and beam collimation mechanism is shown in Figure 15.8b. The sources, each with a unique longitude and latitude, are oriented and collimated by

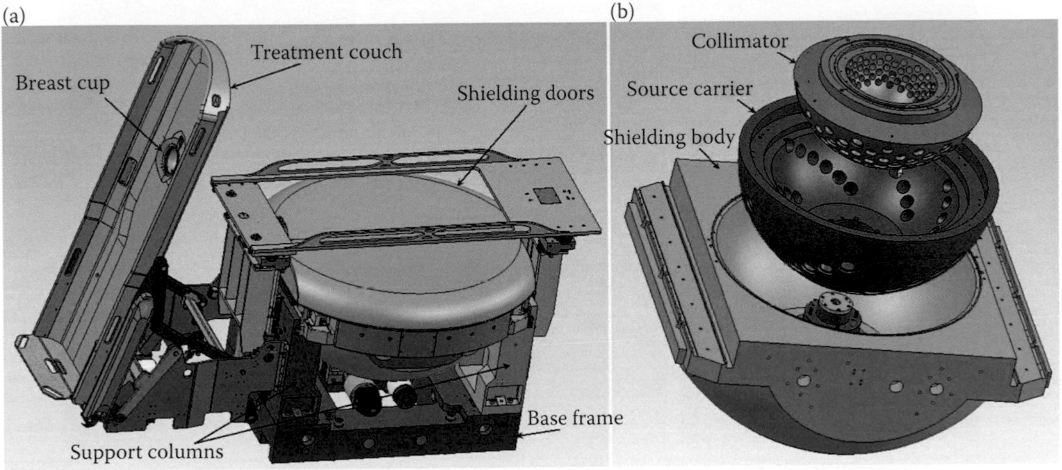

Figure 15.8 (a) The structure of the GammaPod system consists of the irradiation unit, the dynamic treatment couch, and a rotating patient-positioning device for easy prone setup. (b) The concentric structure of the irradiation unit consists of the outer shielding structure, the source carrier, and the innermost collimator. The treatment space is defined by the inner surface of the collimator.

the primary collimators so that they converge at the isocenter at a distance of 38 cm from the sources at the geometrical center of the hemisphere. The secondary collimator is positioned as the final component of the beam-shaping unit on the top of the source carrier and has the same hemispherical bowl-shaped design. Different alignments of the source carrier and the collimator permit three different selections of collimation: 1.5 cm, 2.5 cm, and blocked. When the selected collimator opening is aligned with the source positions, the source carrier and the collimator are rotated together to create 36 non-coplanar conical arcs all converging at the isocenter.

Dosimetric characterization of the GammaPod has shown that the GammaPod is capable of delivering highly conformal dose coverage to the postoperative surgical bed with substantially lower dose spillage outside the target volumes as compared with three-dimensional conformal therapy (Mutaf et al. 2013). This dose-focusing ability provides the potential for preoperative tumor ablation with and without concurrent tumor bed sterilization. Figure 15.9 illustrates the dosimetric feasibility of these three potential modes of treatment.

Comparisons with commonly used single- and multi-lumen high dose rate (HDR) brachytherapy devices were conducted (Oden et al. 2013). In all scenarios investigated in the study, GammaPod delivers more uniform dose distribution to the target, faster dose falloff at a distance away from the target, and lower skin doses as compared with commonly used single- and multi-lumen HDR brachytherapy devices (Oden et al. 2013). Figure 15.10a shows the dose-volume histogram (DVH) for treating a 5-cm clinical target volume (CTV) using different methods. Unlike intracavitary HDR techniques, GammaPod provides uniform dose to the target without the 200% to 250% hot spots. The dose fall off beyond the CTV (Figure 15.10b) is also steeper with GammaPod, reducing the dose to the lung and the heart. Whether or not these dosimetric advantages of GammaPod can translate to clinical advantages require further clinical studies.

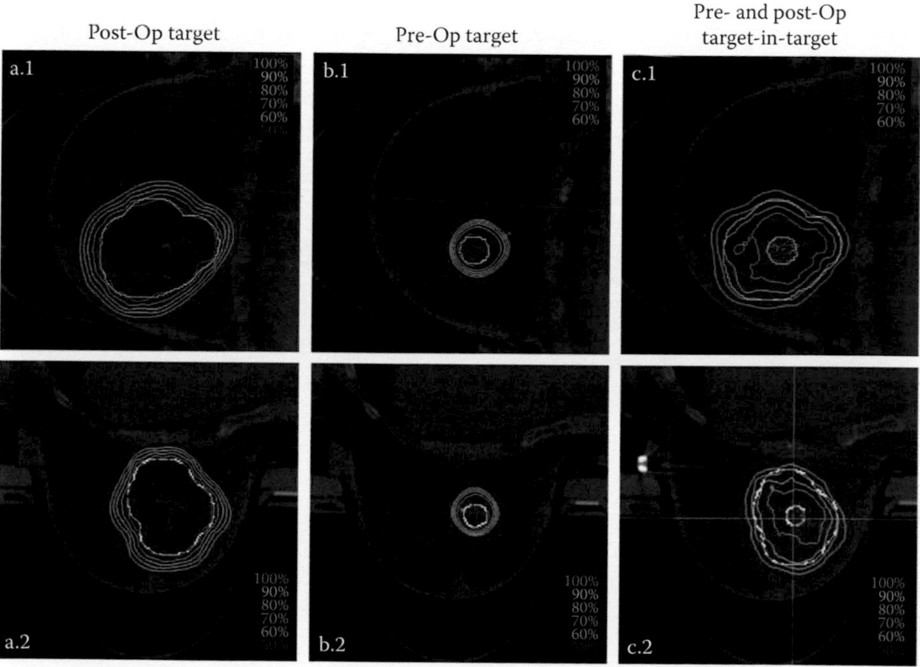

Figure 15.9 Sagittal (top row = a.1, b.1, c.1) and transverse (bottom row = a.2, b.2, c.2) images of GammaPod plans created with different targets (white contours). Post-op surgical cavity with 1.5 cm subclinical extension (left column = a.1 and a.2), preoperative lesion as delineated on pre-op MR images (middle column = b.1 and b.2), and a target-in-target volume with a simultaneous boost approach utilizing both pre-op (full-dose) and post-op (half-dose) targets (right column = c.1 and c.2).

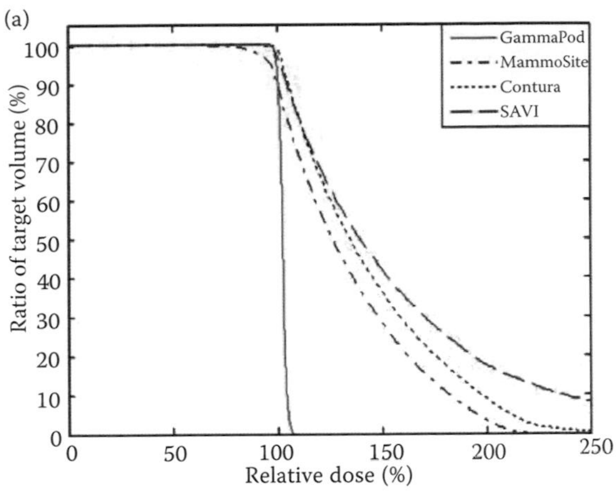

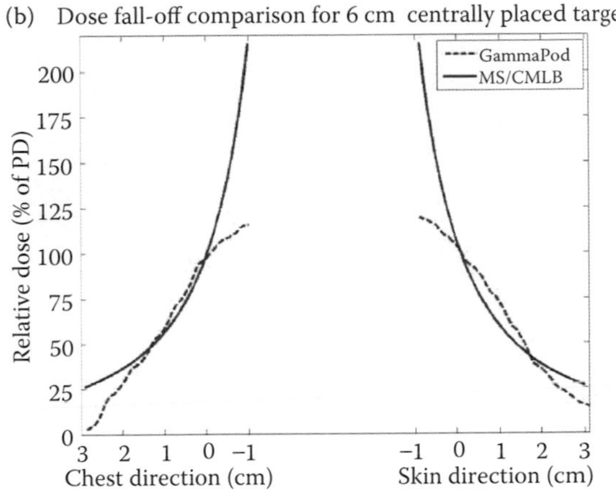

Figure 15.10 Comparison of GammaPod and balloon techniques based on (a) DVH and (b) dose fall-off away from cavity wall normalized to 1 cm from the cavity surface.

15.1.5 MICROBEAM RADIOSURGERY

15.1.5.1 Introduction

A number of scientific studies on small animals have demonstrated the astonishing fact that healthy biological tissue can tolerate an enormous amount of absorbed dose (>500 Gy) when delivered in an array of thin planes of radiation (<500 μm), termed microbeam radiation. Although cells in the direct path of the microbeam radiation are damaged or killed, the adjacent nonirradiated tissues mount a healing response. Studies have also demonstrated that microbeam radiation has a preferential tumoricidal effect. The microbeams damage cancerous tissue in such a fashion that it is difficult for the tumor to recover. Thus, microbeam radiosurgery (MBRS) appears to have tremendous potential to destroy cancerous tumors with little or no toxicity to surrounding healthy tissue.

15.1.5.2 History

An excellent review of the field of MBRS is given by Brauer-Krisch et al. (2010). The study that launched the field was conducted by Curtis (1967). In an effort to understand the effect of cosmic-ray particles on astronauts, the brains of healthy mice were irradiated with a 25-μm-diameter beam of 22-MeV deuterons (Figure 15.11). Histological examination showed that no damage was done to the mouse brain for doses below 4000 Gy. Many years then passed before the biological effect of microbeams of x-rays was studied.

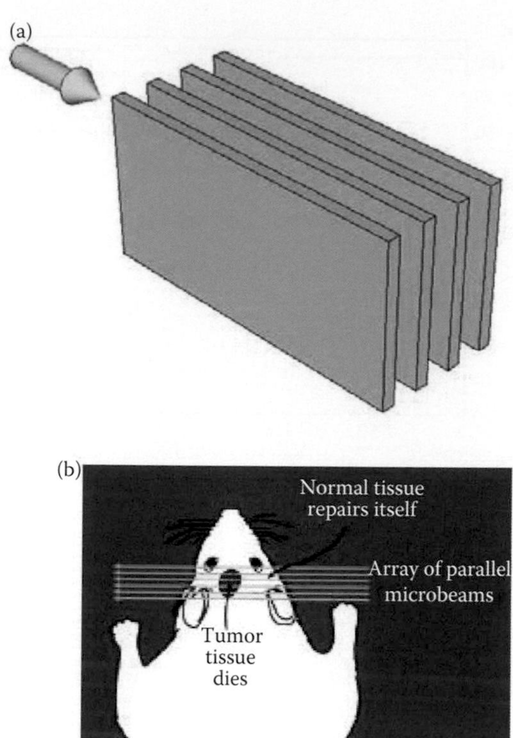

Figure 15.11 (a) Graphical representation of an array of microbeam radiation, a.k.a., microplanar radiation. (b) Schematic showing the effect of microbeam radiation on biological tissues. (From Dilmanian, F. A. et al., *Nuclear Instruments and Methods in Physics Research Section A: Accelerators, Spectrometers, Detectors and Associated Equipment*, 548, 30–37, 2005.)

A key technology needed to be developed: wiggler insertion devices in synchrotrons. In 1995, Slatkin et al. exposed healthy rat brains to microplanar arrays of 50 keV x-rays from a synchrotron source. The microplanes of radiation were 37 μm wide with a 75 μm pitch. Figure 15.12 shows a histological examination of a portion of rat cerebellum exposed in this study. The exposure came in groups of three microplanar beams of alternating dosages, 625 Gy and 2500 Gy. Damage is clearly seen in the tissue

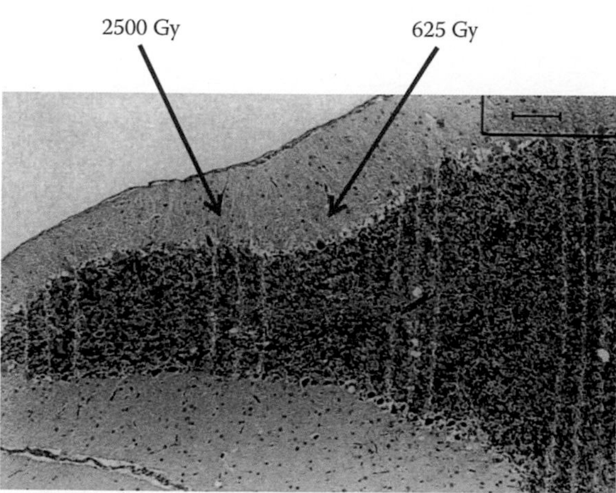

Figure 15.12 Hematoxylin and eosin-stained section of rat cerebellum. Damage results from microplanar arrays of 2500 Gy whereas there is no apparent damage from microplanar arrays of 625 Gy. Scale bar is 100 μm. (From Slatkin, D. N. et al., *Proc. Natl. Acad. Sci. USA*, 92, 19, 8783–87, 1995.)

exposed to 2500 Gy, and no apparent damage results from the 625 Gy exposures. Even though damage is seen in the tissue irradiated at 2500 Gy, the animals showed no behavioral differences compared to nonirradiated animals.

In 2003, Dilmanian et al. performed a comparative study of the efficacy of microbeam versus conventional radiotherapy on mammary tumors inoculated into the hind legs of rats. One set of rats was exposed to 90-μm-wide, 300-μm-pitch, 650-Gy microbeams, and a second set was exposed to a 2 cm × 2 cm broad beam of 45 Gy. The x-ray photon energy was 100 keV. This study showed that tumor ablation was slightly better and that acute and late toxicity effects were significantly lower for the microbeam irradiated rats compared to the broad-beam irradiated rats.

In 2006, Romanelli proposed using MBRS for the treatment of noncancerous brain disorders, such as epilepsy and Parkinson's disease (Dilmanian et al. 2006). In 2011, Romanelli injected rats with kainic acid to induce convulsive epilepsy (Romanelli et al. 2013). A set of the animals were irradiated with 600-μm-wide, 1200-μm-pitch, 360-Gy microbeams. The x-ray photon energy was 100 keV. Seizures disappeared in 2.2 hr for the irradiated animals compared to 40 hr for the nonirradiated animals.

15.1.5.3 Biology

Remarkable though it may be upon first learning, the ability of normal healthy tissue adjacent to microbeam tracks to heal the radiation-induced damage is, to some degree, to be expected. The body has a natural ability to heal small wounds, such as a thin cut to the skin. The biochemistry of wound healing has been studied extensively (http://en.wikipedia.org/wiki/Wound_healing). The processes involved are complicated with events happening on a molecular, cellular, and tissular level over time frames of minutes, hours, days, weeks, and months. The specifics of the healing of microbeam radiation–induced damage are only beginning to be studied (Dilmanian et al. 2007; Smith et al. 2013).

With regard to the preferential tumoricidal effect of microbeams, the evidence amassed so far indicates that differences in the microvasculature of tumors and normal tissue may be the cause. The blood vessels of tumors are faster growing and immature, larger in diameter, fewer in density, and more tortuous in shape than the blood vessels of normal tissue. In an experiment by Serduc et al. (2006), two fluorescent dyes were injected into the blood vessels of normal tissue. One dye had a molecular weight nearly that of a red blood cell. The second dye had a molecular weight approximately 1/100th that of a red blood cell. After irradiation with 1000 Gy in a 25-μm microplanar beam, fluorescence showed no extravasation of the large molecular weight dye. Some extravasation was observed for the smaller molecular weight dye, and this extravasation reduced over time as the blood vessels healed. In a similar experiment by Dilmanian et al. (2002), considerable extravasation of the large molecular weight dye from the blood vessels of a tumor was observed after irradiation with 800 Gy in a 27-μm microplanar beam. Figure 15.13 shows the histological result of Dilmanian et al.

Another possible explanation for the preferential tumoricidal effect of microbeams involves the motility of cancer cells. Crosbie et al. (2010) inoculated the hind legs of rats with mammary cancer cells and irradiated the resultant tumor with an array of 25-μm-wide, 200-μm-pitch microbeams at 560 Gy. Post-irradiation histology showed that the radiation-damaged cancer cells mixed with the nondamaged cancer cells. Within 24 hr, the original location of the radiation tracks could no longer be observed, and the irradiated field was a homogeneous mixture of radiation-damaged and nondamaged cancer cells. As such, there was not a contiguous region of nondamaged cancer cells adjacent to the radiation-damaged cancer cells to mount a healing response. The results of the Crosbie study are shown in Figure 15.14.

15.1.5.4 Problems

Although the advantageous properties of microbeams have been known for some time, there are fundamental problems that have kept this form of radiosurgery from becoming a clinical tool.

First and foremost, effective microbeam arrays of x-ray photons can presently be generated only by synchrotron sources. A synchrotron is required for low beam divergence and HDR. The low beam divergence is necessary to keep the microbeams from spreading to larger widths as they pass through a patient. The HDR is required so that patient motion does not broaden the radiation damage region. Unfortunately, synchrotrons are very large devices with construction and operation costs at levels that only nation states can provide.

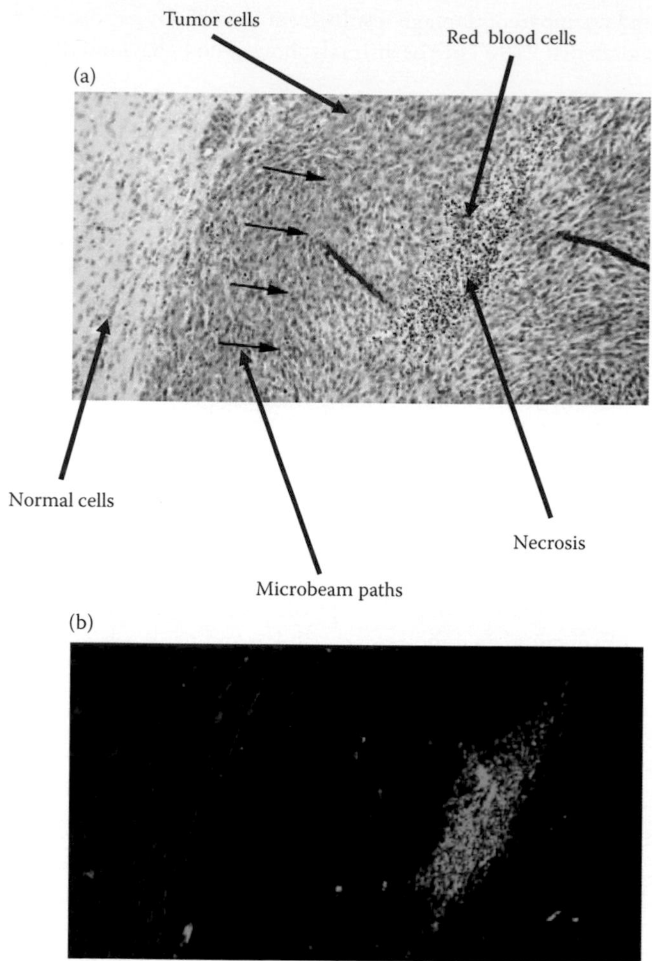

Figure 15.13 (a) Microbeam irradiation induces blood vessel disruption and necrosis in 9L gliosarcoma tumor in rat brain. (From Dilmanian, F. A. et al., *Nat. Synchrotron Light Source Rep.*, Abstract No.: Dilm0599, 2002.) (b) Fluorescence shows extravasation in necrotic region of (a).

A second problem is the low energy of the x-ray photons used in MBRS studies to date. While photons at 100 keV work well for small animals, such photons do not provide sufficient dose at depth for human patients.

A third problem is that all MBRS studies up to now have focused on the brain. To warrant the serious investment in R & D required of a medical device company to bring MBRS to the clinic, microbeams must be shown to be effective in additional anatomy.

15.1.5.5 Solutions

Recently, several groups throughout the world began working to develop a new type of radiation source, which employs inverse Compton scattering (ICS). Figure 15.15 displays the ICS process in which a high-energy electron collides with a low-energy photon to yield a high-energy photon and a reduced-energy electron. Figure 15.16 shows a schematic of an ICS source. A linear accelerator injects high-energy electrons into a small storage ring. The electron beam path in the storage ring intersects an optical cavity into which photons from a visible laser are injected. When the high-energy electrons and low-energy laser photons collide, high-energy x-ray photons are generated. First-generation ICS sources are expected to produce radiation very similar to that of synchrotron radiation with the exception of dose rate. Later generations of ICS sources are expected to match the dose rate of synchrotrons. ICS sources are expected to be a few meters in diameter rather than the kilometer diameter typical of synchrotrons. Also, ICS sources are expected to cost about as much as today's linear accelerators used in conventional radiotherapy.

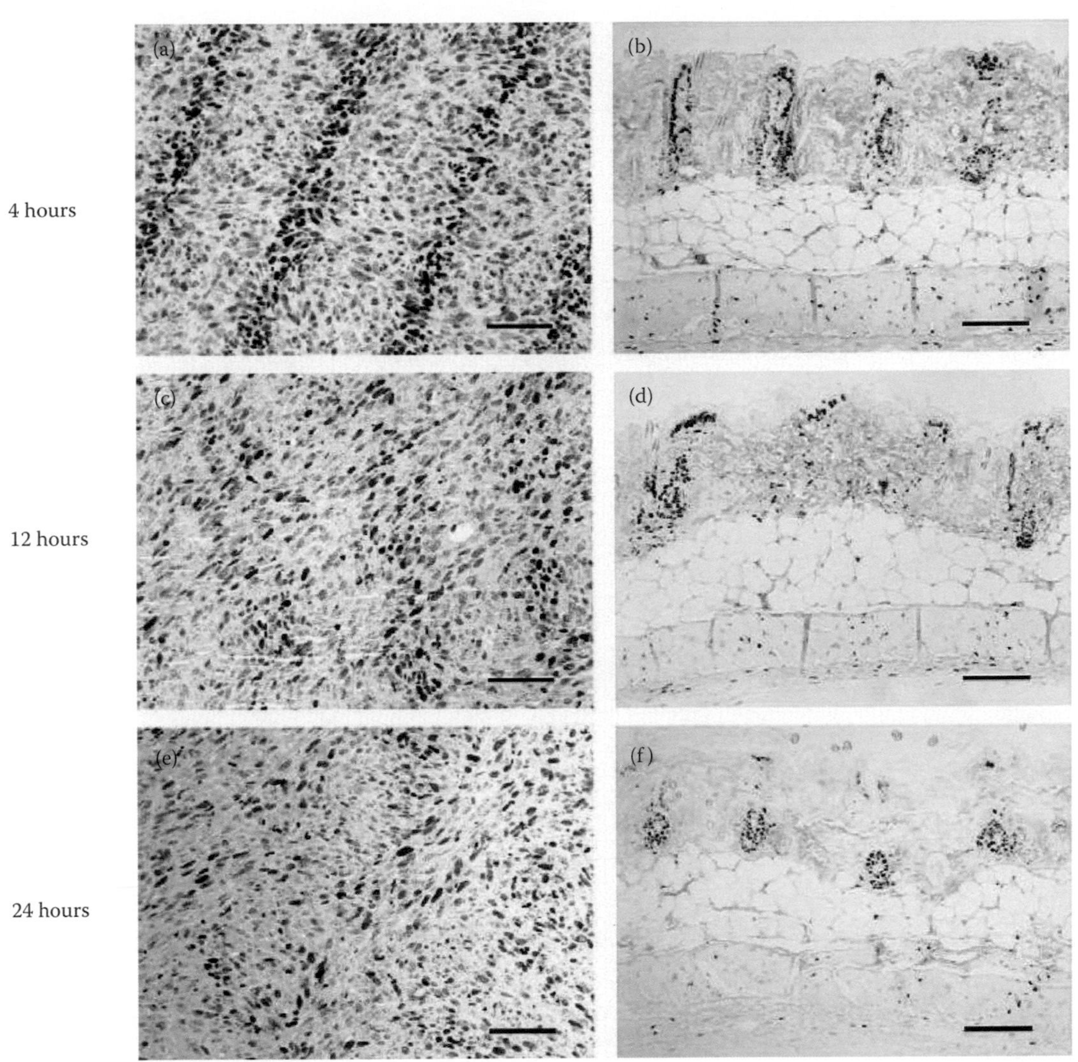

Figure 15.14 γ-H2AX (brown) and BrdU (blue) staining of microbeam irradiated EMT-6.5 mammary tumor (a, c, and e) and normal skin (b, d, and f) of rat as a function of time post-irradiation. The γ-H2AX stain identifies cells with radiation-induced DNA damage. The BrdU stain identifies proliferating cells. Within 24 hours after irradiation, the radiation-damaged and non-damaged cancer cells are completely intermixed. Radiation-damaged skin cells, however, remain localized. Scale bar is 100 μm. (From Crosbie, J. C. et al., *Int. J. Radiat. Oncol. Biol. Phys.*, 77, 3, 886–94, 2010.)

ICS sources will also be capable of producing x-ray photons in the MeV range, which solves the second problem associated with the current state of the art of MBRS. MeV photons will penetrate sufficiently deep into human patients. It must be noted, however, that operating in the MeV range will require careful tailoring of the microplanar beam to account for the scattering that will occur in the patient (Wright 2012).

With regard to the effectiveness of MBRS in tissue other than brain, new studies are underway. For example, a study has been launched at the European Synchrotron Radiation Facility in Grenoble, France, to investigate the effect of microbeams on rodent lungs.

15.1.5.6 Conclusions

Although there are several very challenging difficulties facing the development of MBRS, the potential of MBRS to cure diseases internal to the human body with no toxicity to surrounding healthy tissue warrants the effort to overcome such challenges. A clinically useful device seems feasible.

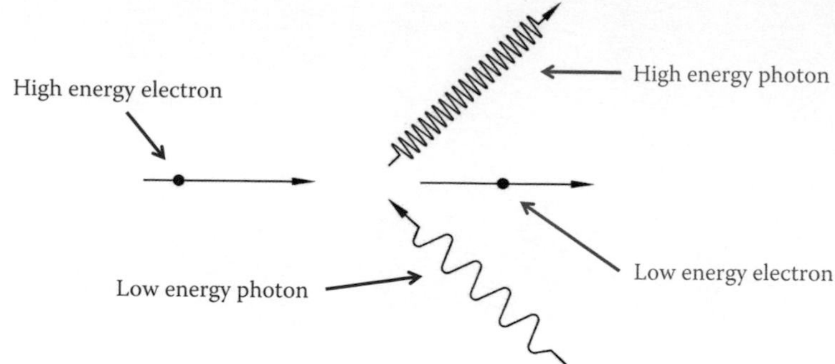

Figure 15.15 Inverse Compton scattering.

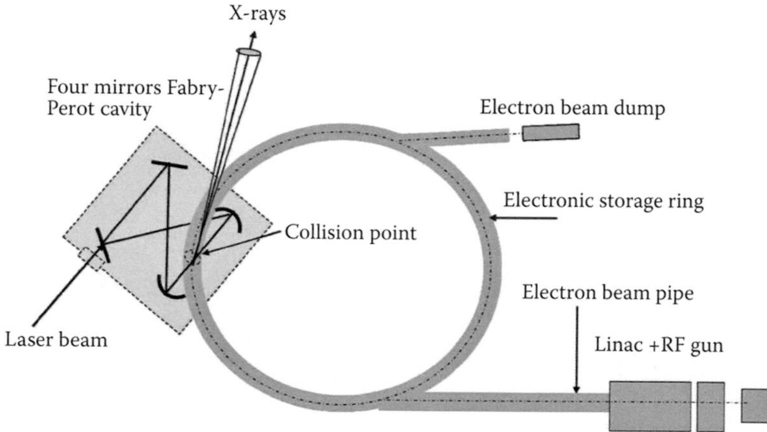

Figure 15.16 Schematic of an ICS source. (From Variola, A., Loulergue, A., and Zomer, F., eds. 2010. THOMX Conceptual Design Report, http://hal.in2p3.fr/docs/00/44/82/78/PDF/ThomXCDRAV.pdf, 2010.)

15.2 RADIATION SENSITIZERS AND ENHANCERS

15.2.1 INTRODUCTION

Stereotactic radiosurgery and stereotactic body radiation are effective treatments for a variety of tumors. Further dose escalation could improve outcomes; however, this would not be without an increased risk of normal tissue toxicity. One method to improve the therapeutic ratio of radiosurgery would be to increase the effectiveness of radiation delivery with a sensitizing or enhancing agent. This could result in improved tumor control and decreased radiation dose needed to achieve control, thus decreasing normal tissue toxicity. Another application of radiosurgery could be not only to target local disease, but also to target systemic disease by combining this treatment with an immunomodulation drug (Figure 15.17).

15.2.2 CYTOTOXIC DRUG THERAPY

Concurrent administration of cytotoxic chemotherapy with fractionated radiation therapy has been shown to improve outcomes in numerous disease sites. This improvement may be due to a radiosensitizing effect and/or impact on clinically occult microscopic metastatic disease. There is little data on the use of concurrent cytotoxic chemotherapy with SRS/SBRT, but the results of two studies using concurrent cytotoxic chemotherapy with SRS in the treatment of recurrent glioma were reported in the late 1990s

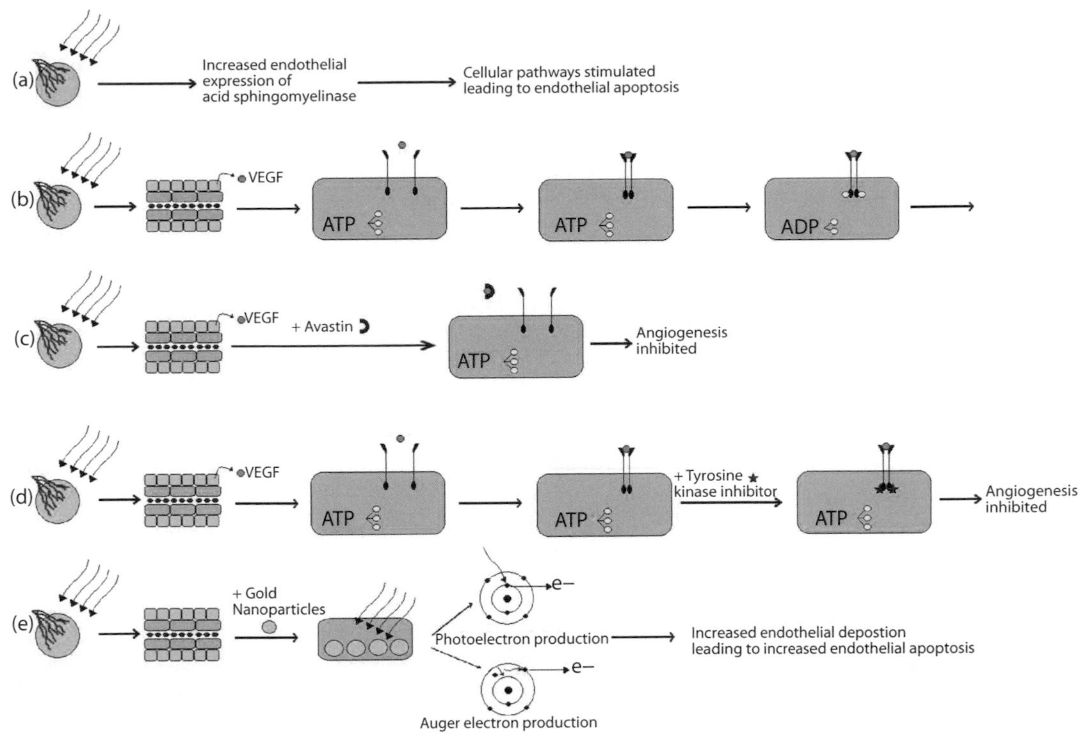

Figure 15.17 Sensitizing the tumor endothelium to the effects of high-dose radiation. (a) High-dose radiation therapy leads to increased expression of acid sphingomyelinase by the tumor endothelium, which results in endothelial apoptosis. The depleted vasculature leads to tumor cell death. (b) In response to high-dose radiation, tumor cells (green) may increase expression of vascular endothelial growth factor (VEGF), which interacts with the extracellular domain of VEGF receptors on tumor endothelium (pink). This leads to dimerization of the VEGF receptor and phosphorylation of the intracellular domain by ATP. The intracellular domain of the VEGF receptor functions as a tyrosine kinase and transfers the phosphate group to other signaling molecules, leading to promotion of angiogenesis and blocking apoptosis. (c) Bevacizumab binds to VEGF and prevents it from interacting with the VEGF receptor on endothelial cells. This inhibits the signal from being received by the endothelium. (d) Tyrosine kinase inhibitors bind to the intracellular domain of the VEGF receptor and prevent phosphorylation by ATP. This inhibits the downstream signaling cascade from VEGF binding to its receptor. (e) Gold nanoparticles concentrate in the vascular endothelium. Short-range photoelectrons and Auger electrons enhance the dose delivery to the endothelium, leading to improved vascular damage and tumor vasculature depletion.

(Glass et al. 1997; Lederman et al. 1998). Glass et al. (1997) treated 20 patients with fractionated stereotactic radiosurgery and concurrent cisplatinum at a dose of 40 mg/m^2 weekly for recurrent, progressive, or persistent malignant astrocytomas and reported a median survival of 55 weeks (Glass et al. 1997). Lederman et al. (1998) treated 14 patients with recurrent glioblastoma multiforme using a course of fractionated stereotactic radiosurgery with concurrent paclitaxel (median dose 120 mg/m^2) given immediately prior to radiation therapy and reported a median survival of 14.2 months (Lederman et al. 1998). While interesting, these reports do not help to determine if the addition of concurrent cytotoxic drug therapy to radiosurgery provided a benefit over radiation therapy alone. The addition of cytotoxic drug therapy in the neoadjuvant or adjuvant setting could improve outcomes with radiosurgery by addressing clinically occult metastatic disease as well as potentially initiating cellular damage that potentiates the effects of hypofractionated radiation therapy. Neoadjuvant and adjuvant cytotoxic drug therapy has been added to SBRT for treatment of locally advanced pancreatic cancer with the stated purpose of addressing distant metastatic disease. This has included a regimen of neoadjuvant gemcitabine followed by SBRT (25 Gy/1 fraction) and then adjuvant gemcitabine (Schellenberg et al. 2008; Schellenberg et al. 2011) as well as SBRT (8–12 Gy × 3 fractions) with adjuvant gemcitabine

(Mahadevan et al. 2010) and neoadjuvant gemcitabine followed by SBRT (Mahadevan et al. 2011). In a retrospective review, Chen et al. (2008) reported findings suggesting increased overall survival in patients treated with adjuvant cisplatin after SBRT for lung cancer. It is not clear from these studies that adding cytotoxic drug therapy in the neoadjuvant, concurrent, or adjuvant setting has an impact on control of the treated lesion. Also, the use of cytotoxic chemotherapy can suppress the immune system, which could be an important component of cancer treatment in the future as will be described later in this section.

15.2.3 TARGETED AGENTS

The addition of cytotoxic chemotherapy agents that are used as radiosensitizers for fractionated radiation therapy does not take into consideration that the cellular response to hypofractionated radiation therapy and mechanisms of cell death may differ from conventional radiobiology. The use of targeted agents to exploit these cellular processes may result in improved outcomes in radiosurgery.

Endothelial damage may play an important role in tumor control in response to hypofractionated radiation therapy. A histopathological analysis of arteriovenous malformations after Gamma Knife radiosurgery demonstrated progressive changes starting with endothelial damage and leading to narrowing and obliteration of the vascular lumen (Schneider, Eberhard, and Steiner 1997). Hypoxia-inducible factor (HIF-1) and vascular endothelial growth factor (VEGF) have been shown to be involved in evasion of endothelial cell apoptosis in response to radiation therapy (Gorski et al. 1999; Moeller et al. 2004), and use of an anti-VEGF treatment was shown in vitro to potentiate radiation-induced lethality in human umbilical vein endothelial cells (Gorski et al. 1999). Endothelial apoptosis was shown to have an impact on tumor response to radiation therapy in a mouse model, in which cells lines with mutations in acid sphingomyelinase or Bax had reduced endothelial apoptosis and grew faster that wild-type tumors (Garcia-Barros et al. 2003). Importantly, these authors showed that evasion of endothelial apoptosis had an impact on tumor response with doses up to 20 Gy in a single fraction.

Bevacizumab is a monoclonal antibody that targets VEGF and inhibits angiogenesis. It has been used in combination with hypofractionated radiation therapy for treatment of recurrent grade III or IV glioma. In a report of 25 patients with recurrent glioblastoma multiforme (GBM) or anaplastic astrocytoma, concurrent and adjuvant bevacizumab with hypofractionated radiation therapy to 30 Gy/5 fractions was showed to be safe and well tolerated with an overall response rate of 50% and median survival of 12.5 months (Gutin et al. 2009). The use of bevacizumab as adjuvant treatment after stereotactic radiosurgery has also been reported. In a case control study of 11 patients with recurrent GBM, the use of bevacizumab after radiosurgery resulted in improved progression-free survival (15 vs. 7 months) and overall survival (18 vs. 12 months) compared to patients receiving radiosurgery alone (Park et al. 2012). The delivery of bevacizumab after SRS for recurrent grade III or IV glioma has been shown to result in improved median survival when compared to delivery prior to SRS (11.2 vs. 2.5 months) (Cuneo et al. 2012). This finding corresponds with the biological data showing upregulation of VEGF after exposure to radiation therapy (Gorski et al. 1999; Moeller et al. 2004).

Tyrosine kinase inhibitors may also be employed to inhibit the VEGF pathway. Vandetanib is a tyrosine kinase inhibitor with activity against vascular endothelial growth factor receptor (VEGFR) and epidermal growth factor receptor and has been used in a phase I study concurrently with hypofractionated radiation therapy for recurrent malignant glioma (Fields et al. 2012). A study of 106 patients with metastatic renal cell carcinoma showed that concurrent administration of sunitinib or sorafenib (tyrosine kinase inhibitors with activity against VEGFR) with single-fraction SBRT to disease in the spine or brain was safe and effective with a local control at 15 months of 98% (Staehler et al. 2008).

Radiosurgery typically involves the treatment of a tumor with high doses of radiation in a single or few total fractions, which may have an impact due to lack of reoxygenation and resistance of hypoxic regions. Carlson et al. (2011) developed a model to evaluate effects of hypoxia in hypofractionated radiation therapy and predicted a loss of up to three logs of cell kill as the dose per fraction increases from conventional sizes (2–2.2 Gy) to a large single fraction (18.3–23.8 Gy). The model also predicted that hypoxic cell radiosensitizers would be most effective when delivered with a single or a few large doses of radiation. RTOG 95-02 was a phase I study evaluating the use the hypoxic cell sensitizer etanidazole with radiosurgery in previously irradiated brain tumors in 51 patients (Drzymala et al. 2008). The combination was well tolerated, but there was no report of tumor control or survival. A randomized trial using external beam radiation therapy (40 Gy/20 fractions) preceded by intraoperative radiation therapy to 25 Gy ± the hypoxic cell sensitizer doranidazole for the treatment of locally advanced pancreatic cancer showed an improvement in response as assessed by computed tomography (47% versus 18%) and in 3-year overall survival (23% versus 0%) (Karasawa et al. 2008). This data suggests combining a hypoxic cell radiosensitizer with hypofractionated radiation therapy results in improved outcomes.

15.2.4 DRUG DELIVERY

Some patients undergoing radiosurgery may not be healthy enough to tolerate systemic drug therapy—particularly cytotoxic chemotherapy—in addition to radiation therapy. Additionally, a drug may be discovered that improves the effects of radiation therapy but is too toxic to deliver systemically. As described earlier in this section, ablative doses of radiation therapy lead to vascular changes, which could impair drug delivery to the tumor in the adjuvant setting. One method to avoid these potential limitations would be to deliver the drug directly to the tumor. This could be accomplished by utilizing fiducial markers—already in use for lung and liver radiosurgery—and coating them with drugs. Dattatri Nagesha et al. (2010) have described the application of drug-eluting nanocoatings to gold fiducial markers and the diffusion of the drugs into a buffer medium. In this work, two systems were evaluated. One system consisted of a hydrophilic drug loaded into a nondegradable polymer coating placed onto a gold fiducial. The diffusion characteristics of this system were an initial high volume of drug release into the buffer medium from the outside of the coating, followed by a slow and sustained release of drug over the period of 40 days. They found that by applying additional coating to the fiducial, the initial rapid release of drug could be controlled. The second system used a hydrophobic drug embedded into a biodegradable polymer to form nanoparticles, which were then mixed into a hydrogel matrix and coated onto the surface of the gold fiducial. As the hydrogel matrix degraded, the nanoparticles were released into the buffer, and as the polymer degraded, the free drug was released. This two-step method was found to allow for a more controlled and sustained release of drug. Robert Cormack et al. (2010) developed an analytic function to describe the concentration of a drug versus distance from the source with diffusion-elimination properties of the drug in tissue taken into consideration. They predicted that four radiosensitizer-loaded fiducials could provide adequate radiosensitization for a 4-cm-diameter lung tumor. The potential advantage of this technology would be the delivery of the drug directly to the tumor with release characteristics that could exploit biological changes occurring at specific points in time as a result of exposure to high-dose radiation therapy. This could also allow for more patients to be eligible for drug therapy, particularly patients with comorbid conditions that would make them poor candidates for systemic therapy.

15.2.5 TIMING OF DRUG DELIVERY

The optimal timing of drug delivery in relation to treatment with high-dose radiation therapy depends on the cellular changes occurring in both the tumor and the surrounding tissue in response to radiation therapy. As described earlier in this section, high-dose radiation therapy induces endothelial cell apoptosis (Garcia-Barros et al. 2003), but levels of pro-angiogenic factors, such as VEGF, have been demonstrated to increase in response to radiation therapy (Moeller et al. 2004). The initiation and duration of this increased expression of VEGF could impact the optimal time and duration of drug exposure to achieve maximum effect. As described in a mouse model, the administration of an anti-VEGF antibody three hours before exposure to ionizing radiation therapy resulted in a greater reduction in tumor volume compared with an anti-VEGF antibody alone or radiation therapy alone (Gorski et al. 1999). In a retrospective review, Cuneo et al. (2012) noted worse outcomes in terms of overall survival in patients who only received bevacizumab before delivery of radiation therapy. This suggests that the timing may be important—the VEGF inhibition may need to be present shortly before or after the inciting factor (radiation therapy) induces increased expression of the growth factor. Further work on the timing of molecular events occurring in response to hypofractionated radiation therapy will help determine additional pathways to target. In some cases, it may be beneficial to deliver a drug before treatment, in effect "priming" the target or depleting a particular substrate. In other cases, the drug would need to be present at the moment of treatment or at a certain point afterward. Another consideration would be the impact of endothelial damage in response to ablative doses of radiation therapy, which could affect the delivery of a drug to the target tissue.

Another important question is whether maximum effect is achieved in a dose-dependent or time-dependent manner. Is a high dose of a drug at a particular time point sufficient to disrupt or exploit a disturbed cellular pathway? Or is the best effect achieved by sustaining a steady concentration of the drug over time while targeted cells are modulating gene expression or metabolic pathways in response to radiation therapy?

15.2.6 IMMUNE MODULATION

The immune response has been implicated in the therapeutic effect of ablative doses of radiation therapy. Lee et al. (2009) reported on the effects of ablative radiation therapy doses on lung, breast, and melanoma cell lines in an animal model. They found that ablative doses of radiation (15–25 Gy/1 fraction) generated a strong CD8+ T-cell immune response, which led to a reduction or eradication of the primary tumor and distant metastases in some cases. They noted increased T-cell priming in draining lymphatics, suggesting that high doses of radiation therapy activates dendritic cells in the tumor, which then migrate and present antigens to T-cells in the draining lymphoid tissue. They also showed that delivery of chemotherapy diminished CD8+ cell priming and response to radiation therapy. This data suggests a potential role for immune modulation to improve the outcomes in radiosurgery and that the use of agents with an immunosuppressive effect could have a negative impact on tumor control.

Immune modulation may help *improve* the control of disease locally, but another application that is gaining interest is in the combination with SRS/SBRT to elicit a response in distant sites of disease. There have been three recent case reports in the literature of the abscopal effect in patients treated for melanoma with high-dose radiation therapy and ipilimumab (Hiniker, Chen, and Knox 2012; Postow et al. 2012; Stamell et al. 2013). The abscopal effect describes a phenomenon in which disease separate from a targeted volume regresses after treatment with radiation therapy. This process is believed to be due to an immune response at the distant sites of disease. Ipilimumab is a monoclonal antibody that blocks a receptor involved in suppressing the activity of cytotoxic T-cells, thus allowing them to respond to antigens without the presence of a coexisting inhibitory molecule. It appears that the treatment of a tumor with high-dose radiation therapy, and thus release of tumor antigens due to cell death, in combination with a therapy targeted at up-regulating the activity of cytotoxic T-cells can result in eradication of distant disease due to modulation of the immune response. There are currently two clinical trials evaluating the efficacy of adding ipilimumab to radiation therapy for treatment of prostate cancer (NCT00861614) and metastatic melanoma (NCT01449279). Interleukin-2 (IL-2) is a cytokine involved in the immune response and has been used in the treatment of metastatic melanoma and metastatic renal cell carcinoma. A recent phase I study evaluated the combination of this drug with SBRT for metastatic melanoma and renal cell carcinoma and reported feasibility of the treatment and promising finding in regards to response to therapy (Seung et al. 2012). Other work on augmenting the effects of radiation therapy with immunomodulation have included mouse models using a listeria monocytogenes–based PSA vaccine for a prostate cancer cell line (Hannan et al. 2012) and a toll-like receptor agonist for a lung adenocarcinoma cell line (Zhang et al. 2012). Both of these studies showed improved tumor responses when combined with high-dose radiation therapy when compared to radiation therapy or immunomodulatory therapy alone.

15.2.7 ENHANCING DOSE DEPOSITION WITH GOLD NANOPARTICLES

Gold nanoparticles have been evaluated to enhance the effects of radiation therapy, mostly with kilovoltage x-rays. Berbeco, Ngwa, and Makrigiorgos (2011) performed a dosimetric study to estimate the dose enhancement to tumor endothelial cells with a 6-MV photon beam. Under the assumption that gold nanoparticles can be concentrated in the tumor endothelium and that the photoelectrons generated in gold have a short range, they predicted that a localized dose enhancement to the vascular endothelium could be achieved with a 6-MV photon beam. This suggests that if damage to the vascular endothelium is a key factor in the control of tumors treated with radiosurgery, a lower dose of radiation may be able to be delivered to spare adjacent normal tissue while providing the desired effect in the tumor vasculature. This would be especially useful if it was determined that there was an inherent difference in the radiosensitivity of tumor cells and the microvasculature.

15.2.8 THE FUTURE

The biology of hypofractionation is likely different from conventional fractionated radiation therapy. This, in itself, does not necessarily mean that agents used for radiosensitization with fractionated radiation therapy will not work in radiosurgery, but delivery and timing may be different to achieve the desired effect. There is likely a role for targeting the tumor microvasculature with agents to enhance the effects of radiosurgery. The optimal timing and duration of drug delivery to maximally enhance the effects of radiosurgery on the microvasculature remains to be determined. Further investigation into the molecular changes occurring in response to ablative doses of radiation will help define new targets for radiosensitization. Using multiple targeted agents at advantageous times in relation to delivery of radiosurgery could result in a coordinated attack on the tumor and improved outcomes. This could include hypoxic cell sensitizers, inhibitors of VEGF, or other targets up-regulated by ablative doses of radiation, immunomodulators, or gold nanoparticles (Figure 15.17).

Conceivably, multiple drugs could be incorporated onto a fiducial marker that would be designed to release the drugs into the tumor at certain points in time. The toxicity of drug delivery could be decreased by diffusion of the drug from a fiducial implanted directly within the tumor rather than by systemic administration. The use of enhancing agents with radiosurgery may achieve improvement in tumor control and allow radiation oncologists to decrease the dose of radiation, thus decreasing treatment-related normal tissue toxicity.

The use of the immune system to improve the local control of SRS/SBRT could be an exciting development in the treatment of localized cancer. However, the use of SRS/SBRT in combination with drugs to modulate the immune response could be a paradigm-shifting development in the treatment of metastatic disease. Currently, patients with metastatic disease are typically treated with combinations of cytotoxic therapy and/or targeted agents, and cures with this treatment approach are extraordinarily rare. In many patients with metastatic disease, comorbid conditions and poor performance status preclude them from treatment with these toxic drugs. The infrastructure is already in place to accurately deliver highly conformal, high-dose radiation therapy to nearly any part of the body. In the future, SRS/SBRT could be combined with new drugs targeting the immune system to extend treatment to more patients with metastatic disease, which may offer more patients prolonged survival and potentially cure of their disease.

15.3 HYPOFRACTIONATED PALLIATIVE RADIOTHERAPY

Most cancer deaths involve extensive loco-regional tumors or metastatic disease to brain, lung, liver, or bone causing pain, disability, and decreased quality of life. Radiotherapy is an important treatment for the alleviation of pain and suffering for cancer patients and up to 30%–50% of all radiation oncology referrals are for palliative radiation (McCloskey et al. 2007).

The skeleton is one of the most common sites of metastatic disease and is often the first site affected by metastases and the most common origin of cancer-related pain (Coleman 2006; Schulman and Kohles 2007). It was estimated that in 2004, 250,000 cancer patients were afflicted with metastatic bone disease (Schulman and Kohles 2007). Metastatic bone disease causes considerable morbidity in cancer patients resulting in pain, hypercalcemia, pathologic fractures, compression of the spinal cord or cauda equina, and spinal instability (Coleman 2006).

In the United States, the most common palliative dose fractionation schedule for metastatic bone disease is 30 Gy/10 treatments delivered over 2 weeks (Fairchild et al. 2009), which results in a complete or partial resolution of pain in about 60% of patients (Chow et al. 2007; Wu et al. 2003). Adding the common 1-week pretreatment planning process to the 2 weeks of treatment delivery results in an overall duration of 3 weeks for completion of palliative treatment. In a retrospective study of end-stage cancer patients receiving palliative radiotherapy, Gripp et al. (2010) found that half of the patients received treatment for >60% of their final days of life. Thus, these prolonged treatment courses subject patients to repeated visits to treatment centers and consume precious time and energy for ill patients and their families.

More than a dozen randomized controlled trials have demonstrated that a single fraction of 8 Gy offers equivalent pain relief to 10 fractions of 3 Gy for the treatment of metastatic bone disease (Lutz et al. 2011). European countries, Australia, and Canada have widely adopted the use of single fraction treatment, and countries with larger reimbursements for higher numbers of treatments routinely offer longer treatment courses (Fairchild et al. 2009). The American Society for Radiation Oncology consensus guidelines established that 8 Gy in a single fraction for patients with uncomplicated bone metastases offers similar rates of pain relief to that achieved with 2-week treatment courses. These single-fraction treatments have no increase in toxicity and maximize patient and caregiver convenience for the treatment of uncomplicated bone metastases (Lutz et al. 2011). Clearly, more work must be done to encourage the use of single-fraction palliative treatments and to design efficient same-day treatment workflows that result in highly conformal plans that aim to minimize acute toxicity.

Innovative academic centers are working toward solutions for efficient conformal palliative workflows. Already the radiation oncology group at Princess Margaret Hospital in Toronto has developed a workflow for the delivery of palliative radiation treatments with simple beam geometry (1–2 beams) in approximately 35 min. With their workflow, the patient lies on a linear-accelerator treatment couch and receives cone-beam CT. While the patient continues to lie on the couch, the contouring and planning is done on the cone-beam CT image set, followed by quality-assurance checks. Then the patient receives a second cone-beam CT to verify his or her position, and the radiation treatment is delivered (Wong et al. 2012). Although this is an excellent start to maximizing efficiency and convenience for patients, we believe that, in the near future, ever more sophisticated plans can be safely delivered in an even shorter amount time.

15.3.1 STAT RAD: A RAPID PALLIATIVE RADIOTHERAPY WORKFLOW IN CLINIC DEVELOPMENT

At the University of Virginia, we are piloting a new workflow called "STAT RAD" to rapidly deliver advanced radiotherapy to patients with metastatic disease on an internal review board–approved clinical trial. This STAT RAD workflow offers same-day palliation in an approximately 6-hr time frame similar to a standard Gamma Knife® (Elekta, Stockholm, Sweden) workflow. STAT RAD is a highly coordinated conventional workflow that includes kVCT simulation, treatment planning, treatment plan quality assurance (QA), and delivery of conformal hypofractionated radiotherapy in a single day. All treatments are planned and delivered on FDA-approved systems, including a TomoHD (Madison, WI) system. With the STAT RAD program, we are now able to offer a unique workflow that delivers rapid, effective, and efficient palliative radiotherapy that is cost-effective, less toxic, and more convenient for cancer patients and their families.

15.3.2 SCAN-PLAN-VERIFY-TREAT STAT RAD WORKFLOW: THE FUTURE OF SBRT

With recent advances in software and technology, we plan to further condense the STAT RAD workflow into the Scan-Plan-Verify-Treat workflow, a 30-min process in which all steps (megavoltage CT [MVCT] simulation, diagnostic image coregistration, treatment planning, and treatment delivery with real-time QA) are performed on a TomoTherapy unit. This advanced workflow will eliminate the need for the patients to undergo a kVCT simulation on a separate unit as well as make it unnecessary for the patient to leave the treatment table between the simulation and treatment delivery.

Requirements for the clinical implementation of the Scan-Plan-Verify-Treat STAT RAD workflow are envisioned as follows:

1. Scan: MVCT simulation image acquisition (10 min) then rigid or deformable image coregistration of existing diagnostic image sets with precontoured target and organ at risk (OAR) volumes to the MVCT simulation scan for contour transfer (3–5 min).
2. Plan: Rapid inverse treatment planning (3–5 min).
3. Verify: Real-time patient-specific QA using CT detectors during treatment delivery (10 min). A rapid secondary independent Monte Carlo dose calculation is another QA possibility.
4. Treat: Patient motion tracking to monitor patient position in real-time during the entire process (Figure 15.18).

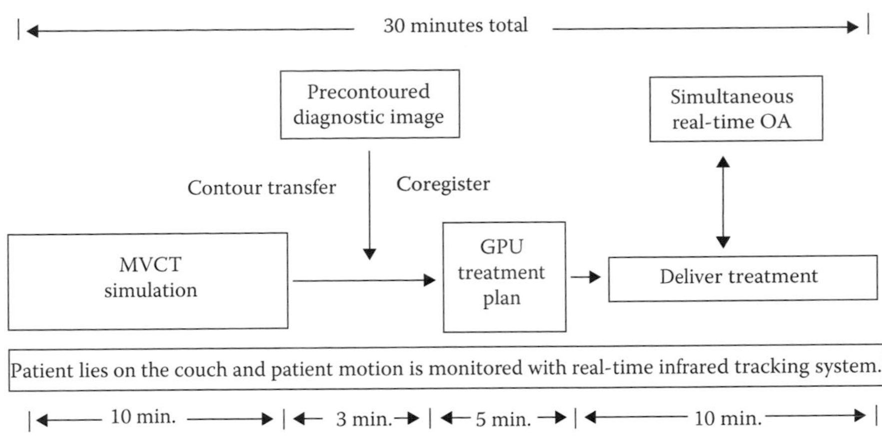

Figure 15.18 Schematic of the Scan-Plan-Verify-Treat STAT RAD workflow.

15.3.3 NEW IMAGE COREGISTRATION WORKFLOW

A kVCT simulation scan is currently used for simulation in the conventional workflow for both palliative and curative radiation planning in most clinics. Compared to MVCT scans, it has higher resolution and allows for administration of iodinated IV and/or GI contrast, which makes it easier to identify soft tissues for treatment planning. However, contrast agents are not generally given for kVCT simulations of patients for palliative treatment of metastases because the soft tissue and bone windows are adequate. Soft tissue and bone windows of MVCT scans have quite reasonable resolution and can easily be coregistered to higher resolution diagnostic studies for contour transfer for the treatment of bone metastases. Tomotherapy image-guidance software has been used clinically since 2004 to automatically coregister kVCT simulation images and daily MVCT scans for image guidance on a daily basis.

Our preliminary unpublished data confirms that the MVCT scan has sufficient resolution, particularly of bone anatomy, for accurate coregistration to contoured diagnostic images and that this one-step coregistration process yields comparable agreement to the conventional two-step image coregistration workflow with ±2–3 mm differences. This level of agreement is consistent with results reported from image coregistration studies performed in a multi-institutional pediatric clinical trial with coregistration data of 51 patients from 45 institutions using 11 different image-software systems. They reported an inherent uncertainty of 2 mm for MRI to CT coregistration (Ulin, Urie, and Cherlow 2010). Thus, preliminary data suggests that the optimization of this one-step image coregistration workflow of diagnostic image sets to a MVCT simulation scan will be clinically similar to the conventional two-image coregistration workflow for bone targets.

15.3.4 RAPID INVERSE TREATMENT PLANNING ON MVCT SCANS

CT image sets are used for radiation treatment planning because the electron density of tissues, which is required for calculating dosage, is easily determined based on the Hounsfield units. The tissue electron density determination is essentially the same for MVCT and kVCT scans. It has previously been reported that as far as the dose calculations are concerned, treatment planning on either a kVCT simulation image set or an MVCT simulation image set yields treatment plans that are within 1% of each other (Langen et al. 2005).

We have published that the TomoTherapy STAT RT treatment-planning module can calculate SBRT plans in just a few minutes (Dunlap et al. 2010). The computing speed of radiation treatment-planning systems is about to take a quantum leap forward with the incorporation of new algorithms that will take advantage of the processing power of graphics processing units (GPUs) whose more rapid and parallel calculating potential can improve treatment-planning speed by 10–20 times (Hissoiny, Ozell, and Després 2009, 2010). Same-day inverse treatment planning of IMRT or 3-D TomoTherapy plans has not been a problem for patients treated with STAT RAD to date. Highly conformal 3-D or IMRT plans can now be generated in a few minutes with FDA approved GPU-based algorithms.

15.3.5 NOVEL CT DETECTOR–BASED QUALITY ASSURANCE METHODOLOGY

Volumetric dose reconstruction from information collected during treatment became possible with the incorporation of radiation imaging detectors, such as electronic portal imaging devices (EPIDs) on linear accelerators and CT detector arrays on TomoTherapy. Dose reconstruction using in-line EPID was first described by McNutt et al. (McNutt, Mackie, and Paliwal 1997; McNutt et al. 1996). The EPID, when deployed during treatment, collects exit fluence from the patient and then back-projects this to x-ray fluence before entering the patient; then, the dose in the patient is recomputed using this entrance fluence and the planning CT images. However, there are many limitations to EPID-based dose verification. For example, the EPID was originally designed for semiquantitative portal imaging, and for the purpose of dose reconstruction, it suffers from a narrow dynamic range, short lifespan, nonlinearity in the dose response, ghost artifacts from low temporal resolution, and cross-plane scatter photon contribution to the measured fluence (Mijnheer 2008). Investigators are currently working on methods to overcome these challenges.

The TomoTherapy unit has an in-line source-patient-detector geometry with CT ion chamber detectors that are used for daily MVCT scan image guidance for accurate patient positioning that remain in place during both imaging and treatment. These CT detectors can also be used to measure the patient exit dose fluence and back-project this onto a planning CT scan for volumetric or 3-D dose reconstruction. Dose verification on TomoTherapy was first studied by Kapatoes et al. (2001), who calculated the entrance fluence from the exit dose using a transfer matrix, which is calculated based on the radiological path length from the source to the detector. The use of a CT ion chamber array has multiple advantages over EPID for exit fluence measurement. It is more durable and has a much longer lifespan. It has a wider dynamic range and doesn't limit treatment positions. Finally, it is less sensitive to the noise from cross-plane scatter photons that complicate EPID-based dose reconstruction (Siewerdsen and Jaffray 2001).

Our preclinical evaluation of the CT detector–based exit radiation dose verification algorithm has been retrospectively studied by Sheng et al. (2012) using in-development software. We compared planned and delivered doses with the conventional phantom QA measurements for 24 patients and 347 treatment fractions. The concordance of planned-to-delivered dose calculated by the in-development software was shown to be ±5% (Sheng et al. 2012). This tolerance is within the standard of care of other current clinically available QA methods. In addition to CT-detector based approaches, a pretreatment independent dose verification calculation is another option for QA.

15.3.6 OPTICAL TRACKING METHODS FOR PATIENT INTRAFRACTIONAL MOTION MONITORING

Consistent patient positioning from planning CT image acquisition through the end of treatment is critical to ensure accurate dose delivery. Physical immobilization devices, such as external body frames, *Aquaplast* masks, and other body molds, and vac-lock vacuum bags are commonly used to ensure patient positioning reproducibility. X-ray or CT image guidance prior to radiation delivery on the treatment unit is routinely employed in the clinic. Methods for optical tracking of the patient surface with or without markers are available to ensure consistent patient positioning after image guidance and throughout treatment (Wagner et al. 2007; Wiersma et al. 2010). This provides a method without ionizing radiation for confirming patient position that can be used in real time during treatment delivery. A mechanism to ensure that the patient's position doesn't change between MVCT simulation and treatment completion would obviate the need for a repeat image guidance MVCT scan just prior to delivery in the Scan-Plan-Verify-Treat workflow.

We have recently developed an in-house optical tracking system using multiple OptiTrack FLEX:V100 cameras (Natural Point, Corvallis, OR). The camera utilizes 26 infrared light-emitting diodes and a charge coupled device to capture the reflective light from markers with special coating. By using multiple cameras, the 3-D position of each reflective marker can be determined precisely. Multiple markers can be placed on a patient and monitored simultaneously. Through strategic positioning of the markers, movements of the head, neck, and extracranial locations can be closely monitored. Other commercially available systems exist for surface optical tracking.

15.3.7 FUTURE DIRECTIONS FOR SPINAL SBRT

The Scan-Plan-Verify-Treat STAT RAD workflow could be incorporated into the treatment of spinal SBRT patients. We have previously reported that treatment-planning algorithms currently exist that can create highly conformal spinal SBRT plans in just a few minutes (Dunlap et al. 2010) and that the CT detector–based quality assurance algorithms can measure exit dose to within ±5% (Sheng et al. 2012). Real-time spinal SBRT simulation, planning, and delivery would eliminate the need for patients to be repositioned between simulation and treatment. With this proposed workflow, the patient is treated in the planning position, which eliminates repositioning error issues. In the near future, workflows, such as this Scan-Plan-Verify-Treat workflow, may revolutionize radiation therapy by making it a real-time process that could have tremendous benefit to patients needing rapid palliation.

REFERENCES

Ahn, P. H., A. I. Ahn, C. J. Lee et al. 2009. Random positional variation among the skull, mandible, and cervical spine with treatment progression during head-and-neck radiotherapy. *Int. J. Radiat. Oncol. Biol. Phys.* 73(2): 626–33.

Berbeco, R. I., W. Ngwa, and G. M. Makrigiorgos. 2011. Localized dose enhancement to tumor blood vessel endothelial cells via megavoltage x-rays and targeted gold nanoparticles: New potential for external beam radiotherapy. *Int. J. Radiat. Oncol. Biol. Phys.* 81(1): 270–6.

Brauer-Krisch, E., R. Serduc, E. A. Siegbahn et al. 2010. Effects of pulsed, spatially fractionated, microscopic synchrotron x-ray beams on normal and tumoral brain tissue. *Mutat. Res.* 704: 160–6.

Buzurovic, I., K. Huang, Y. Yu, and T. K. Podder. 2011. A robotic approach to 4D real-time tumor tracking for radiotherapy. *Phys. Med. Biol.* 56(5): 1299–318.

Cadman, P., and K. Bzdusek. 2011. Co-60 tomotherapy: A treatment planning investigation. *Med. Phys.* 38(2): 556–64.

Cadman, P. F., B. R. Paliwal, and C. G. Orton. 2010. Point/counterpoint: Co-60 tomotherapy is the treatment modality of choice for developing countries in transition toward IMRT. *Med. Phys.* 37(12): 6113–5.

Cao, M., F. Lasley, A. Fakiris, C. Desrosiers, and I. J. Das. 2011. SU-E-T-553: Evaluation of rotational errors in treatment setup of stereotactic body radiotherapy (SBRT) of lung cancer. *Med. Phys.* 38(6): 3616.

Carlson, D. J., P. J. Keall, B. W. Loo Jr., Z. J. Chen, and J. M. Brown. 2011. Hypofractionation results in reduced tumor cell kill compared to conventional fractionation for tumors with regions of hypoxia. *Int. J. Radiat. Oncol. Biol. Phys.* 79(4): 1188–95.

Chen, Y., W. Guo, Y. Lu, and B. Zou. 2008. Dose-individualized stereotactic body radiotherapy for T1-3N0 non-small cell lung cancer: Long term results and efficacy of adjuvant chemotherapy. *Radiother. Oncol.* 88(3): 351–8 (http://www.ncbi.nlm.nih.gov/pubmed/18722684).

Chow, E., K. Harris, G. Fan, M. Tsao, and W. M. Sze. 2007. Palliative radiotherapy trials for bone metastases: A systematic review. *J. Clin. Oncol.* 25(11): 1423–36.

Coleman, R. E. 2006. Clinical features of metastatic bone disease and risk of skeletal morbidity. *Clin. Cancer Res.* 12(20 Pt 2): 6243s–9s.

Constantin, D. E., R. Fahrig, and P. J. Keall. 2011. A study of the effect of in-line and perpendicular magnetic fields on beam characteristics of electron guns in medical linear accelerators. *Med. Phys.* 38(7): 4174–85.

Cormack, R. A., S. Sridhar, W. W. Suh, A. V. D'Amico, and G. M. Makrigiorgos. 2010. Biological in situ dose painting for image-guided radiation therapy using drug-loaded implantable devices. *Int. J. Radiat. Oncol. Biol. Phys.* 76(2): 615–23.

Crosbie, J. C., R. L. Anderson, K. Rothkamm et al. 2010. Tumor cell response to synchrotron microbeam radiation therapy differs markedly from cells in normal tissues. *Int. J. Radiat. Oncol. Biol. Phys.* 77(3): 886–94.

Cuneo, K. C., J. J. Vredenburgh, J. H. Sampson et al. 2012. Safety and efficacy of stereotactic radiosurgery and adjuvant bevacizumab in patients with recurrent malignant gliomas. *Int. J. Radiat. Oncol. Biol. Phys.* 82(5): 2018–24.

Curtis, H. J. 1967. The use of a deuteron microbeam for simulating the biological effects of heavy cosmic-ray particles. *Radiat. Res. Supplement* 7: 250–7.

Dempsey, J., D. Benoit, J. R. Fitzsimmons et al. 2005. A device for realtime 3D image-guided IMRT. *Int. J. Radiat. Oncol. Biol. Phys.* 63(Suppl 1): S202.

Dilmanian, F. A., J. F. Hainfeld, C. A. Kruse et al. 2002. Biological mechanisms underlying the preferential destruction of gliomas by x-ray microbeam radiation. *Nat. Synchrotron Light Source Rep.* Abstract No.: Dilm0599.

Dilmanian, F. A., G. M. Morris, N. Zhong et al. 2003. Murine EMT-6 carcinoma: High therapeutic efficacy of microbeam radiation therapy. *Radiat. Res.* 159: 632–41.

Dilmanian, F. A., Y. Qu, L. E. Feinendegen et al. 2007. Tissue-sparing effect of x-ray microplanar beams particularly in the CNS: Is a bystander effect involved? *Exp. Hematol.* 35(4 Suppl 1): 69–77.

Dilmanian, F. A., Y. Qu, S. Liu et al. 2005. X-ray microbeams: Tumor therapy and central nervous system research. In *Nuclear Instruments and Methods in Physics Research Section A: Accelerators, Spectrometers, Detectors and Associated Equipment*, 548: 30–37.

Dilmanian, F. A., Z. Zhong, T. Bacarian et al. 2006. Interlaced x-ray microplanar beams: A radiosurgery approach with clinical potential. *Proc. Natl. Acad. Sci. USA* 103: 9709–14.

Drzymala, R. E., T. H. Wasserman, M. Won et al. 2008. A phase I-B trial of the radiosensitizer: Etanidazole (SR-2508) with radiosurgery for the treatment of recurrent previously irradiated primary brain tumors or brain metastases (RTOG Study 95-02). *Radiother. Oncol.* 87(1): 89–92.

D'Souza, W. D., S. A. Naqvi, and C. X. Yu. 2005. Real-time intra-fraction-motion tracking using the treatment couch: A feasibility study. *Phys. Med. Biol.* 50(17): 4021–33.

Dunlap, N., A. McIntosh, K. Sheng et al. 2010. Helical tomotherapy-based STAT stereotactic body radiation therapy: Dosimetric evaluation for a real-time SBRT treatment planning and delivery program. *Med. Dosim.* 35(4): 312–9.

Fairchild, A., E. Barnes, S. Ghosh et al. 2009. International patterns of practice in palliative radiotherapy for painful bone metastases: Evidence-based practice? *Int. J. Radiat. Oncol. Biol. Phys.* 75(5): 1501–10.

Fallone, B. G., M. Carlone, B. Murray et al. 2007. Development of a linac-MRI system for real-time ART. *Med. Phys.* 34(6): 2547.

Fields, E. C., D. Damek, L. E. Gaspar et al. 2012. Phase I dose escalation trial of vandetanib with fractionated radiosurgery in patients with recurrent malignant gliomas. *Int. J. Radiat. Oncol. Biol. Phys.* 82(1): 51–7.

Fox, C., H. E. Romeijn, B. Lynch, C. Men, D. M. Aleman, and J. F. Dempsey. 2008. Comparative analysis of ^{60}Co intensity-modulated radiation therapy. *Phys. Med. Biol.* 53(12): 3175–88.

Garcia-Barros, M., F. Paris, C. Cordon-Cardo et al. 2003. Tumor response to radiotherapy regulated by endothelial cell apoptosis. *Science* 300(5622): 1155–9.

Glass, J., C. L. Silverman, R. Axelrod, B. W. Corn, and D. W. Andrews. 1997. Fractionated stereotactic radiotherapy with cis-platinum radiosensitization in the treatment of recurrent, progressive, or persistent malignant astrocytoma. *Am. J. Clin. Oncol.* 20(3): 226–9.

Gorski, D. H., M. A. Beckett, N. T. Jaskowiak et al. 1999. Blockade of the vascular endothelial growth factor stress response increases the antitumor effects of ionizing radiation. *Cancer Res.* 59: 3374–8.

Gripp, S., S. Mjartan, E. Boelke, and R. Willers. 2010. Palliative radiotherapy tailored to life expectancy in end-stage cancer patients: Reality or myth? *Cancer* 116(13): 3251–6.

Gutfeld, O., A. E. Kretzler, R. Kashani, D. Tatro, and J. M. Balter. 2009. Influence of rotations on dose distributions in spinal stereotactic body radiotherapy (SBRT). *Int. J. Radiat. Oncol. Biol. Phys.* 73(5): 1596–601.

Gutin, P. H., F. M. Iwamoto, K. Beal et al. 2009. Safety and efficacy of bevacizumab with hypofractionated stereotactic irradiation for recurrent malignant gliomas. *Int. J. Radiat. Oncol. Biol. Phys.* 75(1): 156–63.

Hannan, R., H. Zhang, A. Wallecha et al. 2012.Combined immunotherapy with listeria monocytogenes-based PSA vaccine and radiation therapy leads to a therapeutic response in a murine model of prostate cancer. *Cancer Immunol. Immunother.* 61(12): 2227–38.

Hiniker, S. M., D. S. Chen, and S. J. Knox. 2012. Abscopal effect in a patient with melanoma. *N. Engl. J. Med.* 366(21): 2035; author reply 2035–6.

Hissoiny, S., B. Ozell, and P. Després. 2009. Fast convolution-superposition dose calculation on graphics hardware. *Med. Phys.* 36(6): 1998–2005.

Hissoiny, S., B. Ozell, and P. Després. 2010. A convolution-superposition dose calculation engine for GPUs. *Med. Phys.* 37(3): 1029–37.

Kaiser, A., T. E. Schultheiss, J. Y. Wong et al. 2006. Pitch, roll, and yaw variations in patient positioning. *Int. J. Radiat. Oncol. Biol. Phys.* 66(3): 949–55.

Kapatoes, J. M., G. H. Olivera, J. P. Balog, H. Keller, P. J. Reckwerdt, and T. R. Mackie. 2001. On the accuracy and effectiveness of dose reconstruction for tomotherapy. *Phys. Med. Biol.* 46(4): 943–66.

Karasawa, K., M. Sunamura, A. Okamoto et al. 2008. Efficacy of novel hypoxic cell sensitiser doranidazole in the treatment of locally advanced pancreatic cancer: Long-term results of a placebo-controlled randomised study. *Radiother. Oncol.* 87(3): 326–30.

Kim, S., H. Jin, H. Yang, and R. J. Amdur. 2009. A study on target positioning error and its impact on dose variation in image-guided stereotactic body radiotherapy for the spine. *Int. J. Radiat. Oncol. Biol. Phys.* 73(5): 1574–9.

Kirkby, C., B. Murray, S. Rathee, and B. G. Fallone. 2010. Lung dosimetry in a linac-MRI radiotherapy unit with a longitudinal magnetic field. *Med. Phys.* 37(9): 4722–32.

Kirkby, C., T. Stanescu, S. Rathee, M. Carlone, B. Murray, and B. G. Fallone. 2008. Patient dosimetry for hybrid MRI-radiotherapy systems. *Med. Phys.* 35(3): 1019–27.

Kron, T., D. Eyles, S. L. John, and J. Battista. 2006. Magnetic resonance imaging for adaptive cobalt tomotherapy: A proposal. *J. Med. Phys.* 31(4): 242–54.

Lagendijk, J. J., B. W. Raaymakers, A. J. Raaijmakers et al. 2008. MRI/linac integration. *Radiother. Oncol.* 86(1): 25–9.

Langen, K. M., S. L. Meeks, D. O. Poole et al. 2005. The use of megavoltage CT (MVCT) images for dose recomputations. *Phys. Med. Biol.* 50(18): 4259–76.

Lederman, G., E. Arbit, M. Odaimi, E. Lombardi, M. Wrzolek, and M. Wronski. 1998. Fractionated stereotactic radiosurgery and concurrent taxol in recurrent glioblastoma multiforme: A preliminary report. *Int. J. Radiat. Oncol. Biol. Phys.* 40(3): 661–6.

Lee, Y., S. L. Auh, Y. Wang et al. 2009. Therapeutic effects of ablative radiation on local tumor require CD8+ T cells: Changing strategies for cancer treatment. *Blood* 114(3): 589–95.

Linthout, N., D. Verellen, K. Tournel, T. Reynders, M. Duchateau, and G. Storme. 2007. Assessment of secondary patient motion induced by automated couch movement during on-line 6-dimensional repositioning in prostate cancer treatment. *Radiother. Oncol.* 83(2): 168–74.

Lutz, S., L. Beck, E. Chang et al. 2011. Palliative radiotherapy for bone metastases: An ASTRO evidence-based guideline. *Int. J. Radiat. Oncol. Biol. Phys.* 79(4): 965–76.

Mahadevan, A., S. Jain, M. Goldstein et al. 2010. Stereotactic body radiotherapy and gemcitabine for locally advanced pancreatic cancer. *Int. J. Radiat. Oncol. Biol. Phys.* 78(3): 735–42.

Mahadevan, A., R. Miksad, M. Goldstein et al. 2011. Induction gemcitabine and stereotactic body radiotherapy for locally advanced nonmetastatic pancreas cancer. *Int. J. Radiat. Oncol. Biol. Phys.* 81(4): e615–22.

McCloskey, S. A., M. L. Tao, C. M. Rose, A. Fink, and A. M. Amadeo. 2007. National survey of perspectives of palliative radiation therapy: Role, barriers, and needs. *Cancer J.* 13(2): 130–7.

McNutt, T. R., T. R. Mackie, and B. R. Paliwal. 1997. Analysis and convergence of the iterative convolution/superposition dose reconstruction technique for multiple treatment beams and tomotherapy. *Med. Phys.* 24(9): 1465–76.

McNutt, T. R., T. R. Mackie, P. J. Reckwerdt, N. Papanikolaou, and B. R. Paliwal. 1996. Calculation of portal dose using the convolution/superposition method. *Med. Phys.* 23(4): 527–35.

Mijnheer, B. 2008. State of the art of in vivo dosimetry. *Radiat. Prot. Dosimetry* 131(1): 117–22.

Moeller, B. J., Y. Cao, C. Y. Li, and M. W. Dewhirst. 2004. Radiation activates HIF-1 to regulate vascular radiosensitivity in tumors: Role of reoxygenation, free radicals, and stress granules. *Cancer Cell* 5(5): 429–41.

Murphy, M. J. 2007. Image-guided patient positioning: If one cannot correct for rotational offsets in external-beam radiotherapy setup, how should rotational offsets be managed? *Med. Phys.* 34(6): 1880–3.

Mutaf, Y. D., J. Zhang, C. X. Yu et al. 2013. Dosimetric and geometric evaluation of a novel stereotactic radiotherapy device for breast cancer: The GammaPod™. *Med. Phys.* 40(4): 041722.

Nagesha, D. K., D. B. Tada, C. K. Stambaugh et al. 2010. Radiosensitizer-eluting nanocoatings on gold fiducials for biological in-situ image-guided radio therapy (BIS-IGRT). *Phys. Med. Biol.* 55(20): 6039–52.

Oden, J., I. Toma-Dasu, C. X. Yu, S. J. Feigenberg, W. F. Regine, and Y. D. Mutaf. 2013. Dosimetric comparison between intra-cavitary breast brachytherapy techniques for accelerated partial breast irradiation and a novel stereotactic radiotherapy device for breast cancer: GammaPod™. *Phys. Med. Biol.* 58(13): 4409–21.

Park, K. J., H. Kano, A. Iyer et al. 2012. Salvage gamma knife stereotactic radiosurgery followed by bevacizumab for recurrent glioblastoma multiforme: A case-control study. *J. Neurooncol.* 107(2): 323–33.

Peng, J. L., C. Liu, Y. Chen, R. J. Amdur, K. Vanek, and J. G. Li. 2011. Dosimetric consequences of rotational setup errors with direct simulation in a treatment planning system for fractionated stereotactic radiotherapy. *J. Appl. Clin. Med. Phys.* 12(3): 3422.

Postow, M. A., M. K. Callahan, C. A. Barker et al. 2012. Immunologic correlates of the abscopal effect in a patient with melanoma. *N. Engl. J. Med.* 366(10): 925–31.

Raaijmakers, A. J., B. W. Raaymakers, and J. J. Lagendijk. 2005. Integrating a MRI scanner with a 6 MV radiotherapy accelerator: Dose increase at tissue-air interfaces in a lateral magnetic field due to returning electrons. *Phys. Med. Biol.* 50(7): 1363–76.

Raaijmakers, A. J., B. W. Raaymakers, and J. J. Lagendijk. 2008. Magnetic-field-induced dose effects in MR-guided radiotherapy systems: Dependence on the magnetic field strength. *Phys. Med. Biol.* 53(4): 909–23.

Romanelli, P., E. Fardone, G. Battaglia et al. 2013. Synchrotron-generated microbeam sensorimotor cortex transections induce seizure control without disruption of neurological functions. *PLoS One* 8(1): e53549.

Santos, D. M., J. St. Aubin, B. G. Fallone, and S. Steciw. 2012. Magnetic shielding investigation for a 6 MV in-line linac within the parallel configuration of a linac-MR system. *Med. Phys.* 39(2): 788–97.

Schellenberg, D., K. A. Goodman, F. Lee et al. 2008. Gemcitabine chemotherapy and single-fraction stereotactic body radiotherapy for locally advanced pancreatic cancer. *Int. J. Radiat. Oncol. Biol. Phys.* 72(3): 678–86.

Schellenberg, D., J. Kim, C. Christman-Skieller et al. 2011. Single-fraction stereotactic body radiation therapy and sequential gemcitabine for the treatment of locally advanced pancreatic cancer. *Int. J. Radiat. Oncol. Biol. Phys.* 81(1): 181–8.

Schneider, B. F., D. A. Eberhard, and L. E. Steiner. 1997. Histopathology of arteriovenous malformations after gamma knife radiosurgery. *J. Neurosurg.* 87(3): 352–7.

Schreibmann, E., T. Fox, and I. Crocker. 2011. Dosimetric effects of manual cone-beam CT (CBCT) matching for spinal radiosurgery: Our experience. *J. Appl. Clin. Med. Phys.* 12(3): 3467.

Schreiner, L. J., C. P. Joshi, J. Darko, A. Kerr, G. Salomons, and S. Dhanesar. 2009. The role of Cobalt-60 in modern radiation therapy: Dose delivery and image guidance. *J. Med. Phys.* 34(3): 133–6.

Schulman, K. L., and J. Kohles. 2007. Economic burden of metastatic bone disease in the U.S. *Cancer* 109(11): 2334–42.

Serduc, R., P. Vérant, J. C. Vial et al. 2006. In vivo two-photon microscopy study of short-term effects of microbeam irradiation on normal mouse brain microvasculature. *Int. J. Radiat. Oncol. Biol. Phys.* 64(2): 1519–27.

Seung, S. K., B. D. Curti, M. Crittenden et al. 2012. Phase 1 study of stereotactic body radiotherapy and interleukin-2—tumor and immunological responses. *Sci. Transl. Med.* 4(137): 137–74.

Sheng, K., R. Jones, W. Yang et al. 2012. 3D dose verification using tomotherapy CT detector array. *Int. J. Radiat. Oncol. Biol. Phys.* 82(2): 1013–20.

Siewerdsen, J. H., and D. A. Jaffray. 2001. Cone-beam computed tomography with a flat-panel imager: Magnitude and effects of x-ray scatter. *Med. Phys.* 28(2): 220–31.

Slatkin, D. N., P. Spanne, F. A. Dilmanian, J. O. Gebbers, and J. A. Laissue. 1995. Subacute neuropathological effects of microplanar beams of x-rays from a synchrotron wiggler. *Proc. Natl. Acad. Sci. USA* 92(19): 8783–7.

Smith, R. W., J. Wang, E. Schültke et al. 2013. Proteomic changes in the rat brain induced by homogenous irradiation and by the bystander effect resulting from high energy synchrotron x-ray microbeams. *Int. J. Radiat. Biol.* 89(2): 118–27.

Staehler, M., N. Haseke, K. Zilinberg, T. Stadler, A. Karl, and C. G. Stief. 2008. Systemic therapy of metastasizing renal cell carcinoma. *Urologe A.* 47(10): 1357–67.

Stamell, E. F., J. D. Wolchok, S. Gnjatic, N. Y. Lee, and I. Brownell. 2013. The abscopal effect associated with a systemic anti-melanoma immune response. *Int. J. Radiat. Oncol. Biol. Phys.* 85(2): 293–5.

St. Aubin, J., D. M. Santos, S. Steciw, and B. G. Fallone. 2010. Effect of longitudinal magnetic fields on a simulated in-line 6 MV linac. *Med. Phys.* 37(9): 4916–23.

St. Aubin, J., S. Steciw, and B. G. Fallone. 2010. Effect of transverse magnetic fields on a simulated in-line 6 MV linac. *Phys. Med. Biol.* 55(16): 4861–9.

Ulin, K., M. M. Urie, and J. M. Cherlow. 2010. Results of a multi-institutional benchmark test for cranial CT/MR image registration. *Int. J. Radiat. Oncol. Biol. Phys.* 77(5): 1584–9.

Variola, A., A. Loulergue, and F. Zomer, eds. 2010. *THOMX Conceptual Design Report.* Available at http://hal.in2p3.fr/docs/00/44/82/78/PDF/ThomXCDRAV.pdf.

Wagner, T. H., S. L. Meeks, F. J. Bova et al. 2007. Optical tracking technology in stereotactic radiation therapy. *Med. Dosim.* 32(2): 111–20.

Wang, H., A. Shiu, C. Wang et al. 2008. Dosimetric effect of translational and rotational errors for patients undergoing image-guided stereotactic body radiotherapy for spinal metastases. *Int. J. Radiat. Oncol. Biol. Phys.* 71(4): 1261–71.

Wiersma, R. D., Z. Wen, M. Sadinski, K. Farrey, and K. M. Yenice. 2010. Development of a frameless stereotactic radiosurgery system based on real-time 6D position monitoring and adaptive head motion compensation. *Phys. Med. Biol.* 55(2): 389–401.

Wong, R. K., D. Letourneau, A. Varma et al. 2012. A one-step cone-beam CT-enabled planning-to-treatment model for palliative radiotherapy-from development to implementation. *Int. J. Radiat. Oncol. Biol. Phys.* 84(3): 834–40.

Wright, M. D. 2012. High energy microbeam radiosurgery. U. S. Patent Application No. 13/492,412.

Wu, J. S., R. Wong, M. Johnston, A. Bezjak, and T. Whelan. 2003. Meta-analysis of dose-fractionation radiotherapy trials for the palliation of painful bone metastases. *Int. J. Radiat. Oncol. Biol. Phys.* 55(3): 594–605.

Yu, C. X., X. Shao, J. Zhang et al. 2013. GammaPod—A new device dedicated for stereotactic radiotherapy of breast cancer. *Med. Phys.* 40(5): 1703–10.

Zhang, H., L. Liu, D. Yu et al. 2012. An in situ autologous tumor vaccination with combined radiation therapy and TLR9 agonist therapy. *PLoS One* 7(5): e38111.

Index

Page numbers followed by f and t indicate figures and tables, respectively.